ADVANCES IN CYCLIC NUCLEOTIDE RESEARCH
VOLUME 3

Advances in Cyclic Nucleotide Research

Series Editors:

Paul Greengard, New Haven, Conn., U.S.A.
G. Alan Robison, Houston, Texas, U.S.A.

International Advisory Board

Bruce Breckenridge, New Brunswick, N.J., U.S.A.
R. W. Butcher, Worcester, Mass., U.S.A.
E. Costa, Washington, D.C., U.S.A.
George I. Drummond, Vancouver, British Columbia, Canada
Nelson Goldberg, Minneapolis, Minn., U.S.A.
Joel G. Hardman, Nashville, Tenn., U.S.A.
Oscar Hechter, Chicago, Ill., U.S.A.
David M. Kipnis, St. Louis, Mo., U.S.A.
Edwin G. Krebs, Davis, Cal., U.S.A.
Thomas A. Langan, Denver, Col., U.S.A.
Joseph Larner, Charlottesville, Va., U.S.A.
Grant W. Liddle, Nashville, Tenn., U.S.A.
Yasutomi Nishizuka, Kobe, Japan
Ira H. Pastan, Bethesda, Md., U.S.A.
Th. Posternak, Geneva, Switzerland
Theodore W. Rall, Cleveland, Ohio, U.S.A.
Martin Rodbell, Bethesda, Md., U.S.A.
Charles G. Smith, New Brunswick, N.J., U.S.A.

Advances in Cyclic Nucleotide Research

Volume 3

EDITORS:

Paul Greengard, Ph.D.
Professor of Pharmacology
Yale University School of Medicine
New Haven, Connecticut

G. Alan Robison, Ph.D.
Chairman, Department of Pharmacology
University of Texas Medical School
Houston, Texas

Raven Press, Publishers ■ New York

© 1973 by Raven Press Books, Ltd. All rights reserved. This book is protected by copyright. No part of it may be duplicated or reproduced in any manner without written permission from the publisher.

Made in the United States of America

International Standard Book Number 0-911216-38-3
Library of Congress Catalog Card Number 71-181305

Preface

Can a productive scientist keep abreast of a scientific literature that doubles in size every fifteen years and shows evidence of continued exponential growth during this decade? Dr. Martin M. Cummings, the Director of the National Library of Medicine, has expressed the belief that it is no longer possible to do so, even in a limited field or discipline (*American Scientist,* 61:163, 1973). The question posed by Dr. Cummings is especially relevant to scientists interested in cyclic nucleotide research, since the literature in this field seems to be expanding at a faster rate than in most, and shows no sign of leveling off.

Thoughtful review articles covering selected portions of the literature can be an invaluable means of helping scientists cope with this problem, and one of the main purposes of *Advances in Cyclic Nucleotide Research* is to provide a forum for reviews of this nature. Our hope is that these reviews will not only make the literature relating to cyclic nucleotides more meaningful, but may also stimulate further productive research. With these aims in mind, we are especially pleased to introduce the third volume of this series. The authors of the six reviews in this volume have set a high standard for the authors of future reviews in this series.

Paul Greengard
G. Alan Robison

Contents

1	Adenyl Cyclase *John P. Perkins*
65	Cyclic Nucleotide Phosphodiesterases *M. M. Appleman, W. J. Thompson, and T. R. Russell*
99	Protein Kinases and Protein Kinase Substrates *Thomas A. Langan*
155	Cyclic GMP *Nelson D. Goldberg, Robert F. O'Dea, and Mari K. Haddox*
225	The Chemistry and Biological Properties of Nucleotides Related to Nucleoside 3',5'-Cyclic Phosphates *Lionel N. Simon, Dennis A. Shuman, and Roland K. Robins*
355	Clinical Studies and Applications of Cyclic Nucleotides *Ferid Murad*
385	Author Index
393	Subject Index

List of Contributors

M. M. APPLEMAN
Department of Biological Sciences, University of Southern California, Los Angeles, California 90007

NELSON D. GOLDBERG
Department of Pharmacology, University of Minnesota, Minneapolis, Minnesota 55455

MARI K. HADDOX
Department of Pharmacology, University of Minnesota, Minneapolis, Minnesota 55455

THOMAS A. LANGAN
Department of Pharmacology, University of Colorado, Denver, Colorado 80220

FERID MURAD
Division of Clinical Pharmacology, Departments of Internal Medicine and Pharmacology, University of Virginia, School of Medicine, Charlottesville, Virginia 22903

ROBERT F. O'DEA
Department of Pharmacology, University of Minnesota, Minneapolis, Minnesota 55455

JOHN P. PERKINS
Department of Pharmacology, University of Colorado School of Medicine, Denver, Colorado 80220

ROLAND K. ROBINS
ICN Nucleic Acid Research Institute, 2727 Campus Drive, Irvine, California 92664

T. R. RUSSELL
Laboratory of Molecular Biology, National Cancer Institute, National Institutes of Health, Bethesda, Maryland 20014

DENNIS A. SHUMAN
ICN Nucleic Acid Research Institute, 2727 Campus Drive, Irvine, California 92664

LIONEL N. SIMON
ICN Nucleic Acid Research Institute, 2727 Campus Drive, Irvine, California 92664

W. J. THOMPSON
University of Washington, Department of Medicine, Division of Endocrinology, Seattle, Washington 98195

Adenyl Cyclase

John P. Perkins*

OUTLINE

I. Introduction 2
II. Assay Procedures 3
III. Characteristics of the Adenyl Cyclase Enzyme System 6
    A. General Properties 6
        1. Species distribution 6
        2. Tissue distribution in mammals 7
        3. Subcellular distribution 8
        4. A working model 11
    B. Properties of the Catalytic Component 13
        1. General considerations 13
        2. The reaction: Stoichiometry and thermodynamics 16
        3. Reaction requirements 17
        4. Stimulation by NaF 23
        5. Properties of detergent-treated adenyl cyclases 30
    C. Characteristics of Hormonal Regulation of Adenyl Cyclase 32
        1. General comments 32
        2. Factors affecting adenyl cyclase activity and its sensitivity to hormones 36
        3. Correlation of hormone binding and enzyme activation 41
        4. Ontogenetic development and differentiation 47
IV. Speculation on the Structure of the Adenyl Cyclase System 51
V. Acknowledgments 56
VI. References 56

---

*Recipient of a Research Career Development Award KO4-CA-70466

I. INTRODUCTION

The enzymatic conversion of adenosine triphosphate (ATP) to adenosine 3',5'-cyclic monophosphate (cAMP) (Fig. 1) was first described in detail by Sutherland and Rall and their co-workers in a series of papers in 1962 (Sutherland, Rall, and Menon, 1962; Rall and Sutherland, 1962; Murad, Chi, Rall, and Sutherland, 1962; Klainer, Chi, Freidberg, Rall, and Sutherland, 1962). It was shown that the enzyme, which they named adenyl cyclase,[1] required ATP and Mg^{++}, was particulate bound, and was present in tissues from at least four phyla. In every case tested, enzyme activity

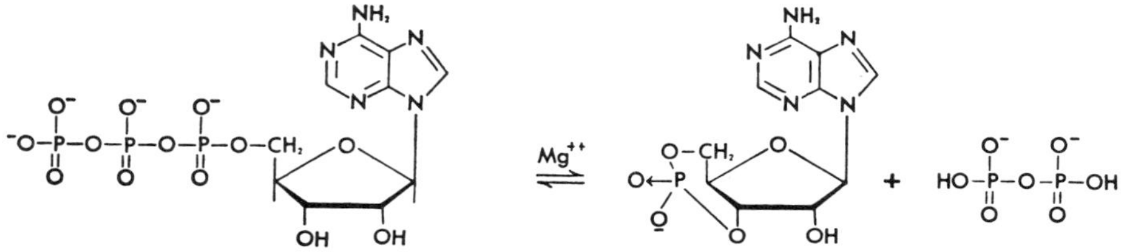

FIG. 1. The reaction catalyzed by adenyl cyclase.

was stimulated by NaF and in a tissue-specific manner was stimulated by different hormones. Their suggestion (Sutherland and Rall, 1960; Rall and Sutherland, 1961; Sutherland, Øye, and Butcher, 1965) that the action of many hormones might be mediated by cAMP as a result of stimulation of adenyl cyclase activity has been confirmed and extended by a myriad of subsequent studies (for reviews, see Robison, Butcher, and Sutherland, 1968, 1971; Sutherland, Robison, and Butcher, 1968; Greengard and Costa, 1970; Jost and Rickenberg, 1971; Robison, Nahas and Triner, 1971; Greengard and Robison, 1972).

Although the adenyl cyclase system cannot yet be described in accurate molecular terms, a large amount of information concerning its properties has been accumulated, and a review of the current status of the problem seems appropriate now. Aspects of the topic have been reviewed previously (Birnbaumer, Pohl, Krans, and Rodbell, 1970; Weiss, 1970; Rall, 1971; Rodbell, 1971; Robison, Butcher, and Sutherland, 1971), but much of the

---

[1]The enzyme should more appropriately be named adenylate cyclase or adenylyl cyclase based on the reaction catalyzed. Out of respect for the original discoverers, because of its common usage, and because it is easier to say and to write, I have chosen in this article to abbreviate the name of the enzyme as adenyl cyclase.

work dealing directly with the properties of the enzyme system in broken cell preparations has appeared only within the last two years.

In this chapter an attempt is made to discuss the important facts and concepts related to the structure and the regulation of function of adenyl cyclases in general. If the review of the literature is at all comprehensive, it is from this point of view since no attempt has been made to compile a comprehensive bibliography of the studies which relate to adenyl cyclase. The discussion is not directed toward the more complex question of how the level of cAMP in cells is regulated, but only to the question of how the activity of the enzyme which synthesizes cAMP is regulated. Within this limitation I will consider primarily those studies which have examined adenyl cyclase activity in broken cell preparations. Only in this experimental situation can one, with any assurance, determine directly the results of enzyme activity, i.e., the production of cAMP. Even then, factors in the crude enzyme preparation other than adenyl cyclase often become significant determinants of how effectively enzyme activity can be measured. With less confidence, studies involving the determination of cAMP content of intact cell preparations will be considered as reflecting adenyl cyclase activity. Studies wherein physiological effects of hormones have been measured will not be considered because of the probability of complicating intervening steps between adenyl cyclase activity and the physiological event measured.

The writing of this review has been aided by the products of two reference gathering organizations. The two volumes (1957–1969; 1970) of the *Cyclic AMP* bibliography generously made available by E. R. Squibb and Sons, Inc.[2] were especially useful because of their completeness and extensive cross-referencing. Search of the more recent literature[3] was aided by the monthly bibliography, *Cyclic AMP*, produced by the University of Sheffield Biomedical Information Project.

II. ASSAY PROCEDURES

Early work on adenyl cyclase relied on biological assays for cAMP (Rall and Sutherland, 1958, 1962). These techniques were tedious and usually involved extensive purification of cAMP samples since they involved multicomponent enzyme systems whose activity could be influenced by a number of adenine nucleotides and certain hexoses. Nonetheless, when properly controlled, such assays are quite sensitive and provide an accurate

---

[2] My appreciation is extended to Dr. N. S. Semenuk who also made available the 1971 volume of *Cyclic AMP* prior to publication.
[3] The literature search was completed in June, 1972.

measure of cAMP. However, simpler methods for the measurement of cAMP have been reported and are described in Volume 2 of this series.

With the availability of radioactively labeled ATP and the advent of quick, simple methods for the purification of cAMP, investigators were provided with a direct radiometric assay procedure for adenyl cyclase activity in broken cell preparations (Rabinowitz, Desalles, Meisler, and Lorand, 1965; Streeto and Reddy, 1967; Krishna, Weiss, and Brodie, 1968; Bär and Hechter, 1969b).

Adenyl cyclase from most tissues has not been readily purified, primarily because it is membrane bound and is relatively unstable. Thus, most investigators utilize whole homogenates or washed membrane fractions as enzyme preparations. In such preparations, ATPase and cAMP-phosphodiesterase (PDE) activities often are significant contaminants and their presence must be considered. Theoretically, linear reaction rates could be obtained if the concentration of ATP used was of such a magnitude that even though destruction occurred, the level of ATP would not drop below concentrations that were saturating for adenyl cyclase activity. The K_m for ATP in the presence of excess Mg^{++} is usually in the range of 0.08 to 0.50 mM. Thus, as long as the concentration of ATP does not drop below about 1.5 mM, cAMP would be produced at a reasonably constant rate. However, two major problems have been encountered. First, ATPase activity as high as 0.6 μmole/min/mg protein has been observed (Drummond and Duncan, 1970). Thus, even relatively high levels of ATP can be rapidly destroyed. Second, since the experimental blank is usually directly proportional to the concentration of labeled ATP (at constant specific activity), in tissues with low adenyl cyclase activity, the small amount of labeled cAMP formed becomes difficult to distinguish from the blank without extensive purification (see Bär and Hechter, 1969b; Krishna and Birnbaumer, 1970).

An alternate means of attaining linear reaction rates has been to include an enzymatic ATP-regenerating system in the assay mixture. In this manner even low concentrations of ATP can be maintained constant throughout the assay period. This technique is especially applicable to the assay of tissues with low adenyl cyclase activity but is useful in most circumstances when properly applied. However, the adequacy of the recycling system often has been verified only for standard assay conditions and not reverified for each new set of assay conditions. The most commonly used regenerating systems have been phosphocreatine and creatine phosphokinase or phosphoenolpyruvate and pyruvate kinase. Drummond and Duncan (1970) have pointed out some potential pitfalls in the use of the latter system.

Rodbell, Birnbaumer, Pohl, and Krans (1971) solved this technical problem in a unique manner by the use of an analog of ATP, 5'-adenylyl imidodiphosphate, labeled with ^{32}P at the α-position. This compound was shown to be an effective substrate for adenyl cyclase of rat liver plasma membranes, but a poor substrate for ATPase.

Phosphodiesterase activity also is often present in tissue homogenates at levels one to two orders of magnitude higher than adenyl cyclase. In most studies PDE activity has been at least partially inhibited by inclusion of a methylxanthine in the adenyl cyclase assay mixture. However, even high concentrations of theophylline (5 to 20 mM) probably do not inhibit PDE activity completely (Drummond and Duncan, 1970) and in some cases can inhibit adenyl cyclase activity (Sheppard, 1970; Weinryb and Michel, 1971).

The inclusion of a high concentration (1 to 6 mM) of nonlabeled cAMP in the assay mixture with or without a methylxanthine has been shown to be an adequate means of protecting the labeled cAMP formed during the reaction (Weiss and Costa, 1968; Birnbaumer, Pohl, and Rodbell, 1969). A measure of the cAMP (absorbance 260 nm) in the purified sample compared to that in no-enzyme blanks allows a correction to be made not only for losses of labeled cAMP during purification but for losses due to PDE activity during the assay. This was found to amount to a significant correction in our studies of adenyl cyclase activity in brain homogenates because of the high PDE activity (Perkins and Moore, 1971). Such considerations are especially important when variations in assay conditions can lead to variation in PDE activity; e.g., Triton X-100 added to adenyl cyclase assays can increase PDE activity twofold (Cheung, 1967). The addition to assay mixtures of both an ATP recycling system and excess unlabeled cAMP could lead to significant reduction in the specific activity of the substrate (labeled ATP) if PDE activity coupled with adenylate kinase activity is adequate to convert significant amounts of unlabeled cAMP to ADP. The conversion of the ADP to ATP by the recycling system would result in a dilution of the specific activity of ATP.

The most widely used method for the isolation of cAMP from adenyl cyclase reaction mixtures is that of Krishna, Weiss, and Brodie (1968). The procedure involves chromatography of the samples on columns of Dowex 50 followed by the precipitation of contaminating nucleotides with $ZnSO_4$ and $Ba(OH)_2$. When properly carried out (see Krishna and Birnbaumer, 1970), the method is adequate for the determination of even small amounts of labeled cAMP. Two points concerning the procedure not pointed out by the authors merit consideration. First, the $ZnSO_4$-$Ba(OH)_2$ precipitation should not be carried out in the presence of high concentrations of labeled ATP since formation of cAMP by chemical reaction can occur in the presence of $Ba(OH)_2$ and ATP (Cook, Lipkin, and Markham, 1957). Second, an adequate separation of ATP and cAMP on Dowex 50 under the conditions described (Krishna et al., 1968) occurs only when cAMP is present at high concentrations. This problem can be eliminated by the addition of carrier cAMP to adenyl cyclase reaction mixtures prior to chromatography on the Dowex columns. Alternatively, low concentrations of cAMP can be separated from ATP using the modification described by Otten, Johnson, and Pastan (1971). Essentially, the modification involves elution of ATP with dilute (0.1 N) HCl followed by elution of cAMP with water. Other methods for purification of cAMP are discussed in Volume 2 of this series.

An often used criterion for the identity of cAMP in purified unknown samples is the absence of "cAMP" after incubation of the sample with preparations containing PDE activity. Such a determination is clearly not definitive. The PDE preparations used are invariably impure and thus could contain other enzyme activities that could convert any cAMP-like substances to other structures with different purification properties. Two variations seem more appropriate: (1) the quantitative formation of 5'-AMP (or adenosine) should be demonstrated after incubation with PDE, or (2) authentic cAMP, labeled with a different isotope, should be added to the sample, at approximately equal concentrations, and the *rate* as well as the extent of loss of both forms of "cAMP" determined.

In general, the reaction requirements for adenyl cyclase from a number of tissues are similar. Tris·HCl or glycyl-glycine buffer systems have been used in most instances. The pH profile of the reaction rate exhibits a fairly broad maximum between 7.0 and 8.5 with a rapid decline in reaction rate below pH 7.0. The ionic strength of the reaction mixture is usually not critical with certain exceptions which will be discussed in later sections. A divalent cation, usually Mg^{++}, is required for the reaction and adequate concentrations for maximal activity should be determined for the enzyme from each different source. A complex relationship exists between Mg^{++} and ATP concentrations, which is discussed below.

Much of the misinformation in the literature concerning the properties of adenyl cyclase can be traced to inadequate assay techniques, usually involving one or more of the points considered above.

III. CHARACTERISTICS OF THE ADENYL CYCLASE ENZYME SYSTEM

A. General Properties

1. *Species Distribution*

Cyclic AMP appears to be omnipresent in animal species as well as in at least 10 different strains of bacteria (Ide, 1971). Adenyl cyclase also is assumed to be present in such cases and has been demonstrated to be in every metazoan species examined to date (for extensive listing of occurrence, see Weiss, 1970). There are definite differences in the properties of mammalian and bacterial adenyl cyclases which may be related to the apparently different modes of regulation of enzyme activity. In mammalian cells synthesis of cAMP is regulated by hormonal influence on adenyl cyclase activity, the hormone being produced in a cell type distinct from that which it affects. In bacteria the regulation of cAMP synthesis appears to be related to the nutritional state of the bacteria, and there is no evidence to suggest a mechanism involving cell to cell communication (Pastan and Perlman, 1970). From a teleological point of view, the role of cAMP in the two phyla appears to be somewhat different. In mammals the adenyl cyclase of a particular cell type responds to specific extracellular signals to produce cAMP, with a resultant change in cellular function, for the good of the organism as a whole. In bacteria the rise in cAMP occurs in the cell to satisfy a nutritional need for that same cell. A study of the phylogenetic development of hormone-sensitive adenyl cyclases would be of interest. Based on the above rationale it is predicted that hormonal-type regulation should have developed concomitantly with the appearance of heterocellular organisms. To date there has been no report of an effect of a metazoan hormone on the activity of adenyl cyclase of unicellular organisms.

2. *Tissue Distribution in Mammals*

In every tissue examined, with the possible exceptions of dog erythrocytes (Sutherland, Rall, and Menon, 1962) and a line of rat hepatoma cells (Granner, Chase, Aurbach, and Tomkins, 1968), the presence of either cAMP or adenyl cyclase activity *per se* has been demonstrated. Consistent with the 2nd messenger hypothesis (Sutherland, Øye, and Butcher, 1965), the hormonal responsiveness of broken cell preparations of adenyl cyclase mirrors the hormonal responsiveness of the tissue in terms of physiological effects. However, an orientation related to tissue specificity can be misleading when the tissue is comprised of different cell types responding to different hormones. This is clearly the case in the kidney where the enzyme from the renal medulla is activated by vasopressin while the renal cortex enzyme is responsive to parathyroid hormone (Chase and Aurbach, 1968).

Although the level of basal enzyme activity varies considerably from tissue to tissue (see Sutherland, Rall, and Menon, 1962; Rall, 1971), the general kinetic properties of the catalytic aspect of adenyl cyclase are similar in most cases. Thus, the primary distinguishing feature of the enzyme in different tissues is its pattern of responsiveness to hormones.

Recently a role for cAMP in the regulation of mammalian cell growth has been suggested by work from a number of laboratories. The basic observations are that cAMP (Ryan and Heidrick, 1968), analogues of cAMP (Heidrick and Ryan, 1970; Johnson, Friedman, and Pastan, 1971; Sheppard, 1971; and Macintyre, Perkins, Wintersgill, and Vatter, 1972), activators of adenyl cyclase (Johnson and Pastan, 1971), and inhibitors of PDE activity (Bürk, 1968; Prasad and Sheppard, 1972) can decrease the rate and/or extent of growth of malignant cells in culture. These findings fit well with the inverse correlation of growth rate and cAMP content of a number of fibroblast lines in culture and the observation that the level of cAMP in normal (nonmalignant) cells increases when such cells reach the contact-inhibited state where growth ceases (Otten, Johnson, and Pastan, 1971). Taken together, these observations imply that malignant transformation might involve either decreased ability to synthesize cAMP or an increased ability to destroy it. However, the observations to date do not suggest any general alteration of adenyl cyclase that can be related to the lower cAMP content of malignant cells. Basal enzyme activity of malignant cells can be increased over the normal tissue (Trauton, Roth, and Pastan, 1969) or decreased (Emmelot and Bos, 1971), and in one case apparently was absent (Granner et al., 1968). Hormonal sensitivity can be lost (Allen, Munshower, Morris, and Weber, 1971; Schimmer, 1972) or retained (Schimmer, Ueda, and Sato, 1968; Pastan, Pricer, and Blanchette-Mackie, 1970; Emmelot and Bos, 1971; Rosen, Hirsch, and Goren, 1971). The specificity of responsiveness to hormones may be different in tumors as compared to their tissue of origin (Dexter and Allen, 1970; Schorr, Rathnam, Saxena, and Ney, 1971). Thus, no valid generalizations can be made relating the properties of adenyl cyclase and the characteristics of the growth of mammalian cells.

3. *Subcellular Distribution*

The early studies of Sutherland, Rall, and Menon (1962) demonstrated that adenyl cyclase activity was tightly associated with sedimentable fractions of tissue homogenates. Davoren and Sutherland (1963) showed that adenyl cyclase of pigeon erythrocyte ghosts sedimented with the $600 \times g$ "nuclear" fraction. However, a predominantly nuclear localization of the enzyme was excluded by their demonstration that nuclei and adenyl cyclase activity could be separated by centrifugation of cell lysates in 20% glycerol solutions. Similar techniques were used by these workers to exclude a predominantly mitochondrial localization of the enzyme as well. It was concluded that adenyl cyclase activity was probably localized in the plasma membrane of the pigeon erythrocyte. More recent work in a number of laboratories suggests that in most, if not all, mammalian tissues this enzyme exists predominantly in the plasma membrane.

Under appropriate conditions erythrocytes and isolated adipocytes are amenable to lysis with retention of a great deal of plasma membrane integrity. Such preparations appear under the light microscope to be composed of empty sacs (ghosts) relatively devoid of nuclei or other internal structures. Rodbell (1964) first described preparations of isolated adipocytes, and subsequent studies from his and other laboratories have established that the activity of the many lipolytic hormones are reflected in the hormonal responsiveness of the adenyl cyclase of adipocyte ghosts (Butcher, Baird, and Sutherland, 1968; Bär and Hechter, 1969a; Birnbaumer, Pohl, and Rodbell, 1969; Birnbaumer and Rodbell, 1969).

Rosen and Rosen (1969) have described in some detail the properties of adenyl cyclase of partially purified frog erythrocyte ghosts and fragments of these ghosts. Phase microscopic examination failed to detect nuclei or other internal structures within the ghosts. The preparation contained approximately 70% of the enzyme activity of the total lysate. The enzyme of the 150-fold purified membrane fragments was still stimulated by NaF and epinephrine.

Wolff and Jones (1971) purified bovine thyroid membranes approximately 100-fold on the basis of TSH-stimulated adenyl cyclase activity. The adenyl cyclase activity co-purified with Na^+-K^+-ATPase, K^+-stimulated, p-nitrophenyl phosphatase, and 5'-nucleotidase which are all thought to be valid plasma membrane marker enzymes.

The adenyl cyclase of rat brain was found to be of highest specific activity in subcellular fractions rich in synaptosomes (DeRobertis, Arnaiz, Alberici, Butcher, and Sutherland, 1967). Although such structures have specialized functions they are nonetheless extensions of the external cell membrane of neurons.

Recently, Rodbell and his co-workers (Pohl, Birnbaumer, and Rodbell, 1971; Birnbaumer, Pohl, and Rodbell, 1971, 1972; Rodbell, Krans, Pohl, and Birnbaumer, 1971a,b; Rodbell, Birnbaumer, Pohl, and Krans, 1971) have extensively characterized the adenyl cyclase of rat liver plasma membranes prepared by the method of Neville (1968). Both 5'-nucleotidase and adenyl cyclase were purified 15- to 20-fold in this preparation. Electron micrographic examination revealed numerous paired membrane sheets with typical intercellular junctions characteristic of the plasma membrane of hepatic parenchymal cells, as well as numerous vesicles of uncertain origin.

The idea that the adenyl cyclase system exists predominantly in the plasma membrane is based primarily on studies such as those outlined above, but also has received strong support from observations that hormones, bound to large, insoluble support materials, are effective in eliciting normal physiological responses. Schimmer, Ueda, and Sato (1968) first demonstrated that ACTH or the synthetic eicosapeptide analogue of ACTH, diazotized to *p*-amino-benzyl-cellulose, was active in the stimulation of steroidogenesis in cultures of adrenal tumor cells. Recently, Johnson, Blecher, and Giorgio (1972) demonstrated that glucagon diazotized to agarose beads could stimulate adenyl cyclase activity of rat liver plasma membranes, but more importantly could activate lipolysis in intact adipocytes. Norepinephrine diazotized to agarose also was an effective lipolytic agent. As will be discussed later, a number of investigators have demonstrated directly the binding of polypeptide and amine hormones to partially purified plasma membrane preparations (Lefkowitz, Roth, and Pastan, 1971; Lefkowitz and Haber, 1971; Rodbell, Krans, Pohl, and Birnbaumer, 1971; Schramm, Feinstein, Naim, Lang, and Lasser, 1972).

Certainly, observations such as those described above do not exclude the possibility that adenyl cyclase could be present in certain internal membranous structures. Seraydarian and Mommaerts (1965) and Rabinowitz, Desalles, Meisler, and Lorand (1965) reported that cyclase activity was distributed largely in mitochondrial and microsomal fractions of rabbit skeletal muscle homogenates. However, recent studies of Severson, Drummond, and Sulakhe (1972) provide convincing evidence that the major portion of enzyme activity can be demonstrated in highly purified sarcolemma. They suggest that at least a portion of adenyl cyclase activity observed in other cell fractions could be due to contamination by fragments of sarcolemma. Entman, Levey, and Epstein (1969) reported the presence of adenyl cyclase of high specific activity in microsomal fractions of cardiac muscle homogenates. However, the percentage of the total enzyme activity recovered in such fractions was not presented, and contamination with fragments of sarcolemma could not be discounted.

Recently, work from two laboratories has shown the presence of adenyl cyclase activity in purified preparations of nuclei from rat ventral prostate (Liao, Lin, and Tymoczko, 1971) and rat liver (Soifer and Hechter, 1971). The enzyme of rat liver nuclei was shown to have properties that distinguished it from the enzyme obtained from membrane fractions of liver devoid of nuclei. The possibility of adenyl cyclase activity in nuclei poses the interesting question as to whether it might play a role in steroid hormone function. In these studies cortisol and dexamethasone (0.1 to 20 μg/ml) were without effect on the enzyme activity of rat liver nuclei. However, in view of the role of a protein steroid-receptor in steroid action (Jensen, Numata, Brecher, and DeSombre, 1971), a study of the effect of the steroid-receptor complex on the activity of the nuclear adenyl cyclase would be of interest.

The possibility that cAMP might play a role in steroid hormone action has been discussed in detail by Hechter and Soifer (1972).

Tao and Lipmann (1971) demonstrated that adenyl cyclase activity of *Escherichia coli*, although associated with particulate subcellular fractions, was readily obtained in a soluble form by washing the particles with aqueous buffers. Ide (1971) was able to divide 10 strains of bacteria into two classes based on whether most of the adenyl cyclase activity was found in supernatant or particulate fractions after disruption of the bacteria in a French press. Thus, in bacteria the association of adenyl cyclase with membrane structures is much more tenuous than in mammalian tissues.

At present it appears that in most mammalian cells the adenyl cyclase system exists primarily in the plasma membrane, but it also may exist and function in other cellular membranous structures.

4. A Working Model

There has long been indirect evidence that the adenyl cyclase system was composed of at least two components: a receptor or regulatory component and a catalytic component. First, although the properties of the basal catalytic reactions are quite similar for the enzyme from different tissues, the hormonal regulation of enzyme activity can be highly selective. Second, for most enzyme preparations the hormonal sensitivity is readily lost under a variety of mildly stressful conditions while little or no change in catalytic activity may occur. As a working hypothesis, Robison, Butcher, and Sutherland (1967) proposed the model shown in Fig. 2. The model assumes the existence of two subunits with a specific directional orientation. Hormone interaction with the receptor unit causes an alteration in the receptor which in turn alters the catalytic unit in a way that leads to increased enzyme activity. It is logical and consistent with available evidence that the two subunits should be oriented as shown; for example, hormones bound to agarose beads can elicit normal responses in intact cells, *intracellular* ATP is the substrate of choice of adenyl cyclase of intact cells, and cAMP accumulates *within* cells upon hormone stimulation.

An alternate model has been suggested (Hechter and Halkerston, 1964; Hechter, 1965) differing in essence only by the proposal that intermediate moieties function between the hormone receptor and the catalytic unit. Recently, Rodbell and co-workers (Birnbaumer, Pohl, Krans, and Rodbell, 1970; Rodbell, 1971) have described in detail a model of the adenyl cyclase system composed of a hormone discriminator (receptor), a transducer, and an amplifier (catalytic unit) (Fig. 3). The term "transducer" is used by these workers to denote "that element which couples events occurring at the

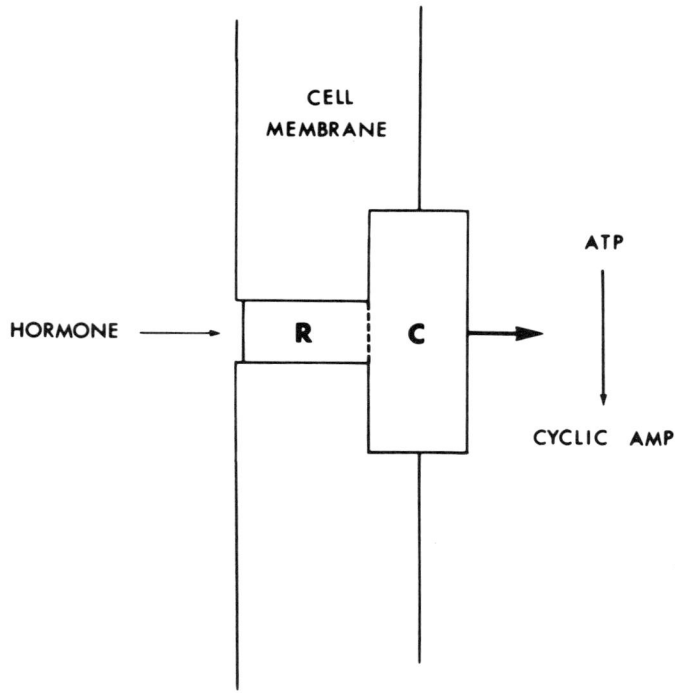

FIG. 2. Hypothetical, two-component model of the adenyl cyclase system (taken from Robison, Butcher, and Sutherland, 1967).

discriminator to events taking place at the amplifier" (Rodbell, 1972). In its simplest form, the adenyl cyclase system could be a regulatory enzyme with the discriminator being an allosteric site for hormone interaction on a regulatory subunit and the transducer being alterations in enzyme conformation which result in changes in the activity of the catalytic subunit (amplifier). However, evidence to be presented below is also compatible with an adenyl cyclase system composed of receptor and catalytic moieties which are distinct and physically separated but capable of interaction as a result of their location in a dynamic lipid matrix.

The properties of the adenyl cyclase system are discussed below, first in relation to the catalytic component of the system, then in terms of the receptor-transducer components. Finally, the enzyme system is considered in terms of current ideas of membrane structure.

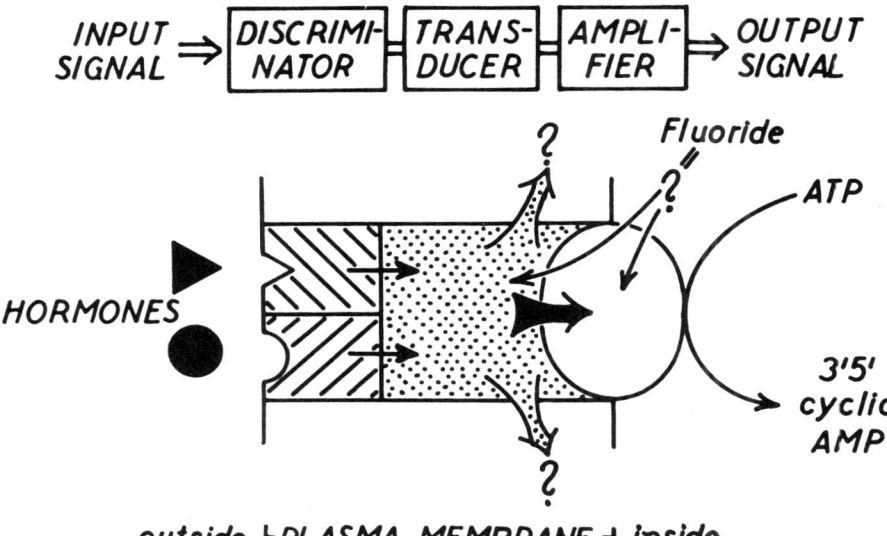

FIG. 3. Hypothetical, three-component model of the adenyl cyclase system (modified from Birnbaumer, Pohl, Krans, and Rodbell, 1970).

B. Properties of the Catalytic Component

1. *General Considerations*

There are numerous indications that the adenyl cyclase activity observed in broken cell preparations only partially reflects the properties of the enzyme inferred from studies of intact cells. For example, the order of potency as well as the absolute potencies of norepinephrine, epinephrine, and isoproterenol are different when the responsiveness of adenyl cyclase of fat-cell ghosts is compared to effects of these amines on whole cell preparations of adipose tissue (see Himms-Hagen, 1967). It has been a general finding that broken cell preparations are less sensitive to hormonal stimulation than whole-cell preparations (see Robison, Butcher, and Sutherland, 1971). Also, virtually all broken-cell preparations of the enzyme are stimulated by NaF, but NaF does not stimulate the adenyl cyclase activity of intact cells even though it probably enters most cells readily. Such observations suggest that significant alteration of the enzyme system occurs upon fragmentation of the cell membrane. Furthermore, since the enzyme is probably oriented within the membrane with receptors exposed on the outside and catalytic components exposed on the inside, the formation of vesicles could result in the expression of artifactual properties. "Inside-out" vesicles could reduce or prevent hormone interaction and "right-side-out" vesicles might effectively reduce total enzyme activity if the vesicles were impervious to the exogenous, labeled substrate. The interpretation of the effects of substances that alter membrane integrity (detergents, Ca^{++}, EDTA, lipolytic or proteolytic enzymes, etc.) on adenyl cyclase activity should include consideration of such structural artifacts.

The nonstimulated or basal adenyl cyclase activity/gm wet weight observed in homogenates varies considerably from tissue to tissue (see Sutherland, Rall, and Menon, 1962). However, there is no obvious correlation between such a measure of basal enzyme activity and the tissue content of cAMP; values usually range between 0.5 and 1.5 μmoles cAMP/kg tissue. As pointed out by Schramm and Naim (1970), we have little understanding of the significance of the so-called basal activity of adenyl cyclase in broken-cell preparations. They suggested that basal activity could either be representative of a true hormone-independent activity involved in the maintenance of basal cAMP content or could be an artifact due either to perturbation of the enzyme system during homogenization or to the carry-over of small amounts of hormone into the assay mixture. The latter possibility is apparently not the case in many instances since specific hormone antagonists usually do not alter basal enzyme activity. Furthermore, as will be discussed later, there are a number of substances that inhibit hormonal effects but do not change basal activity.

Since virtually all mammalian cells maintain significant intracellular levels in cAMP (about 1 μM), and since a significant amount of cAMP is excreted in urine (1 to 2 μmoles/day for man), continuous, limited synthesis of cAMP is a likely occurrence. Within the normal organism regulation of the synthesis of cAMP is a highly dynamic process. That is, cells are constantly exposed to slightly fluctuating levels of hormone and what is observed as the basal cAMP content is determined in part by "tonic" hormonal influence. However, in whole-cell tissue preparations (or cells in culture) the content of cAMP usually remains constant for hours in chemically defined media devoid of hormones or in the presence of hormone antagonists. Thus, in such instances either cAMP must be maintained at a steady state level by a balance of synthesis and destruction or the cell content of cAMP must be in a stable form not susceptible to destruction (O'Dea, Haddox, and Goldberg, 1971).

We have examined in two systems the rate of incorporation of radioactive label (^{3}H-adenine) into cAMP during incubation in defined media in the absence of hormones and/or in the presence of hormone antagonists. In experiments with cultured astrocytoma cells and with slices of rat cerebral cortex, the specific activity of cAMP was found to increase slowly to a constant value while total cAMP content did not change. After reaching the point where the specific activity of cAMP was constant, the addition of norepinephrine resulted in marked increases in total cAMP content (brain slices, six- to eightfold; astrocytes, 30-fold), but no change in the specific activity of cAMP was observed. Barring the possibility of reversal of the adenyl cyclase reaction (see below), these experiments indicate a measurable turnover of the basal cAMP pool. The observation that the cAMP formed under the influence of norepinephrine was of the same specific activity as that attained prior to its addition suggests that both the basal cAMP and the effector-induced cAMP were formed from the same ATP substrate pool. Although these studies support the existence of significant basal activity of adenyl cyclase in whole cells, it is not clear if this activity can be equated with the basal enzyme activity observed in broken cell preparations.

If hormonal control of adenyl cyclase is stringent in nature, the catalytic component can be considered to exist in two forms, one inactive and the other fully active. Thus, in whole cells any basal synthesis of cAMP would be in proportion to the amount of enzyme in the active form in the absence of hormones. Basal activity of broken-cell preparations could be greater or less than that in whole cells, depending on the effect of membrane fragmentation on the distribution of the two forms of the enzyme. Consistent with such a model, the *catalytic* properties of basal activity should be the same as those of hormone-induced activity since only one form of active enzyme exists.

Alternatively, if the so-called inactive form of the enzyme was simply *less* active, then basal activity could be ascribed to the inherent activity of this form. In such a model the fully active form of the catalytic unit need only be formed in the presence of hormone(s) or other activators. In this model the form of the enzyme responsible for basal activity might well express catalytic properties different from those of the hormone-induced, fully active form of the enzyme.

To date there are few indications of differences in the catalytic properties of basal vs. hormone- or NaF-activated forms of the enzyme. It will be pointed out later that activity in the presence of hormones can readily be inhibited by factors which do not alter basal activity. However, such differences could be due to requirements for the activation process and not to any basic difference in the properties of the catalytic unit.

There are two reasons for this expression of concern over the existence and identity of a basal enzyme activity. First, it appears that an inability to maintain adequate basal levels of cAMP may be involved in the loss of growth control observed in transformed cells in culture (Johnson and Pastan, 1971). Second, such considerations will eventually be of importance in the process of unraveling the complex structure-function relationship of adenyl cyclase.

2. The Reaction: Stoichiometry and Thermodynamics

In 1962 Rall and Sutherland first reported the details of the reaction (see Fig. 1) catalyzed by adenyl cyclase. The reaction involves the conversion of one molecule of ATP (probably as the ATP-Mg^{++} complex) to a molecule of cAMP and one of pyrophosphate. Thus, only ATP and Mg^{++} are required for the expression of basal enzyme activity in broken-cell preparations.

There has been some controversy over the potential reversibility of the adenyl cyclase reaction and the functional significance of the high energy of hydrolysis of the 3'-ester bond of cAMP. It has been shown in convincing fashion that under appropriate conditions the reversibility of the adenyl cyclase reaction is readily demonstrable (Greengard, Hayaishi, and Colowick, 1969; Hayaishi, Greengard, and Colowick, 1971; Takai, Kurashina, Suzuki, Ckamoto, Ueki, and Hayaishi, 1971; Khandelwal and Hamilton, 1971). The equilibrium constant of the reaction [Kobs = (cAMP)(PPI)/(ATP) = 0.065 M] favors the formation of ATP under standard conditions, i.e., reactants and products at 1.0 M. The equilibrium constant indicates a free energy of hydrolysis ($\Delta G°_{obs}$) for the 3'-ester bond of cAMP of -11.9 kcal/mole. The enthalpy of hydrolysis of cAMP to 5'-AMP was shown by Greengard, Rudolph, and Sturtevant (1969) to be -14.1 kcal/mole.

The inability of others (Cheung and Chiang, 1971; Tao and Lipmann, 1969; and Rosen and Rosen, 1969) to observe reversibility of the reaction can probably be attributed to the use of unfavorable reaction conditions, e.g., low concentrations of cAMP and PPi, low enzyme concentration and short reaction time. However, as has been pointed out by Hayaishi, Greengard, and Colcwick (1971), at the concentrations of cAMP, PPi, and ATP normally prevailing in mammalian cells, significant reversal of the adenyl cyclase reaction is highly unlikely.

Based on the high energy of hydrolysis of the 3'-ester bond Greengard, Hayaishi, and Colowick (1969) have suggested that cAMP might serve as an adenylating agent for proteins, e.g., for cAMP-dependent protein kinase. To date there is no evidence to support the occurrence of such a reaction.

3. Reaction Requirements

In Fig. 4 the relation of enzyme activity to ATP concentration is shown for adenyl cyclase of rat cerebral cortex. It is clear that ATP in excess of Mg^{++} is inhibitory. Such a relationship has been demonstrated in virtually every adenyl cyclase system examined (see, for example, Birnbaumer, Pohl, and Rodbell, 1969; Drummond, Severson and Duncan, 1971; Pohl, Birnbaumer, and Rodbell, 1971).

Since under optimal reaction conditions (pH 7.0 to 8.0; ATP above 3 mM; Mg^{++} 5 to 10 mM) ATP is predominantly in the form of the ATP-Mg complex, this complex is probably the substrate form of ATP. However, Hirata and Hayaishi (1967) have proposed that the substrate for the enzyme of *Brevibacterium liquefaciens* is the $ATP(Mg)_2$ complex.

Where it has been examined, the substrate specificity of adenyl cyclases

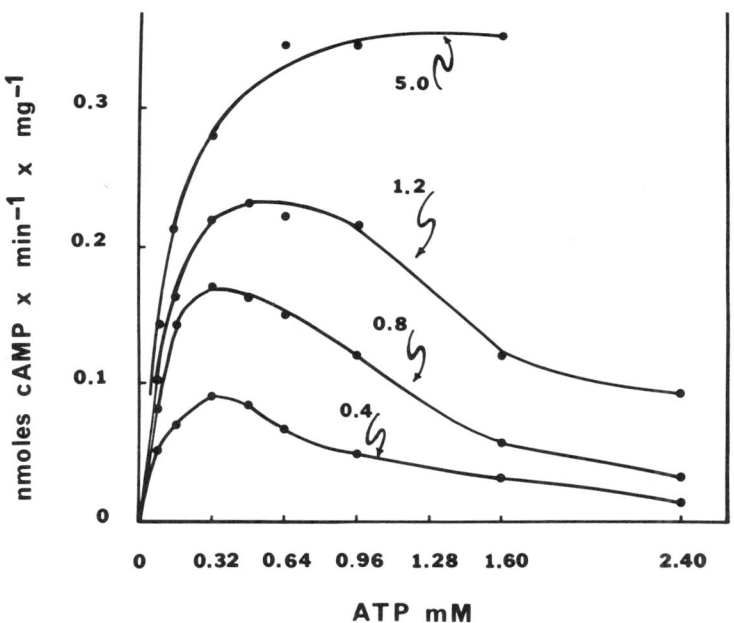

FIG. 4. The relationship of adenyl cyclase activity to ATP concentration. The enzyme preparation was a washed membrane fraction from rat cerebral cortex. The concentration of Mg^{++} is indicated in the figure. The details of the assay procedure were as described by Perkins and Moore (1971).

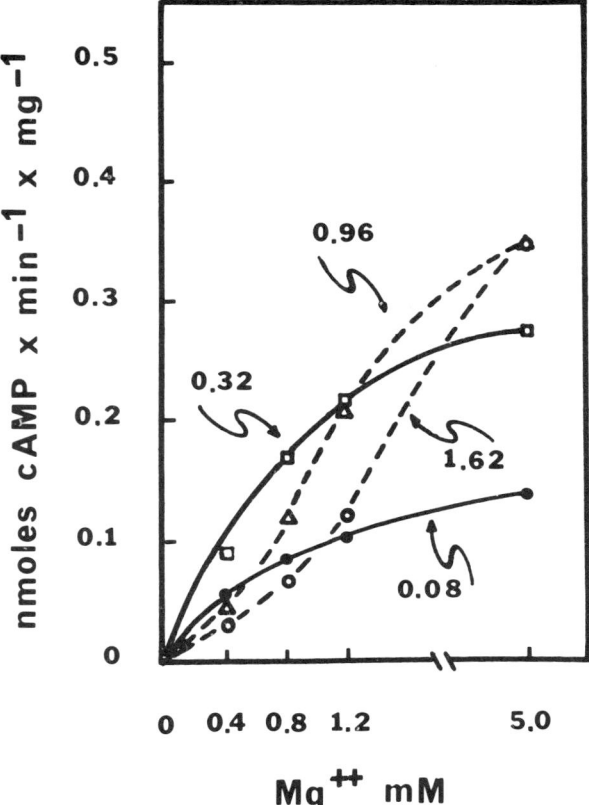

FIG. 5. The relationship of adenyl cyclase activity to Mg^{++} concentration. The enzyme preparation was a washed membrane fraction from rat cerebral cortex. The concentrations of ATP are indicated in the figure. The details of the assay procedure were as described by Perkins and Moore (1971).

for ATP is quite high. Only deoxy-ATP has been shown to act as an alternate substrate for the enzyme of *B. liquefaciens* (Hirata and Hayaishi, 1967), frog erythrocyte (Rosen and Rosen, 1969), sheep thyroid gland (Burke, 1970c), and rat fat cells (Bär and Hechter, 1969d). Other nucleoside triphosphates (TTP, CTP, dTTP, UTP) have been reported to inhibit the utilization of ATP but only to a limited extent.

In Fig. 5 the relation of enzyme activity to Mg^{++} concentration is shown for the enzyme of rat cerebral cortex. Two points can be made. First, at the two higher concentrations of ATP (0.96 and 1.62 mM), the relationship is sigmoidal as is to be expected since it is clear that ATP in excess of Mg^{++}

is inhibitory (see Fig. 4). Second, Mg^{++} increases enzyme activity over a range of concentrations that exceed the concentration necessary to convert most of the ATP to the ATP-Mg complex. This phenomenon was first described by Birnbaumer, Pohl, and Rodbell (1969) for adenyl cyclase of rat adipocyte ghosts and was confirmed by Drummond and Duncan (1970) for the enzyme from rabbit heart membranes. These workers concluded that Mg^{++} was interacting with the enzyme at a binding site other than the catalytic site to increase the V_{max} of the reaction.

Recently an alternate interpretation of such data has been proposed (C. DeHaen, *personal communication*). DeHaen pointed out that relationships similar to those shown in Figs. 4 and 5 are to be expected if free ATP is inhibitory and ATP-Mg is the true substrate. The concentration of Mg^{++} required in any adenyl cyclase system to bring about maximum activity would be related to the relative affinities of ATP and ATP-Mg for the enzyme; i.e., the higher the affinity of the enzyme for free ATP, the higher the Mg^{++} required for maximum activity. Calculations based on the results of Birnbaumer, Pohl, and Rodbell (1969) and Drummond and Duncan (1970) indicate that DeHaen's proposed mechanism is feasible if the affinity of the enzyme for free ATP is at least 100 times that of its affinity for ATP-Mg. Although this formulation is simpler than a mechanism requiring a second binding site for Mg^{++}, both possibilities are consistent with the available data.

The divalent cation requirement of most adenyl cyclases can be satisfied by Mn^{++} as well as by Mg^{++}, and Co^{++} also is partially active in some instances. Ca^{++}, Zn^{++}, and Cu^{++} are usually inhibitory. However, in most reports only a single concentration of Mn^{++} was tested and the results range from less activity (Sutherland, Rall, and Menon, 1962; Marcus and Aurbach, 1971) to equal activity (Sutherland, Rall, and Menon, 1962) to more activity (Birnbaumer, Pohl and Rodbell, 1969; Burke, 1970*a*). It is clear from the data in Fig. 6 why such variable results might have been obtained. Depending on the concentration of divalent cation at which the comparison is made, the ratio of Mn^{++}/Mg^{++} activity varies from 5 to 0.5 with the heart preparation (also see Drummond, Severson, and Duncan, 1971) and from 100 to 2 with the brain enzyme. With adenyl cyclase from at least five different tissues (brain, heart, erythrocyte, salmon testis, and thyroid), enzyme activity is greater at low concentrations of Mn^{++} (up to 1 to 5 mM) than at equal concentrations of Mg^{++}. The reason for the increased activity in the presence of Mn^{++} cannot be explained solely on the basis of the greater affinity of Mn^{++} for free ATP (Sillen and Martel, 1964) since the enzyme from rabbit heart (Drummond, Severson, and Duncan, 1971) and rat cerebral cortex (Fig. 6) is stimulated by Mn^{++} to a greater degree than

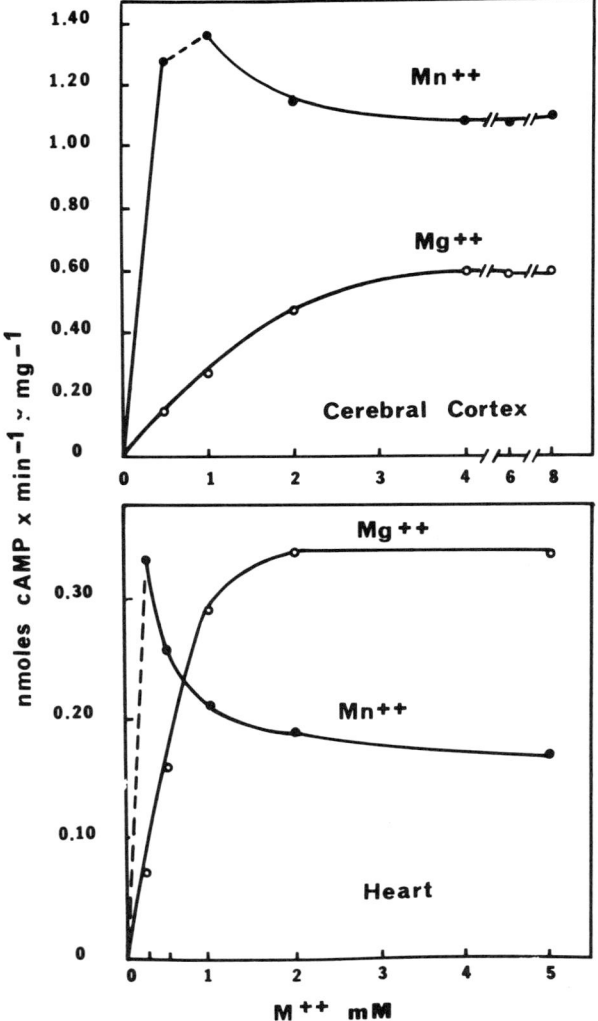

FIG. 6. The relation of adenyl cyclase activity to Mg^{++} and Mn^{++} concentrations. The enzyme preparations were washed membrane fractions from rat cerebral cortex or rabbit ventricular myocardium. Both enzyme preparations were "pre-activated" with NaF, prior to assay, as described for Fig. 8 (from Perkins and Moore, 1971).

at any concentration of Mg^{++}. Thus, it would appear that the V_{max} of the enzymic reaction is greater with ATP-Mn than with ATP-Mg. If a second cation binding site is involved, then apparently Mn^{++} not only has greater affinity for the site but its interaction results in a greater increase in the

V_{max} of the reaction than does that of Mg^{++}. The precise mechanism by which Mn^{++} and Mg^{++} interact with adenyl cyclases remains to be elucidated. The problem is worthy of continued study in view of the interesting effect of Mn^{++} on hormonal stimulation of the enzyme from thyroid, liver, renal cortex, toad bladder, and fat cells, to be discussed below.

The reason for the decrease in enzyme activity as Mn^{++} concentration is increased (Fig. 6) is not clear, although it could be related to competition with bound Ca^{++} which has been postulated to be required for activity of the enzyme of bovine cerebral cortex (Bradham, Holt, and Sims, 1970; Bradham, 1972). The evidence of a role of Ca^{++} in adenyl cyclase function is indirect and based primarily on the inhibition of enzyme activity by EGTA. As shown in Fig. 7, increasing concentrations of EGTA lead to partial inhibition of the enzyme of bovine cerebral cortex. The apparent partial inhibition could be due to complete inhibition of one of two or more species of enzyme, perhaps derived from the different cell types composing the cerebrum. The activity can be restored upon addition of Ca^{++} as shown by Bradham (1972). However, the activity also was restored by Sr^{++}, and Johnson and Sutherland (1973) have shown that the effect of EGTA on the activity of solubilized adenyl cyclase of rat cerebellum could be reversed by Mn^{++} as well as Ca^{++}. These latter workers concluded that in the presence of EGTA, activity was lost due to chelation of a required divalent cation other than Mg^{++}. The data were insufficient to determine the identity of the cation. It should be pointed out that the primary effect of Ca^{++} added to enzyme assay mixtures is inhibition. Furthermore, extensive dialysis of adenyl cyclase of rat cerebral cortex does not effect a requirement for exogenous Ca^{++}.

The relevance of these studies to adenyl cyclases from sources other than brain has yet to be determined. However, a role for Ca^{++} in the stimulation of adenyl cyclase of adrenal cortex by ACTH has been reported (see below).

In general, monovalent cations have no striking effects on the basal activity of adenyl cyclases. Lithium ion appears to have the most pronounced effects, causing inhibition of basal activity of the enzyme of rat renal cortex (Marcus and Aurbach, 1971), inhibition of basal and NaF-stimulated activity of the rat brain enzyme (Forn and Valdecas, 1971), and inhibition of TSH stimulation of the thyroid enzyme without alteration of basal activity (Burke, 1970a; Wolff and Jones, 1971). Birnbaumer, Pohl, and Rodbell (1969) reported that Li^+ stimulated basal activity of the fat-cell enzyme but inhibited both NaF- and ACTH-stimulated activities. Dousa and Hechter (1970) have demonstrated that Li^+ exerts a marked inhibition on both basal activity and vasopressin-stimulated activity of the enzyme of rabbit renal medulla. Forn and Valdecasas (1971) also demonstrated that Li^+ inhibited the norepinephrine-induced formation of cAMP in slices of rat cerebral

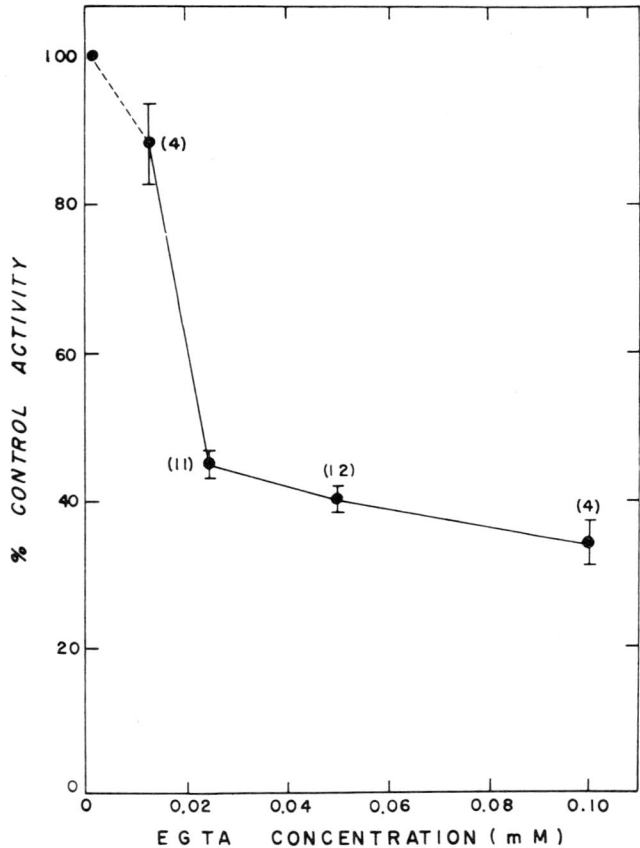

FIG. 7. The effect of EGTA on the activity of adenyl cyclase of bovine cerebral cortex (taken from Bradham et al., 1970).

cortex, and histamine-induced formation of cAMP in slices of rabbit cerebral cortex.

The effect of increased tonicity of assay mixtures on the activity of adenyl cyclase of synaptosome fractions of rat brain has been studied by de Robertis, Arnaiz, Alberici, Butcher, and Sutherland (1967). The synaptosome fraction isolated in 0.32 M sucrose was assayed in reaction mixtures ranging from 0.032 to 0.5 M sucrose. The activity was markedly increased as the tonicity due to either sucrose or salt was decreased. As mentioned earlier, the presence of adenyl cyclase in vesicular structures could confer special, but perhaps artifactual properties on the enzyme. De Robertis et al. (1967)

interpreted their findings as indicating that most of the enzyme is compartmentalized, perhaps within the synaptosomes, under isotonic or hypertonic conditions.

4. *Stimulation by NaF*

NaF was found, somewhat serendipitously, to stimulate adenyl cyclase activity (Rall and Sutherland, 1958; Sutherland, Rall, and Menon, 1962). Subsequently, it has been found to activate the enzyme from virtually all sources studied with the exception of certain bacteria (Tao and Lipmann, 1970; Ide, 1971; also see Ray, Tomasi, and Marinetti, 1970; Pennington, Brown, Chattopadhyay, Conaway, and Morris, 1970). Recently, a soluble, purified adenyl cyclase from *Streptococcus salivarius* has been shown to respond to NaF with increased activity (Khandelwal and Hamilton, 1971). As pointed out earlier, since NaF does not cause accumulation of cAMP in whole-cell preparations (Robison, Butcher, and Sutherland, 1971), it is assumed not to activate adenyl cyclase in intact cells. Thus, it appears reasonable to assume that the process of membrane fragmentation alters the enzyme system rendering it susceptible to activation by NaF. The mechanism of NaF activation is still unknown, but a number of the characteristics of its interaction with the enzyme system have been examined.

Since ATPase activity is significantly reduced by 5 to 15 mM NaF, a mechanism involving conservation of substrate was a possibility. However, such a mechanism is not sufficient to explain the effects of NaF since initial reaction rates are increased even at high ATP concentrations, and activation is also observed in the presence of ATP-regenerating systems where ATP concentrations are maintained constant. Also, the apparent K_m for ATP is not changed upon activation by NaF (Bär and Hechter, 1969b; Drummond and Duncan, 1970; however, see Hepp, Edel, and Wieland, 1970).

The magnitude of the effect of NaF varies markedly with enzymes from different tissues (Table 1). The maximally effective concentration of NaF ranges from between 3 and 4 to between 12 and 15 mM. However, the effective concentration range of NaF may be influenced by the concentration of Mg^{++} in the assay mixture. Birnbaumer, Pohl, and Rodbell (1969) have demonstrated that the eventual decrease in the enzyme activity of fat-cell ghosts at high NaF concentrations can be reversed by increasing the Mg^{++} concentration. Mg^{++} and F^- are known to interact to form MgF_2 complexes (stability constant $10^{8.2}$; Sillen and Martell, 1964) and MgF_2PO_4 complexes are also postulated to occur (Warburg and Christian, 1941).

TABLE 1. *Stimulation of adenyl cyclases from various sources by NaF*

Enzyme source	Basal activity (pmoles cAMP/min/mg)	[NaF][a] (mM)	NaF activity (pmoles cAMP/min/mg)	Fold increase	Reference
Toad bladder epithelium	2	10	15	7.5	Hynie and Sharp, 1971
Rat parotid	100	10	1400	14	Schramm and Naum, 1970
Rat fat cell ghosts	68	4	270	4	Birnbaumer, Pohl, and Rodbell, 1969
Rat cerebral cortex	200	4	460	2.3	Perkins and Moore, 1971
Rat liver	2	10	13	6.5	Bär and Hahn, 1971
Rat liver (purified)	20	15	200	10	Pohl, Birnbaumer, and Rodbell, 1971
Rat pineal	97	10	423	4	Weiss, 1969
Rat renal cortex	12	7	65	5	Marcus and Aurbach, 1971
Mouse liver	9	6	85	9	Hepp, Edel, and Wieland, 1970
Rabbit skeletal muscle (purified)	34	12	636	19	Severson, Drummond, and Sulakhe, 1972
Guinea pig ventricle	40	8	260	6.5	Drummond and Duncan, 1970
Frog erythrocyte (purified)	600	10	12,000	20	Rosen and Rosen, 1969
Bovine thyroid	20	10	100	5	Burke, 1970a
Streptococcus salivarious (purified)	20,000	10	45,000	2	Khandelwal and Hamilton, 1971
Human astrocytoma (1181N1)	200	10	295	1.5	Perkins et al., 1971
Salmon testis	0	7	12	∞	Menon and Smith, 1971

[a] For those studies where it was determined, the minimal concentration of NaF causing maximal stimulation has been listed.

Adenyl cyclases from most tissues are not responsive to hormones in the presence of maximally effective concentrations of NaF. Since enzyme activity can usually be activated to a greater degree by NaF than by hormones, it has been assumed by some workers that NaF causes full expression of enzyme activity. However, there are some exceptions to this generalization. For example, the enzyme from rat liver is stimulated to a greater degree by glucagon than by NaF (Birnbaumer, Pohl, and Rodbell, 1971). Also, Hynie and Sharp (1971) have shown the enzyme from frog bladder epithelium to be more responsive to vasopressin than to NaF. These workers found that theophylline had a marked inhibitory effect on the vasopressin-induced activity but less of an effect on NaF-induced activity. The suggestion was made that theophylline might show a similar preferential inhibition of hormone-induced activity of the enzyme from other tissues. Perkins and Moore (1971) demonstrated that Triton X-100 activated the enzyme of rat cerebral cortex to twice the extent brought about by NaF. Thus, there are numerous observations that NaF added to the assay mixture does not necessarily cause full expression of enzyme activity.

The general observation that maximally effective concentrations of NaF and hormone do not result in an additive increase in activity leads to the suggestion that the activated states produced by these effectors are either the same or mutually exclusive. Conversely, Weiss (1969) observed partial additivity of maximal effects of norepinephrine and NaF on the enzyme of rat pineal gland, and Burke (1970b) observed that TSH and NaF were additive in their stimulation of the enzyme of bovine thyroid gland. However, a possible explanation for these exceptions may lie in the fact that simple addition of NaF to assay mixtures may not result in the full expression of the NaF-activated state of the enzyme (Perkins and Moore, 1971; also see below).

The active molecular species involved in stimulation by NaF appears to be the fluoride ion. Drummond and Duncan (1970) examined the effect of a large number of anions (as the Na salts) as well as a number of organic fluorine compounds on the activity of cardiac muscle adenyl cyclase. Only F^- was effective, added either as NaF or $NH_4F \cdot HF$. Other halides were minimally stimulatory or ineffective; IO_3^- was a strong inhibitor of the basal activity.

Recently it has been demonstrated that the activation of adenyl cyclase by NaF is not readily reversed for the enzyme from rat parotid gland (Schramm and Naim, 1970), mouse adrenal cortex tumor (Pastan, Pricer, and Blanchette-Mackie, 1970), beef adrenal cortex (Kelley and Koritz, 1971), rat cerebral cortex (Perkins and Moore, 1971), and rabbit skeletal muscle (Severson, Drummond and Sulakhe, 1972). Such demonstrations are based

TABLE 2. *Irreversible nature of the stimulation of adenyl cyclase by NaF*

Components added to the preliminary incubation	Activity in subsequent assay (pmoles cAMP/min/mg)		Reference
	−NaF	+NaF	
Rat cerebral cortex			
MgSO$_4$	200	460	Perkins and Moore, 1971
NaF	520	620	Perkins and Moore, *unpublished results*
MgSO$_4$, NaF	820	840	
ATP, MgSO$_4$, NaF	830	815	
Rat parotid gland			
MgCl$_2$	60	1,500	Schramm and Naim, 1970
NaF	100	1,400	
MgCl$_2$, NaF	600	1,900	
ATP, MgCl$_2$, NaF	1,300	2,200	
Rabbit skeletal muscle			
None	34	636	Severson, Drummond, and Sulakhe, 1972
MgSO$_4$	28	−	
NaF	253	−	
MgSO$_4$, NaF	245	−	

on experiments wherein the enzyme preparation is first incubated with NaF alone or in combination with Mg^{++} or ATP or both, then diluted, and washed or dialyzed prior to assay in the presence and absence of NaF. Results from three laboratories are summarized in Table 2. Prior treatment of the enzyme of rat parotid gland with NaF alone had little effect on activity, but NaF in the presence of $MgCl_2$ or $MgCl_2$ and ATP caused a marked increase in activity that was not lost upon washing the enzyme in NaF-free medium. Basically similar results were obtained with the enzyme of rat cerebral cortex except that NaF alone during the preliminary incubation had a significant stimulatory effect. NaF alone was as effective as NaF plus $MgSO_4$ in activating the skeletal muscle enzyme, but in both cases the activity retained after dialysis was only 40% of the activity of the enzyme in the presence of NaF. Only in the case of the enzyme from rat cerebral cortex did the activation appear to be completely irreversible; i.e., NaF had no further effect in the final assay. However, the various experiments are difficult to compare on a quantitative basis since the preliminary reactions were carried out under quite different conditions. Only in one case (Perkins and Moore, 1971) is there assurance that the preliminary reaction had gone to completion (see Fig. 8). Schramm and Naim (1970) incubated the samples at 25°C for 5 min while Severson, Drummond and Sulakhe (1972) incubated samples at 4°C for 30 min. Based on the results of Kelley and Koritz (1971) and of Perkins

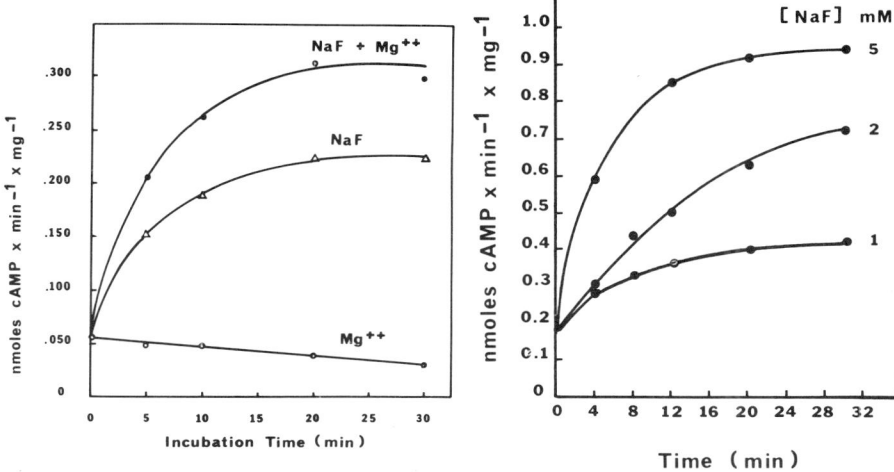

FIG. 8. The time course of activation of adenyl cyclase by NaF. *Left:* The enzyme preparation was a washed membrane fraction of rabbit ventricular myocardium. MgSO$_4$ or NaF or both were included in the activation incubation at a concentration of 5 mM as indicated. Incubation time refers to the time of exposure of the enzyme to the additives prior to washing in a NaF-free buffer and assay in the absence of NaF. *Right:* Washed membrane fraction of rat cerebral cortex. The concentrations of NaF in the activation reaction are indicated in the figure. Experimental detail was as described by Perkins and Moore, (1971).

and Moore (1971), it is unlikely that the activation would have reached completion under such conditions.

The results of Pastan, Pricer, and Blanchette-Mackie (1970) are of interest because of two peculiarities. First, NaF was found to be required for the "solubilization" of adenyl cyclase of mouse adrenal tumors by fragmentation in a French pressure cell in the presence of phospholipids. After this treatment, significant enzyme activity was found in the 105,000 × g supernate. Subsequent dialysis in NaF-free buffer did not reduce enzyme activity, but re-addition of NaF to dialyzed enzyme resulted in a further twofold increase. It is of interest that the enzyme activity of the 600 × g precipitate was reduced to about 25% by dialysis but could be restored to nearly original activity by re-addition of NaF. Second, both particulate and "soluble" fractions were responsive to ACTH after removal of NaF by dialysis. A possible explanation for the apparent lack of complete activation by the initial exposure to NaF could involve the temperature, 1°C, at which the exposure occurred. As mentioned above, the activation process is temperature dependent. Severson, Drummond, and Sulakhe (1972) have demonstrated the marked temperature sensitivity of adenyl cyclase activity measured in the presence of NaF.

When the kinetics of activation of adenyl cyclase of rat brain or rabbit heart were analyzed in greater detail (Fig. 8), it became apparent that the reaction was dependent on time, NaF concentration, and temperature. The applicability of these observations to adenyl cyclases of other tissues is not known, but the importance of an awareness of this phenomenon is illustrated by the observation (Perkins and Moore, 1971) that the effect on the enzyme of 5-day rat cerebral cortex of 10 mM NaF added to the assay mixture amounted to only a 35% increase. By contrast, a 30-min incubation with 5 mM NaF followed by assay in the presence of 5 mM NaF resulted in a 200% increase in activity. The inability of NaF to stimulate adenyl cyclase of young rat brain when simply added to assay mixtures has also been observed by Schmidt, Palmer, Dettbarn, and Robison (1970).

Rodbell and co-workers (Birnbaumer, Pohl, and Rodbell, 1969; Birnbaumer, Pohl, Krans, and Rodbell, 1970) have suggested that the mechanism of action of NaF stimulation of the fat-cell ghost enzyme could be explained by a reduction in the apparent $K_{Dissoc.}$ for Mg^{++} at a hypothesized second site on the enzyme. The corollary of such a hypothesis is that high concentrations of Mg^{++} should substitute for NaF, i.e., should fully activate the enzyme. This mechanism does not appear to offer a complete explanation for the effects of NaF on the enzyme from other tissues. First, it is clear that for the enzymes of a variety of tissues increased Mg^{++} does not substitute for NaF stimulation. This has been shown to be the case for heart (Drummond and Duncan, 1970; Drummond, Severson, and Duncan, 1971), skeletal muscle (Severson, Drummond, and Sulakhe, 1972), liver (Pohl, Birnbaumer, and Rodbell, 1971; Hepp, Edel, and Wieland, 1971), toad bladder (Hynie and Sharp, 1971), and thyroid (Wolff and Jones, 1971). Second, the results of Drummond, Severson, and Duncan (1971) indicate that the primary effect of NaF on the enzyme of cardiac muscle is on the V_{max} of the reaction and cannot be explained by an increase in the affinity of the enzyme for Mg^{++}. However, these studies examined the effect of NaF added directly to the enzyme assay mixture. As mentioned earlier, Mg^{++} and F^- are known to interact; thus secondary equilibria involving these two ions could become an important consideration in the interpretation of the effect of F^- on the interaction of the enzyme with Mg^{++}. Since the activation of adenyl cyclase by F^- was enhanced in the presence of Mg^{++} (Table 2), it is clear that some interaction occurs during the activation process. In an attempt to avoid these complex interactions of F^- and Mg^{++}, we first activated the enzyme by exposure to NaF and $MgSO_4$ at 30°C for 30 min, washed the particulate enzyme free of these ions, and then studied the effect of Mg^{++} on its activity

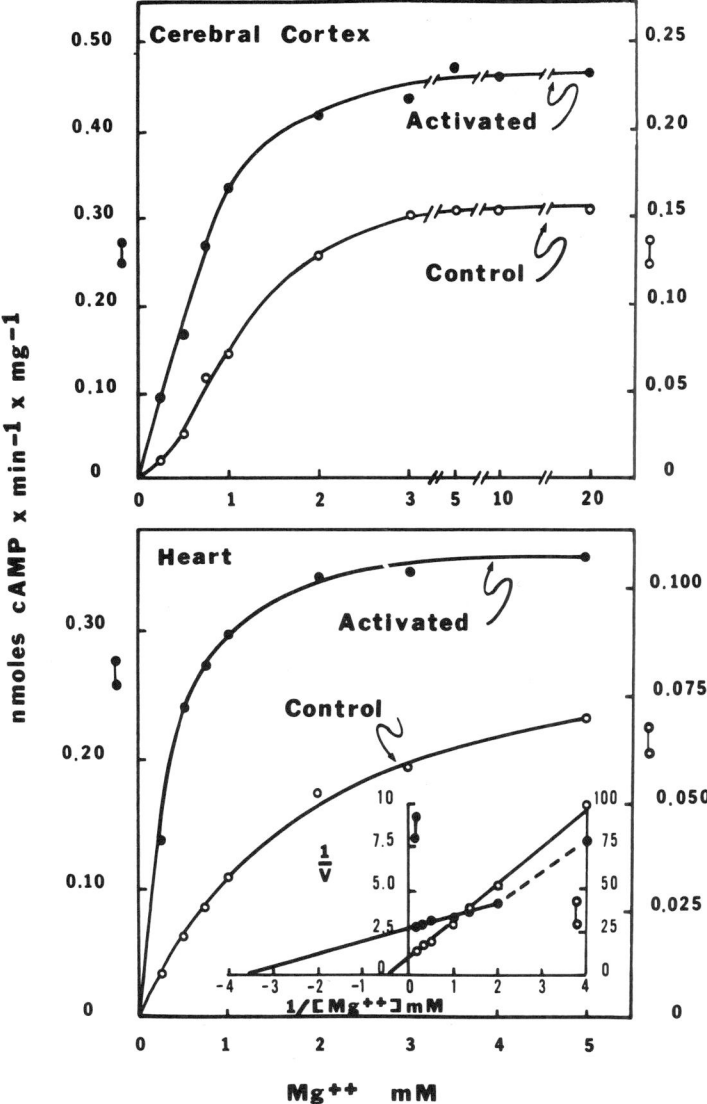

FIG. 9. The effect of Mg^{++} on the activity of adenyl cyclase "pre-activated" by exposure to NaF. The enzyme preparations used were washed membrane fractions of rat cerebral cortex and rabbit ventricular myocardium. The enzymes were activated by pretreatment with NaF as described for Fig. 8, Table 2, and in Perkins and Moore (1971).

(Fig. 9). Activation by F^- effects two changes in the enzyme from rabbit heart: an increase in V_{max} (fourfold) and a decrease (ninefold) in the apparent $K_{Dissoc.}$ for Mg^{++}. With the enzyme of rat cerebral cortex, the primary effect was on the V_{max} (threefold), only a twofold decrease in the apparent $K_{Dissoc.}$ for Mg^{++} being observed.

It is apparent from the discussion above that activation of the enzyme by F^- changes the concentration-effect relationship of Mg^{++}. This mechanism of activation could result in a marked increase in enzyme activity (e.g., about 16-fold for the heart enzyme at 0.5 mM Mg^{++}; Fig. 9) if the intracellular Mg^{++} concentration is limiting and if hormones cause a similar change in the concentration-effect relationship for Mg^{++}. There is evidence to support the latter suggestion (Birnbaumer, Pohl, and Rodbell, 1969; Drummond, Severson, and Duncan, 1971).

5. Properties of Detergent-treated Adenyl Cyclases

A number of attempts have been made to solubilize mammalian adenyl cyclases but these have met with limited success. Sutherland, Rall, and Menon (1962) described the solubilization with Triton X-100 of adenyl cyclase from bovine cerebral cortex, dog skeletal muscle, heart, and liver. These preparations were not sensitive to hormones and were no longer soluble when the Triton was removed.

Pastan, Pricer, and Blanchette-Mackie (1970) "solubilized" mouse adrenal tumor adenyl cyclase in the presence of NaF and phospholipids by the use of a French pressure cell. The enzyme which was not sedimented at $105,000 \times g$, after dialysis in NaF-free media, was sensitive to stimulation by ACTH and NaF. The molecular weight was estimated to be 3 to 7×10^6 daltons by filtration of the solubilized preparation on columns of Bio-gel A-15m. The kinetic properties of the soluble preparation were not reported.

Recently Levey (1970, 1971a,b,c; see also Levey and Klein, 1972) has reported his studies with soluble cat heart adenyl cyclase. The adenyl cyclase of cat left ventricle myocardium was reported to be greater than 90% solubilized when the tissue was homogenized in Lubrol PX, an ethylene oxide condensate of dodecanol. The solubilized enzyme was no longer responsive to stimulation by hormones (catecholamines, glucagon, or histamine) but was stimulated by NaF to the same extent as the particulate enzyme. The detergent could be removed by chromatography on DEAE-cellulose as determined with the use of Lubrol PX labeled with ^{14}C in the ethylene oxide moiety. The enzyme remained in a soluble form after removal of the Lubrol. Chromatography on Sephadex G-200 of the detergent-free, soluble enzyme indicated a molecular weight of 100,000 to 200,000 daltons. An analysis of the kinetic properties of the solubilized enzyme is yet to be reported.

A Lubrol PX-solubilized preparation of rat cerebellum adenyl cyclase has been characterized in some detail in regard to cation dependency (Johnson and Sutherland, 1973). The enzyme was markedly activated by treatment with detergent, and NaF had no further stimulatory effect. However, upon removal of most of the detergent by chromatography on Sephadex G-200 the enzyme eluted in a turbid fraction. This form of the enzyme was sedimentable and could now be stimulated by NaF. The detergent-dispersed enzyme did not show significant alteration in its kinetic properties in regard to stimulation by Mn^{++} and Mg^{++} and inhibition by Ca^{++}. It was of interest, however, that the enzyme was not inhibited by Ca^{++} if Mn^{++} was present. Unlike particulate preparations of rat and bovine cerebral cortex, which are inhibited only 40 to 60% by EGTA, the solubilized enzyme of rat cerebellum was completely inhibited (>90%) by this chelating agent. The

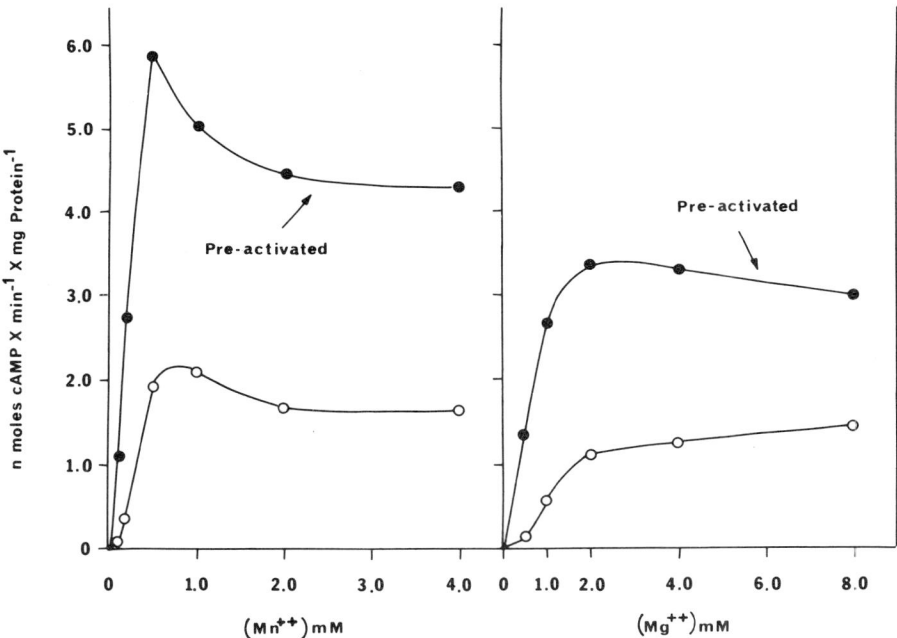

FIG. 10. The effect of Mg^{++} and Mn^{++} concentrations on the activity of Lubrol PX-solubilized adenyl cyclase of rat cerebral cortex before and after activation with NaF. The experimental procedure is described in the text.

effect of EGTA on particulate preparations of the enzyme of rat cerebellum has not been reported.

We have recently examined the effect of varying concentrations of Mg^{++} and Mn^{++} on the activity of Lubrol PX-solubilized preparations of rat cerebral cortex. The enzyme was prepared in two ways. Lyophilized, washed membranes were resuspended in 0.05 M Tris-HCl buffer, pH 7.5, and one-half the preparation was incubated with 5 mM NaF and 5 mM $MgSO_4$ for 30 min at 30°C. After washing to remove NaF, the preparations were made 2% in Lubrol PX, and the 28,000 $\times$ g supernate was collected after 6-min incubation at 30°C. In Fig. 10 is shown a comparison of the effects of Mg^{++} and Mn^{++} on both NaF-treated and untreated, detergent-solubilized enzyme. These data compared with those in Fig. 6 suggest that solubilization with Lubrol PX does not significantly alter the concentration-effect relationship of these two cations. It should be mentioned that after solubilization (without prior exposure to NaF) the enzyme is not responsive to NaF. However, prior exposure of the enzyme to NaF results in increased activity after solubilization. It is clear from the studies of Johnson and Sutherland (1973) and from our work that the activity of the detergent-dispersed enzyme of rat brain is unstable; complete loss of activity occurs after 60 min at 30°C. Effects of solubilization on hormonal responsiveness will be discussed in detail in a later section.

C. Characteristics of Hormonal Regulation of Adenyl Cyclase

1. *General Comments*

By far the most interesting property of adenyl cyclase is its responsiveness to hormones. In fact, the primary motivation for conducting the kinds of studies presented above is the hope of gaining insight into the mechanism(s) whereby hormones increase and apparently decrease, as well, enzyme activity. However, primarily because the enzyme system has not been amenable to purification, details of the mechanism of hormonal effects have not evolved rapidly.

The model proposed by Robison, Butcher, and Sutherland (1967) as shown in Fig. 2 is probably the simplest formulation consistent with the available data, i.e., a single receptor subunit interacting directly with a catalytic subunit. In this case the coupling of hormone binding to catalytic activity is accomplished by a change in subunit conformation through direct interaction. This model seems adequate for cells that respond to a single hormone (monovalent regulation) such as is apparently the case for the adrenal cortex cell. However, even in this example the organization of the adenyl cyclase system may not be as simple as it appears (see discussion below of the work of Schorr, Rathnam, Saxena, and Ney, 1971). For those cells whose adenyl cyclase responds to a number of hormones (multivalent regulation), this model seems less feasible.

The adenyl cyclase of rat adipocytes has been shown to respond to six hormones: epinephrine, glucagon, ACTH, LH, TSH, and secretin. There are at least three reasonable relationships of receptor and catalytic components to explain such multivalent regulation: (1) six separate receptor-catalytic units each specific for one hormone; (2) a single type of receptor-catalytic unit with a nonspecific receptor; and (3) six specific and separate receptors associated with a single type of catalytic unit. The studies supporting the latter relationship have been reviewed recently (Birnbaumer, Pohl, Krans, and Rodbell, 1970; Rodbell, 1971). Briefly, the evidence is based on experiments with intact cells (Butcher, Baird, and Sutherland, 1968) and fat-cell ghosts (Bär and Hechter, 1969a; Birnbaumer, Pohl, and Rodbell, 1969; Birnbaumer and Rodbell, 1969; Rodbell, Birnbaumer, and Pohl, 1970) demonstrating that: (1) maximally effective concentrations of individual hormones when combined do not have additive effects; (2) certain antagonists are hormone specific, e.g., propranolol blocks only the effect of catecholamines; (3) responsiveness to the six hormones exhibits differential lability when the enzyme preparation is treated with trypsin; and (4) there is differential dependence of hormone responsiveness on Ca^{++}.

Recently Schorr, Rathnam, Saxena, and Ney (1971) have described the multivalent regulation of the adenyl cyclase of a mouse adrenal cortex tumor. The properties of this system are similar to those of the rat adipocyte ghosts in that a number of hormones (ACTH, catecholamines, TSH, LH, and FSH) apparently influence the same catalytic component through interaction at distinct receptors.

Studies of the regulation of cAMP content of brain slices indicate a strikingly complex picture also apparently involving multivalent regulation of adenyl cyclase activity. These studies have been reviewed recently (Rall and Sattin, 1970; Shimizu, Creveling, and Daly, 1971; Krishna, Forn, Voigt, Paul, and Gessa, 1971; Rall and Gilman, 1971) and will not be discussed in detail here. Briefly, the level of cAMP in brain slices is increased by catecholamines, serotonin, histamine, and adenosine; K^+, veratridine, batrachotoxin, and ouabain apparently act indirectly by causing the release of adenosine. The unique observation from these studies is the synergism of action that occurs when maximally effective concentrations of two of the agonists are combined, e.g., histamine plus norepinephrine or adenosine plus norepinephrine (Shimizu, Creveling, and Daly, 1970; Huang, Shimizu, and Daly, 1971). This is to be contrasted with the less than additive effects of two or more agonists on the activity of rat fat-cell or mouse adrenal-tumor adenyl cyclases. Unfortunately adenyl cyclase activity of broken-cell preparations of most brain regions responds only minimally to catecholamines (20 to 100% increase) (Klainer, Chi, Freidberg, Rall, and Sutherland, 1962; McCune, Gill, von Hungen, and Roberts, 1971; Kebabian, Petzold, and Greengard, 1972) and not at all to histamine, serotonin, or adenosine. Thus, a direct measure of the interaction of these agents with brain adenyl cyclase activity is not possible.

The cellular heterogeneity of brain slices, composed primarily of neurons and glia, complicates the study of this phenomenon in the already less satisfactory intact-cell system. In an attempt to simplify the analysis of the regulation of adenyl cyclase activity in central nervous system tissue, we have been studying, in culture, several clonal human astrocytoma cell lines

TABLE 3. *Effect of norepinephrine, adenosine, histamine, and prostaglandin E_1, alone and in combination, on the cyclic AMP content of two lines (118132, 1181N1) of human astrocytoma cells in culture*

	Cyclic AMP Content (cpm/mg protein)			
	118132		1181N1	
Additives	Observed	Calculated based on additivity	Observed	Calculated based on additivity
Experiment A				
None	3,000	—	4,400	—
Norepinephrine (0.1 mM)	78,700	—	24,700	—
Adenosine (0.1 mM)	47,000	—	12,300	—
Norepinephrine + adenosine	80,000	122,700	24,500	32,600
Experiment B				
None	—	—	3,500	—
Norepinephrine (0.1 mM)	—	—	23,600	—
Histamine (0.1 mM)	—	—	14,200	—
Norepinephrine + histamine	—	—	27,500	34,300
Experiment C				
None	3,200	—	4,000	—
Norepinephrine (0.1 mM)	77,400	—	22,600	—
Prostaglandin E_1 (0.001 mM)	40,200	—	116,200	—
Norepinephrine + prostaglandin E_1	111,100	114,400	123,900	134,800

Cells, in 60-mm plastic Petri dishes, were exposed to ^{3}H-adenine for 1 hr, the medium removed and replaced with adenine-free, agonist-containing medium. Exposure to the agonists was for 10 min at 37°C in a CO_2:air, 5:95% atmosphere. The incubations were terminated by aspiration of the medium and addition of 1.0 ml of 5% TCA to extract ^{3}H-cAMP. The ^{3}H-cAMP was purified by Dowex 50 and paper ion-exchange chromatography as described by Perkins and Moore (1973).

(Perkins, Macintyre, Riley, and Clark, 1971; Clark and Perkins, 1971, 1972). The cells were originally obtained from a cerebral glioma (Ponten and Macintyre, 1968). Of interest to the topic under discussion is the observation that these cells respond to four different effectors—catecholamines, histamine, adenosine and prostaglandin E_1 (PGE_1)—with a rise in cAMP content. Since the cells represent a clonal population, it is probable that the adenyl cyclase of each cell is responsive to each of the four agonists. In Table 3 are shown the results of adding together maximally effective concentrations of various combinations of the four effectors. In contrast to the effects on brain slices, combinations of the agonists do not result in synergistic effects on the increase in cAMP content. On the contrary, the effects are less than additive, suggesting a circumstance analogous to that observed with rat fat cells, i.e., multiple hormone influence on the same catalytic component. Evidence that the receptors are distinct and specific for each agonist is summarized in Table 4. Propranolol inhibits only the effect of norepinephrine whereas theophylline blocks only the effect of adenosine (see Sattin and Rall, 1970). We have shown previously (Clark and Perkins, 1971) that responsiveness to catecholamines varies markedly during the growth cycle whereas the response to histamine is essentially

TABLE 4. *The effect of propranolol and theophylline on the increase in cyclic AMP content of human astrocytoma cells (118132) by norepinephrine, adenosine, and prostaglandin E_1*

Additives	Cyclic AMP content (cpm/mg protein)	% inhibition
None	3,000	
Norepinephrine (0.1 mM)	78,700	
Norepinephrine (0.1 mM) plus propranolol (0.001 mM)	4,200	97
Norepinephrine (0.1 mM) plus theophylline (0.1 mM)	74,900	5
Adenosine (0.1 mM)	47,000	
Adenosine (0.1 mM) plus propranolol (0.001 mM)	45,200	0
Adenosine (0.1 mM) plus theophylline (0.1 mM)	4,700	96
Prostaglandin E_1 (0.001 mM)	41,100	
Prostaglandin E_1 (0.001 mM) plus propranolol (0.01 mM)	49,400	−17
Prostaglandin E_1 (0.001 mM) plus theophylline (0.1 mM)	60,600	−50

For experimental detail see Table 3, Clark and Perkins (1971), and Perkins and Moore (1973).

unchanged, supporting the idea that different receptors mediate the responses.

It is not clear to what extent the properties of the adenyl cyclase system of normal glia are analogous to those of these tumor glial cells. However, if the analogy is complete, our results suggest that the synergistic effects of combinations of agonists observed in brain slices are a result of the interaction of different cell types and may not reflect a further complexity in the regulation of adenyl cyclase activity *per se*. At least these observations provide a description of another type of cell exhibiting independent, multihormone regulation of a common adenyl cyclase.

2. *Factors Affecting Adenyl Cyclase Activity and Its Sensitivity to Hormones*

In general the responsiveness of adenyl cyclases to hormones is a much more labile property than is the responsiveness to NaF or the expression of basal catalytic activity. Thus, treatment of enzyme preparations with various substances can markedly reduce hormonal effects with little or no alteration of basal or NaF-induced activity. However, it should be recognized that the common use of a second agent (e.g., proteolytic enzymes, phospholipases, detergents, etc.) to alter hormone-induced enzyme activity does not unequivocally demonstrate a property of the receptor component of the system. Such effects could be due to an alteration of the receptor, the coupling component (transducer), or even an alteration of the hormone-induced active form of the catalytic unit. The control for such experiments is usually the demonstration that the agent has no effect on basal activity. However, if the form of the catalytic component responsible for basal activity is different from the hormone-induced form, then such a control may not be valid. As mentioned above it is probable that the same, active form of the catalytic component of the enzyme is induced by either NaF or hormone. If this is in fact the case, the site of action of agents which inhibit hormone effects but not the effect of NaF is probably the receptor-transducer aspect of the system. To differentiate between an action on the receptor and one on the coupling system would require concomitant analysis of the effect on hormone binding and on enzyme activity. As will be discussed below, to date we have an inadequate understanding of the relation of hormone binding to change in adenyl cyclase activity, in part due to the difficulty of differentiating non-functional *vs* functional binding.

(a) *Agents causing nonspecific alteration of membrane structure:* Sutherland, Rall, and Menon (1962) were the first to demonstrate that non-ionic detergents can abolish hormone responsiveness. Ionic detergents usually eliminate all aspects of enzyme activity (Sutherland, Rall, and Menon, 1962; Perkins and Moore, 1971). Levey (1970, 1971a,b,c,; Levey and Klein, 1972) has recently shown that cat heart adenyl cyclase can be solubilized by treatment with Lubrol PX with resultant loss of sensitivity to hormones but maintenance of sensitivity to NaF. Also the enzyme of rat renal cortex was partially solubilized by Triton X-100 or Lubrol PX (Marcus and Aurbach, 1971) with retention of NaF-induced activity but complete loss of sensitivity to parathyroid hormone. Pohl, Birnbaumer, and Rodbell (1971) have studied the effects of a variety of agents on the activity of adenyl cyclase of rat liver membranes. Digitonin (0.1 to 0.4%) did not solubilize the enzyme but selectively prevented glucagon stimulation whereas the effect of NaF was slightly increased. Treatment of liver membranes with 2 M urea or with phospholipase A also caused selective loss of glucagon responsiveness but maintenance of NaF sensitivity. Digitonin has been shown to prevent the stimulation of the fat-cell ghost enzyme by epinephrine, ACTH, secretin, and glucagon (Birnbaumer, Pohl, and Rodbell, 1971).

Trypsinization of erythrocytes (Øye and Sutherland, 1966) and rat adipocytes (Rodbell, Birnbaumer, and Pohl, 1970) has little effect on basal or NaF-induced enzyme activity observed in subsequent assay of ghosts. This is consistent with an intracellular localization of the catalytic component. However, hormonal responsiveness of adipocyte ghosts was markedly and differentially altered. Conversely, direct treatment of ghosts with trypsin leads to a simultaneous loss of basal and NaF- and hormone-induced activity (Birnbaumer, Pohl, and Rodbell, 1969).

It is to be expected that even agents acting by nonspecific mechanisms should exhibit some degree of selectivity since the properties of plasma membranes should be somewhat cell specific. This expectation is borne out by the results of Wolff and Jones (1970) who have shown that chlorpromazine is an inhibitor of the response of thyroid adenyl cyclase to TSH and PGE_1, the adrenal cortex enzyme to ACTH, and the liver enzyme to epinephrine and to a lesser degree to glucagon. On the other hand, the effect of parathyroid hormone on the kidney cortex enzyme was not altered by chlorpromazine. Basal activity was not affected in any case and NaF-induced activity was either not changed or increased. The effect of TSH on the thyroid enzyme also was blocked by thymol, cobramine B, and gramicidin S (Wolff and Jones, 1971).

At least two conclusions can be drawn from observations such as these. First, relatively nonspecific alteration of membrane structure can lead to loss of hormone-responsiveness with maintenance of basal and/or NaF-induced activity. Thus, it is clear that hormonal stimulation involves structures not required for expression of basal activity or for stimulation of activity by NaF. Second, agents that affect both protein and lipid structure can alter the hormone-responsiveness of the enzyme suggesting the involvement of both chemical species. Based on his extensive studies, Rodbell (1972) has suggested that the receptor component is a protein or lipoprotein and that lipids are probably involved in the coupling mechanism as well. The observations that hormonal sensitivity can be restored to detergent-treated adenyl cyclase of heart (Levey, 1971b,c; Levey and Klein, 1972), to the phospholipase A-treated enzyme from liver (Pohl, et al, 1971), and to the digitonin-treated enzyme of liver membranes (Pohl, 1971) by the addition of specific phospholipids support such suggestions.

(b) *Cations:* As mentioned previously, Mn^{++} markedly stimulates the basal and NaF-induced activities of a number of adenyl cyclases. Thus, it is of interest that this ion does not stimulate ACTH-induced activity of rat adipocyte ghosts (Birnbaumer, Pohl, and Rodbell, 1969) or TSH-induced activity of sheep thyroid membranes (Burke, 1970a). Concentrations of Mn^{++} above 5 mM inhibit glucagon-induced activity of liver membranes but not NaF-induced activity (Birnbaumer, Pohl, and Rodbell, 1971). Mn^{++} increases basal activity of the enzyme from rat renal cortex but prevents further stimulation of the enzyme by parathyroid hormone; NaF-induced activity is also somewhat reduced (Marcus and Aurbach, 1971). Hynie and Sharp (1971) have demonstrated a striking inhibition by Mn^{++} of the vasopressin-induced activity of the enzyme from toad bladder. The interpretation of these observations in terms of the structure of the enzyme system must await further investigation.

The responses of both adipose and adrenal cortex adenyl cyclases to ACTH have been shown to be dependent on Ca^{++} (Bär and Hechter, 1969c; Birnbaumer and Rodbell, 1969), although Ca^{++} at concentrations above 1 mM causes inhibition. In the case of the adrenal cortex enzyme, evidence has been presented that Ca^{++} is not required for binding of ACTH to the receptor (Lefkowitz, Roth, and Pastan, 1970). Since Ca^{++} is not required for basal or NaF-induced activity the effects of Ca^{++} would appear to be on the coupling process (see Sayers, Beall, and Seelig, 1972, for an alternate interpretation of these observations).

Such a role for Ca^{++} appears to be unique for ACTH-stimulated enzyme systems since in most cases the addition of Ca^{++} is inhibitory or the addition of calcium chelators such as EGTA usually increases hormone effects or has no effect. For example, the effect of vasopressin on frog bladder adenyl cyclase can be completely blocked at concentrations of Ca^{++} that have no effect on NaF-induced activity (Hynie and Sharp, 1971). Also, both EDTA (Pohl, Birnbaumer, and Rodbell, 1971) and EGTA (Hepp, Edel, and Wieland, 1970) stimulate glucagon-induced activity whereas Ca^{++} inhibits it.

Although monovalent cations usually have not been shown to affect hormone responsiveness, there are such examples. K^+ stimulated (20% increase) the effect of ACTH on fat-cell ghosts (Birnbaumer, Pohl, and Rodbell, 1969) and reversed the inhibitory effect of ouabain on the activation of the thyroid enzyme by TSH (Burke, 1970a). The effect of TSH on adenyl cyclase of purified bovine plasma membranes was stimulated by K^+ but the effect of NaF was not changed (Wolff and Jones, 1971). TSH-induced enzyme activity of the thyroid was inhibited by Na^+ and Li^+ (Burke, 1970; Wolff and Jones, 1971). Li^+ was shown also to inhibit the catecholamine-induced increase in cAMP levels of brain slices (Forn and Valdecasas, 1971). A sex difference was observed (Marcus and Aurbach, 1971) in the response of the rat renal cortex enzyme to monovalent cations. The PTH-induced activity of the enzyme from male but not female rats was increased by K^+ and Rb^+.

(c) *Agents that act through specific receptor mechanisms:* It is well established that the effects of hormones can be blocked by specific substances which appear, either by their structural analogy or the competitive nature of their antagonism, to act at the hormone receptor. Such antagonism of the effect of hormones on the activity of adenyl cyclase in broken-cell preparations has been demonstrated for catecholamines (Murad, Chi, Rall, and Sutherland, 1962), glucagon (Rodbell, Birnbaumer, Pohl, and Sundby, 1971), ACTH (Lefkowitz, Roth, Pricer, and Pastan, 1970), and prostaglandins (see Kuehl and Humes, 1972). When used at appropriate concentrations such antagonists are specific for a single hormone and usually have no effect on basal or NaF-induced activities. High concentrations of catecholamine antagonists such as propranolol and phentolamine may alter adenyl cyclase by a nonspecific mechanism (see Levey, Roth, and Pastan, 1969; Himms-Hagen, 1970).

These observations are of pharmacological interest, but such agents do not play a role in the normal regulation of adenyl cyclase activity. However, there are at least three circumstances wherein normal, physiologically important agents may inhibit enzyme activity through interaction at specific receptors. Robison, Butcher, and Sutherland (1967, 1971) have reviewed the evidence in support of an inhibitory role for the interaction of catecholamines with the adrenergic α-receptor. The hypothesis suggests that interaction of catecholamines with the β-receptor leads to activation of adenyl cyclase whereas interaction with the α-receptor has an inhibitory effect. Although there is evidence to support such a mechanism in some tissues, in brain interaction with α-like receptors has been shown to result in an increase in intracellular cAMP content (Chasin, Rivkin, Mamrak, Samaniego, and Hess, 1971; Perkins and Moore, 1973; Palmer, Sulser, and Robison, 1973).

The mechanism of the antilipolytic effect of insulin in fat cells and the "anti-glycogenolytic" effect in liver remains obscure, but recent studies of Illiano and Cuatrecasas (1972) and Hepp (1971) suggest that insulin may block hormone-induced activation of adenyl cyclase in fat cells and liver.

Finally, the antilipolytic effect of PGE_1 in fat cells may result from blockade of hormone-induced activation of adenyl cyclase. Butcher and Baird (1968) have shown that the increased level of cAMP in fat cells brought about by epinephrine, rapidly declines to basal levels if PGE_1 is added. However, the mechanism involved is not clear since these results could involve either inhibition of the effect of epinephrine on adenyl cyclase or stimulation by PGE_1 of PDE activity. It is clear, however, that prostaglandins can *stimulate* some adenyl cyclases (see Butcher, 1970; Kuehl, Humes, Cirillo, and Ham, 1972) apparently through a receptor-mediated mechanism (Kuehl and Humes, 1972).

(d) *Miscellaneous effects:* Cholera toxin has recently been shown to activate irreversibly adenyl cyclase from intestine (Kimberg, Field, Johnson, Henderson, and Gershon, 1971) and from liver (Gorman and Bitensky, 1972). In liver homogenates of animals pretreated with cholera toxin, basal activity was markedly increased, an effect of epinephrine was not apparent, but glucagon still stimulated enzyme activity. Thus, in liver, cholera toxin appears, like NaF, to act at a step beyond the hormone receptor to activate fully the catecholamine-sensitive adenyl cyclase. The lack of an effect on glucagon-induced activity may be explained by the extensive studies of Bitensky (Bitensky, Russell, and Robertson, 1969; Cohen and Bitensky, 1969; Bitensky, Russell, and Blanco, 1970; Bitensky, Gorman, Neufeld, and King, 1971) showing that, in liver, catecholamine- and glucagon-responsive adenyl cyclases are probably separate entities.

Rosen and Rosen (1970) described a toxin from preparations of clostridial neuraminidase that increases basal activity of frog erythrocyte adenyl cyclase and potentiates the effects of catecholamines. It is of interest that the toxin had no effect on the enzyme from tadpole erythrocytes, which does not respond to catecholamines.

The stimulation of intestinal mucosal adenyl cyclase by cholera toxin has been studied in some detail by Kimberg, Field, Johnson, Henderson, and Gershon (1971). Stimulation occurred whether the toxin was applied *in vivo* or *in vitro,* but in either case 1 to 2 hr was required for development of the response; i.e., toxin added directly to the enzyme assay mixture just prior to a short-term assay had no effect. The response of the enzyme to

PGE_1 and NaF was still apparent in toxin treated preparations. A slow onset of action of cholera toxin also was observed by Sharp and Hynie (1971). The site of action of the toxin is not clear since in one case hormone responsiveness was lost and in another PGE_1 effects were preserved.

3. *Correlation of Hormone Binding and Enzyme Activation*

It has long been assumed but only recently demonstrated that the first step in hormone action is the binding of the hormone to structure-discriminating receptors. Consistent with this assumption, the 2nd messenger hypothesis, and the current models of adenyl cyclase structure, glucagon, ACTH, vasopressin, and catecholamines can be shown to bind to plasma membranes at specific receptor sites. This has been accomplished utilizing hormones covalently linked to agarose beads (beads too large to enter cells) or directly by the use of purified plasma membrane fragments to measure the binding of radioisotopically labeled hormones. The thrust of the latter type of study is to demonstrate the equivalence of the kinetics and stoichiometry of binding and activation of adenyl cyclase. To date success has been only partial.

Lefkowitz, Roth, and Pastan (1971) have shown the binding of ^{125}I-ACTH to a lipid extract of a mouse adrenal tumor containing "partially solubilized" adenyl cyclase (Pastan, Pricer, and Blanchette-Mackie, 1970). Scatchard plots of the data demonstrated two classes of binding sites (Lefkowitz, 1971), but binding was highly specific for ACTH; other peptide hormones or biologically inactive derivatives of ACTH did not bind. Partially active fragments and derivatives of ACTH were bound to a degree consistent with their partial activity. Unfortunately, the binding studies were not carried out under the conditions used for assay of adenyl cyclase activity. Further, a critical comparison of the kinetics of binding and activation have yet to be carried out in this system.

Lefkowitz and Haber (1971) have recently reported their studies of the binding of catecholamines to a microsomal fraction of dog ventricular myocardium. ^3H-norepinephrine, bound to the particles, was displaced by other catecholamines in the order of their effectiveness on the force and rate of cardiac contraction, i.e., isoproterenol > epinephrine > norepinephrine > dopamine. Moreover, 50% inhibition of binding occurred at concentrations similar to those resulting in half-maximal activation of dog heart adenyl cyclase (see Murad, Chi, Rall, and Sutherland, 1962). The α-adrenergic receptor antagonist phentolamine was ineffective in displacing ^3H-norepinephrine whereas 0.5 mM propranolol, a β-receptor antagonist, caused 50% displacement. It is surprising that propranolol, a potent antagonist of the effects of catecholamines on contractility or adenyl cyclase activity, should be so impotent in displacing bound ^{3}H-norepinephrine. Hopefully, subsequent studies with this system will be carried out under conditions similar to those used for adenyl cyclase assay and will include a demonstration that the bound radioactivity is still norepinephrine.

Schramm, Feinstein, Naim, Land, and Lasser (1972) have demonstrated the binding of ^{3}H-epinephrine to turkey erythrocyte membranes. The concentration required for half-maximal binding (30 μM) was the same as that required for half-maximal stimulation of the adenyl cyclase of the same membrane preparation. Again the conditions for binding were not the same as for assay of adenyl cyclase. Binding was rapid at 37°C, being maximal by 1 min, the first time measured. The bound radioactivity was extracted from the membrane, purified, and determined to be epinephrine. That displacement of bound ^{3}H-epinephrine was more effective with propranolol than with phentolamine suggests a β-receptor-like interaction. It was of interest that whereas 20 μM propranolol caused 30% inhibition of binding, the same concentration caused 90% inhibition of the effect of epinephrine on adenyl cyclase activity. The explanation for such a result is not clear unless only a portion of the sites binding epinephrine are involved in activation of the enzyme.

In a series of recent papers (Pohl et al., 1971a,b; Birnbaumer et al., 1971, 1972; Rodbell et al., 1971a,b,c,d; Rodbell, 1972), Rodbell and his associates have studied glucagon binding to purified rat liver membranes and have attempted to relate this to stimulation of the adenyl cyclase of such membranes. A number of interesting observations were made. First, good agreement was found for the relationship of glucagon concentration vs. binding or activation of adenyl cyclase. However, binding and effects on enzyme activity were not measured under identical conditions. Apparently, the important difference was the absence of ATP in the binding assay (see Rodbell et al., 1971b,c). ATP at the concentration (3.2 mM) used for the assay of enzyme activity was shown to cause a 50% decrease in the binding of submaximal concentrations of glucagon to the membrane fragments. GTP at low concentrations (0.05 to 500 μM) was found to influence both the binding of glucagon and its effect on enzyme activity. However, the relationship could not be readily interpreted since, on one hand, GTP reduced the affinity of binding of glucagon whereas on the other it increased markedly the effect of glucagon on enzyme activity without apparent change in the concentration vs. activation relationship. Based on their studies Rodbell has suggested that GTP plays a physiologically significant, obligatory role in the action of glucagon on adenyl cyclase activity. The finding by others that GTP also augments the effects of PGE$_1$ on adenyl cyclase activity of platelets (Krishna and Harwood, 1972) lends support for the general nature of the GTP effect. However, the picture is still not clear and awaits resolution by further experimentation.

First, the apparently contradictory effects of GTP on hormone binding and on adenyl cyclase need to be clarified. Second, the physiological significance of this phenomenon needs further support. For example, it is of interest to determine whether GTP acts on the hormone receptor or the coupling factor and whether it acts extracellularly or intracellularly. If GTP acts at an extracellular site, it should potentiate the effect of the hormone on cAMP accumulation in intact cells. If it acts at an intracellular site, the concentration relationship of the effect would indicate that the GTP concentration of most cells would usually be at saturating levels for this process. Even if not, the level of ATP (which can substitute for GTP at millimolar concentrations) should be sufficient to saturate the system involved. Of course, the existence of specific nucleotide pools in association with adenyl cyclase could obviate the latter argument. Shimizu, Creveling, and Daly (1970) have suggested the existence of unique substrate pools of ATP for adenyl cyclase in brain.

These same workers have carried out a comparative analysis of the kinetics of binding of ^{125}I-glucagon and of activation of adenyl cyclase (Rodbell et al., 1971b,c; Birnbaumer et al., 1972; Rodbell, 1972). If the assumption is made that activation of adenyl cyclase is a direct function of the number of receptor sites occupied, then the time course for binding is much too slow to explain the essentially instantaneous effect of glucagon on the rate of cAMP formation. The explanation for the lack of agreement of binding and activation rates remains obscure but may well involve effects on the binding-activation process by GTP or ATP.

Bitensky (see Bitensky, Gorman, Neufeld, and King, 1971) was first to report the apparently irreversible activation of liver adenyl cyclase by glucagon. This observation can now be put into proper perspective due to the recent work of Rodbell et al. (1971b,c) and Birnbaumer et al. (1972). Glucagon in the absence of nucleotides or EDTA binds essentially irreversibly to liver membranes, but does not activate the enzyme if assayed at low ATP concentrations and in the absence of GTP. The glucagon-treated enzyme can be washed free of unbound hormone and added to an assay mixture containing high concentrations of ATP (or presumably low ATP plus GTP) whereupon a stimulated rate of activity will be observed (Bitensky et al., 1971). The primary effect of EDTA, ATP, and GTP appears to be to increase the rate of dissociation of bound glucagon and thus decrease the apparent affinity constant. It is of interest that deshistidyl-glucagon completely inhibits activation of adenyl cyclase at a concentration of the analogue that displaces only 10% of the bound glucagon (Rodbell, 1972). This is similar to the observation of Schramm et al. (1972) that propranolol fully blocks activation of the erythrocyte ghost enzyme by epinephrine under conditions where only 30% of the bound catecholamine is displaced.

Studies concerning the relationship of hormone binding to enzyme activation are in their infancy but already it appears that it may be naive to assume tacitly that there will be a direct relation between the number of receptors occupied and the extent of enzyme activation. Two basic anomalies indicate the problem. First, the rate of activation of the enzyme is much faster than the rate of saturation of the receptors. Second, specific hormone antagonists thought to act by competition for the receptor can completely block activation at concentrations that only partially prevent binding of the hormone.

In view of these apparent discrepancies, it may be worthwhile to consider briefly two modifications of classical receptor theory. The concept of "spare receptors" has been discussed by Stephenson (1956) and Furchgott (1966) and may be applicable to hormone-receptor interactions. The assumption made for our purposes is that all receptors for a given hormone are identical in terms of binding and activation properties but that only a small fraction of the total receptor population needs to be occupied to induce a maximal effect.[4] Two predictions can be made on a theoretical basis concerning the relation of binding to activation; these are illustrated in Fig. 11. First, under identical conditions the concentration-effect curve should lie to the left of the concentration-binding curve to a degree inversely related to the percentage of total receptors necessary to elicit a maximal response. Second, at hormone concentrations which at equilibrium will bind more receptors than required for maximal effect, the time for occupation of the number of receptors required for full activity should be less than the time necessary to attain equilibrium of binding with the total population. However, at concentrations of the hormone which at equilibrium will occupy a number of receptors equal to or less than the number required for total activity, the rates of binding and activation should be similar.

The observation (Rodbell et al., 1971a; Schramm et al., 1972) that binding and activation occur over identical ranges of hormone concentration would appear to be incompatible with the predictions of the spare receptor model. However, as pointed out before, these hormone binding and activation experiments were not carried out under identical conditions. It appears evident from the results of Rodbell et al. (1971b) that if binding experiments

---

[4]This idea is not unique since there appear to be spare receptors involved in the regulation of the rat adipocyte enzyme; they just happen to be different types of receptors. However, the concept is the same in terms of the mechanism involved since in either the fat cell or the liver cell only a portion of the total receptor population is required to activate the enzyme maximally.

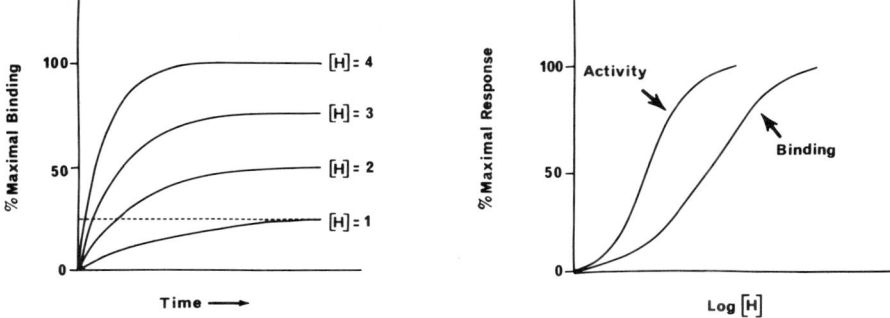

FIG. 11. Spare receptor theory: relation of hormone binding to enzyme activation. *Left:* The assumption is made that [H] = 1 is the concentration of hormone necessary to elicit maximal enzyme activation, but that binds only 25% of the receptor population. It is clear that the time for maximal activation at [H] = 2 to 4 is markedly less than the time required to reach equilibrium of binding. *Right:* The concentration-activation curve lies to the left of the concentration-binding curve to a degree inversely related to the percentage of the receptor population necessary for maximal activation.

were carried out in the presence of maximally effective concentrations of GTP or ATP, the concentration-binding curve for glucagon would be displaced to a higher range of hormone concentrations. Thus, in this regard, the incompatibility with a spare receptor model may be more apparent than real.[5]

The spare receptor model also predicts that the rate of binding and activation should be equal at concentrations of the hormone that produce a submaximal activation of the enzyme. However, at concentrations of glucagon giving half-maximal activation of the rat liver enzyme, Rodbell et al. (1971*b*) and Birnbaumer et al. (1972) observed a significant difference in the rates of binding and activation under comparable albeit not identical conditions. Unfortunately the binding assay used in these studies was not readily

---

[5] It is of interest that fat-cell ghosts bind much less glucagon per milligram protein than do liver cell membranes (Rodbell et al., 1971*a*). Furthermore, the concentration of glucagon required for half-maximal stimulation of the fat-cell ghost enzyme is 25 times that observed for the liver enzyme. Since one might expect the glucagon receptor to have similar binding properties independent of its tissue of origin, this observation is somewhat strange. However, it could be explained if in fat cells there are no spare receptors for glucagon. Thus, even if the $K_{Disocc.}$ for the fat cell and liver receptor for glucagon binding were identical, the concentration of glucagon required for half-maximal activation would be greater in the absence of spare receptors. Since [Receptor-Hormone] $\times$ $K_{Disocc.}$ = [R] $\times$ [H], for an equal amount of R·H to form in the presence of a lower concentration of R, the concentration of H would have to be increased. Of course, other explanations come to mind but this one is consistent with the presence or absence of spare receptors.

applicable to a rapid kinetic analysis. It is clear that such discrepancies can only be resolved by accurate, rapid kinetic analysis of binding and enzyme activation under identical conditions. The rate-limiting step appears to be the development of adequate methods.

Although a consideration of the existence of spare hormone receptors may be of some help in understanding the relationship of hormone binding to enzyme activation, it is not clear how this concept could account for the discrepancy in the effect of competitive antagonists on binding and activity. Rodbell (1972) has suggested that such results might be explained on the basis of two populations of receptors; however, these workers have not been able to detect more than one type of receptor for glucagon by their kinetic analysis (Rodbell, 1971; Birnbaumer et al., 1972). The suggestion proposed is that two receptor populations might exist, one which binds glucagon and leads to activation and another which binds glucagon but has no effect on activity. However, in such a circumstance deshistidyl-glucagon would be allowed to compete only at the "active" receptor to explain the observation of 10% displacement of bound hormone leading to 100% inactivation of the enzyme. This seems somewhat unlikely since it has been demonstrated that deshistidyl-glucagon can displace 100% of half-maximal levels of bound glucagon (Rodbell et al., 1971b). Suffice it to say, more work needs to be done.

Another modification of receptor theory should be mentioned before leaving this subject, namely, the "rate" theory of Paton (1961). Essentially, the theory states that the response is not related to the number of receptors occupied but to the rate of occupation or, in other words, to the rate of turnover of the drug-receptor complex. Thus, an increment of response occurs upon the act of binding, but a second increment cannot occur until the drug dissociates and then reassociates. In a review on the subject, Waud (1968) likened the mechanism to that producing sound with a piano: depression of a key produces sound but unless the key is released and then depressed again the sound fades. In terms of hormone-receptor interaction, an essentially irreversibly bound hormone would elicit an initial quantum of response, upon the act of binding, but then would stabilize the receptor as an inactive complex. The addition to such a complex of a substance which decreased the affinity of the receptor for the hormone would elicit a response. Although there are reasons to question the feasibility of such a model (see Waud, 1968), it would appear to fit, at least in part, the observation that irreversibly bound glucagon does not stimulate adenyl cyclase but agents that accelerate dissociation (EDTA, GTP, ATP) allow expression of the hormonal response. The rate theory also would predict a much faster rate of receptor response than of receptor occupation (see Ariëns, Simonis, and Van Rossum, 1964, p. 190, Figs. 30 and 31).

The studies of Oka and Topper (1971) are of interest with respect to the applicability of Paton's rate theory to hormone-receptor interactions. These studies suggest that insulin must be released from its binding sites on mammary epithelial cells in order to stimulate uptake by the cells of α-aminoisobutyric acid.

As a final comment in this section, the recent observations of Walter, Schwartz, Hechter, Douša and Hoffman (1972) should be mentioned. This report describes the synthesis and properties of bromoacetyl-oxytocin, an irreversible inhibitor of neurohypophyseal hormone-stimulated adenyl cyclase. They suggest the use of this analogue and others like it as a possible affinity label for hormone receptors. It is of further interest that compounds that bind irreversibly with receptors and thus reduce the total population of available receptors have been useful in the demonstration of spare receptors in certain biological systems (see Furchgott, 1966). It seems feasible to use such hormone analogues not only as affinity labels for the isolation and purification of receptors but for the kinetic demonstration of the existence of spare receptors.

4. *Ontogenetic Development and Differentiation*

Rosen and Rosen (1968) first observed that the expression of hormone sensitivity of adenyl cyclase can occur at a stage of development later than the expression of catalytic activity. Whether such an observation can be attributed to the absence of the receptor component or to a lack of the coupling component is not known. Schmidt, Palmer, Dettbarn, and Robison (1970) have reported a distinct difference in the time of development of the responsiveness of adenyl cyclase of brain slices to catecholamines and the appearance of basal catalytic activity. We have carried out similar studies using slices of rat cerebrum and have evidence to suggest that the catecholamine receptor may be present in the developing rat cerebral cortex before it is coupled to the catalytic component. In Fig. 12 is shown the development of catalytic activity as measured in broken cell preparations. From Fig. 13 it is clear that the cAMP content of slices from a 10-day cerebrum is not increased by norepinephrine. However, in the presence of adenosine, norepinephrine markedly increases cAMP content over and above the level observed in the presence of adenosine alone. At day 11 or 12 there is an abrupt increase in the responsiveness of the slices to norepinephrine alone, but the synergistic effect of norepinephrine and adenosine in combination is still observed. These observations suggest that adrenergic receptors are present in the tissue prior to the 11th day. The fact that norepinephrine alone has no effect suggests that either the receptor does not bind the amine in the absence of adenosine or that the coupling function is

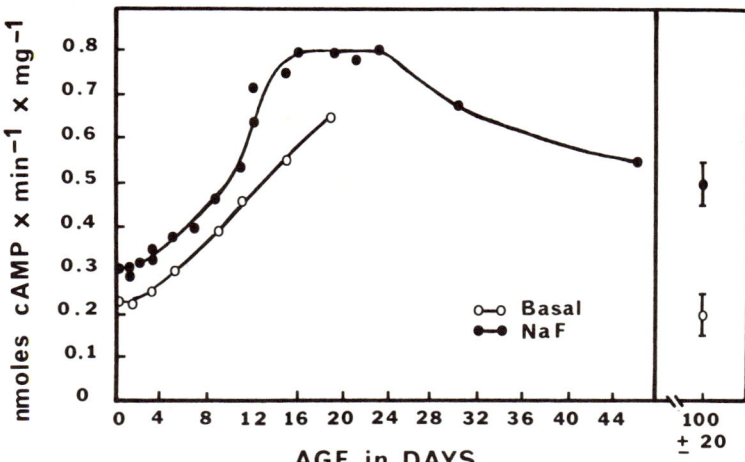

FIG. 12. Ontogenetic development of adenyl cyclase activity in rat cerebral cortex. Enzyme activity was measured in homogenates in the presence or absence of 10 mM NaF. The details of the assay procedure were as described by Perkins and Moore (1971).

not fully operative in the absence of adenosine. Schultz and Daly (1973) have made analogous observations using slices of adult guinea pig cerebral cortex. In this tissue catecholamines alone have no effect on cAMP content but act synergistically with adenosine or histamine. Direct binding studies using radioactive catecholamines may resolve the question as to whether amine binding occurs in the absence of adenosine.

Braun and Hechter (1970) have made the interesting observation that upon adrenalectomy of rats the responsiveness of adipose tissue to ACTH is lost, but other hormones still activate adenyl cyclase activity. Sensitivity to ACTH is regained by administration of cortisol to the adrenalectomized animals or is increased over normal if control animals are given cortisol. Again, direct binding studies might reveal if the ACTH receptor or the coupling factor is involved. However, since responsiveness to the other hormones is not altered, if the transducer is involved, one would have to imagine that there are six distinct coupling structures in rat adipocytes.

Finally, I would like to mention some observations made using tumor cells which provide some insight into the organization of the adenyl cyclase system and which suggest a means whereby this organization might be manipulated experimentally. Ney et al. (1969) observed that ACTH did not increase the cAMP content or the rate of steroidogenesis of a rat adrenocortical carcinoma. However, in homogenates ACTH was more effective in activating the adenyl cyclase from the tumor than from the normal adrenal gland.

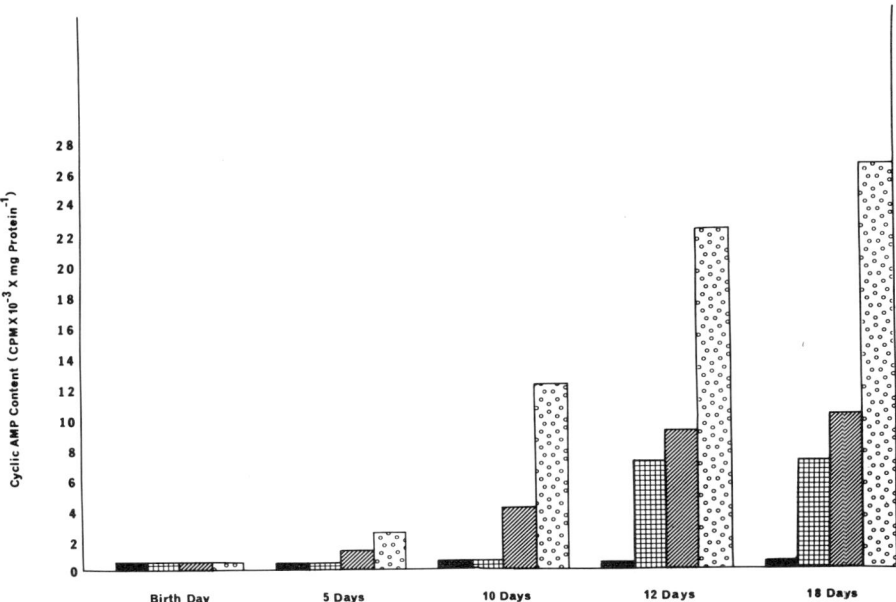

FIG. 13. Ontogenetic development of the response of adenyl cyclase of rat cerebral cortex to norepinephrine and adenosine. Slices of rat cerebral cortex were first incubated with ^{14}C-adenine (Shimizu, Creveling, and Daly, 1970), then incubated in the absence or presence of 30 μM norepinephrine, 100 μM adenosine, or both (respectively from left to right in the figure). Cyclic AMP was isolated as described by Perkins and Moore (1973).

One of the possible explanations for such an observation is the existence of a mechanism for the masking of hormone receptors. In this case the malignant transformation could have resulted in the masking of a receptor normally expressed. More recent work (Schorr et al., 1971) suggests that, conversely, malignant transformation might lead to the unmasking of hormone receptors normally not expressed. These workers observed a mouse adrenal tumor to have an adenyl cyclase which was responsive not only to ACTH but to catecholamines, TSH, LH, and FSH as well. Whether all of these receptors are present but not expressed in the adrenal cortex or whether the malignant transformation resulted in abnormal gene expression is not clear. However, it is clear that an adenyl cyclase normally operating under monovalent regulation can be converted into an enzyme responsive to five different hormones.

The concept of masked receptors has been invoked (Burger and Goldberg, 1967; Burger, 1969) to explain the appearance of agglutinability to wheat germ agglutinin in normal 3T3 fibroblasts after treatment of the cells with trypsin. Malignant transformation of the normal fibroblasts (e.g., with polyoma virus) results in a marked increase in agglutinability without any necessity of prior exposure to trypsin. It would be of interest to test the effects of hormones that are usually noneffective on tissues after they have been exposed to a mild trypsinization.

Schimmer (1972) has studied the adenyl cyclases of three isogenetic adrenal cortex tumor cell lines in culture. All three lines produce steroids, and the rate of steroidogenesis can be stimulated by dibutyryl-cAMP in each line, but by ACTH in only one of the lines. The kinetic properties and stability of the basal enzyme activity is similar in all of the cell lines which leads to the suggestion that the ACTH receptor or the coupling function has been lost in the two unresponsive lines. Direct binding studies should distinguish between these two possibilities.

Observations such as these suggest that it might be possible to obtain a spectrum of clonal cell lines with selective deletions of different components of the adenyl cyclase system. Based on such a rationale we have been examining the regulation of cAMP metabolism in a number of astrocytoma cell lines in culture (Perkins et al., 1971; Clark and Perkins, 1971, 1972). In Table 5 is shown the effector sensitivity of line 1181N1, line 118132 (cloned from routinely overgrown cultures of line 1181N1), and line EH1181N1, a Rous sarcoma virus-transformed line derived from the 1181N1 cells. It is clear that each cell type responds to the four effectors in a different pattern. The 118132 cells have essentially lost their sensitivity to histamine but retain responsiveness to catecholamine, adenosine, and PGE_1. In no case does there appear to be a change with respect to only one effector; quantitatively, alterations occur in the response to each of the agonists. The catalytic properties of the enzyme in broken-cell preparations are similar for

TABLE 5. *Comparison of the effects of norepinephrine, adenosine, histamine, and prostaglandin E_1, on the cAMP content of three lines of human astrocytoma cells in culture*

Agonist	Concn. (mM)	Cyclic AMP content (cpm/mg protein)		
		1181N1	118132	EH1181N1
None	—	4,400	3,000	4,500
Norepinephrine	0.1	24,700	78,700	11,300
Adenosine	0.1	12,300	47,000	11,600
Histamine	0.1	13,500	3,200	8,000
Prostaglandin E_1	0.001	115,600	41,100	13,600

For experimental detail, see Table 3, Clark and Perkins (1971), and Perkins and Moore (1973).

each cell line, but there is variation in the basal activity (per mg protein) and the enzymes exhibit different stability properties. These studies are in their early stages but provide support for the suggestion that the adenyl cyclase system may be amenable to manipulation at the genetic level.

IV. SPECULATION ON THE STRUCTURE OF THE ADENYL CYCLASE SYSTEM

In a recent review Singer and Nicolson (1972) discussed the evidence suggesting that the plasma membrane has certain dynamic properties. The membrane is visualized (Fig. 14) as being composed of a matrix of oriented phospholipids in which are embedded integral membrane proteins. The proteins are maintained within the lipid matrix in stable association because of their structural asymmetry, i.e., the ionic portion being in contact with the aqueous phase of the membrane surfaces and the nonpolar region being em-

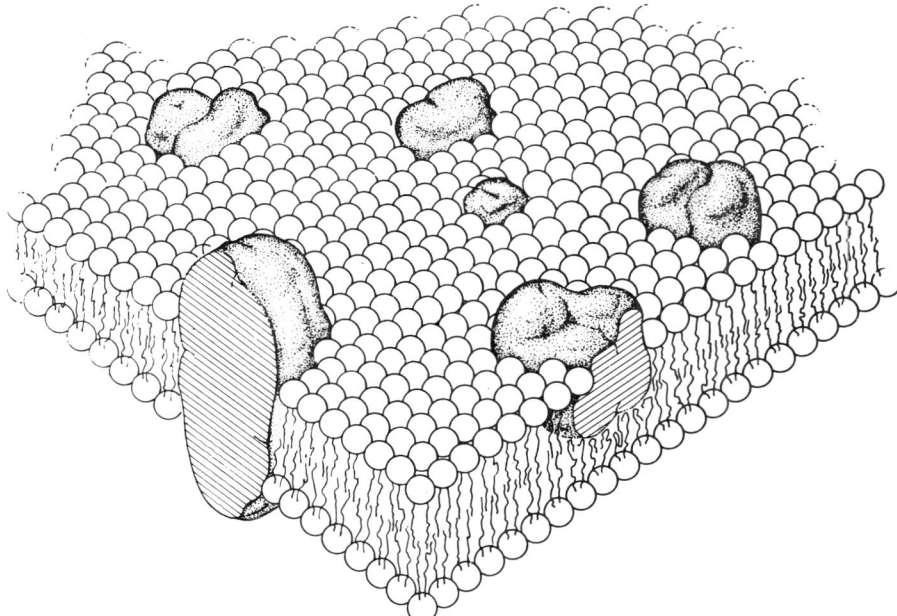

FIG. 14. The lipid-globular protein mosaic model with a lipid matrix (the fluid mosaic model). The solid bodies with stippled surfaces represent the globular integral proteins. The open circles represent the ionic and polar head groups of the phospholipid molecules, in a bilayer, which make contact with water. The wavy lines represent the fatty acid chains (taken from Singer and Nicolson, 1972).

FIG. 15. The lipid-globular protein mosaic model of membrane structure: schematic cross-sectional view. The phospholipids are as depicted in Fig. 14. The integral globular proteins, with the heavy lines representing folded peptide chains, are represented with their ionic residues exposed on the membrane surfaces and their nonpolar residues embedded in the hydrophobic interior of the membrane (taken from Singer and Nicolson, 1972).

bedded in the hydrophobic interior of the membrane (Fig. 15). An important feature of the model is the postulated fluid nature of the phospholipid matrix, with the resultant possibility of translocation of the integral proteins within the plane of the membrane. The lipids comprising the matrix are predominantly arranged in an interrupted bilayer. However, the model does not exclude the possibility that some portion of the membrane lipids could be in specific association with the integral proteins. Such a membrane structure is compatible with either a two-component model of adenyl cyclase as proposed by Robison, Butcher, and Sutherland (1967) or a three component model as suggested by Hechter and Halkerston (1964) and by Birnbaumer et al. (1970).

These models have a number of features in common which seem to be justified by the studies presented above. First, both propose that hormone receptors are probably protein or lipoprotein in nature and are probably integral components of the membrane exposed on its outer surface. Second, both propose that the catalytic component is protein or lipoprotein in nature and is an integral component of the membrane exposed on its inner surface. Further, this entity probably has a common structure and similar properties in all types of cells. Third, both propose that the interaction of a hormone with a receptor results, through alteration in the structure of the components of the system, in increased catalytic activity. The basic difference in the mechanisms is that the two-component model suggests that functional alteration in the catalytic (C) component is effected by a change in the structure of the regulatory subunit (R) through direct interaction of the R and C units. The three component model allows for the involvement of a third structure in the coupling or transducing of changes in the receptor to changes in catalytic function.

Any useful model of the adenyl cyclase system should provide a means for explaining (1) basal activity, (2) activation by NaF, (3) activation by agents such as detergents, (4) activation by hormones, (5) desensitization to hormones without alteration of basal or NaF-induced activity, (6) univalent regulation, (7) multivalent regulation, and (8) inhibition of hormone activation by agents acting through receptor-mediated mechanisms. Such a model should also be compatible with the demonstrations that R and C components can be present but apparently "uncoupled" (Birnbaumer et al., 1972; Schultz and Daly, 1973; and Fig. 13).

Adenyl cyclase systems involving univalent hormonal regulation, in the simplest case, would be composed of R and C units in a 1:1 stoichiometry. To be compatible with the properties of the liver enzyme system, one must consider the participation of a third component, lipid in nature, which is involved in coupling changes in the R unit to changes in the C unit. Even so, the univalent system could simply be a two-subunit lipoprotein existing as an integral membrane protein. However, as proposed, the model involving a 1:1 stoichiometric complex of R and C units, is not compatible with the existence of spare receptors. Since in this model every R unit is coupled to a C unit, the rate of synthesis of cAMP would be a direct function of the percentage of the total receptor population bound to hormone. The alternative is to postulate more R units than C units, the ratio to be compatible with the degree or redundancy of receptors established by experimentation. However, if the degree of redundancy is high, it is not reasonable to think in terms of a strict stoichiometric relationship of multiple R units in direct contact with a single C unit; e.g., $R_{20}C$ is structurally untenable. On the other hand, it is possible to imagine that the forces generated by the interaction of a hormone with one of a number of spare receptors could result in a conformational change in the R unit favoring association with the C unit. The mechanism could involve translational movement of the R or C unit to a juxtaposition favoring complex formation. Such a model is consistent with a fluid mosaic membrane structure.[6] Thus, this model predicts that increased catalytic activity results from the interaction of R and C units in 1:1 com-

---

[6] Singer and Nicolson (1972) predict that effects involving translational movement of the integral proteins could require minutes to complete. The activation of adenyl cyclase appears to require only a few seconds. However, *localized* movement might be compatible with the time scale of enzyme activation.

plexes but is still compatible with the existence of spare receptors. The coupling or transducing function would be a function of the fluidity of the lipid matrix and be accomplished by the migration of R or C units to form the complex. The model is applicable to cells exhibiting multivalent regulation if the assumption is made that the degree of activation of the C unit in the R-C complex is independent of the nature of the R unit. It follows then that the percent of maximal activation elicited by a hormone would be a function of the ratio of the number of those hormone-specific R units to the total number of C units. These assumptions are necessary if the model is to explain the observations with fat-cell ghosts; i.e., the six hormones do not compete when added in combination but any combination of effects always sum to a ceiling value, supposedly the value due to activation of all the C units.

If, on the other hand, different R-C complexes lead to different degrees of activation of the C unit then the model of adenyl cyclase under discussion is not compatible with the observations, since in this case hormones eliciting different maximal responses would be competitive when added in combination.

I would like to describe another possible model based on the assumption that direct, stoichiometric interaction of R and C units is not involved. The model assumes that the change in the structure of the R unit (upon interaction with a hormone) in turn alters the structure of the lipid matrix of the membrane. The degree to which the matrix can be functionally altered by hormone-receptor interaction has a ceiling, and maximal alteration occurs by summation of the effects of hormone-altered R units. The model is compatible with the idea that different hormone-receptor interactions could lead to a different degree of alteration in the lipid matrix, and is compatible as well with the existence of spare receptors. Activation of the C units would occur as a result of the change in their lipid environment.

There appears to be no real basis for a choice between these two models or others which have been postulated. Models are put forth primarily to stimulate experimentation which will eliminate, alter, or verify them. The models proposed above are different conceptually in that the first attempts to explain enzyme activation on the basis of the interaction of protein subunits in a stoichiometric complex. The coupling mechanism involves translocation of the subunits, within the plane of the membrane, to juxtaposition. The second suggestion does not require direct, stoichiometric subunit interaction but rather relies on a receptor-induced "field effect" on the lipid environment of the catalytic component of the system.

Either model provides reasonable mechanisms for the eight properties of the adenyl cyclase system listed earlier. (1) Basal activity of intact cells could be due to the intrinsic activity of the constrained state of the C unit which exists in the absence of hormones. If regulation is stringent, the basal activity could be due to a small proportion of the active form of the enzyme existing even in the absence of hormones. (2) NaF could activate the enzyme by interaction at any point in the chain of events distal to receptor alteration. However, the fact that NaF stimulates the soluble, purified enzyme of *Streptococcus salivarius* suggests that its action may be directly on the C unit. (3, 4) Since the C unit, in the absence of hormones, is assumed to exist in a constrained state, any agent or treatment which nonspecifically (detergents) or specifically (hormones) alters the structures involved in that constraint could lead to activation of the enzyme. (5) Desensitization to hormone effects will occur by any means that disrupts either the receptor or the coupling structure. (6, 7) Univalent and multivalent regulation occur by means discussed above. (8) Competitive hormone antagonists can be considered to act by occupation of the receptor without causing the change in the receptor necessary for enzyme activation. Hormones, such as insulin, catecholamine α-agonists, and prostaglandins, should inhibit only hormone activation and not basal activity. The mechanism could involve interaction with membrane receptors which, in the first model, compete for the available C units but do not form active complexes, or, in the second model, would have a stabilizing effect on the basal state of the lipid matrix.

The final elucidation of the structure of the adenyl cyclase system awaits the successful application of the classical biochemist's technique of isolation and characterization of the component parts, followed by reassembly into a functional unit. There are indications that the protein components of the system may be amenable to solubilization and purification. The use of binding assays for specific receptors coupled with affinity chromatography offers some promise that receptors can be separated from the catalytic component and purified (e.g., see Krug, Desbuquois, and Cuatrecasas, 1971). The purification of the catalytic component from mammalian sources could present a problem but may not be necessary if certain bacterial enzymes are similar enough to the mammalian enzyme.

To date, Levey (1971*a,b,c;* Levey and Klein, 1972) apparently is the only investigator to have been successful in the preparation of a mammalian adenyl cyclase that is soluble after removal of detergents. It is clear that in these preparations receptors still exist since hormone-sensitive enzyme activity is regained by the addition of specific phospholipids to the soluble enzyme. Lefkowitz and Levey (1972) have shown that the phospholipids are not required for hormone binding, at least not for binding of catecholamines. Thus, the phospholipid appears to restore the coupling capacity of the soluble enzyme system. Hopefully, these techniques can be applied to the enzymes from other tissues with equal success.

Finally, it is apparent that if the powerful tool of genetic manipulation can be applied to the study of the adenyl cyclase system in mammalian cells, progress should be greatly accelerated. The possibility of obtaining a variety of clonal cell lines with selective deletions of different components of the system should open a whole new avenue of approach to the characterization of the structure and function of this complex regulatory enzyme.

V. ACKNOWLEDGMENTS

Experimental work from the author's laboratory reported in this article was supported by U.S. Public Health Service grants AM 13236, NS 10233, and NS 09199.

VI. REFERENCES

Allen, D. O., Munshower, J., Morris, H. P., and Weber, G. (1971): Regulation of adenyl cyclase in hepatomas of different growth rates. *Cancer Research*, 31:557–560.

Ariëns, E. J., Simonis, A. M. and Van Rossum, J. M. (1964): *Molecular Pharmacology: The Mode of Action of Biologically Active Compounds,* Vol. I, edited by E. J. Ariëns, p. 190. Academic Press, New York.

Bär, H. P., and Hahn, P. (1971): Development of rat liver adenyl cyclase. *Canadian Journal of Biochemistry*, 49:85–89.

Bär, H. P., and Hechter, O. (1969a): Adenyl cyclase and hormone action. I. Effects of adrenocorticotropic hormone, glucagon and epinephrine on the plasma membrane of rat fat cells. *Proceedings of the National Academy of Sciences*, 63:350–356.

Bär, H. P., and Hechter, O. (1969b): Adenyl cyclase assay in fat cell ghosts. *Analytical Biochemistry*, 29:476–489.

Bär, H. P., and Hechter, O. (1969c): Adenyl cyclase and hormone action. III. Calcium requirement for ACTH stimulation of adenyl cyclase. *Biochemical and Biophysical Research Communications*, 35:681–686.

Bär, H. P., and Hechter, O. (1969d): Substrate specificity of adenyl cyclase from rat fat cell ghosts. *Biochimica et Biophysica Acta*, 192:141–144.

Birnbaumer, L., Pohl, S. L., Krans, M. L., and Rodbell, M. (1970): The actions of hormones on the adenyl cyclase system. *Advances in Biochemical Psychopharmacology*, 3:185–208.

Birnbaumer, L., Pohl, S. L., and Rodbell, M. (1969): Adenyl cyclase in fat cells. I. Properties and the effects of adrenocorticotropin and fluoride. *Journal of Biological Chemistry*, 244:3468–3476.

Birnbaumer, L., Pohl, S. L., and Rodbell, M. (1971): The glucagon-sensitive adenyl cyclase system in plasma membranes of rat liver. II. Comparison between glucagon and fluoride-stimulated activities. *Journal of Biological Chemistry*, 246:1857–1860.

Birnbaumer, L., Pohl, S. L., Rodbell, M., and Sundby, F. (1972): The glucagon-sensitive adenylate cyclase system in plasma membranes of rat liver. VII. Hormonal stimulation: Reversibility and dependence on concentration of free hormone. *Journal of Biological Chemistry*, 247:2038–2043.

Birnbaumer, L., and Rodbell, M. (1969): Adenyl cyclase in fat cells. II. Hormone receptors. *Journal of Biological Chemistry*, 244:3477–3482.

Bitensky, M. W., Gorman, R. E., Neufeld, A. H., and King, R. (1971): A specific, reversible, macromolecular inhibitor of hepatic glucagon responsive adenyl cyclase. *Endocrinology*, 89:1242–1249.

Bitensky, M. W., Russell, V., and Blanco, M. (1970): Independent variation of glucagon and epinephrine responsive components of hepatic adenyl cyclase as a function of age, sex and steroid hormones. *Endocrinology*, 86:154–159.

Bitensky, M. W., Russell, V., and Robertson, W. (1968): Evidence for separate epinephrine and glucagon responsive adenyl cyclase systems in rat liver. *Biochemical and Biophysical Research Communications*, 31:706–712.
Bradham, L. S. (1972): Comparison of the effects of Ca^{2+} and Mg^{2+} on the adenyl cyclase of beef brain. *Biochimica et Biophysica Acta*, 276:434–443.
Bradham, L. S., Holt, D. A., and Sims, M. (1970): The effect of Ca^{2+} on the adenyl cyclase of calf brain. *Biochimica et Biophysica Acta*, 201:250–260.
Braun, T., and Hechter, O. (1970): Glucocorticoid regulation of ACTH sensitivity of adenyl cyclase in rat fat cell membranes. *Proceedings of the National Academy of Sciences*, 66:995–1001.
Burger, M. M. (1969): A difference in the architecture of the surface membrane of normal and virally-transformed cells. *Proceedings of the National Academy of Sciences*, 62:994–1001.
Burger, M. M., and Goldberg, A. R. (1967): Identification of a tumor-specific determinant on neoplastic cell surfaces. *Proceedings of the National Academy of Sciences*, 57:357–366.
Bürk, R. R. (1968): Reduced adenyl cyclase activity in a polyoma virus transformed cell line. *Nature*, 219:1272–1275.
Burke, G. (1970a): Effects of cations and ouabain on thyroid adenyl cyclase. *Biochimica et Biophysica Acta*, 220:30–41.
Burke, G. (1970b): Comparison of thyrotropin and sodium fluoride effects on thyroid adenyl cyclase. *Endocrinology*, 86:346–352.
Burke, G. (1970c): Substrate specificity of thyroid adenyl cyclase. *Life Sciences*, Part I, 9:789–795.
Butcher, R. W. (1970): Prostaglandins and cyclic AMP. *Advances in Biochemical Psychopharmacology*, 3:173–183.
Butcher, R. W., and Baird, C. E. (1968): Effects of prostaglandins on adenosine 3',5'-monophosphate levels in fat and other tissues. *Journal of Biological Chemistry*, 243:1713–1717.
Butcher, R. W., Baird, C. E., and Sutherland, E. W. (1968): Effects of lipolytic and antilipolytic substances on adenosine 3',5'-monophosphate levels in isolated fat cells. *Journal of Biological Chemistry*, 243:1705–1712.
Chase, L. R., and Aurbach, G. D. (1968): Renal adenyl cyclase: Anatomically separate sites for parathyroid hormone and vasopressin. *Science*, 159:545–547.
Chasin, M., Rivkin, I., Mamrak, F., Samaniego, S. G., and Hess, S. M. (1971): α- and β-adrenergic receptors as mediators of accumulation of cyclic adenosine 3',5'-monophosphate in specific areas of guinea pig brain. *Journal of Biological Chemistry*, 246:3037–3041.
Cheung, W. Y. (1967): Properties of cyclic 3',5'-nucleotide phosphodiesterase from rat brain. *Biochemistry*, 6:1079–1087.
Cheung, W. Y., and Chiang, M.-H. (1971): Adenyl cyclase. The extent of reversibility of the reaction. *Biochemical and Biophysical Research Communications*, 43:868–874.
Clark, R. B., and Perkins, J. P. (1971): Regulation of adenosine 3',5'-cyclic monophosphate concentration in cultured human astrocytoma cells by catecholamines and histamine. *Proceedings of the National Academy of Sciences*, 68:2757–2760.
Clark, R. B., and Perkins, J. P. (1972): The effect of adenosine on the formation of cyclic AMP (cAMP) in cultured human astrocytoma cells. *Federation Proceedings*, 31:513.
Cohen, K. L., and Bitensky, M. W. (1969): Inhibitory effects of alloxan on mammalian adenyl cyclase. *Journal of Pharmacology and Experimental Therapeutics*, 169:80–86.
Cook, W. H., Lipkin, D., and Markham, R. (1957): The formation of a cyclic dianhydrodiadenylic acid (I) by the alkaline degradation of adenosine-5'-triphosphoric acid (II). *Journal of the American Chemical Society*, 79:3607–3608.
Davoren, P. R., and Sutherland, E. W. (1963): The cellular location of adenyl cyclase in the pigeon erythrocyte. *Journal of Biological Chemistry*, 238:3016–3023.
de Robertis, E., Arnaiz, G. R. D. L., Alberici, M., Butcher, R. W., and Sutherland, E. W. (1967): Subcellular distribution of adenyl cyclase and cyclic phosphodiesterase in rat brain cortex. *Journal of Biological Chemistry*, 242:3487–3496.
Dexter, R. N., and Allen, D. O. (1970): A glucagon-sensitive adenyl cyclase in pheochromocytoma. *Clinical Research*, 18:601.

Doušá, T., and Hechter, O. (1970): The effect of NaCl and LiCl on vasopressin-sensitive adenyl cyclase. *Life Sciences*, Part I, 9:765-770.

Doušá, T., Hechter, O., Schwartz, I. L., and Walter, R. (1971): Neurohypophyseal hormone-responsive adenylate cyclase from mammalian kidney. *Proceedings of the National Academy of Sciences*, 68:1693-1697.

Drummond, G. I., and Duncan, L. (1970): Adenyl cyclase in cardiac tissue. *Journal of Biological Chemistry*, 245:976-983.

Drummond, G. I., Severson, D. L., and Duncan, L. (1971): Adenyl cyclase: Kinetic properties and nature of fluoride and hormone stimulation. *Journal of Biological Chemistry*, 246:4166-4173.

Emmelot, P., and Bos, C. J. (1971): Studies on plasma membranes. XIV. Adenyl cyclase in plasma membranes isolated from rat and mouse livers and hepatomas, and its hormone sensitivity. *Biochimica et Biophysica Acta*, 249:285-292.

Entman, M. L., Levey, G. S., and Epstein, S. E. (1969): Demonstration of adenyl cyclase activity in canine cardiac sarcoplasmic reticulum. *Biochemical and Biophysical Research Communications*, 35:728-733.

Forn, J., and Valdecasas, F. G. (1971): Effects of lithium on brain adenyl cyclase activity. *Biochemical Pharmacology*, 20:2773-2779.

Furchgott, R. F. (1966): The use of β-haloalkylamines in the differentiation of receptors and in the determination of dissociation constants of receptor-agonist complexes. In: *Advances in Drug Research*, Vol. 3, edited by N. J. Harper and A. B. Simmonds, pp. 21-55. Academic Press, London.

Gorman, R. E., and Bitensky, M. W. (1972): Selective effects of cholera toxin on the adrenaline responsive component of hepatic adenyl cyclase. *Nature*, 235:439-440.

Granner, D., Chase, L. R., Aurbach, G. D., and Tomkins, G. M. (1968): Tyrosine aminotransferase: Enzyme induction independent of adenosine-3',5'-monophosphate. *Science*, 162:1018-1020.

Greengard, P., and Costa, E., editors (1970): *Advances in Biochemical Psychopharmacology, Vol. 3, The Role of Cyclic AMP in Cell Function*. Raven Press, New York.

Greengard, P., Hayaishi, O., and Colowick, S. P. (1969): Enzymatic adenylation of pyrophosphate by 3',5'-cyclic AMP; reversal of the adenyl cyclase reaction. *Federation Proceedings*, 28:467.

Greengard, P., and Robison, G. A., editors (1972): *Advances in Cyclic Nucleotide Research, Vol. 1, Physiology and Pharmacology of Cyclic AMP*. Raven Press, New York.

Greengard, P., Rudolph, S. A., and Sturtevant, J. M. (1969): Enthalpy of hydrolysis of the 3' bond of adenosine 3',5'-monophosphate and guanosine 3',5'-monophosphate. *Journal of Biological Chemistry*, 244:4798-4800.

Hayaishi, O., Greengard, P., and Colowick, S. P. (1971): On the equilibrium of the adenylate cyclase reaction. *Journal of Biological Chemistry*, 246:5840-5843.

Hechter, O. (1965): In: *Mechanisms of Hormone Action*, edited by P. Karlson. Academic Press, New York.

Hechter, O., and Halkerston, I. D. K. (1964): On the action of mammalian hormones. In: *The Hormones*, edited by G. Pinkus, K. V. Thimann, and E. B. Astwood. Academic Press, New York.

Hechter, O., and Soifer, D. (1971): *Basic Actions of Sex Steroids on Target Organs*, edited by P. O. Hubinont, F. Leroy, and P. Galand. Karger, Basel.

Heidrick, M. L., and Ryan, W. L. (1970): Cyclic nucleotides in cell growth *in vitro*. *Cancer Research*, 30:376-378.

Hepp, K. D. (1971): Inhibition of glucagon-stimulated adenyl cyclase by insulin. *FEBS Letters*, 12:263-266.

Hepp, K. D., Edel, R., and Wieland, O. (1970): Hormone action on liver adenyl cyclase activity: The effect of glucagon and fluoride on a particulate preparation from rat and mouse liver. *European Journal of Biochemistry*, 17:171-177.

Himms-Hagen, J. (1967): Sympathetic regulation of metabolism. *Pharmacological Reviews*, 19:367-461.

Himms-Hagen, J. (1970): Adrenergic receptors for metabolic responses in adipose tissue. *Federation Proceedings,* 29:1388-1401.
Hirata, M., and Hayaishi, O. (1967): Adenyl cyclase of *Brevibacterium liquefaciens. Biochimica et Biophysica Acta,* 149:1-11.
Huang, M., Shimizu, H., and Daly, J. (1971): Regulation of adenosine cyclic 3',5'-phosphate formation in cerebral cortical slices: Interaction among norepinephrine, histamine, serotonin. *Molecular Pharmacology,* 7:155-162.
Hynie, S., and Sharp, G. W. G. (1971a): Adenyl cyclase in the toad bladder. *Biochimica et Biophysica Acta,* 230:40-51.
Hynie, S., and Sharp, G. W. G. (1971b): Inhibition by manganese of the action of antidiuretic hormone on adenyl cyclase in toad bladder. *Journal of Endocrinology,* 50:231-235.
Ide, M. (1971): Adenyl cyclase of bacteria. *Archives of Biochemistry and Biophysics,* 144:262-268.
Illiano, G., and Catrecasas, P. (1972): Modulation of adenylate cyclase activity in liver and fat cell membranes by insulin. *Science,* 175:906-908.
Jensen, E. V., Numata, M., Brecher, P. I., and DeSombra, E. R. (1971): *The Biochemistry of Steroid Hormone Action,* edited by R. M. S. Smellie. Academic Press, London.
Johnson, C. B., Blecher, M., and Giorgio, J. (1972): Hormone receptors. I. Activation of rat liver plasma membrane adenyl cyclase and fat cell lipolysis by agarose-glucagon. *Biochemical and Biophysical Research Communications,* 46:1035-1041.
Johnson, G. S., Friedman, R. M., and Pastan, I. (1971): Restoration of several morphological characteristics of normal fibroblasts in sarcoma cells treated with adenosine 3',5'-cyclic monophosphate and its derivatives. *Proceedings of the National Academy of Sciences,* 68:425-429.
Johnson, G. S., and Pastan, I. (1971): Change in growth and morphology of fibroblasts by prostaglandins. *Journal of the National Cancer Institute,* 47:1357-1364.
Johnson, R. A., and Sutherland, E. W. (1973): Detergent dispersed adenylate cyclase from rat brain. Effects of fluoride, cations and chelators. *Journal of Biological Chemistry* (in press).
Jost, J.-P., and Rickenberg, H. V. (1971): Cyclic AMP. *Annual Review of Biochemistry,* 70:741-744.
Kebabian, J. W., Petzold, G., and Greengard, P. (1972): Dopamine-sensitive adenylate cyclase in caudate nucleus of rat brain, and its similarity to the "dopamine receptor." *Proceedings of the National Academy of Sciences,* 69:2145-2149.
Kelley, L. A., and Koritz, S. B. (1971): Bovine adrenal cortical adenyl cyclase and its stimulation by adrenocorticotropic hormone and NaF. *Biochimica et Biophysica Acta,* 237:141-155.
Khandelwal, R. L., and Hamilton, I. R. (1971): Purification and properties of adenyl cyclase from *Streptococcus salivarius. Journal of Biological Chemistry,* 246:3297-3304.
Kimberg, D. V., Field, M., Johnson, J., Henderson, A., and Gershon, E. (1971): Stimulation of intestinal mucosal adenyl cyclase by cholera enterotoxin and prostaglandins. *Journal of Clinical Investigation,* 50:1218-1230.
Klainer, L. M., Chi, Y. M., Freidberg, S. L., Rall, T. W., and Sutherland, E. W. (1962): Adenyl cyclase. IV. The effects of neurohormones on the formation of adenosine 3',5'-phosphate by preparations from brain and other tissues. *Journal of Biological Chemistry,* 237:1239-1243.
Krishna, G., and Birnbaumer, L. (1970): On the assay of adenyl cyclase. *Analytical Biochemistry,* 35:393-397.
Krishna, G., Forn, J., Voigt, K., Paul, M., and Gessa, G. L. (1970): Dynamic aspects of neurohormonal control of cyclic 3',5'-AMP synthesis in brain. *Advances in Biochemical Psychopharmacology,* 3:155-173.
Krishna, G., and Harwood, J. P. (1972): Requirement for guanosine triphosphate in the prostaglandin activation of adenylate cyclase of platelet membranes. *Journal of Biological Chemistry,* 247:2253-2254.
Krishna, G., Weiss, B., and Brodie, B. B. (1968): A simple, sensitive method for the assay of

adenyl cyclase. *Journal of Pharmacology and Experimental Therapeutics*, 163:379–385.

Krug, F., Desbuquois, B., and Cuatrecasas, P. (1971): Glucagon affinity absorbents: Selective binding of receptors of liver cell membranes. *Nature New Biology*, 234:268–270.

Kuehl, F. A., and Humes, J. L. (1972): Direct evidence for a prostaglandin receptor and its application to prostaglandin measurements. *Proceedings of the National Academy of Sciences*, 69:480–484.

Kuehl, F. A., Humes, J. L., Cirillo, V. J., and Ham, E. A. (1972): In: *Advances in Cyclic Nucleotide Research*, Vol. 1: *Physiology and Pharmacology of Cyclic AMP*, edited by P. Greengard, and G. A. Robison. Raven Press, New York.

Lefkowitz, R. J. (1971): *Colloquium on the Role of Adenyl Cyclase and Cyclic 3',5'-AMP in Biological Systems*, edited by P. Condliffe and M. Rodbell, pp. 88–95. Fogarty International Center, Government Printing Office.

Lefkowitz, R. J., and Haber, E. (1971): A fraction of the ventricular myocardium that has the specificity of the cardiac beta-adrenergic receptor. *Proceedings of the National Academy of Sciences*, 68:1773–1777.

Lefkowitz, R. J., and Levey, G. S. (1972): Norepinephrine: Dissociation of β-receptor binding from adenylate cyclase activation in solubilized myocardium. *Life Sciences*, Part II, 11:821–828.

Lefkowitz, R. J., Roth, J., and Pastan, I. (1970): Effects of calcium on ACTH stimulation of the adrenal: Separation of hormone binding from adenyl cyclase activation. *Nature*, 228:864–866.

Lefkowitz, R. J., Roth, J., and Pastan, I. (1971): ACTH-receptor interaction in the adrenal: A model for the initial step in the action of hormones that stimulate adenyl cyclase. *Annals of the New York Academy of Sciences*, 185:195–209.

Lefkowitz, R. J., Roth, J., Pricer, W., and Pastan, I. (1970): ACTH receptors in the adrenals: Specific binding of ACTH-^{125}I and its relation to adenyl cyclase. *Proceedings of the National Academy of Sciences*, 65:745–752.

Levey, G. S. (1970): Solubilization of myocardial adenyl cyclase. *Biochemical and Biophysical Research Communications*, 38:86–92.

Levey, G. S. (1971a): Solubilization of myocardial adenyl cyclase: Loss of hormone responsiveness and activation by phospholipids. *Annals of the New York Academy of Sciences*, 185:449–457.

Levey, G. S. (1971b): Restoration of glucagon responsiveness of solubilized myocardial adenyl cyclase by phosphatidylserine. *Biochemical and Biophysical Research Communications*, 43:108–113.

Levey, G. S. (1971c): Restoration of norepinephrine responsiveness of solubilized myocardial adenylate cyclase by phosphatidylinositol. *Journal of Biological Chemistry*, 246:7405–7407.

Levey, G. S., and Klein, I. (1972): Solubilized myocardial adenylate cyclase: Restoration of histamine responsiveness by phosphatidylserine. *Journal of Clinical Investigation*, 51:1578–1582.

Levey, G. S., Roth, J., and Pastan, I. (1969): Effect of propranolol and phentolamine on canine and bovine responses to TSH. *Endocrinology*, 84:1009–1015.

Liano, S., Lin, A. H., and Tymoczko, J. L. (1971): Adenyl cyclase of cell nuclei isolated from rat ventral prostate. *Biochimica et Biophysica Acta*, 230:535–538.

Macintyre, E. H., Perkins, J. P., Wintersgill, C. J., and Vatter, A. E. (1972): The responses in culture of human astrocytes and neuroblasts to N^6,O^2-dibutyryl cyclic adenosine monophosphate. *Journal of Cell Science* (in press).

Marcus, R., and Aurbach, G. D. (1971): Adenyl cyclase from renal cortex. *Biochimica et Biophysica Acta*, 242:410–421.

McCune, R. W., Gill, T. H., von Hungen, K., and Roberts, S. (1971): Catecholamine-sensitive adenyl cyclase in cell-free preparations from rat cerebral cortex. *Life Sciences*, Part II, 10:443–450.

Menon, K. M. J., and Smith, M. (1971): Characterization of adenyl cyclase from the testis of chinook salmon. *Biochemistry*, 10:1186–1190.

Murad, F., Chi, Y. M., Rall, T. W., and Sutherland, E. W. (1962): Adenyl cyclase. III. The effect of catecholamines and choline esters on the formation of adenosine 3′,5′-phosphate by preparations from cardiac muscle and liver. *Journal of Biological Chemistry,* 237:1233-1238.

Neville, D. M. (1968): Isolation of an organ specific protein antigen from cell surface membrane of rat liver. *Biochimica et Biophysica Acta,* 154:540-552.

Ney, R. L., Hochella, N. J., Grahame-Smith, D. G., Dexter, R. N., and Butcher, R. W. (1969): Abnormal regulation of adenosine 3′,5′-monophosphate and corticosterone formation in an adrenocortical carcinoma. *Journal of Clinical Investigation,* 48:1733-1739.

O'Dea, R. F., Haddox, M. K., and Goldberg, N. D. (1971): Interaction with phosphodiesterase of free and kinase-complexed cyclic adenosine 3′,5′-monophosphate. *Journal of Biological Chemistry,* 246:6183-6190.

Otten, J., Johnson, G. S., and Pastan, I. (1971): Cyclic AMP levels in fibroblasts: relationship to growth rate and contact inhibition of growth. *Biochemical and Biophysical Research Communications,* 44:1192-1199.

Øye, I., and Sutherland, E. W. (1966): The effect of epinephrine and other agents on adenyl cyclase in the cell membrane of avian erythrocytes. *Biochimica et Biophysica Acta,* 127:347-354.

Palmer, G. C., Sulser, F., and Robison, G. A. (1973): Effects of neurohumoral and adrenergic agents on cyclic AMP levels in various areas of the rat brain *in vitro. Neuropharmacology (in press).*

Pastan, I., and Perlman, R. (1970): Cyclic adenosine monophosphate in bacteria. *Science,* 169:339-344.

Pastan, I., Pricer, W., and Blanchette-Mackie, J. (1970): Studies on an ACTH-activated adenyl cyclase from a mouse adrenal tumor. *Metabolism,* 19:809-817.

Paton, W. D. M. (1961): A theory of drug action based on the rate of the drug-receptor combination. *Proceedings of the Royal Society B,* 154:21-69.

Pennington, S. N., Brown, H. D., Chattopadhyay, S., Conaway, C., and Morris, H. P. (1970): Effect of sodium fluoride on the epinephrine response of liver and hepatoma adenyl cyclase. *Experientia,* 26:139-140.

Perkins, J. P., Macintyre, E. H., Riley, W. D., and Clark, R. B. (1971): Adenyl cyclase, phosphodiesterase and cyclic AMP dependent protein kinase of malignant glial cells in culture. *Life Sciences,* Part I, 10:1069-1080.

Perkins, J. P., and Moore, M. M. (1971): Adenyl cyclase of rat cerebral cortex: Activation by sodium fluoride and detergents. *Journal of Biological Chemistry,* 246:62-68.

Perkins, J. P., and Moore, M. M. (1973): Characterization of the adrenergic receptors mediating a rise in adenosine 3′,5′-monophosphate in rat cerebral cortex. *Journal of Pharmacology and Experimental Therapeutics, (in press).*

Pohl, S. L. (1971): *Colloquium on the Role of Adenyl Cyclase and Cyclic AMP in Biological Processes,* p. 85. Fogarty International Center, Government Printing Office.

Pohl, S. L., Birnbaumer, L., and Rodbell, M. (1971): The glucagon-sensitive adenyl cyclase system in plasma membranes of rat liver. I. Properties. *Journal of Biological Chemistry,* 246:1849-1856.

Ponten, J., and Macintyre, E. H. (1968): Long term culture of normal and neoplastic human glia. *Acta Pathologica et Microbiologica Scandinavica,* 74:465-486.

Prasad, K. N., and Sheppard, J. R. (1972): Inhibitors of cyclic nucleotide phosphodiesterase induce morphological differentiation of mouse neuroblastoma cell culture. *Experimental Cell Research,* 73:436-440.

Rabinowitz, M., Desalles, L., Meisler, J., and Lorand, L. (1965): Distribution of adenyl cyclase activity in rabbit muscle fractions. *Biochimica et Biophysica Acta,* 97:29-36.

Rall, T. W. (1971): Mammalian adenyl cyclase. In: *Colloquium on the Role of Adenyl Cyclase and Cyclic 3′,5′-AMP in Biological Systems,* edited by P. Condliffe and M. Rodbell. Fogarty International Center, Government Printing Office.

Rall, T. W., and Gilman, A. G. (1970): Formation and disposition of cyclic AMP in nervous system tissue. *Neurosciences Research Progress Bulletin,* 8:239-266.

Rall, T. W., and Sattin, A. (1970): Factors influencing the accumulation of cyclic AMP in brain tissue. *Advances in Biochemical Psychopharmacology,* 3:113–133.

Rall, T. W., and Sutherland, E. W. (1958): Formation of cyclic adenine nucleotide by tissue particles. *Journal of Biological Chemistry,* 232:1065–1076.

Rall, T. W., and Sutherland, E. W. (1961): The regulatory role of adenosine 3′,5′-phosphate. *Cold Spring Harbor Symposium on Quantitative Biology,* 26:347–354.

Rall, T. W., and Sutherland, E. W. (1962): In: *Methods in Enzymology,* Vol. V, p. 386, edited by S. P. Colowick, and N. P. Kaplan. Academic Press, New York.

Rall, T. W., and Sutherland, E. H. (1962): The enzymatically catalyzed formation of adenosine 3′,5′-phosphate and inorganic pyrophosphate from adenosine triphosphate. *Journal of Biological Chemistry,* 237:1228–1232.

Ray, T. K., Tomasi, V., and Marinetti, G. V. (1970): Hormone action at the membrane level. I. Properties of adenyl cyclase in isolated plasma membranes of rat liver. *Biochimica et Biophysica Acta,* 211:20–30.

Robison, G. A., Butcher, R. W., and Sutherland, E. W. (1967): Adenyl cyclase as an adrenergic receptor. *Annals of the New York Academy of Sciences,* 139:703–723.

Robison, G. A., Butcher, R. W., and Sutherland, E. W. (1968): Cyclic AMP. *Annual Review of Biochemistry,* 37:149–174.

Robison, G. A., Butcher, R. W., and Sutherland, E. W. (1971): *Cyclic AMP.* Academic Press, New York.

Robison, G. A., Nahas, G. G., and Triner, K. (1971): Cyclic AMP and cell function. *Annals of the New York Academy of Sciences,* Vol. 185.

Rodbell, M. (1964): Metabolism of isolated fat cells. I. Effects of hormones on glucose metabolism and lipolysis. *Journal of Biological Chemistry,* 239:375–380.

Rodbell, M. (1971): Hormones, receptors and adenyl cyclase activity in mammalian cells. In: *Colloquium on the Role of Adenyl Cyclase and Cyclic 3′,5′-AMP in Biological Systems,* edited by P. Condliffe and M. Rodbell, pp. 88–95. Fogarty International Center, Government Printing Office.

Rodbell, M. (1972): Cell surface receptor sites. In: *Current Topics in Biochemistry,* edited by C. B. Anfinsen, R. F. Goldberger, and A. N. Schechter. Academic Press, New York.

Rodbell, M., Birnbaumer, L., and Pohl, S. L. (1970): Adenyl cyclase in fat cells. III. Stimulation by secretin and the effects of trypsin on the receptors for lipolytic hormones. *Journal of Biological Chemistry,* 245:718–722.

Rodbell, M., Krans, M. J., Pohl, S. L., and Birnbaumer, L. (1971a): The glucagon-sensitive adenyl cyclase system in plasma membranes of rat liver. III. Binding of glucagon: Method of assay and specificity. *Journal of Biological Chemistry,* 246:1861–1871.

Rodbell, M., Krans, M. J., Pohl, S. L., and Birnbaumer, L. (1971b): The glucagon-sensitive adenyl cyclase system in plasma membranes of rat liver. IV. Effects of guanyl nucleotides in binding of ^{125}I-glucagon. *Journal of Biological Chemistry,* 246:1872–1876.

Rodbell, M., Birnbaumer, L., Pohl, S. L., and Krans, M. J. (1971c): The glucagon-sensitive adenyl cyclase system in plasma membranes of rat liver. V. An obligatory role of guanyl nucleotides in glucagon action. *Journal of Biological Chemistry,* 246:1877–1882.

Rodbell, M., Birnbaumer, L., Pohl, S. L., and Sundby, F. (1971d): The reaction of glucagon with its receptor: Evidence for discrete regions of activity and binding in the glucagon molecule. *Proceedings of the National Academy of Sciences,* 68:909–913.

Rosen, O. M., Hirsch, A. H., and Goren, E. N. (1971): Factors which influence cyclic AMP formation and degradation in an islet cell tumor of the Syrian hamster. *Archives of Biochemistry and Biophysics,* 146:660–663.

Rosen, O. M., and Rosen, S. M. (1968): The effect of catecholamines on the adenyl cyclase of frog and tadpole hemolysates. *Biochemical and Biophysical Research Communications,* 31:82–91.

Rosen, O. M., and Rosen, S. M. (1969): Properties of an adenyl cyclase partially purified from frog erythrocytes. *Archives of Biochemistry and Biophysics,* 131:449–456.

Rosen, O. M., and Rosen, S. M. (1970): A bacterial activator of frog erythrocyte adenyl cyclase. *Archives of Biochemistry and Biophysics,* 141:346–352.

Ryan, W. L., and Heidrick, M. L. (1968): Inhibition of cell growth *in vitro* by adenosine 3',5'-monophosphate. *Science,* 162:1484-1485.

Sattin, A., and Rall, T. W. (1970): The effect of adenosine and adenine nucleotides on the cyclic adenosine 3',5'-phosphate content of guinea pig cerebral cortex slices. *Molecular Pharmacology,* 6:13-23.

Sayers, G., Beall, R. J., and Seelig, S. (1972): Isolated adrenal cells: Adrenocorticotropic hormone, calcium, steroidogenesis and cyclic adenosine monophosphate. *Science,* 175:1131-1133.

Schimmer, B. P. (1972): Adenylate cyclase activity in adrenocorticotropic hormone-sensitive and mutant adrenocortical tumor cell lines. *Journal of Biological Chemistry,* 247:3134-3138.

Schimmer, B. P., Ueda, K., and Sato, G. H. (1968): Site of action of adrenocorticotropic hormone (ACTH) in adrenal cell cultures. *Biochemical and Biophysical Research Communications,* 32:806-810.

Schmidt, M. J., Palmer, E. C., Dettbarn, W.-D., and Robison, G. A. (1970): Cyclic AMP and adenyl cyclase in the developing rat brain. *Developmental Psychobiology,* 3:53-67.

Schorr, I., Rathnam, P., Saxena, B. B., and Ney, R. L. (1971): Multiple specific hormone receptors in the adenylate cyclase of an adrenocortical carcinoma. *Journal of Biological Chemistry,* 246:5806-5811.

Schramm, M., Feinstein, H., Naim, E., Lang, M., and Lasser, M. (1972): Epinephrine binding to the catecholamine receptor and activation of the adenylate cyclase in erythrocyte membranes. *Proceedings of the National Academy of Sciences,* 69:523-527.

Schramm, M., and Naim, E. (1970): Adenyl cyclase of rat parotid gland: Activation by fluoride and norepinephrine. *Journal of Biological Chemistry,* 245:3225-3231.

Schultz, J., and Daly, J. W. (1973): Cyclic adenosine 3',5'-monophosphate in guinea pig cerebral cortical slices: I. *Journal of Biological Chemistry,* 248:843-866.

Seraydarian, K., and Mommaerts, W. F. H. M. (1965): Density gradient separation of sarcotubular vesicles and other particulate constituents of rabbit muscle. *Journal of Cell Biology,* 26:641-656.

Severson, D. L., Drummond, G. I., and Sulakhe, P. V. (1972): Adenyl cyclase in skeletal muscle: Kinetic properties and hormonal stimulation. *Journal of Biological Chemistry,* 247:2949-2958.

Sharp, G. W. G., and Hynie, S. (1971): Stimulation of intestinal adenyl cyclase by cholera toxin. *Nature,* 229:266-269.

Sheppard, H. (1970): Inhibition of norepinephrine stimulated adenyl cyclase by theophylline. *Nature,* 228:567-568.

Sheppard, J. R. (1971): Restoration of contact-inhibited growth to transformed cells by dibutyryl adenosine 3',5'-cyclic monophosphate. *Proceedings of the National Academy of Sciences,* 68:1316-1320.

Shimizu, H., Creveling, C. R., and Daly, J. (1970a): Stimulated formation of adenosine 3',5'-cyclic phosphate in cerebral cortex: synergism between electrical activity and biogenic amines. *Proceedings of the National Academy of Sciences,* 65:1033-1040.

Shimizu, H., Creveling, C. R., and Daly, J. W. (1970b): Effect of membrane depolarization and biogenic amines on the formation of cyclic AMP in incubated brain slices. *Advances in Biochemical Psychopharmacology,* 3:135-154.

Sillen, L. G., and Martel, A. E. Eds. (1964): *Stability Constants of Metal-Ion Complexes.* The Chemical Society, Burlington House, London, p. 257.

Singer, J. J., and Nicolson, G. S. (1972): The fluid mosaic model of the structure of cell membranes. *Science,* 175:720-731.

Soifer, D., and Hechter, O. (1971): Adenyl cyclase activity in rat liver nuclei. *Biochimica et Biophysica Acta,* 230:539-542.

Stephenson, R. P. (1956): A modification of receptor theory. *British Journal of Pharmacology,* 11:379-393.

Streeto, J. M., and Reddy, W. J. (1967): An assay for adenyl cyclase. *Analytical Biochemistry,* 21:416-426.

Sutherland, E. W., Øye, I., and Butcher, R. W. (1965): The action of epinephrine and the role

of the adenyl cyclase system in hormone action. *Recent Progress in Hormone Research,* 21:623–646.
Sutherland, E. W., and Rall, T. W. (1960): The relation of adenosine 3′,5′-phosphate and phosphorylase to the actions of catecholamines and other hormones. *Pharmacological Reviews,* 12:265–299.
Sutherland, E. W., Rall, T. W., and Menon, T. (1962): Adenyl cyclase. I. Distribution, preparation and properties. *Journal of Biological Chemistry,* 237:1220–1227.
Sutherland, E. W., Robison, G. A., and Butcher, R. W. (1968): Some aspects of the biological role of adenosine 3′,5′-monophosphate (cyclic AMP). *Circulation,* 37:279–306.
Takai, K., Kurashina, Y., Suzuki, C., Okamoto, H., Ueki, A., and Hayaishi, O. (1971): The reversibility of the adenylate cyclase reaction. *Journal of Biological Chemistry,* 246:5843–5845.
Tao, M., and Lipmann, F. (1969): Isolation of adenyl cyclase from *Escherichia coli. Proceedings of the National Academy of Sciences,* 63:86–92.
Taunton, O. D., Roth, J., and Pastan, I. (1969): Studies on the adrenocorticotropin hormone-activated adenyl cyclase of a functional adrenal tumor. *Journal of Biological Chemistry,* 244:247–253.
Walter, R., Schwartz, I. L., Hechter, O., Doušа, T., and Hoffman, P. L. (1972): Bromoacetyl-oxytocin, an irreversible inhibitor of neurohypophyseal hormone-stimulated adenylate cyclase, and a possible affinity label for hormone receptors. *Endocrinology,* 91:39–48.
Warburg, O., and Christian, W. (1941): Isolierung und kristallisation des gärungsferments enolase. *Biochemische Zeitschrift,* 310:384–410.
Waud, D. R. (1968): Pharmacological receptors. *Pharmacological Reviews,* 20:49–88.
Weinryb, I., and Michel, M. (1971): Divergent effects of theophylline on adenylate cyclase preparations from guinea-pig heart and lung. *Experientia,* 27:1386.
Weiss, B. (1969): Similarities and differences in the norepinephrine- and sodium fluoride-sensitive adenyl cyclase system. *Journal of Pharmacology and Experimental Therapeutics,* 166:330–338.
Weiss, B. (1970): Factors affecting adenyl cyclase activity and its sensitivity to biogenic amines. In: *Biogenic Amines as Physiological Regulators,* edited by J. J. Blum. Prentice-Hall, New Jersey.
Weiss, B., and Costa, E. (1968): Selective stimulation of adenyl cyclase of rat pineal gland by pharmacologically active catecholamines. *Journal of Pharmacology and Experimental Therapeutics,* 161:310–319.
Wolff, J., and Jones, A. B. (1971): The purification of bovine thyroid plasma membranes and the properties of membrane-bound adenyl cyclase. *Journal of Biological Chemistry,* 246:3939–3947.
Wolff, J., and Jones, A. B. (1970): Inhibition of hormone-sensitive adenyl cyclase by phenothiazines. *Proceedings of the National Academy of Sciences,* 65:454–459.

Cyclic Nucleotide Phosphodiesterases

M. M. Appleman, W. J. Thompson, and T. R. Russell

OUTLINE

I. Introduction	66
II. Methods in General	66
A. Phosphodiesterase Assays	66
B. Substrate Specificity	68
III. Enzyme Characterization	68
A. Multiple Forms	68
B. Kinetic Regulation	72
C. Liver Phosphodiesterase — Kinetic Complexity	74
D. Brain Phosphodiesterase — Endogenous Activators	75
E. Subcellular Distribution	76
F. Lower Organisms and Plants	78
G. Tissue Culture	78
IV. Pharmacology	79
A. Methylxanthines	79
B. Puromycin	81
C. 4-Benzyl-2-Imidazolidinones	81
D. Papaverine	82
E. Other Inhibitors	82
F. Activators	84
V. Hormonal Effects	84
A. General	84
B. Insulin	85
C. Other Hormones	87
VI. Summary	89
VII. Acknowledgments	89
VIII. References	90

I. INTRODUCTION

Regulation of cyclic nucleotide levels in biological systems is a function chiefly of both the rate of synthesis and the rate of degradation. This was recognized by Sutherland and his co-workers from the first discovery of the involvement of cyclic AMP in physiological control, and the report of an enzymatic activity capable of the hydrolysis of the cyclic nucleotide to 5'-AMP (Sutherland and Rall, 1958). Since that finding there has been a considerable amount of work on both the characterization of the cyclic nucleotide phosphodiesterase enzymes and the significance of their activities in physiological regulation. The recent demonstration of a variety of distinct forms of phosphodiesterases in animal tissues has, however, introduced new complexity into this field of biomedical investigation. In this review an attempt will be made to present a current picture of cyclic nucleotide phosphodiesterases which will be of value to those concerned with biochemical enzymology and those whose interest is in pharmacology and endocrinology. A good historical review including phosphodiesterase is in the book by Robison et al. (1971).

There are two major problems which have been encountered in preparing this review. The first of these is that cyclic nucleotide phosphodiesterases have been implicated in a number of biological phenomena by *in vivo* experiments which show a response to caffeine or other inhibitors of the enzyme. In view of the potential for alternative or indirect effects for these drugs, further experimental evidence is required for these findings to be related to phosphodiesterases for purposes of this review. The second problem is that until very recently most investigators have assayed cyclic nucleotide phosphodiesterase under conditions which could not differentiate between the various activities. Where this appeared to be the case, the enzymatic activity is referred to in this review as "total phosphodiesterase." The activity will be described in terms of relative substrate specificity or affinity when possible.

II. METHODS IN GENERAL

A. Phosphodiesterase Assays

As is the case with most enzymes associated with cyclic nucleotides, a major problem in measuring the activity of phosphodiesterase has been the development of assays capable of determining very small amounts of product. Many of the current methods are derived from systems initially utilized in measuring the cyclic nucleotides themselves or in providing phosphodiesterase-based blank values for the nucleotide measurements.

Phosphodiesterase products are directly measured in the methods of Cheung (1967a, 1969b) and Mandel and Kuehl (1967) by utilizing radioactive substrate and chromatographic separation of the 5'-AMP. A method based on titrimetric determination of the proton released during hydrolysis has been reported (Cheung, 1969b). In all other methods the product is reacted upon by other enzymes leading to more easily separated biochemically stable compounds. The most common secondary reaction is the conversion of the 5'-nucleotide to the nucleoside and inorganic phosphate through reaction with an excess of nucleotidase. In the original assay of Butcher and Sutherland (1962) inorganic phosphate was measured colorimetrically. This method has been displaced in popularity by assays using tritiated or ^{14}C-labeled substrates and separation of the nucleoside products by a number of methods: anion exchange using columns (Hynie et al., 1966; DeLange et al., 1968; Beavo et al., 1970a) or direct resin addition (Brooker et al., 1968; Thompson and Appleman, 1971a), paper chromatography (Cheung, 1967a; Rosen, 1970; Gulyassay and Oken, 1971; Schroder and Plageman, 1972), and precipitation (Sobel et al., 1968; Pöch, 1971; Schonhofer et al., 1972). These have been modified by using ^{32}P-cyclic AMP and measuring either decrease in (Pöch, 1971; Schonhofer et al., 1972) or release of (Schonhofer et al., 1972) ^{32}P-labeled substrate. A microassay using enzyme cycling techniques has been reported (Weiss et al., 1972) as well as high-speed liquid chromatography (Pennington, 1971). Although other secondary systems have been proposed, e.g., myokinase (Cheung, 1966) and adenylic acid deaminase (Drummond and Perrot-Yee, 1961), the use of highly specific radioactive cyclic nucleotides and the snake venom nucleotidase is the most sensitive, specific, and convenient assay for the phosphodiesterase available.

The optimal conditions for the assay of phosphodiesterase vary with the source of the enzyme, but in general the pH optima are rather broad with maxima between 7.5 and 8.5. There is a divalent metal requirement, best satisfied by $^{2+}$Mg, although $^{2+}$Mn can substitute (Cheung, 1967b). The concentration of the metal can be critical; it must be greater than the concentration of potential chelators in the assay solution but can become inhibitory above 10 mM. Early assays routinely included imidazole (Nair, 1966), but the value of this addition is questionable. Caution must be observed in including secondary enzyme systems directly in the phosphodiesterase reaction mixture as they may contain factors which affect activity. Cheung (1967a) has shown activation by some snake venoms that were intended as a source of nucleotidase. Substrate concentrations for routine assays vary from millimolar (total phosphodiesterase) to less than 0.1 μM.

Histological assays for cyclic nucleotide phosphodiesterase have been described by Goren et al. (1971), using lead to detect inorganic phosphate released from 5'-AMP by alkaline phosphatase, and by Christiansen and Monn (1971), using the coupled enzyme system of myokinase, etc. Although they provide a good means of revealing a number of phosphodiesterase activities in polyacrylamide and starch gels, these systems have not detected the high-affinity enzyme activity shown to be present in most tissues.

B. Substrate Specificity

The specificity of total phosphodiesterase from a variety of mammalian tissues has been well tabulated (Cheung, 1970a). The activities of major interest are those which are specific for the naturally occurring purine base 3',5'-cyclic nucleotides cyclic AMP and cyclic GMP. These enzymes can also hydrolyze the 3',5'-cyclic phosphate bond of cyclic IMP but show little if any activity against 3',5'-cyclic pyrimidine nucleotides (Hardman and Sutherland, 1965), 2',3'-nucleotides, or nucleic acids. For a more detailed discussion of enzyme specificity, see below.

III. ENZYME CHARACTERIZATION

A. Multiple Forms

Research on cyclic AMP phosphodiesterase in the 10 years following its discovery depended upon assays requiring substrate concentrations to be in the vicinity of 1 mM. The work of Butcher and Sutherland (1962) on beef heart, Drummond and Perrott-Yee (1961) on rabbit tissue, Nair (1966) on dog heart, and Cheung (1967b) on rat brain indicated that most mammalian tissues contain a soluble phosphodiesterase with a K_m of approximately 10^{-4} M for cyclic AMP, which is inhibited by methylxanthines. There was some question as to the efficiency of an enzyme with such a low affinity for its substrate, the Michaelis constant being two orders of magnitude higher than reported tissue cyclic AMP levels. This activity was, however, generally accepted as the biological mechanism for the destruction of cyclic nucleotides.

Brooker et al. (1968) used a more sensitive isotopic phosphodiesterase assay during the development of the isotope dilution method for measuring cyclic nucleotides. It was found using very low substrate concentrations that one could detect kinetic behavior in rat brain preparations suggesting that more than a single type of cyclic AMP phosphodiesterase activity was present. Extrapolation of the data yielded two apparent K_m values of 100 μM and 1 μM for cyclic AMP. The rat brain preparation also hydrolyzed cyclic GMP, but this nucleotide did not interfere with cyclic AMP hydrolysis at low substrate concentrations. Rosen (1970) showed chromatographically that frog erythrocytes contain two distinct phosphodiesterase activities.

As will be seen below, kinetic evidence is only suggestive of the multiple forms of a particular enzyme activity. Confirmation requires physical separation followed by characterization of the isolated enzymes. This was achieved

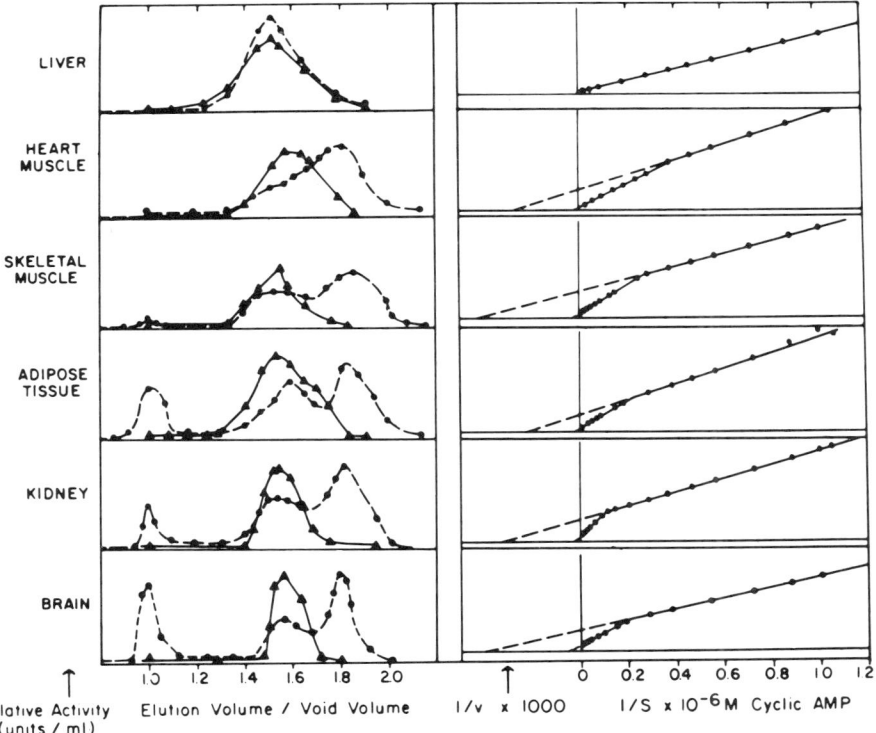

FIG. 1. Cyclic nucleotide phosphodiesterase activity profiles on agarose gel filtration columns (left) and Lineweaver-Burk kinetic plots (right) of preparations from various rat tissues. The enzymatic activity of each fraction is normalized relative to that for the fraction with the maximum activity in each profile. Gel filtration profiles show assays of phosphodiesterase activity at cyclic GMP concentrations of 10 μM (▲——▲) and cyclic AMP concentrations of 0.1 μM (●– – –●). Kinetic plots are determined by linear least square analysis to a correlation coefficient of 0.995 (Thompson and Appleman, 1971b).

TABLE 1. Michaelis-Menten constants of cyclic nucleotide phosphodiesterase

Source	Cyclic AMP High K_m – μM	Cyclic AMP Low K_m – μM	Cyclic GMP K_m – μM	References
Beef heart	100–500	—	100	Butcher and Sutherland, 1962; Goren and Rosen, 1971; Therriault and Winters, 1970; Goodsell et al., 1971.
Dog heart	490	—	—	Nair, 1966.
Rat heart	87[a]	3.9[a]	22[a]	Thompson and Appleman, 1971b.
Bovine heart	12–25[a]	0.8[a]	Low: 1.6–3[a] High: 20[a]	Beavo, et al., 1970a; O'Dea, et al., 1971.
Guinea pig heart	130–210	—	—	Pöch, 1971.
Rat liver	20–90[a]	0.7–5[a]	Low: 5[a] High: 15–20[a]	Menahan et al., 1969; Thompson and Appleman, 1971b; Russell et al., 1972a; Hemmington and Dunn, 1971; Beavo et al., 1971; Thompson et al., 1973.
Rat hepatoma	100–200	1–1.5	—	Schroder et al., 1972.
Rat brain (cortex)	100–200[a]	1–2[a]	12.9[a]	Cheung, 1967b; Brooker et al., 1968; Thompson and Appleman, 1971a; Kakiuchi et al., 1970, 1971; Weiss et al., 1972.
Mouse brain	260	—	—	Breckenridge and Johnson, 1969.
Trout brain	90	—	—	Yamamoto and Massey, 1969.
Bovine brain	140	—	—	Cheung, 1970b.
Rat skeletal muscle	57[a]	2.2[a]	27[a]	Thompson and Appleman, 1971b.
Rabbit skeletal muscle	20	4[a]	15	Huang and Kemp, 1971.
Rat fat cells	38–500[a]	0.9–3[a]	12–38[a]	Hepp et al., 1969; Thompson and Appleman, 1971b; Loten and Sneyd, 1970; Solomon, 1972; Klotz et al., 1972a; Schonhofer et al., 1972; Blecher et al., 1968.
Human adipose tissue	400	0.4	—	Solomon, 1972.
Rat kidney	30–300[a]	3–5[a]	9[a]	Dousa and Rychlik, 1970; Jard and Bernard, 1970; Thompson and Appleman, 1971b.
Rat adrenal	100	—	100	Klotz et al., 1972b.
Mouse skin	90	—	—	Mier and Urselman, 1972.
Frog gastric mucosa	250	1–2.1	—	Sung et al., 1972.
Human jejunum	460	—	—	Greene and Herman, 1972.
Rat thyroid	30–50	—	—	Bastomsky et al., 1971.
Bovine corpus luteum	250	—	—	Stansfield et al., 1971.
Mouse islets	500	3.5	—	Ashcroft et al., 1972.
Guinea pig islets	30	3.0	—	Sams and Montague, 1972.

Hamster tumor islets	—	2	—	Goldfine et al., 1971.
Rod outer segments	7,000	—	180	Pannbacker et al., 1972.
Rat thymic lymphocytes	8	0.9	—	Franks and MacManus, 1971.
Frog erythrocytes	300[a]	—	—	Rosen, 1970.
Rat erythrocyte	11	—	—	Sheppard and Wiggan, 1970.
Human malignant glial cells	15	not present	—	Perkins et al., 1972.
3T3 Cells	71	2.5	—	Armiento et al., 1972.
SV40–3T3 Cells	not present	2.5	—	Armiento et al., 1972.
L-929 Fibroblasts	300	0.5	—	Manganiello and Vaughan, 1972a.
Mouse L	80–170	1–1.5	—	Schroder et al., 1972.
HeLa	100–400	1–2.0	—	Schroder et al., 1972.
Slime mold	2,000	15	—	Chang, 1968; Chassey, 1972.

[a] K_m determined on separated fraction.

for rat brain cortex phosphodiesterase preparations by Thompson and Appleman (1971a). Using agarose gel filtration columns, it was found that a higher molecular weight enzyme with a K_m of 100 μM for cyclic AMP could be separated from a lower molecular weight enzyme with an apparent K_m of 5 μM. Only the former enzyme could carry out the hydrolysis of cyclic GMP and it actually had a slightly higher affinity for this nucleotide than for cyclic AMP. The low molecular weight enzyme on the other hand was quite specific for cyclic AMP and showed kinetic behavior suggesting negative cooperativity (Conway and Koshland, 1968). Subsequently, separation and confirmation of multiple enzymes was achieved for rat adipose tissue, heart, skeletal muscle, and kidney (Thompson and Appleman, 1971b), (Fig. 1).

The existence of multiple forms of cyclic nucleotide phosphodiesterase in brain was confirmed by Kakiuchi et al. (1971). Kinetic analysis of a variety of tissues and organisms suggests that the occurrence of multiple enzyme activities is a widespread phenomenon (Table 1). In a number of tissues the isolated low K_m cyclic AMP hydrolytic activity exhibits kinetic behavior suggesting negative cooperativity. It should be emphasized that confirmation of this type of allosteric behavior requires evidence from a number of separation procedures that only a single molecular species is present, as well as extensive kinetic analysis (Russell et al., 1973). Molecular weights have not been included in Table 1 because in most cases they are estimated. Thompson and Appleman (1971b) have reported values of approximately 400,000 for the high K_m phosphodiesterase and approximately 200,000 for the low K_m enzyme from the mammalian tissues tested. Slightly lower molecular weights have been reported for frog erythrocytes (Rosen, 1970), and much lower estimates for the rat kidney enzyme molecular weights have been reported by Jard and Bernard (1970). Using agarose gel filtration, the multiplicity of adipose tissue phosphodiesterases and approximate molecular weights reported by Thompson and Appleman (1971b) were confirmed by Klotz et al. (1972a).

B. Kinetic Regulation

The biphasic plots obtained upon graphical analysis of the kinetics of cyclic AMP hydrolysis by crude phosphodiesterase preparations from a number of tissues were readily explained by the identification and separation of the multiple forms of this enzyme activity (Thompson and Appleman, 1971b), but there remained a troublesome kinetic anomaly exhibited by the separated high affinity, cyclic AMP-specific enzyme. It was of some interest to determine if this behavior indicated the presence of yet more multiple activities or if it represented a single enzyme with negatively cooperative allosteric sites (Conway and Koshland, 1968). Further purification on gel filtration (Sephadex G-200) and DEAE-cellulose ion exchange columns gave no evidence for contamination by separable enzyme activities (Russell et al., 1972a). Computer modeling studies, undertaken to deal in a more quantitative manner with the observed kinetic anomalies, also tended to support the suggestion of allosteric regulation. The data fit a negative cooperative model with cooperativity constants of 0.2 to 0.5 for the enzyme from the various rat tissues examined (a constant greater than 1.0 indicates positive cooperativity, equal to 1.0 indicates Michaelis-Menten behavior, less than 1.0 indicates negative cooperativity). The data could also be fitted to a model based on contamination by a lower affinity enzyme, but only if it was assumed that 80 to 90% of the total activity was that of the contaminating enzyme—an assumption far out of line with the physical separation evidence. On the basis of this evidence Russell et al. (1972) advocated a negative-cooperative allosteric phosphodiesterase in which binding at a first high affinity cyclic AMP site induces a conformational change in the enzyme leading to a lower affinity at a second site. Proof of this model will require studies on a highly purified preparation of phosphodiesterase.

A mechanism coupling the negative cooperative phosphodiesterase with adenylate cyclase and protein kinase for optimal cell function can be derived. A hormone, acting as a positive effector, activates adenylate cyclase which raises the level of cyclic AMP in the cell to that necessary to stimulate the protein kinase. Phosphodiesterase at first acts in concert with the cyclase to establish the new steady-state level of cyclic AMP and then, following the removal of the hormonal stimulus, acts to return the cyclic nucleotide to the basal level. Since cyclic AMP must exist intact for a finite period of time to activate the protein kinase, the concentration should not be returned to the basal level too rapidly. A positive cooperative system cannot achieve this delicate control since its velocity increases too steeply over a small change in substrate concentration. Even an enzyme obeying classical Michaelis-Menten kinetics may have a change in velocity too great to allow cyclic AMP levels to remain sufficiently high. A negative cooperative phosphodiesterase may, however, accomplish this "kinetic buffering" of cyclic AMP levels, especially if its cooperativity constant, although less than 1.0, is not so low that the cell is "burned out" by prolonged exposure to elevated cyclic nucleotide concentrations.

The regulation of cellular physiology as a result of a negative-cooperative allosteric type of phosphodiesterase could be brought about by two enzymes of differing properties if these enzymes had the proper affinities for the substrate and if they were located together within the cell. This latter requirement does not seem to hold true in those cases where the subcellular localization has been examined (see below). The positively cooperative cyclic nucleotide phosphodiesterase activity reported in liver by Beavo et al. (1970a) and Russell et al. (1973) must, if the above mechanism holds true, serve some quite distinct physiological function other than the negatively cooperative high affinity enzyme.

C. Liver Phosphodiesterase—Kinetic Complexity

Phosphodiesterase activity from liver indicates the presence of only a single enzyme species when fresh homogenates are analyzed kinetically or on gel filtration columns (Thompson and Appleman, 1971b). If, however, this preparation is exposed to high salt conditions (Beavo et al., 1970a), DEAE-cellulose chromatography or trypsin treatment (Russel et al., 1973), or even simple storage at 4°C (Hemington and Dunn, 1971), changes appear in both the kinetic and physical properties of the phosphodiesterase activity. DEAE-cellulose chromatography of rat liver extracts reveals at least three cyclic nucleotide phosphodiesterase fractions (Fig. 2).

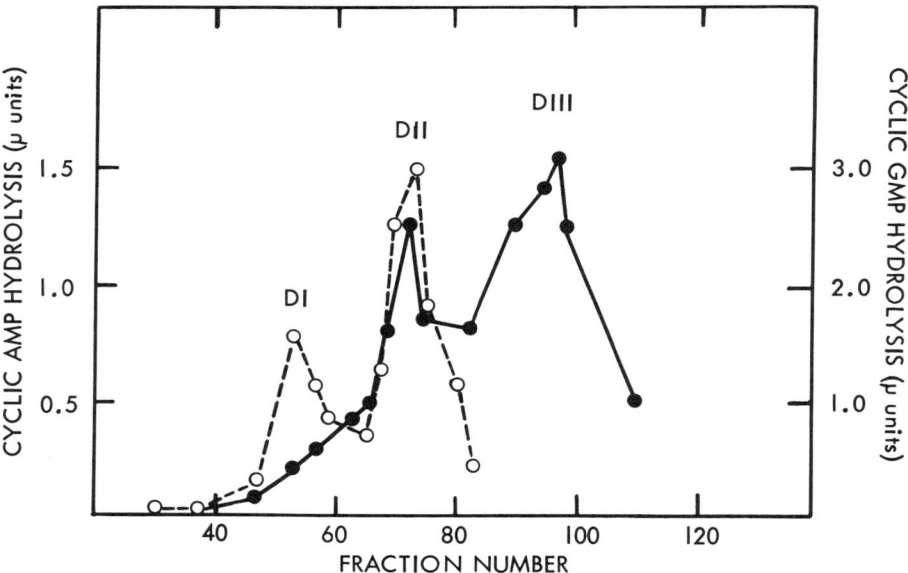

FIG. 2. DEAE-cellulose profile of liver phosphodiesterase from a sonicated extract. Buffer 0.05 M Tris-acetate pH 6.0, gradient from 0 to 1 M sodium acetate from fraction 30 to fraction 130. O----O = Cyclic GMP (2.5×10^{-7} M). ●——● = Cyclic AMP (1.25×10^{-7} M) (Russell et al., 1973).

Fraction I hydrolyzes cyclic GMP, is unable to hydrolyze cyclic AMP, and is not inhibited by the latter nucleotide. Thus far, this tissue appears to be unique in containing an enzyme capable of hydrolyzing cyclic GMP independently of the more prevalent nucleotide cyclic AMP. Liver extracts contain a heat-stable, trypsin-sensitive factor which stimulates the phosphodiesterase activity of fraction I only (Terasaki, *unpublished*). Fraction II hydrolyzes both cyclic AMP and cyclic GMP with relatively low substrate affinities. This activity, which is the kinetically dominant activity in the fresh liver homogenate is similar to the "high K_m" enzyme of other tissues. It differs from most tissue preparations, however, in that the hydrolysis of cyclic AMP is activated by low concentrations of cyclic GMP at pH 7.4. This behavior has been reported previously for other preparations from liver (Beavo et al., 1971), thymic lymphocytes (Franks and McManus, 1971), and adipose tissue (Klotz and Stock, 1972). If this evidence is considered in conjunction with the facts that at high concentrations each nucleotide is a noncompetitive inhibitor of the hydrolysis of the other and that at pH 7.4 the hydrolysis of each nucleotide shows kinetic behavior indicative of positive cooperativity (Beavo et al., 1970a; Russell et al., 1972a), a model can be proposed of a polymeric enzyme composed of cyclic AMP phosphodiesterase subunits and cyclic GMP phosphodiesterase subunits. Substrate bound to one type of subunit can influence the binding at the other type and all subunits would bear a similar allosteric activation site with a greater affinity for cyclic GMP than for cyclic AMP. The third fraction of the liver phosphodiesterase is the high affinity enzyme, specific for cyclic AMP hydrolysis, and displaying negatively cooperative kinetics. It is inhibited by cyclic GMP in a hyperbolic fashion. A rational physiological basis for the kinetic complexity and latent activities of the liver phosphodiesterases must await further investigation. It would appear that cyclic GMP must have a significant role in biological regulation: there are at least two high affinity sites in liver phosphodiesterases for it, one catalytic and one regulatory.

D. Brain Phosphodiesterase—Endogenous Activators

The cyclic nucleotide phosphodiesterase activity from a variety of sources appears to be subject to endogenous factors capable of activation or inhibition of the enzyme. The greatest progress in this area has been on activation by brain preparations. Cheung (1970b, 1971) measured total cyclic nucleotide phosphodiesterase and found that the considerable activity lost during purification of the bovine brain enzyme could be restored by addition of a factor present in crude fractions. This factor was characterized as a protein by trypsin sensitivity, had a molecular weight of approximately 40,000, was quite stable to heat, and was resistant to nuclease attack. Although stimulation of the activity could also be achieved by trypsin, pronase, or a component of snake venom, the endogenous factor itself did not appear to act catalytically. The millimolar substrate concentrations used in these activator studies suggest that the observed effects were primarily on the high K_m enzyme (Thompson and Appleman, 1971a).

Kakiuchi has carried out an extensive investigation of brain phosphodiesterase (Kakiuchi et al., 1970, 1971) and has confirmed the presence of two distinct activities using gel filtration. The cyclic nucleotide specificities reported are broader than those of Thompson and Appleman (1971a), who found little cyclic GMP hydrolysis by the low K_m enzyme. Kakiuchi discovered that the lower K_m enzyme from brain, unlike the high K_m activity, was greatly stimulated by ^{2+}Ca with a maximum effect at about 10 μM; both enzymes require ^{2+}Mg. Of potential physiological importance was the discovery that the sensitivity to ^{2+}Ca of the ^{2+}Ca plus ^{2+}Mg dependent phosphodiesterase was under the control of a heat-stable, nondialyzable factor (PAF) present in brain extracts. This factor, with a molecular weight of approximately 60,000, reduced the required ^{2+}Ca concentration for activation; it did not act catalytically and did not serve as a protective protein. The physical similarity to the activating factor described by Cheung (1970b) indicates that the two may be identical, although some differences in their metal requirements and in the enzyme preparations upon which they act must be resolved before this is confirmed. Rat brain phosphodiesterase has also been suggested as possessing an inhibitory protein unit which may be regulated by ^{2+}Ca (Miki and Yoshida, 1972).

Activating factors possibly related to that from brain have been separated from phosphodiesterase preparations from beef heart (Goren and Rosen, 1971; Wang et al., 1972), bullfrog gastric mucosa (Ray and Forte, 1972), and kidney and adipose tissue (Thompson, 1971). A specific physiological function has not been presented for any of these endogenous activation systems.

E. Subcellular Distribution

If one can generalize on the basis of a large amount of data derived from random comments about phosphodiesterase preparative methods and specific examination of the subcellular distribution of this enzymatic activity, it can be stated that in most tissues the low K_m cyclic AMP phosphodiesterase is localized in a particulate or membranous fraction whereas the less specific enzyme with a lower affinity appears to be soluble. This could explain many of the early findings, especially in brain preparations, that in assays measuring the total enzyme activity there was a distribution between soluble and particulate phases (e.g., Cheung and Salganicoff, 1967). The particulate activity was shown to be associated with nerve endings (De Robertis et al., 1967; Gaballah and Popoff, 1971) and particles of synaptosomal dimensions (Thompson and Appleman, 1971a), and it could be released partially by treatment with Triton X-100 or osmotic shock, and totally by sonication. Recent studies on frog gastric mucosa phosphodiesterases indicate an interesting enzyme distribution. Cyclic AMP phosphodiesterase and cyclic GMP phosphodiesterase are soluble enzymes that are localized in various cell types (Sung et al., 1972). Rat cerebellum phosphodiesterase was found to be most active in gray matter (Weiss and Costa, 1968). Studies on the distribution of high K_m cyclic AMP phosphodiesterase activity of rabbit brain showed variability among brain regions but no correlation with monoamine content or cellular density (Breckenridge and Johnson, 1969; Shanta et al., 1966). Phosphodiesterase activity of rat cortex that is detectable by cytochemical localization is thought to be located postsynaptically (Florendo et al., 1971).

Other tissues, although having lower total phosphodiesterase activity than brain, have a similar distribution (Dousa and Rychlik, 1968). Hepp et al. (1969) solubilized a particulate, rat fat cell activity by sonic disruption. Thompson and Appleman (1971b) found that sonication can release into a 30,000 × g supernatant greater than 90% of the activity from rat brain, kidney, adipose tissue, liver, heart, and skeletal muscle. Combined with kinetic analysis of the phosphodiesterase activity of these tissues, the hypothesis was presented that low K_m cyclic AMP specific phosphodiesterase was membrane bound and that the high K_m cyclic AMP phosphodiesterase and cyclic GMP phosphodiesterases are soluble enzymes. Sonication can also partially solubilize a low K_m cyclic AMP phosphodiesterase from a relatively pure preparation of plasma membrane of chicken embryo fibroblasts grown in tissue culture (Russell and Pastan, *in preparation*). The work of Beavo et al. (1971), Russell et al. (1973), and Thompson et al. (1973) has shown that in rat liver the low K_m enzyme is particulate, appearing to be distributed among a number of membranous fractions, not including mitochondria. House et al. (1972) have isolated relatively pure liver plasma membranes containing an insulin-sensitive phosphodiesterase which is probably the low K_m enzyme. The low affinity cyclic AMP-cyclic GMP phosphodiesterase and a cyclic GMP-specific enzyme from rat liver are both soluble (Russell et al., 1973; Thompson et al., 1973).

F. Lower Organisms and Plants

A cyclic AMP phosphodiesterase has been found in *E. coli* which appears to require three components for complete activity. Component I is a high molecular weight catalytic unit, component II is a low molecular weight protein, and component III is a dialyzable activator (Monard et al., 1969). Cyclic AMP is required for the synthesis of inducible enzymes in *E. coli*, and, under severe catabolite repression (high glucose), cyclic AMP levels are lowered (Perlman and Pastan, 1968). It appears that component II is present in higher concentrations under severe catabolite repression. The dialyzable activator, component III, could be a metabolic product of glucose since glucose-6-phosphate, fructose-6-phosphate, and other sugar phosphates can substitute for component III in stimulating phosphodiesterase activity.

A cyclic AMP phosphodiesterase has been found in the extracellular medium of the cellular slime mold *Dictyostelium discoideum* (Chang, 1968; Pannbacker and Bravard, 1972). The enzyme has a molecular weight of 300,000 and a K_m of 2 mM. The enzyme, which is not inhibited by caffeine, appears to be involved in creating extracellular cyclic AMP gradients, which in turn cause the individual amebae cells to aggregate (Bonner et al., 1969). An inhibitor of this enzyme has been found in the medium of aggregating cells but not in the medium of cells which fail to aggregate (Riedel and Gerisch, 1971).

Potato contains phosphodiesterase activity in the 105,000 g supernatant. Two active peaks elute from agarose columns but these fractions have not been characterized (Shimoyama et al., 1972). Wood et al. (1972) found two cyclic AMP phosphodiesterases in crown-gall tumor tissue of *Vinca rosea* that showed inhibition by the cell division-promoting substance cytokinesin I.

G. Tissue Culture

The relationship of cyclic AMP to cell growth in normal and transformed fibroblasts is currently under intensive study (Ryan and Heidrick, 1968; Johnson et al., 1971; Hsie and Puck, 1971). Cyclic AMP phosphodiesterase has been shown to be present in a number of normal and transformed fibroblasts (Manganiello and Vaughan, 1972a; Armiento et al., 1972; Heidrick and Ryan, 1971). Evidence has shown biphasic kinetic plots in normal fibroblast lines similar to those found for other tissues. A human glial cell tumor (Perkins et al., 1971) and SV40 transformed 3T3 cells have been shown to contain only the low K_m enzyme (Armiento et al., 1972). The untransformed

3T3 cells contain a high and low K_m activity. L cells treated with prostaglandin E_2 (Manganiello and Vaughan, 1972a) and 3T3 and SV40-3T3 cells (Armiento et al., 1972) treated with dibutyryl cyclic AMP and theophylline show an increase in phosphodiesterase activity. This increase in activity can be blocked by treatment with cycloheximide and actinomycin D, suggesting that cyclic AMP can regulate the synthesis of phosphodiesterase. Recent studies of Grimm and Frank (1972) indicate that fetal calf serum can cause activation of cyclic AMP phosphodiesterase of cultured embryonic rat cells and may be related to cyclic AMP and cell growth.

IV. PHARMACOLOGY

A considerable variety of chemicals has been examined as possible effectors of cyclic nucleotide phosphodiesterase. Although these studies have not yet made great progress in the development of new drugs, they have made valuable contributions toward the characterization of the active site of the enzyme as well as providing some knowledge of the role of the various forms of phosphodiesterase in the complex systems involved in the metabolism and utilization of cyclic nucleotides. It is to be hoped that this basic information will eventually lead to a rational approach to the treatment of metabolic and endocrine disorders.

In the discussion of a number of drugs which follows, it should be recalled that, with the possible exception of the methylxanthines, no class of compounds has been proven to act physiologically through effects on cyclic nucleotide phosphodiesterase. The reader is also warned that whereas an attempt is made to define effects in terms of one or the other of the primary forms of phosphodiesterase, this is not always possible with the data available.

A. Methylxanthines

The pharmacological agents most often employed for inhibition of phosphodiesterase are the methylxanthines (Sutherland and Rall, 1958; Butcher and Sutherland, 1962). Caffeine, theophylline, theobromine, and aminophylline have been used frequently to inhibit undesirable cyclic nucleotide degradation in assays of adenylate or guanylate cyclase, or to serve as synergistic agents for *in vivo* and *in vitro* studies of the actions of exogenous cyclic AMP or cyclic GMP. Although the effects of theophylline, the most widely used of the methylxanthines, are frequently attributed to inhibition of phosphodiesterase, it should be remembered that these compounds have been shown to affect a number of potentially unrelated phenomena, e.g., calcium efflux, ATP depletion, protein synthetic effects, oxygen consumption and heart rate, and endocrine secretions (Brisson et al., 1969; Fleischer et al., 1969; McNeill et al., 1969; Rall and West, 1963; Halkerston et al., 1966; Ensinck et al., 1970; Cerasi and Luft, 1969; Strubelt et al., 1979). Conversely, it is not uncommon for investigation to reveal no increase in the level of cyclic AMP in a biological system under theophylline treatment at concentrations adequate to inhibit a major portion of the phosphodiesterase activity. Schwabe and Ebert (1972) suggest that effects of theophylline in fat cells are not due to phosphodiesterase inhibition. Ideally, criteria for the implication of cyclic AMP and its phosphodiesterase in a biological mechanism should include the use of at least one other inhibitor of the enzyme now that more active and specific compounds have been developed (Simon et al., *this volume*).

In vitro studies of the theophylline inhibition of phosphodiesterase activity, measured at cyclic AMP concentrations in the millimolar range, indicate that this compound is a competitive inhibitor of the enzymes from beef heart (Butcher and Sutherland, 1962), rat brain (Cheung, 1967b; Nakano and Ishii, 1970b), kidney (Senft et al., 1968; Dousa and Rychlik, 1970), skeletal muscle (Huang and Kemp, 1971), adipose tissue (Schwabe et al., 1972; Pöch, 1971; Schonhofer et al., 1972), adrenal (Klotz et al., 1972b), and pancreatic islets (Sams and Montague, 1972). K_i values are close to 1 mM. Reports have also been published describing the methylxanthine as a noncompetitive inhibitor of the enzyme from dog heart (Nair, 1966; Therreault and Winters, 1970), a rather surprising finding from the structural viewpoint. Where measurements have been made at low enough substrate concentrations to measure the low K_m enzyme, theophylline has also been shown to be a competitive inhibitor: such studies were carried out with brain (Goldberg et al., 1970), skeletal muscle (Huang and Kemp, 1971), and adipose tissue (Schwabe et al., 1972; Schonhofer et al., 1972). Phosphodiesterase of rat hepatoma, mouse L, and HeLa cells show competitive or noncompetitive inhibition by theophylline depending on the substrate concentration used (Schroder and Plageman, 1972). Structural requirements for the xanthines to inhibit cyclic nucleotide phosphodiesterase have been extensively investigated on the activity from adipose tissue (Beavo et al., 1970b) and beef heart (Goodsell et al., 1971). Inhibition was correlated with small nonpolar substitutions at positions 1 and 3 of the xanthine rings, longer side chains at the 8 position, but no modification of the 7-N position; the structurally related imidazopyrazines and triazolopyrimidines also inhibited enzyme activity (Mandel, 1971; Davies et al., 1971).

In vivo confirmation of the ability of methylxanthines to inhibit phosphodiesterase has been attempted. Although *in vitro* experiments had indicated inhibition of the enzyme in most tissues, only the anterior pituitary phosphodiesterase was inhibited in rats 30 min after injection of 2 to 5 mg of theophylline or caffeine per 100 g body weight (Vernikos-Harris, 1968). A greater degree of correlation was achieved by Beavo et al. (1970*b*) between inhibition of phosphodiesterase by various methylxanthine derivatives and their lipolytic potency in rat adipocytes. Relatively slight inhibition seemed to be adequate to increase lipolytic rates. Good correlation was also demonstrated by Mandel (1971) between potentiation of norepinephrine-stimulated lipolysis and phosphodiesterase inhibition by imidazopyrazines.

B. Puromycin

The antibiotic puromycin, which contains a purine base, has been shown to be an effective competitive inhibitor of cyclic AMP phosphodiesterase from rat diaphragm muscle (Appleman and Kemp, 1966). This finding provides a possible explanation for the glycogenolytic or insulin antagonistic activities of this compound, which are independent of its accepted function of inhibiting protein biosynthesis (Hofert and Boutwell, 1963). Puromycin provided some of the early evidence that the separated low K_m rat brain phosphodiesterase might possess a regulatory site. This antibiotic was a competitive inhibitor at high cyclic AMP concentrations but appeared as noncompetitive at substrate concentrations below the K_m for the high affinity site (Goldberg et al., 1970; Thompson and Appleman, 1971*c*).

C. 4-Benzyl-2-imidazolidinones

An interesting group of phosphodiesterase inhibitors has been described by Sheppard and his colleagues (Dalton et al., 1970; Sheppard and Wiggan, 1970, 1971). Although the assay conditions do not permit a choice to be made of which type of enzyme is the target or of what type is the inhibition, it is clear that the activity from rat erythrocytes is severely inhibited by Ro 7-2956 [4-(3,4-dimethoxybenzyl)-2-imidazolidinone]. The 3 butoxy-4-methoxy derivative (Ro 20-1724) was a 1,000-fold better inhibitor than theophylline. Although these compounds had lipolytic and hypotensive activity, there was not as direct a correlation between pharmacological potency and phosphodiesterase inhibition as was observed by Beavo et al. (1970*b*) for the xanthine derivatives. The nature of some of these inhibitors provides a basis for discussion of a possible hydrophobic region in the active site of the phosphodiesterase. Further information must be obtained, however, before the relative importance of hydrophobic sites and the problems of permeability in relation to these drugs can be determined.

D. Papaverine

A proposal that cyclic AMP is involved in the regulation of smooth muscle tone and contractility (Moore et al., 1968) is supported by the finding that the isoquinoline opium alkaloid papaverine, a smooth muscle relaxant, is a potent inhibitor of phosphodiesterase preparations from uterine, vascular, and striated muscle (Kukovetz and Pöch, 1970; Triner et al., 1970; Rufeger et al., 1971; Toson and Carpenedo, 1972), and adrenals (Klotz et al., 1972b). Although papaverine with a K_i of approximately 10^{-5} M is from 10 to 100 times as potent an inhibitor as theophylline, it still requires high concentrations for complete suppression of the enzyme, even *in vitro*. Both the activity measured at low substrate concentrations, in preparations from brain (Goldberg et al., 1970), heart and platelets (Smith and Mills, 1970), and the high K_m enzyme, from bovine rod outer segments (Pannbacher et al., 1972), are inhibited competitively by papaverine. The type of inhibition by papaverine seen in heart muscle and coronary artery preparations is competitive or noncompetitive, seemingly dependent on the degree of enzyme purity (Pöch and Kukovetz, 1971). These findings should cause concern and problematic awareness to the investigators of the kinetics of phosphodiesterase activity. Caution must always be observed in applying isolated enzyme data to physiological situations. The work of Schwabe et al. (1972) has shown that, despite their potent ability to inhibit adipose tissue phosphodiesterase *in vitro*, papaverine and some other inhibitors not only do not stimulate lipolysis by isolated fat cells, but also can actually block the stimulation by norepinephrine. Similarly, Klotz et al. (1972b) were unable to correlate adrenal phosphodiesterase inhibition with effects on steroidogenesis.

E. Other Inhibitors

Several other types of compound have potential pharmacological importance as inhibitors of phosphodiesterase. Although many reports are of such a nature as to make analysis difficult, the high K_m cyclic AMP phosphodiesterase appears to be the common target. Penothiazine-type tranquilizers and reserpine derivatives, at concentrations of 10^{-5} M, inhibit enzyme preparations from beef heart and brain (Honda and Imamura, 1968). The inhibition was potentiated by halogenation at ring 2 of the phenothiazine. Although these workers reported imipramine, desmethylimipramine, and amitriptyline as ineffective in their system (see also Nakano and Ishii, 1970a), Schwabe et al. (1972) found imipramine to be a competitive inhibitor of the low K_m enzyme from rat adipocytes. McNeill and Muschek (1970) reported that the tricyclic antidepressants do inhibit phosphodiesterase activity of rat brain and suggested a role for cyclic AMP in the reuptake of norepinephrine into adrenergic nerve terminals. Beer et al. (1972) have shown a good correlation between anxiety reduction and inhibition of rat brain cyclic AMP phosphodiesterase for a variety of drugs measured at low substrate concentrations.

An extensive study has been made of a series of inhibitors of phosphodiesterase containing pyrimidine bases (Simon et al., *this volume*). In general, these compounds, unlike the inhibitors containing purine bases, were not competitive with cyclic AMP. By manipulation of the functional groups on the ring, potent inhibitors with some tissue or enzyme selectivity were obtained.

Inhibition of cyclic nucleotide phosphodiesterase in the kidney has been suggested as a mode of action for a number of drugs. Senft et al. (1968a) reported that furosemide and hydrochlorothiazide inhibited the renal enzyme activity not only *in vitro* but also *in vivo*. When administered i.v., these compounds were effective phosphodiesterase inhibitors in the regions of the kidney known to respond physiologically to their actions. More recently the antidiuretic effectiveness of chloropropamide in the treatment of diabetes insipidus was suggested to be due to inhibition by this drug of renal phosphodiesterase (Fichman and Brooker, 1970). Further investigation revealed that whereas both chloropropamide and tolbutamide were inhibitors of the kidney phosphodiesterase, the concentrations required were so high (millimolar) as to put in question the notion that this represented the major mechanism of these drugs (Brooker and Fichman, 1971; Chaudhuri and Wines, 1971). Tolbutamide has been shown to be a competitive inhibitor of phosphodiesterase activity in isolated rat islets and a number of other tissues from a variety of animal sources (Goldfine et al., 1971; Roth, 1971; Brown et al., 1972), but the significance of these findings is as yet unclear. Studies of pancreatic islet tissue have indicated that alpha adrenergic stimulation can decrease cyclic AMP levels (Turtle and Kipnis, 1967), and that the alpha adrenergic antagonist dihydroergotamine inhibits the low K_m phosphodiesterase of rat adipocytes (Ward and Fain, 1971) and cat brain (Iwangoff and Enz, 1971), although it also inhibits adenylate cyclase. McNeill et al. (1972) have shown brain cyclic AMP phosphodiesterase to be inhibited by phentolamine, another α-adrenergic antagonist. A relationship between inhibition of phosphodiesterase and potency in stimulating catecholamine action on muscle contraction has been reported for a number of synthetic beta adrenergic agonists (Nott, 1971).

F. Activators

Imidazole can stimulate cyclic nucleotide phosphodiesterase activity *in vitro*. This has been reported for high K_m enzyme preparations from heart (Butcher and Sutherland, 1962), brain (Cheung, 1967; Nakano and Ishii, 1970b; McNeill et al., 1972; Miki and Yoshida, 1972), and corpora lutea (Stansfield et al., 1971), and a low K_m phosphodiesterase from skeletal muscle (Huang and Kemp, 1971). On the other hand, little or no effect of imidazole was reported on enzyme preparations from adipose tissue (Allen and Clark, 1971) and on the low K_m activity from brain (Goldberg et al., 1970). The variability of imidazole stimulation and the pH optimum shift frequently associated with it may indicate that its effect is nonspecific, perhaps involving chelation of metals or displacement of inhibitory buffering ions. Imidazole can, however, counteract the cyclic AMP-mediated stimulation of lipolysis *in vivo,* and this has been suggested to be due to activation of phosphodiesterase (Goodman, 1969a). A similar mechanism has been proposed for histamine action (Goodman, 1968a). The antilipolytic activity of nicotinic acid has led to the suggestion that this compound also might be a phosphodiesterase activator, but the concentrations required for *in vitro* stimulation appear to be excessive (Kupiecke and Marshall, 1968; Blecher et al., 1968; Nakano, 1970; Skidmore et al., 1971; Peterson et al., 1968; Sams and Montague, 1972; Abdulla, 1969).

V. HORMONAL EFFECTS

A. General

Although there has been considerable effort in recent years to find direct effects of hormones on cyclic nucleotide phosphodiesterase, the results do not compare with those in the area of adenylate cyclase. The hypothesis has been proposed that hormones, for which the physiological response is paralleled by depressed cyclic AMP levels, might act through activation or even induction of the synthesis of phosphodiesterase to achieve this effect. Conversely, a hormone might inhibit the phosphodiesterase in a complementary action to activation of adenylate cyclase.

There are other aspects of phosphodiesterase enzymology which must be taken into account before one can evaluate this hypothesis. It is apparent now that mammalian tissues contain two, three, or more distinct forms of cyclic nucleotide phosphodiesterase. Until the relationship of each of these enzymes to the possible cyclic nucleotide pools, and, thereby, to various cell functions is defined, it is difficult to predict hormonal interrelationships.

Most investigations have been carried out on homogenates of target tissues that were assayed under conditions which could not even kinetically resolve the various forms. Recent work utilizing better kinetic analysis, fractionation of cellular components, or even enzyme purification shows more promise of interpreting the hormone-phosphodiesterase relationships. Some researchers are concerned that hormonal activation of phosphodiesterase activity of a cell, without simultaneous inhibition of cyclase activity, would, in effect, be an ATPase. Although this might be true, rough calculations of the activity of these enzymes indicate that the magnitude of this energy loss would be insufficient to affect the economics of cell metabolism. The thermodynamics of cyclic AMP synthesis and degradation (Hayaishi et al., 1971) are also opposed to an ATPase-like function.

B. Insulin

The primary candidate for a hormone acting through modification of phosphodiesterase activity is insulin. This hormone has been known for decades to have many effects in adipose and hepatic tissue which were antagonistic to those of epinephrine and glucagon. When these latter effects were related to elevated cyclic AMP levels in response to the stimulation of adenylate cyclase, an intensive effort was made to connect the mechanism of insulin action with cyclic nucleotide metabolism. This effort was stimulated by the finding of Butcher et al. (1966, 1968) that under certain conditions insulin could be shown to reduce adipose tissue cyclic AMP levels and by the correlation of diabetic states to elevated cyclic AMP levels in liver (Jefferson et al., 1968). The antagonistic effects of insulin with glucagon and catecholamines on cyclic AMP levels have been well supported by the work of Exton and co-workers (Exton et al., 1971) in liver. Unfortunately, it has proven to be a difficult step from demonstrating an action of insulin on cyclic AMP levels to positively identifying an enzymatic mechanism for this action. There is some evidence that insulin can depress the activity of hormonally stimulated adenylate cyclase in liver and adipose tissue (Ray et al., 1970; Hepp, 1971; Illiano and Cuatrecasas, 1972).

Despite the fact that reports correlating insulin action with elevated phosphodiesterase activity appeared as long ago as 1966, there is as yet no absolute confirmation or even general acceptance of the mechanism. Schultz et al. (1966) reported that total cyclic AMP phosphodiesterase activity was depressed in liver preparations from acutely alloxan diabetic rats and that *in vivo* insulin treatment could restore the activity. More detailed studies followed confirming this phenomenon in other tissues (Senft et al., 1968a,b,c). The phosphodiesterase assays used in those investigations were carried out at cyclic AMP concentrations such that the total activity of all forms of the enzyme rather than that of any single form would be measured. The time element for the insulin action was 15 to 45 min and the hormone levels used were very high (0.5 U/kg body weight).

Subsequent detailed studies of insulin-phosphodiesterase interactions have produced both positive and negative results with no readily apparent correlation to the choice of tissue or any other recognizable variable.

Adipose tissue phosphodiesterase activity showed no dramatic response to the diabetic state or insulin treatment in a number of studies (Blecher et al., 1968; Müller-Oerlinghausen et al., 1968; Hepp et al., 1969; Vaughan and Murad, 1969). Most of these investigations were carried out using phosphodiesterase assays at high cyclic AMP concentrations, capable of masking subtle enzyme changes. This point was emphasized by Loten and Sneyd (1970) and Loten (1970) who found significant effects of *in vitro* insulin treatment on adipose tissue enzymes. This important work utilized kinetic analysis to show that the action of the hormone could be interpreted as an increase in the maximal velocity of a low K_m phosphodiesterase and/or an increase in the affinity of a high K_m enzyme. This *in vitro* response has been confirmed by Clark et al. (1972) but has not been demonstrable in cell free systems (Allen and Clark, 1971). Another negative study of insulin and adipose tissue phosphodiesterase used substrate concentrations in the low K_m range, but used enzyme preparations insufficient to maintain the low K_m phosphodiesterase (Fain and Rosenberg, 1972). Experiments testing cyclic AMP phosphodiesterase activity and response to insulin under conditions where insulin is maximally antilipolytic, however, are yet to be completed. Indeed, the basic premise that activation of phosphodiesterase will decrease the level of cellular cyclic AMP has not been established.

The story of insulin effects on hepatic cyclic AMP phosphodiesterase is similar. Menahan et al. (1969), Müller-Oerlinghausen et al. (1968), and others could find little if any correlation of enzyme activity with insulin levels but Kupiecki (1969) reported depressed phosphodiesterase in livers from alloxan diabetic rats and restoration with insulin treatment. Working with a subfraction of rat liver plasma membranes, House et al. (1972) found stimulation of phosphodiesterase activity after a very brief exposure of this cell free system to insulin at low concentration. The target enzyme for this hormonal effect is probably the cyclic AMP specific, high affinity phosphodiesterase, which has been shown to be associated with a particulate fraction in liver homogenates (Russell et al., 1973). Using sucrose gradient fractionation of rat liver homogenates, Thompson et al. (1973) have shown that membrane-bound high affinity cyclic AMP phosphodiesterase of diabetic rats is stimulated by insulin administration. The cyclic GMP phosphodiesterase and low affinity cyclic AMP phosphodiesterases, which are separate and apparently soluble enzymes, were unaffected. As with adipose tissue phosphodiesterase, the activated enzyme had an increased V_{max} and no change in the apparent Michaelis-Menten constant. Although these experiments cannot be interpreted as direct insulin action, they do give support to the above hypothesis. These investigators were unable to confirm any *in vitro* action of insulin on isolated low K_m rat liver cyclic AMP phosphodiesterase.

It should be recalled here that there is considerable literature demonstrating that there are insulin receptors on the external side of target cell membranes (e.g., Kono, 1972, and Cuatrecasas, 1972). The sum total of these different approaches would tend to suggest, but not confirm, that at least some of the metabolic actions of insulin are due to the activation of a cyclic AMP phosphodiesterase activity contained in plasma membranes, that is perhaps associated with an adenylate cyclase and able to reduce the nucleotide concentration in at least some intracellular pool if not in the entire cell. As emphasized by Robison et al. (1971), the possibility that even this enzymatic response to insulin may be secondary to some other biological effect has not yet been eliminated.

C. Other Hormones

It is quite difficult to support direct effects of hormones other than insulin on cyclic nucleotide phosphodiesterase because the time element required for the demonstration of physiological effects of hormone depletion or hormone excess is sufficient for many secondary actions to occur. For example, treatment of fibroblasts with prostaglandin E_1 leads to the induction of synthesis of low K_m phosphodiesterase over a period of many hours (Manganiello and Vaughan, 1972a). But this stimulation of specific enzyme synthesis, which can be blocked by cycloheximide or actinomycin D, is believed not to be the direct result of the hormone but rather of the elevated cyclic AMP levels which it produces; other agents which can elevate intracellular cyclic AMP levels also lead to higher phosphodiesterase activity (Armiento et al., 1972). This may be the case in fat cells as preliminary data show increased low K_m cyclic AMP phosphodiesterase activity after 10 min of exposure to epinephrine (Lavis, Thompson, and Williams, *unpublished observations*). Although the actions of other hormones have not been as clearly defined, it is obvious that evidence from long-term *in vivo* experiments must be handled with caution.

The physiological effects of growth hormone are many and not well-defined, but one of these effects, the potentiation of theophylline-stimulated lipolysis in adipose tissue without strong effects on basal lipolytic activity (Goodman, 1968b, 1969b; Fain, 1968), has been suggested as occurring through the suppression of the synthesis of cyclic nucleotide phosphodiesterase. Indeed, hypophysectomized rats do show elevated levels of the high affinity phosphodiesterase in adipose tissue (Clark et al., 1972) and heart (Thompson and Williams, *unpublished observations*) but these effects are not dramatic and are not reversed by growth hormone administration. Thompson et al. (1973) have shown that administration of high doses of growth hormone into hypophysectomized rats causes stimulation of discontinuous, sucrose-gradient-fractionated, low K_m, membrane-bound cyclic AMP phosphodiesterase after 30 min. It would appear that the elucidation of any relationship between phosphodiesterase activity and growth hormone must await the development of improved experimental systems. This conclusion would also apply to glucocorticoid action where the demonstration of attenuated responsiveness of liver to cyclic AMP in adrenalectomized animals (e.g., Schaeffer et al., 1969) has not been defined in terms of a specific enzyme of cyclic nucleotide metabolism. However, on the basis of decreased cyclic AMP phosphodiesterase activity upon incubation of cultured hepatoma cells with dexamethasone, it has been proposed that some of the "permissive" effects of steroids may involve phosphodiesterase (Manganiello and Vaughan, 1972b).

A more direct involvement of hormone action on cyclic AMP phosphodiesterase is that described by Amer and McKinney (1972) of the gastrointestinal polypeptide hormone cholecystokinin. There appears to be an activation of purified low K_m phosphodiesterase from rabbit liver, dog intestinal mucosa, and other tissues when exposed to the hormone *in vitro*, and the activation may be at the expense of the high K_m enzyme.

Reports of the inhibition of high cyclic AMP phosphodiesterase of rat adipose tissue and partially purified beef heart by 3,3′,5-triiodo-L-thyronine aroused speculation that phosphodiesterase may be involved in the regulation of the action of thyroid hormone (Mandel and Kuehl, 1967). T_3 inhibited beef heart high K_m phosphodiesterase competitively with a K_i of 4×10^{-4} M. Inhibition by T_3 of a relatively pure, high K_m frog erythrocyte phosphodiesterase activity (Rosen, 1970) and human adipose tissue phosphodiesterase (Solomon, 1972) have also been reported. The high concentrations of this compound necessary for inhibition, however, make its physiological significance unclear. More physiological concentrations of T_3 have no effect on high K_m cyclic AMP phosphodiesterase activity of spermatozoa (Casillas and Hoskins, 1970), low K_m cyclic AMP phosphodiesterase activity of adipose tissue, or adipose tissue enzyme prepared from T_3 treated rats (Caldwell and Fain, 1971).

High K_m cyclic AMP phosphodiesterase activity of homogenates of myocardium from thyrotoxic cats has been reported as normal (Sobel et al., 1969), as have beef heart homogenate and particulate high K_m phosphodiesterase from hypothyroid cats (Levey et al., 1969). Recent studies using chronic injection of thyroxin to normal animals (12.5 µg/day for 10 days, 60 µg on day 11) indicate decreased low K_m cyclic AMP phosphodiesterase of heart homogenates and decreased cyclic GMP phosphodiesterase activity of skeletal muscle homogenates (Thompson and Williams, *unpublished observations*). Preliminary investigations as to the relative activity of cyclic AMP and cyclic GMP phosphodiesterase in mammalian tissue of hyperthyroid, hypothyroid, and euthyroid states indicate that significant attention to phosphodiesterase kinetic complexities might reveal some interaction of thyroid hormones with these enzymes (Bastomsky et al., 1971). Changes in tissue levels of cyclic AMP and in sensitivity to catecholamines and theophylline with altered thyroid status indicate a need for further investigation in this area, but presently phosphodiesterase involvement with thyroid hormone action is not established.

An interesting hypothesis that long-acting thyroid stimulator (LATS) may act as an antibody to phosphodiesterase and cause inhibition (McKenzie, 1967) has not been supported by recent experiments (Miyai et al., 1971).

VI. SUMMARY

Cyclic nucleotide phosphodiesterase can be seen from this review to be a class of enzymes rather than a single enzyme. This complexity means that a portion of previous work should be reexamined for the possibility of masked relationships of hormones or other physiological effectors with one or the other of the major phosphodiesterase species. It also provides some additional opportunities for the development of pharmacological agents with more specific sites of action. Future reviews will probably deal with individual enzymes and individual functions.

VII. ACKNOWLEDGMENTS

The authors wish to express their appreciation to Drs. M. S. Amer, W. Y. Cheung, R. Jungus, S. Kakiuchi, V. Lavis, H. Sheppard, L. N. Simon, W. Terasaki, and R. H. Williams for providing materials prior to publication.

VIII. REFERENCES

Abdulla, Y. H. (1969): β-Adrenergic receptors in human platelets. *Journal of Atherosclerosis Research,* 9:171–177.
Allen, D. O., and Clark, J. B. (1971): Effect of various antilipolytic compounds on adenyl cyclase and phosphodiesterase activity in isolated fat cells. *Advances in Enzyme Regulation,* 9:99–112.
Amer, M. S., and Kenney, G. R. (1972): On the mechanism of action of choleceptokinesis: Effect on phosphodiesterase. *Pharmacologist,* 11:291.
Appleman, M. M., and Kemp, G. (1966): Puromycin: A potent metabolic effect independent of protein synthesis. *Biochemical and Biophysical Research Communications,* 24:564–568.
Armiento, M. d', Johnson, G. S., and Pastan, I. (1972): Regulation of adenosine $3':5'$-cyclic monophosphate phosphodiesterase activity in fibroblasts by intracellular concentrations of cyclic adenosine monophosphate. *Proceedings of the National Academy of Sciences,* 69:459–462.
Ashcroft, S. J. H., Randle, P. J., and Taljedal, I. B. (1972): Cyclic nucleotide phosphodiesterase activity in normal mouse pancreatic islets. *FEBS Letters,* 20:263.
Bastomsky, C. H., Zakarija, M., and McKenzie, J. M. (1971): Thyroid hydrolysis of cyclic AMP as influenced by thyroid gland activity. *Biochimica et Biophysica Acta,* 230:286–295.
Beavo, J. A., Hardman, J. G., and Sutherland, E. W. (1970a): Hydrolysis of cyclic guanosine and adenosine $3',5'$-monophosphates by rat and bovine tissues. *Journal of Biological Chemistry,* 245:5649–5655.
Beavo, J. A., Hardman, J. G., and Sutherland, E. W. (1971): Stimulation of adenosine $3',5'$-monophosphate hydrolysis by guanosine $3',5'$-monophosphate. *Journal of Biological Chemistry,* 246:3841–3846.
Beavo, J. A., Rogers, N. L., Crofford, O. B., Hardman, J. G., Sutherland, E. W., and Newman, E. V. (1970b): Effects of xanthine derivatives on lipolysis and on adenosine $3',5'$-monophosphate phosphodiesterase activity. *Molecular Pharmacology,* 6:597–603.
Beer, B., Chasin, M., Clody, D. E., Vogel, J. R., and Horovitz, Z. P. (1972): Cyclic adenosine monophosphate phosphodiesterase in brain: Effect on anxiety. *Science,* 176:428–430.
Blecher, M., Merline, N. S., and Ro'Ane, J. T. (1968): Control of the metabolism and lipolytic effects of cyclic $3',5'$-adenosine monophosphate in adipose tissue by insulin, methyl xanthines, and nicotinic acid. *Journal of Biological Chemistry,* 243:3973–3977.
Bonner, J. J., Barkley, D. S., Hall, E. M., Konijn, T. M., Masow, J. W., O'Keefe, G., and Wolfe, P. B. (1969): Acrasin, acrasinase, and the sensitivity to acrasin in *Dictyostelium discoideum*. *Developmental Biology,* 20:72–87.
Breckenridge, B. McL, and Johnson, R. E. (1969): Cyclic $3',5'$-nucleotide phosphodiesterase in brain. *Journal of Histochemistry and Cytochemistry,* 17:505–510.
Brisson, G. R., Malaisse-Lagae, F., and Malaisse, W. J. (1972): The stimulus-secretion coupling of glucose-induced insulin release. *Journal of Clinical Investigation,* 51:232–241.
Brooker, G., and Fichman, M. (1971): Chlorpropamide and tolbutamide inhibition of adenosine $3',5'$-cyclic monophosphate phosphodiesterase. *Biochemical and Biophysical Research Communication,* 42:824–828.
Brooker, G., Thomas, L., Jr., and Appleman, M. M. (1968): The assay of adenosine $3',5'$-cyclic monophosphate and guanosine $3',5'$-cyclic monophosphate in biological materials by enzymatic radio-isotopic displacement. *Biochemistry,* 7:4177–4181.
Brown, J. D., Steele, A. A., Stone, D. B., and Steele, F. A. (1972): The effect of tolbutamide on lipolysis and cyclic AMP concentration in white fat cells. *Endocrinology,* 90:47–51.
Butcher, R. W., Baird, C. E., and Sutherland, E. W. (1968): Effects of lipolytic and antilipolytic substances on adenosine $3',5'$-monophosphate levels in isolated fat cells. *Journal of Biological Chemistry,* 243:1705–1712.
Butcher, R. W., Sneyd, J. G. T., Park, C. R., and Sutherland, E. W. (1966): Effect of insulin on adenosine $3',5'$-monophosphate in the rat epididymal fat pad. *Journal of Biological Chemistry,* 241:1651–1653.

Butcher, R. W., and Sutherland, E. W. (1962): Adenosine 3',5'-phosphate in biological materials. *Journal of Biological Chemistry*, 237:1244-1250.
Caldwell, A., and Fain, J. N. (1971): Triiodothyronine stimulation of cyclic adenosine 3',5'-monophosphate accumulation in fat cells. *Endocrinology*, 89:1195-1204.
Casillas, E. R., and Hoskins, D. D. (1970): Activation of monkey spermatozoal adenyl cyclase by thyroxine and triiodothyronine. *Biochemical and Biophysical Research Communications*, 40:255-261.
Cerasi, E., and Luft, R. (1969): The effect of an adenosine 3',5'-monophosphate diesterase inhibitor (aminophylline) on the insulin response to glucose infusion in prediabetic and diabetic subjects. *Hormone and Metabolic Research*, 1:162-168.
Chang, Y. Y. (1968): Cyclic 3',5'-adenosine monophosphate phosphodiesterase produced by the slime mold dictyostelium discoideum. *Science*, 161:57-59.
Chassy, B. M. (1972): Cyclic nucleotide phosphodiesterase in dictyostelium discoideum: Interconversion of two enzyme forms. *Science*, 175:1016-1018.
Chaudhuri, T. K., and Winer, N. (1971): Effect of sulfonylureas on renal phosphodiesterase. *Clinical Research*, 19:650.
Cheung, W. Y. (1966): Inhibition of cyclic nucleotide phosphodiesterase by adenosine 5'-triphosphate and inorganic pyrophosphate. *Biochemical and Biophysical Research Communications*, 23:214-219.
Cheung, W. Y. (1967a): Cyclic 3',5'-nucleotide phosphodiesterase: Pronounced stimulation by snake venom. *Biochemical and Biophysical Research Communications*, 29:478-482.
Cheung, W. Y. (1967b): Properties of cyclic 3',5'-nucleotide phosphodiesterase from rat brain. *Biochemistry*, 6:1079-1087.
Cheung, W. Y. (1969a): Cyclic 3',5'-nucleotide phosphodiesterase: Preparation of a partially inactive enzyme and its subsequent stimulation by snake venom. *Biochimica et Biophysica Acta*, 191:303-315.
Cheung, W. Y. (1969b): Cyclic 3',5'-nucleotide phosphodiesterase: A continuous titimetric assay. *Analytical Biochemistry*, 28:182-191.
Cheung, W. Y. (1970a): Cyclic nucleotide phosphodiesterase. *Advances in Biochemical Psychopharmacology*, 3:51-56.
Cheung, W. Y. (1970b): Cyclic 3',5'-nucleotide phosphodiesterase. *Biochemical and Biophysical Research Communications*, 38:533-538.
Cheung, W. Y. (1971): Cyclic 3',5'-nucleotide phosphodiesterase: Evidence for and properties of a protein activator. *Journal of Biological Chemistry*, 246:2859-2869.
Cheung, W. Y., and Salganoff, L. (1967): Cyclic 3',5'-nucleotide phosphodiesterase: Localization and latent activity in rat brain. *Nature*, 214:90-91.
Christiansen, R. O., and Monn, E. (1971): Adenosine 3',5'-monophosphate phosphodiesterase: Multiple molecular forms. *Science*, 173:540-541.
Clark, L. J., Goodman, M., and Jungas, R. (1972): Effects of hypophyseal, adrenal, and thyroid hormones and of insulin on phosphodiesterase activity in adipose tissue. *In press.*
Conway, A., and Koshland, D. E., Jr. (1968): Negative cooperativity in enzyme action. *Biochemistry*, 7:4011-4022.
Cuatrecasas, P. (1972): The nature of insulin-receptor interactions. In: *Insulin Action*, pp. 137-169. Academic Press, New York.
Dalton, C., Quinn, J. B., Burghardt, C. R., and Sheppard, H. (1970): Investigation of the mechanism of action of the lipolytic agent 4-(3,4-dimethoxybenzyl)-2-imidazolidinone (Ro 7-2956). *Journal of Pharmacology and Experimental Therapeutics*, 173:270-276.
Davies, G. E., Rose, F. L., and Somerville, A. R. (1971): An Ne inhibitor of phosphodiesterase with anti-bronchoconstrictor properties. *Nature New Biology*, 234:50-51.
Delange, R. J., Kemp, R. G., Riley, W. D., Cooper, R. A., and Krebs, E. G. (1968): Activation of skeletal muscle phosphodiesterase kinase by adenosine triphosphate and adenosine 3',5'-monophosphate. *Journal of Biological Chemistry*, 243:2200-2208.
De Robertis, E., De Lores Arnaiz, G. R., and Alberici, M. (1967): Subcellular distribution of adenyl cyclase and cyclic phosphodiesterase in rat brain cortex. *Journal of Biological Chemistry*, 242:3487-3493.
Dousa, T., and Rychlik, I. (1968): Adenyl cyclase and adenosine 3',5'-cyclic phosphate phos-

phodiesterase in the receptor tissues of neurohypophysial hormones. *Life Sciences*, 7:1039–1044.
Dousa, T., and Rychlik, I. (1970): Metabolism of adenosine 3',5'-cyclic phosphate. II. Some properties of adenosine-3',5'-cyclic-phosphate phosphodiesterase from the rat kidney. *Biochimica et Biophysica Acta*, 204:11–17.
Drummond, G. I., and Perrott-Yee, S. (1961): Enzymatic hydrolysis of adenosine 3',5'-phosphoric acid. *Journal of Biological Chemistry*, 236:1126–1129.
Ensinck, J. W., Stoll, R. W., Gale, C. C., Santen, R. J., Touber, J. L., and Williams, R. H. (1970): Effect of aminophylline on the secretion of insulin, glucagon, luteinizing hormone and growth hormone in humans. *Journal of Clinical Endocrinology*, 31:153–161.
Exton, J. H., Lewis, S. B., Ho, R. J., Robison, G. A., and Park, C. R. (1971): The role of cyclic AMP in the interaction of glucagon and insulin in the control of liver metabolism. *Annals of the New York Academy of Sciences*, 185:85–99.
Fain, J. N. (1968): Effect of dibutyryl-3',5'-AMP, theophylline and norepinephrine on lipolytic action of growth hormone and glucocorticoid in white fat cells. *Endocrinology*, 82:825–830.
Fain, J. N., and Rosenberg, L. (1972): Antilipolytic action of insulin on fat cells. *Diabetes*, 21, Suppl. 2:414–425.
Fichman, M., and Brooker, G. (1970): Cyclic AMP and the antidiuretic effect of chlorpropamide in diabetes insipidus. *Clinical Research*, 18:121.
Fleischer, N., and Donald, R. A. (1969): Involvement of adenosine 3',5'-monophosphate in release of ACTH. *American Journal of Physiology*, 217:1287–1291.
Florenco, N. T., Barrnett, R. J., and Greengard, P. (1971): Cyclic 3',5'-nucleotide phosphodiesterase: Cytochemical localization in cerebral cortex. *Science*, 173:745–747.
Franks, D. J., and MacManus, J. P. (1971): Cyclic GMP stimulation and inhibition of cyclic AMP phosphodiesterase from thymic lymphocytes. *Biochemical and Biophysical Research Communications*, 42:844–849.
Gaballah, S., and Popoff, C. (1971): Cyclic 3',5'-nucleotide phosphodiesterase in nerve endings of developing rat brain. *Brain Research*, 25:220–222.
Goldberg, N. D., Lust, W. D., O'Dea, R. F., Wei, S., and O'Toole, A. G. (1970): A role of cyclic nucleotides in brain metabolism. *Advances in Biochemical Psychopharmacology*, 3:67–87.
Goldfine, I., Perlman, R., and Roth, J. (1971): Inhibition of cyclic 3',5'-AMP phosphodiesterase in islet cells and other tissues by tolbutamide. *Nature*, 234:295–296.
Goodman, H. M. (1968a): Proposed mode of action of histamine. *Nature*, 219:1053.
Goodman, H. M. (1968b): Effects of growth hormone on the lipolytic response of adipose tissue to theophylline. *Endocrinology*, 81:1027–1034.
Goodman, H. M. (1969a): Metabolic effects of imidazole in adipose tissue. *Biochimica et Biophysica Acta*, 176:60–64.
Goodman, H. M. (1969b): Failure of growth hormone alone to potentiate epinephrine-induced lipolysis. *Proceedings of the Society of Experimental Biology and Medicine*, 132:821–824.
Goodsell, E. B., Stein, H. H., and Wenzke, K. J. (1971): 8-Substituted theophyllines. In vitro inhibition of 3',5'-cyclic adenosine monophosphate phosphodiesterase and pharmacological spectrum in mice. *Journal of Medicinal Chemistry*, 14:1202–1205.
Goren, E. N., Hirsch, A. H., and Rosen, O. M. (1971): Activity stain for the detection of cyclic nucleotide phosphodiesterase separated by polyacrylamide gel electrophoresis and its application to the cyclic nucleotide phosphodiesterase of beef heart. *Analytical Biochemistry*, 43:156–161.
Goren, E. N. and Rosen, O. M. (1971): The effect of nucleotides and a non-dialyzable factor on the hydrolysis of cyclic AMP by a cyclic nucleotide phosphodiesterase from beef heart. *Archives of Biochemistry and Biophysics*, 142:720–723.
Greere, H. L., and Herman, R. H. (1972): Human jejunal adenyl cyclase and phosphodiesterase. *Biochemical Medicine*, 6:19–28.
Grimm, J., and Frank, W. (1972): Stimulation of embryonic rat cells in culture by calf serum. *Zeitschrift für Naturforschung*, 27:562–566.

Gulyassy, P. F. (1971): Inhibition of cyclic 3',5'-nucleotide phosphodiesterase by adenine compounds. *Life Sciences,* 10:451–461.
Gulyassy, P. F., and Oken, R. L. (1971): Assay of cyclic 3',5'-nucleotide phosphodiesterase in tissue homogenates. *Proceedings of the Society of Experimental Biology and Medicine,* 137:361–365.
Halkerston, I. D. K., Feinstein, M., and Hechter, O. (1966): An anomalous effect of theophylline on ACTH and adenosine 3',5'-monophosphate stimulation. *Proceedings of the Society of Experimental Biology and Medicine,* 122:896–900.
Hardman, J. G., and Sutherland, E. W. (1965): A cyclic 3',5'-nucleotide phosphodiesterase from heart with specificity for uridine 3',5'-phosphate. *Journal of Biological Chemistry,* 240:3704–3705.
Hayaishi, O., Greengard, P., and Colowick, S. P. (1971): On the equilibrium of the adenylate cyclase reaction. *Journal of Biological Chemistry,* 246:5840–5845.
Heidrick, M. L., and Ryan, W. L. (1971): Adenosine 3',5'-cyclic monophosphate and contact inhibition. *Cancer Research,* 31:1313–1315.
Hepp, K. D. (1971): Increased glucagon sensitivity of mouse liver adenyl cyclase in streptozotocin diabetes. *Diabetologia,* 1:484.
Hepp, K. C., Menahan, L. A., Wieland, O., and Williams, R. H. (1969): Studies on the action of insulin in isolated adipose tissue cells. *Biochimica et Biophysica Acta,* 184:554–565.
Hemington, J. G., and Dunn, A. (1971): Identification of insulin binding activity and isolation of endogenous insulin from rat liver. *Biochemical and Biophysical Research Communications,* 44:71–77.
Hofert, J. F., and Boutwell, R. K. (1963): Puromycin-induced glycogenolysis as an event independent from inhibited protein synthesis in mouse liver; effects of puromycin analogs. *Archives of Biochemistry and Biophysics,* 103:338–344.
Honda, F., and Imamura, H. (1968): Inhibition of cyclic 3',5'-nucleotide phosphodiesterase by phenothiazine and reserpine derivatives. *Biochimica et Biophysica Act,* 161:267–269.
House, P. D. R., Poulis, P., and Weidemann, M. J. (1972): Isolation of a plasma-membrane subfraction from rat liver containing an insulin-sensitive cyclic-AMP phosphodiesterase. *European Journal of Biochemistry,* 24:429–437.
Hsie, A. W., and Puck, T. T. (1971): Morphological transformation of Chinese hamster cells by dibutynyl adenosine 3',5'-monophosphate and testosterone. *Proceedings of the National Academy of Sciences,* 68:358–361.
Huang, Y. C., and Kemp, R. G. (1971): Properties of a phosphodiesterase with high affinity for adenosine 3',5'-cyclic phosphate. *Biochemistry,* 10:2278–2283.
Hynie, S., Krishna, G., and Brodie, B. B. (1966): Theophylline as a tool in studies of the role of cyclic adenosine 3',5'-monophosphate in hormone-induced lipolysis. *Journal of Pharmacology and Experimental Therapeutics,* 153:90–96.
Illiano, G., and Cuatrecasas, P. (1971): Modulation of adenylate cyclase activity in liver and fat cell membranes by insulin. *Science,* 175:906–908.
Iwangoff, P., and Enz, A. (1971): The effect of dihydroergotamine on the phosphodiesterase activity of cat grey matter. *Experientia,* 27:1258–1259.
Jard, S., and Bernard, M. (1970): Presence of two 3',5'-cyclic AMP phosphodiesterases in rat kidney and frog bladder epithelial cells extracts. *Biochemical and Biophysical Research Communications,* 41:781–788.
Jefferson, L. S., Exton, J. H., Butcher, R. W., Sutherland, E. W., and Park, C. R. (1968): Role of adenosine 3',5'-monophosphate in the effects of insulin and anti-insulin serum on liver metabolism. *Journal of Biological Chemistry,* 243:1031–1038.
Johnson, G. S., Friedman, R. M., and Pastan, I. (1971): Restoration of several morphological characteristics of normal fibroblasts in sarcoma cells treated with adenosine 3',5'-cyclic monophosphate and its derivatives. *Proceedings of the National Academy of Sciences,* 68:425–429.
Kakiuchi, S., and Yamazaki, R. (1970): Calcium dependent phosphodiesterase activity and its activating factor (PAF) from brain. *Biochemical and Biophysical Research Communications,* 41:1104–1110.

Kakiuchi, S., Yamazaki, R., and Teshima, Y. (1971): Cyclic 3',5'-nucleotide phosphodiesterase, IV. *Biochemical and Biophysical Research Communications*, 42:968–974.
Klotz, U., Berndt, S., and Stock, K. (1972a): Characterization of multiple cyclic nucleotide phosphodiesterase activities of rat adipose tissue. *Life Sciences*, 11:7–17.
Klotz, U., and Stock, K. (1972): Influence of cyclic guanosine-3',5'-monophosphate on the enzymatic hydrolysis of adenosine-3',5'-monophosphate. *Naunyn-Schmiedebergs Archiv für Pharmakologie*, 274:54–62.
Klotz, U., Vapaatalo, H., and Stock, K. (1972b): Rat adrenal cyclic nucleotide-phosphodiesterase; inhibition by drugs known to affect steroidogenesis. *Naunyn-Schmiedebergs Archiv für Pharmakologie*, 273:376–385.
Kono, T. (1972): The insulin receptor of fat cells: The relationship between the binding and physiological effects of insulin. In: *Insulin Action*, pp. 171–204. Academic Press, New York.
Kukovetz, W. R., and Pöch, G. (1970): Inhibition of cyclic 3',5'-nucleotide phosphodiesterase as a possible mode of action of papaverine and similarly acting drugs. *Naunyn-Schmiedebergs Archiv für Pharmakologie*, 267:189–194.
Kupiecki, F. P. (1969): Reduced adenosine 3',5'-monophosphate phosphodiesterase activity in the pancreas and adipose tissue of spontaneously diabetic mice. *Life Sciences*, 8:645–649.
Kupiecki, F. P., and Marshall, N. B. (1967): Effects of 5-methylpyrazole-3-carboxylic acid (U-19425) and nicotinic acid on lipolysis *in vitro* and *in vivo* and on cyclic-3',5'-AMP phosphodiesterase. *Journal of Pharmacology and Experimental Therapeutics*, 160:166–170.
Levey, G. S., Skelton, C. L., and Epstein, S. E. (1969): Decreased myocardial adenyl cyclase activity in hypothyroidism. *Journal of Clinical Investigation*, 48:2244–2250.
Loten, E. G. (1970): Activation of adipose tissue cyclic 3',5'-nucleotide phosphodiesterase by insulin. *Proceedings of the University of Otago Medical School*, 48:44–45.
Loten, E. G., and Sneyd, J. G. T. (1970): An effect of insulin on adipose-tissue adenosine 3',5'-cyclic monophosphate phosphodiesterase. *Biochemical Journal*, 120:187–193.
Mandel, L. R. (1971): Inhibition of cyclic 3',5'-adenosine monophosphate phosphodiesterase by substituted imidazopyrazines. *Biochemical Pharmacology*, 20:3412–3431.
Mandel, L. R., and Kuehl, F. A., Jr. (1967): Lipolytic action of 3,3'5-triiodo-L-thyronine, a cyclic AMP phosphodiesterase inhibitor. *Biochemical and Biophysical Research Communications*, 28:13–18.
Manganiello, V., and Vaughan, M. (1972a): Prostaglandin E_1 effects on adenosine 3',5'-cyclic monophosphate concentration and phosphodiesterase activity in fibroblasts. *Proceedings of the National Academy of Sciences*, 69:269–273.
Manganiello, V., and Vaughan, M. (1972b): An effect of dexamethasone on adenosine 3',5'-monophosphate content and adenosine 3',5'-monophosphate phosphodiesterase activity of cultured hepatoma cells. *Journal of Clinical Investigation*, 51:2763–2767.
McKenzie, J. M. (1967): The long-acting thyroid stimulator: Its role in Graves' disease. *Recent Progress in Hormone Research*, 23:1–46.
McNeil, J. H., Lee, C., and Muschek, L. D. (1972): The effect of phentolamine and other drugs in rat brain phosphodiesterase. *Canadian Journal of Physiology and Pharmacology*, 50:840–844.
McNeil, J. H., and Muschek L. D. (1970): Inhibition of brain phosphodiesterase by tricyclic antidepressants. *Clinical Research*, 18:625.
McNeil, J. H., Nassar, M., and Brody, T. M. (1969): The effect of theophylline on amine-induced cardiac phosphorylase activation and cardiac contractility. *Journal of Pharmacology and Experimental Therapeutics*, 165:234–241.
Menahan, L. A., Hepp, K. D., and Wieland, O. (1969): Liver 3',5'-nucleotide phosphodiesterase and its activity in rat livers perfused with insulin. *European Journal of Biochemistry*, 8:435–443.
Mier, P. D., and Urselman, E. (1972): Adenosine 3',5'-cyclic monophosphate phosphodiesterase in skin. *British Journal of Dermatology*, 86:141–149.
Miki, N., and Yoshida, H. (1972): Purification and properties of cyclic AMP phosphodiesterase from rat brain. *Biochimica et Biophysica Acta*, 268:166–174.

Miyai, K., Amino, N., Azukizawa, M., and Kumahara, Y. (1971): *In vitro* effects of LATS and TSH on phosphodiesterase activity in human thyroid homogenates. *Acta Endocrinologica*, 67:209–215.
Monard, D., Janacek, J., and Rickenberg, H. U. (1969): The enzymatic degradation of 3',5'-cyclic AMP in strains of *E. coli* sensitive and resistant to catabolite repression. *Biochemical and Biophysical Research Communications*, 35:584–591.
Monn, E., and Christiansen, R. O. (1971): Adenosine 3',5'-monophosphate phosphodiesterase: Multiple forms. *Science*, 173:540–541.
Moore, P. F., Iorio, L. C., and McManus, J. M. (1968): Relaxation of the guinea-pig tracheal chain preparation by N^6, 2'-O-dibutyryl 3',5'-cyclic adenosine monophosphate. *Journal of Pharmacy and Pharmacology*, 20:368–372.
Müller-Oerlinghausen, B., Schwabe, U., Hasselblatt, A., and Schmidt, F. H. (1968): Activity of 3',5'-AMP phosphodiesterase in liver and adipose tissue of normal and diabetic rats. *Life Sciences*, 7:593–598.
Nair, K. G. (1966): Purification and properties of 3',5'-cyclic nucleotide phosphodiesterase from dog heart. *Biochemistry*, 5:150–157.
Nakano, J. (1970): Effects of nicotinic acid on hormone-, theophylline-, and dibutyryl cyclic AMP-induced lipolysis and on cyclic AMP phosphodiesterase. *Research Communications in Chemical Pathology and Pharmacology*, 1:769–779.
Nakano, J., and Ishii, I. (1970*a*): Effect of desmethylimipramine on hormone-, theophylline-, and dibutyryl cyclic AMP-induced lipolysis in isolated rat fat cells. *Proceedings of the Society of Experimental Biology and Medicine*, 134:210–212.
Nakano, J., and Ishii, T. (1970*b*): Effect of ketone bodies on hormone-, theophylline- and dibutyryl cyclic AMP-induced lipolysis and on cyclic AMP-phosphodiesterase. *Research Communications in Chemical Pathology and Pharmacology*, 1:485–496.
O'Dea, R. F., Haddox, M. K., and Goldberg, N. D. (1971): Interaction with phosphodiesterase of free and kinase complexed cyclic adenosine 3',5'-monophosphate. *Journal of Biological Chemistry*, 246:6183–6190.
Pannbacker, R. G., and Bravard, L. J. (1972): Phosphodiesterase in dictyostelium discoideum and the chemotactic response to cyclic adenosine monophosphate. *Science*, 175:1014–1015.
Pannbacker, R. G., Fleischman, D. E., and Reed, D. W. (1972): Cyclic nucleotide phosphodiesterase: High activity in a mammalian photoreceptor. *Science*, 175:757–758.
Pennington, S. N. (1971): 3',5'-Cyclic adenosine monophosphate phosphodiesterase assay using high speed liquid chromatography. *Analytical Chemistry*, 43:1701–1703.
Perkins, J. P., MacIntyre, E. H., Riley, W. D., and Clark, R. B. (1971): Adenyl cyclase, phosphodiesterase and cyclic AMP dependent protein kinase of malignant glial cells in culture. *Life Sciences*, 10:1069–1080.
Perlman, R. L., and Pastan, J. (1968): Regulation of β-galactosidase synthesis in *Escherichia coli* by cyclic adenosine 3',5'-monophosphate. *Journal of Biological Chemistry*, 243:5420–5427.
Peterson, M. J., Hillman, C. C., and Ashmore, J. (1968): Nicotinic acid: Studies on the mechanism of its antilipolytic action. *Molecular Pharmacology*, 4:1–9.
Pöch, G. (1971): Assay of phosphodiesterase with radioactivity labeled cyclic 3',5'-AMP as substrate. *Naunyn-Schmiedesbergs Archiv für Pharmakologie*, 268:272–299.
Pöch, G., and Kukovetz, R. (1971): Papaverine-induced inhibition of phosphodiesterase activity in various mammalian tissues. *Life Sciences*, 10:133–144.
Rall, T. W., and West, T. C. (1963): The potentiation of cardiac inotropic responses to norepinephrine by theophylline. *Journal of Pharmacology and Experimental Therapeutics*, 139:269–274.
Ray, T. K., and Forte, J. B. (1972): Demonstration of an "activator factor" and an "inhibitor factor" in the cyclic AMP phosphodiesterase from oxyntic cells of bullfrog gastric mucosa. *FEBS Letters*, 20:205–208.
Ray, T. K., Tomasi, V., and Marinetti, G. V. (1970): Hormone action at the membrane level. *Biochimica et Biophysica Acta*, 211:20–30.

Riedel, V., and Gerisch, G. (1971): Regulation of extracellular cyclic-AMP phosphodiesterase activity during development of *Dictyostelium discoideum*. *Biochemical and Biophysical Research Communications*, 42:119-124.

Robison, G. A., Butcher, R. W., and Sutherland, E. W. (1971): *Cyclic AMP*. Academic Press, New York.

Rosen, O. M. (1970): Preparation and properties of a cyclic 3′,5′-nucleotide phosphodiesterase isolated from frog erythrocytes. *Archives of Biochemistry and Biophysics*, 137:435-441.

Roth, J. (1971): Sulfonylureas: Effects *in vivo* and *in vitro*. *Annals of Internal Medicine*, 75:607-621.

Rufeger, U., Tellhelm, B., and Frimmer, M. (1971): Inhibition by papaverine of 3′,5′-nucleotide phosphodiesterase (E.C.3.1.4.1.) from bovine uterus muscle. *Naunyn-Schmiedebergs Archiv für Pharmakologie*, 270:428-430.

Russell, T. R., Terasaki, W. L., and Appleman, M. M. (1973): Separate phosphodiesterases for the hydrolysis of cyclic AMP and cyclic GMP in rat liver. *Journal of Biological Chemistry*, 248:1334-1340.

Russell, T. R., Thompson, W. J., Schneider, F. W., and Appleman, M. M. (1972): 3′,5′-cyclic adenosine monophosphate phosphodiesterase: Negative cooperativity. *Proceedings of the National Academy of Sciences*, 69:1791-1795.

Ryan, W. L., and Heidrick, M. L. (1968): Inhibition of cell growth *in vitro* by adenosine 3′,5′-monophosphate. *Science*, 162:1484-1485.

Sams, D. J., and Montagne, W. (1972): The role of adenosine 3′,5′-cyclic monophosphate in the regulation of insulin release. *Biochemical Journal*, 129:945-952.

Schaeffer, L. D., Chenoweth, M., and Dunn, A. (1969): Adrenal corticosteroid involvement in the control of liver glycogen phosphorylase activity. *Biochimica et Biophysica Acta*, 192:292-303.

Schonhofer, P. S., Skidmore, I. F., Bourne, H. R., and Krishna, G. (1972): Cyclic 3′,5′-AMP phosphodiesterase in isolated fat cells. *Pharmacology*, 7:65-77.

Schroder, J., and Plageman, P. G. W. (1972): Cyclic 3′,5′-nucleotide phosphodiesterases of Novikoff rat hepatoma, Monse L, and HeLa cells growing in suspension culture. *Cancer Research*, 32:1082-1087.

Schultz, G., Senft, G., and Munske, K. (1966): Der Einfluß von Insulin auf die enzymatische Regulation der Glycogenolyse. *Naturwissenschaften*, 53:529.

Schwabe, U., Berndt, S., and Ebert, R. (1972): Activation and inhibition of lipolysis in isolated fat cells by various inhibitors of cyclic AMP phosphodiesterase. *Naunyn-Schmiedebergs Archiv für Pharmakologie*, 273:287-298.

Schwabe, U., and Ebert, R. (1972): Different effects of lipolytic hormones and phosphodiesterase inhibitors on cyclic AMP levels in isolated fat cells. *Naunyn-Schmiedebergs Archiv für Pharmakologie*, 274:287-298.

Senft, G., Munske, K., Schultz, G., and Hoffmann, M. (1968a): The influence of hydrochlorothiazide and other sulfamoyl diuretics on the activity of 3′,5′-AMP phosphodiesterase in rat kidney. *Naunyn-Schmiedebergs Archiv für Pharmakologie*, 259:344-359.

Senft, G., Schultz, G., Munske, K., and Hoffmann, M. (1968b): Influence of insulin on cyclic 3′,5′-AMP phosphodiesterase activity in liver, skeletal muscle, adipose tissue, and kidney. *Diabetologia*, 4:322-329.

Senft, G., Schultz, G., Munske, K., and Hoffmann, M. (1968c): Effects of glucocorticoids and insulin on 3′,5′-AMP phosphodiesterase activity in adrenalectomized rats. *Diabetologia*, 4:330-335.

Shanta, T. R., Woods, W. D., Waitzman, M. B., and Bourne, G. H. (1966): Histochemical method for localization of cyclic 3′,5′-nucleotide phosphodiesterase. *Histochemie*, 7:177-190.

Sheppard, H., and Wiggan, G. (1970): Analogues of 4-(3,4-dimethoxybenzyl)-2-imidazolidinone as potent inhibitors of rat erythrocyte adenosine cyclic 3′,5′-phosphate phosphodiesterase. *Molecular Pharmacology*, 7:111-115.

Sheppard, H., and Wiggan, G. (1971): Different sensitivities of the phosphodiesterases (adenosine-3',5'-cyclic phosphate 3'-phosphohydrolase) of dog cerebral cortex and erythrocytes to inhibition by synthetic agents and cold. *Biochemical Pharmacology,* 20:2128–2130.

Shimoyama, M., Kawai, M., Tanigawa, Y., and Ueda, I. (1972): Evidence for and some properties of a 3',5'-cyclic AMP phosphodiesterase inhibitor in potato. *Biochemical and Biophysical Research Communications,* 47:59–65.

Skidmore, I. R., Schonhofer, P. S., and Kritchevsky, D. (1971): Effects of nicotinic acid and some of its homologues on lipolysis, adenyl cyclase, phosphodiesterase and cyclic AMP accumulation in isolated fat cells. *Pharmacology,* 6:330–338.

Smith, J. B., and Mills, D. C. B. (1970): Inhibition of adenosine 3',5'-cyclic monophosphate phosphodiesterase. *Biochemical Journal,* 120:20.

Sobel, B. E., Dempsey, P. J., and Cooper, T. (1968): Adenyl cyclase activity in the chronically denervated cat heart. *Biochemical and Biophysical Research Communications,* 33:758–762.

Sobel, B. E., Dempsey, P. J. and Cooper, T. (1969): Normal myocardial adenyl cyclase activity in hyperthyroid cats. *Proceedings of the Society of Experimental Biology and Medicine,* 132:6–9.

Solomon, S. S. (1972): Phosphodiesterase activity of rat and human adipose tissue. *Journal of Laboratory and Clinical Medicine,* 79:598–610.

Stansfield, D. A., Horne, J. R., and Wilkinson, G. H. (1971): Adenosine 3',5'-cyclic phosphate phosphodiesterase of corpus luteum. *Biochimica et Biophysica Acta,* 227:413–418.

Strubelt, O., Steffen, J., and Stutz, U. (1970): Der Einfluß der Schilddrüsenfunktion auf die chronotropen und einige metabolische Wirkungen von Theophyllin und Coffein. *Naunyn-Schmiedebergs Archiv für Pharmakologie,* 267:135–154.

Sung, C. P., Wiebelhaus, V. D., Jenkins, B. C., Adlercreutz, P., Hirschowitz, B. I., and Sachs, G. (1972): Heterogeneity of 3',5'-phosphodiesterase of gastric mucosa. *American Journal of Physiology,* 223:648–650.

Sutherland, E. W., and Rall, T. R. (1958): Fractionization and characterization of a cyclic adenine ribonucleotide formed by tissue particles. *Journal of Biological Chemistry,* 232:1077–1091.

Therriault, D. G., and Winters, V. G. (1970): Enzymic hydrolysis of adenosine 3',5'-monophosphate by rat and beef heart supernatant. *Life Science,* 9:1053–1060.

Thompson, W. J. (1971): Studies on cyclic nucleotide phosphodiesterase. Thesis, *University of Southern California.*

Thompson, W. J., and Appleman, M. M. (1971a): Multiple cyclic nucleotide phosphodiesterase activities from rat brain. *Biochemistry,* 10:311–316.

Thompson, W. J., and Appleman, M. M. (1971b): Characterization of cyclic nucleotide phosphodiesterases of rat tissues. *Journal of Biological Chemistry,* 246:3145–3150.

Thompson, W. J., and Appleman, M. M. (1971c): Cyclic nucleotide phosphodiesterase and cyclic AMP. *Annals of the New York Academy of Sciences,* 185:36–41.

Thompson, W. J., Little, S. A., and Williams, R. H. (1973): Effect of insulin and growth hormone on rat liver cyclic nucleotide phosphodiesterase. *In preparation.*

Toson, G. C., and Carpenedo, F. (1972): Inhibition by papaverine and eupaverin of 3',5'-cyclic AMP phosphodiesterase from rabbit skeletal muscle. *Naunyn-Schmiedebergs Archiv für Pharmakologie,* 273:168–171.

Triner, L., Vulliemoz, Y., Schwartz, I., and Nahas, G. G. (1970): Cyclic phosphodiesterase activity and the action of papaverine. *Biochemical and Biophysical Research Communications,* 40:64–69.

Turtle, J. R., and Kipnis, D. M. (1967): An adrenergic receptor mechanism for the control of cyclic 3',5'-adenosine monophosphate synthesis in tissues. *Biochemical and Biophysical Research Communications,* 28:797–802.

Vernikos-Danellis, J., and Harris, C. G., III (1968): The effect of *in vitro* and *in vivo* caffeine, theophylline, and hydrocortisone on the phosphodiesterase activity of the pituitary, median eminence, heart, and cerebral cortex of the rat. *Proceedings of the Society of Experimental Biology and Medicine,* 128:1016–1021.

Wang, J. H. C., Teo, T. S., and Wang, T. H. (1972): Hysteretic substrate activation of bovine heart c-AMP phosphodiesterase. *Biochemical and Biophysical Research Communications,* 46:1306–1311.

Ward, W. F., and Fain, J. N. (1971): The effects of dihydroergotamine upon adenyl cyclase and phosphodiesterase of rat fat cells. *Biochimica et Biophysica Acta,* 237:387–390.

Weiss, B., and Costa, E. (1968): Regional and subcellular distribution of adenyl cyclase and 3′,5′-cyclic nucleotide phosphodiesterase in brain and pineal gland. *Biochemical Pharmacology,* 17:2107–2116.

Weiss, B., Lehne, R., and Strada, S. (1972): Rapid microassay of adenosine 3′,5′-monophosphate phosphodiesterase activity. *Analytical Biochemistry,* 45:222–235.

Wood, H. N., Lin, M. C., and Braun, A. C. (1972): The inhibition of plant and animal adenosine 3′,5′-cyclic monophosphate phosphodiesterases by a cell-division-promoting substance from tissues of higher plant species. *Proceedings of the National Academy of Sciences,* 69:403–406.

Yamamoto, M., and Massey, K. L. (1969): Cyclic 3′,5′-nucleotide phosphodiesterase of fish (Salmo Gairdnerli) brain. *Comparative Biochemistry and Physiology,* 30:941–954.

Protein Kinases and Protein Kinase Substrates

Thomas A. Langan

I. Introduction 99
II. Species Distribution of cAMP-dependent Protein Kinase 102
III. Role of Protein Kinases in Mediating the Effects of cAMP 103
IV. Activation of Protein Kinase by cAMP 108
V. Other Factors Influencing Protein Kinase Activity 113
VI. Multiple Forms and Subunit Structure 119
VII. Subcellular Distribution 124
VIII. Substrates and Substrate Specificity of cAMP-dependent Protein Kinase ... 127
IX. Concluding Remarks 139
X. References 140

I. INTRODUCTION

The activation of protein kinases appears at present to be the major, if not the only, mechanism by which cyclic adenosine 3′,5′-monophosphate (cAMP) carries out its function as second messenger in the transmission of hormonal signals. This review chapter will outline our current knowledge of these cAMP-dependent protein kinases, and attempt to assess our understanding of the role played by protein phosphorylation in linking cAMP to the wide variety of physiological responses it is known to elicit. Emphasis will be on the conceptual state of the field rather than on the provision of an exhaustive bibliography of papers in the area.[1] It should be stated that

---

[1] A complete bibliography of publications concerned with all aspects of cAMP is compiled periodically by E. R. Squibb and Sons, Inc. I thank Dr. N. S. Semenuk for providing portions of the 1971 bibliography prior to publication. Other reviews concerned with protein kinase are

the various assessments made here are attempts to reflect current information and are not intended as predictions of the conclusions that will be reached as investigations continue.

The first observation of protein kinase activity was made by Burnett and Kennedy (1954), who described a liver enzyme that catalyzes the phosphorylation of casein. The finding shortly thereafter that protein phosphorylation is involved in the activation of glycogen phosphorylase (Sutherland and Wosilait, 1955; Fischer and Krebs, 1955; Rall, Sutherland, and Wosilait, 1956; Krebs and Fischer, 1956; Krebs, Kent, and Fischer, 1958) was the key to the demonstration of hormone action in a cell-free system (Rall, Sutherland, and Berthet, 1957) and to the discovery that a heat-stable factor, cAMP, mediated the effects of hormones on the phosphorylation and activation of phosphorylase (Rall, Sutherland, and Berthet, 1957; Rall and Sutherland, 1958). However, this early indication that protein phosphorylation is involved in the action of cAMP was somewhat adventitious, since cAMP does not, of course, act directly on phosphorylase kinase. Determination of the site of action of cAMP required the exhaustive study of muscle phosphorylase activation carried out by Krebs and Fischer and more recently by Krebs and colleagues, the key findings being that phosphorylase kinase itself exists in an active, phosphorylated form and in an inactive, dephosphorylated form and that cyclic AMP stimulates this activation process (Krebs, Graves, and Fischer, 1959; Krebs, Love, Bratvold, Trayser, Meyer, and Fischer, 1964). Among the clues which led to the now well-known two-step cascade in the action of cAMP on phosphorylase activation was the finding that insignificant amounts of cAMP were bound by preparations of phosphorylase kinase, even though phosphorylation and activation of these preparations were readily stimulated by the cyclic nucleotide; the stoichiometry suggested that the component responsible for binding cAMP was present in catalytic amounts (Krebs, DeLange, Kemp, and Riley, 1966; DeLange, Kemp, Riley, Cooper, and Krebs, 1968). This led to the demonstration that a third enzyme, a phosphorylase kinase kinase, is the enzyme directly activated by cAMP (Walsh, Perkins, and Krebs, 1968).

In addition, the Krebs group recognized early that proteins other than phosphorylase kinase could serve as substrates for the cAMP-dependent protein kinase. For example, casein and, in collaboration with G. Dixon and

---

those of Krebs (1972, 1973), Walsh and Krebs (1973), Fischer, Heilmeyer, and Haschke (1971), Larner and Villar-Palasi (1971), Hers, de Wulf, and Stalmans (1970), Villar-Palasi and Larner (1970), Krebs, DeLange, Kemp, and Riley (1966), Krebs and Fischer (1962), and Rabinowitz (1962).

B. Jergil, protamine were found to be substrates (Walsh, Perkins, and Krebs, 1968). Appreciating the significance of the phosphorylation of multiple substrates, Krebs and co-workers adopted the term *protein kinase* rather than *phosphorylase kinase kinase* for the enzyme. Prompted by these observations, I (Langan, 1968a) tested the effect of cAMP on preparations of a liver enzyme which catalyzes the phosphorylation of histones and protamine (Langan and Smith, 1967) and found that cAMP stimulates the phosphorylation of histones (but not protamine) by these enzyme preparations. Eventually it was shown that histone kinase and protein kinase are one and the same enzyme. The studies of Larner and co-workers (Friedman and Larner, 1963; Rosell-Perez and Larner, 1964; Huijing and Larner, 1966a) and those of Belocopitow (1961) had provided evidence that cAMP stimulates phosphorylation of glycogen synthetase in partially purified preparations. Further work showed that phosphorylation of both phosphorylase kinase and glycogen synthetase is catalyzed by the same preparations of cAMP-dependent protein kinase (Schlender, Wei, and Villar-Palasi, 1969) and that the kinase involved is the same in each case (Soderling, Hickenbottom, Reimann, Hunkeler, Walsh, and Krebs, 1970; Villar-Palasi, Larner, and Shen, 1971).

Aside from any potential biological significance of cAMP-dependent histone phosphorylation, the knowledge that protein kinase catalyzes the phosphorylation of histones provided a convenient substrate for the study of this enzyme in many laboratories. In particular, Kuo and Greengard (1969a), using histones as substrates, quickly established that cAMP-dependent protein kinase is widely distributed in animal tissues and organisms. They postulated on this basis that all of the biochemical and physiological effects of cAMP might be mediated through regulation of the activity of protein kinases in various tissues. As noted below, the findings in animal cells are thus far generally consistent with this view, although in bacteria one action of cAMP appears not to involve protein phosphorylation.

Protein phosphorylation activated by cAMP is of the type which occurs on the hydroxyl group of serine and, to a much smaller extent, threonine residues. This type of protein phosphate was first demonstrated in the egg yolk phosphoprotein phosvitin (or vitellinic acid) and in casein by Lipmann and Levene (1932) and Lipmann (1933). Phosphorylation of serine residues in total protein fractions from various tissues was detected in early experiments utilizing ^{32}P-labeled phosphate, either by isolation of phosphoserine from partial acid hydrolysates or by the lability of the protein phosphate in alkali (von Euler, Hevesy, and Solodkowska, 1948; Davidson, Frazer, and Hutchison, 1951; Johnson and Albert, 1953; Kennedy and Smith, 1954;

Agren, De Verdier, and Glomset, 1954). Alkali lability is a characteristic of phosphate esterified to serine and threonine residues and is due to beta-elimination (not hydrolysis) of phosphate which takes place when the amino acids are in peptide linkage (Anderson and Kelley, 1959). These observations of protein phosphorylation in tissues led to Burnett and Kennedy's (1954) work on casein kinase activity in liver. Apparently similar enzymes, acting on phosvitin or casein, have been reported in a number of tissues (Sundararajan, Kumar, and Sandra, 1958; Rabinowitz and Lipmann, 1960; Krane, Stone, and Glimcher, 1965; Jackson, Jackson, and Freeman, 1965; Lorini, Pinna, Moret, and Siliprandi, 1965). In spite of the overlap in some cases with respect to the utilization of casein as a substrate, this type of phosphoprotein kinase is distinct from cAMP-dependent protein kinase (Langan and Smith, 1967; Baggio, Pinna, Moret, and Siliprandi, 1970; Pinna, Clari, and Moret, 1971; Takeda, Yamamura, and Ohga, 1971; Ruddon and Anderson, 1972). It should be emphasized that cAMP-dependent protein phosphorylation appears to comprise only a fraction of the total amount of protein phosphorylation which occurs in tissues. In particular, phosphorylation of a class of highly phosphorylated phosphoproteins found concentrated in the cell nucleus (Langan, 1967; Kleinsmith and Allfrey, 1969) has not been reported to be stimulated by cAMP. It should also be noted that phosphorylation of proteins on residues other than serine and threonine has been demonstrated, e.g., histidine (Ramaley, Bridger, Moyer, and Boyer, 1967; Walinder, 1968; Kundig and Roseman, 1971) and glutamic or aspartic acid (Bader, Post, and Jean, 1967). The extreme acid lability of these phosphorylated amino acids dictates that histidyl or acyl phosphate will not usually be encountered in the normal course of investigating phosphorylation of proteins on hydroxy amino acids.

II. SPECIES DISTRIBUTION OF cAMP-DEPENDENT PROTEIN KINASE

Numerous reports on the occurrence of cAMP-dependent protein kinases in various tissues and organisms have appeared since the original paper of Walsh, Perkins, and Krebs (1968). In addition to those noted in the introduction, early reports which served to confirm the widespread distribution of this enzyme included those of Miyamoto, Kuo, and Greengard (1969a), Corbin and Krebs (1969), Jergil and Dixon (1970), Jard and Bastide (1970), Kuo, Krueger, Sanes and Greengard (1970), and Weller and Rodnight (1970). It now seems clear that cAMP-dependent protein kinase is essentially universally present in animal tissues, and this wide distribution somewhat diminishes the pertinence of commenting at this point on additional reports of its occurrence in any one cell type. However, the extremely high specific activity of protein kinase in mammalian spermatozoa (Hoskins, Casillas, and Stephens, 1972) is of interest in view of the possible role of cAMP in sperm motility. Also, the occurrence of protein kinase in RNA tumor viruses and other budding viruses (Strand and August, 1971; Hatanaka, Twiddy, and Gilden, 1972) is especially interesting. It is not known, however, whether the protein kinase is coded in the viral genome or derived from host-cell material contained in the virus. Insects and other arthropods, as well as certain mammalian tissues, also contain varying amounts of cyclic GMP-dependent protein kinases (Kuo and Greengard, 1970a; Greengard and Kuo, 1970; Kuo, Wyatt, and Greengard, 1971).

In contrast to the wealth of reports of animal cell protein kinase, this reviewer is aware of no reports of cAMP-dependent protein kinase in plants. The published report of cAMP-stimulated protein kinase in *Escherichia coli* (Kuo and Greengard, 1969b) has not been followed by similar findings in other bacteria. A protein (histone) kinase has been detected and partially purified from extracts of *Arthrobacter atrocyoneus*, but the response of the enzyme preparations from this source to cAMP has been variable, and it is not yet clear if this is a cAMP-dependent protein kinase (Li, 1972). It is possible that the detection of cAMP-dependent protein kinase in bacteria is hampered by the availability of a suitable substrate, since it seems unlikely that highly active histone-phosphorylating enzymes would be present in prokaryotes.

III. ROLE OF PROTEIN KINASES IN MEDIATING THE EFFECTS OF cAMP

The possibility that phosphorylase kinase kinase might have multiple physiological functions was suggested by the observations of Walsh, Perkins, and Krebs (1968) that their enzyme preparations phosphorylated several proteins.' The extremely wide species distribution of cAMP-dependent protein kinase led Greengard and co-workers to propose specifically that protein kinases might mediate *all* the varied effects of cAMP in different tissues and organisms (Kuo and Greengard, 1969a; Greengard and Kuo, 1970). Since the wide distribution of protein kinase has been established almost exclusively with the use of histones as substrate, the idea that protein kinases from all sources can act on multiple substrates is implicit in this hypothesis. The unifying simplicity of the hypothesis is useful to the extent that it focuses attention on the importance of establishing the nature of the substrates for protein kinase in cells, and also in that it provides a framework for initial investigation of cyclic AMP action in a given situation. It should not of course serve to discourage tests for effects of cAMP on processes other than protein phosphorylation.

Krebs (1973) (Table 1) has proposed a set of criteria to be satisfied in determining if a particular effect of cAMP is mediated by protein phosphorylation, somewhat analogous to the criteria for establishing if a particular hormone response is mediated by cAMP (Robison, Butcher, and Sutherland, 1971). The first criterion appears to be universally satisfied in animal cells, owing to the wide distribution of protein kinase. As is the case with the criteria for involvement of cAMP in a hormone response, we may not expect each of these criteria to be fully satisfied even in cases where the overall weight of evidence is convincing, since technical factors may interfere. Criteria 3 and 4 would seem especially subject to this possibility. At present there are three effects of cAMP for which the criteria are reasonably satisfied: acceleration of glycogen breakdown (Posner, Stern, and Krebs, 1965; Riley, DeLange, Bratvold, and Krebs, 1968; Walsh, Perkins, and Krebs, 1968; Drummond, Harwood, and Powell, 1969; Mayer and Krebs, 1970; Walsh, Perkins, Brostrom, Ho, and Krebs, 1971), reduction of glycogen synthesis (Belocopitow, 1961; Friedman and Larner, 1963; Traut and Lipmann, 1963; Craig and Larner, 1964; Rosell-Perez and Larner, 1964; Larner, Villar-Palasi, and Brown, 1969; Schlender, Wei, and Villar-Palasi, 1969; Soderling, Hickenbottom, Reimann, Hunkeler, Walsh, and Krebs, 1970), and increased adipose tissue lipolysis (Hollenberg, Raben, and Astwood, 1961; Vaughan, Berger, and Steinberg, 1964; Corbin and Krebs, 1969; Huttunen, Steinberg, and Mayer, 1970*a,b;* Corbin, Reimann, Walsh, and Krebs, 1970; Huttunen and Steinberg, 1971), the protein kinase substrates being, respectively, phosphorylase kinase, glycogen synthetase, and adipose tissue lipase (or possibly an intervening lipase kinase). In fact, all the criteria have been satisfied in these three cases, with the exception that only the modification of enzyme *activity* has been demonstrated *in vivo*, not the actual phosphorylation of the enzymes in response to hormone or cAMP. Other proteins whose phosphorylation may be involved in media-

TABLE 1. *Criteria for mediation of cAMP effect by protein phosphorylation*

1. Cell-type involved contains a cAMP-dependent protein kinase.
2. Protein substrate exists which bears a functional relationship to the cAMP-mediated process.
3. Phosphorylation of the substrate alters its function *in vitro*.
4. Protein substrate is modified *in vivo* in response to cAMP.
5. A phosphoprotein phosphatase exists to reverse the process.

From Krebs (1973) with permission of the author and publisher.

tion of cAMP effects are discussed in the section on protein kinase substrates.

In animal cells, the hypothesis that protein kinase mediates all the effects of cAMP also receives support from the fact that there have been no well-documented reports of a case of cAMP regulation not involving protein kinase. The principal possible exception is the enzyme phosphofructokinase, whose activity is modulated *in vitro* by cAMP by processes that appear to involve binding the cAMP as a ligand rather than phosphorylation of the enzyme (Stone and Mansour, 1967; Kemp and Krebs, 1967). The physiological significance of the effects of cAMP observed *in vitro* for hormonal regulation of phosphofructokinase are difficult to assess at present. The potential role of phosphofructokinase as a site of action for cAMP has recently been reviewed (Mansour, 1972).

Cyclic AMP causes inactivation of phosphorylase phosphatase in partially purified preparations from bovine adrenal cortex (Riley and Haynes, 1963; Merlevede and Riley, 1966) and pigeon breast muscle (Chelala and Torres, 1969, 1970). ATP and magnesium must be present for these effects, so there is ample opportunity for mediation of the response by protein kinase, in keeping with the hypothesis. The systems are complicated, however, by the facts that ATP and magnesium in the absence of cAMP cause activation and that ATP alone causes inactivation of the enzyme (Merlevede and Riley, 1966; Chelala and Torres, 1970). The mechanism of inhibition of calcium-stimulated ATPase in rat heart sarcolemma particles by high concentrations of cAMP is difficult to judge from the data so far available (Dietze and Hepp, 1972), but again the opportunity for protein phosphorylation is present.

Rasmussen (Rasmussen, 1970; Rasmussen and Tenenhouse, 1968; Prince, Berridge, and Rasmussen, 1972) has emphasized the importance of Ca^{++} as a potential mediator of hormone actions, in particular of those hormones interacting with the adenyl cyclase system. He envisages an effect of hormones on cell Ca^{++} by three mechanisms: (1) the release of Ca^{++} chelated to ATP when the ATP is converted to cAMP by adenyl cyclase, (2) the mobilization of Ca^{++} from intracellular stores by cAMP, and (3) the stimulation of extracellular Ca^{++} uptake by hormone-membrane interactions independent of increases in cAMP. As has been pointed out (Rasmussen and Tenenhouse, 1968), the first mechanism, which has only a hypothetical basis, cannot be important in the large number of responses which can be obtained by adding exogenous cAMP. Similarly, the third mechanism cannot be important in these responses. The second mechanism may well occur, but it seems possible that the action of cAMP on release of intracellular Ca^{++} stores might involve protein phosphorylation as the primary step.

The third possibility, that some of the effects of hormones which activate adenyl cyclase are due to effects on Ca^{++} uptake independent of increases in cAMP, is an important one (see also Mayer and Stull, 1971). However, strictly speaking, it does not relate to the hypothesis that all the actions of cAMP are due to activation of protein kinase. It is clear that Ca^{++} may be required for hormone responses at steps subsequent to the action of cAMP-dependent protein kinase, for example, in the action of phosphorylase kinase (Brostrom, Hunkeler, and Krebs, 1971). Further, it should be recognized that some responses brought about by cAMP may also be brought about independently by changes in Ca^{++}, without changes in cAMP, for example, glycogen breakdown during muscular contraction (Posner, Stern, and Krebs, 1965; Drummond, Harwood, and Powell, 1969).

In animal cells, then, it appears that activation of protein kinase is the only well-documented mechanism by which cAMP acts. Whether additional mechanisms are discovered in the future remains to be seen. Nevertheless, it is already clear that protein kinase activation is a mechanism of major importance in cellular regulation by cAMP, regardless of whether other mechanisms also exist.

In *E. coli* and a number of other bacteria, cAMP regulates the transcription of numerous genes coding for inducible enzymes involved in the utilization of energy sources other than glucose (Perlman and Pastan, 1968; Ullman and Monod, 1968; Chambers and Zubay, 1969; Pastan and Perlman, 1970). Several types of evidence indicate that this action of cAMP is not mediated by protein phosphorylation. Cyclic AMP, whose level rises in the absence of glucose (Makman and Sutherland, 1965), acts in this system by binding to a protein designated catabolite gene-activating protein (Zubay, Schwartz, and Beckwith, 1970) or cAMP receptor protein (Emmer, de Crombrugghe, Pastan, and Perlman, 1970) which, when combined with cAMP, causes the attachment of RNA polymerase to the promoter sites of the regulated operons. Transcription of the genes then occurs, provided that substrate for the inducible enzymes is also present to cause dissociation of repressor proteins. Martelo, Woo, Reimann, and Davie (1970) found that phosphorylation of *E. coli* RNA polymerase by mammalian cAMP-dependent protein kinase increases the activity of the enzyme in transcribing T4 DNA template, and suggested that phosphorylation of the polymerase might be the basis for the effects of cAMP on bacterial gene transcription. However, stimulation of specific gene transcription by cAMP can be demonstrated in cell-free systems in which DNA, purified RNA polymerase, and purified cAMP receptor protein are the only macromolecules present (de Crombrugghe, Chen, Anderson, Nissely, Gottesman, Pastan, and Perlman, 1971). Under these conditions, the following evidence against the participation of protein phosphorylation in the action of cAMP has been obtained. First, the purified cAMP receptor protein used in these experiments has no detectable protein kinase activity when measured under conditions appropriate for the assay of mammalian protein kinase (Emmer, de Crombrugghe, Pastan, and Perlman, 1970), nor does addition of mammalian protein kinase have any effect on the gene transcription system (Pastan, 1970). Second, when adenylylimidodiphosphate (AMP-P-N-P) (which can substitute for ATP in the RNA polymerase reaction, but cannot donate a gamma phosphate as required in a protein kinase reaction) is used in place of ATP, the effects of cAMP on specific mRNA synthesis are still obtained (de Crombrugghe, Chen, Anderson, Gottesman, Perlman, and Pastan, 1971). However, the validity of this experiment depends on the complete specificity of any protein phosphorylation system present for ATP, since the other triphosphates present are the normal ones. Third, using the antibiotic rifampicin, which blocks initiation steps in RNA synthesis but not the polymerization of nucleoside triphosphates into RNA strands, the effect of cAMP has been shown to be on the formation of an initiation complex between DNA, cAMP receptor protein, and RNA polymerase. The initiation complex can be formed in the absence of all four nucleoside triphosphates upon addition of cAMP; its presence is demonstrated by subsequently measuring RNA synthesis in the presence of rifampicin (de Crombrugghe, Chen, Anderson, Nissely, Gottesman, Pastan, and Perlman, 1971). In all these experiments, the presence of traces of bound ATP in the purified RNA polymerase or cAMP receptor protein preparations might invalidate the conclusions, since only a very small amount of protein phosphorylation might be necessary to give maximal RNA synthesis. However, genetic considerations pointed out by Burgess (1971) argue against this possibility. Certain mutations in the cAMP receptor protein cause the bacteria to lose the ability to synthesize certain inducible enzymes, but not others. This strongly suggests that the cAMP receptor protein interacts with promotor DNA regions (which presumably differ sufficiently to allow a slightly altered mutant receptor protein to bind only to certain promotors). On the other hand, it is hard to imagine, if cAMP receptor protein acts to phosphorylate RNA polymerase, how a mutation in the cAMP receptor protein could result in a selective alteration in the attachment of RNA polymerase to different promotors. The weight of evidence, then, appears large enough to justify the conclusion that protein phosphorylation is not involved in the action of cAMP on specific gene transcription in bacteria.

Cyclic AMP has also been reported to stimulate RNA synthesis in several mammalian systems (Sharma and Talwar, 1970; Wilson and Wright, 1970; Dokas and Kleinsmith, 1971; Adiga, Murthy, and McKenzie, 1971;

Jost and Sahib, 1971; Varrone, Ambesi-Impiombato, and Macchia, 1972). In the latter system, the maximal effect of cAMP is stated to depend on preincubation with cAMP in the absence of added nucleoside triphosphates, suggesting the possibility that cAMP may stimulate RNA synthesis in eukaryotes by a mechanism analogous to that found in bacteria. More extensive studies are necessary to confirm this possibility. A mechanism for stimulation of RNA synthesis in eukaryotes which involves protein phosphorylation has been proposed (Langan, 1968a, 1969a, 1971a). This will be discussed further below.

If cAMP acts in eukaryotes by mechanisms which do not involve protein phosphorylation, then one would expect to find cAMP binding proteins which are not related to the protein kinase system. Two potential candidates for this type of cAMP binding protein are discussed in the section on protein kinase subunits.

IV. ACTIVATION OF PROTEIN KINASE BY cAMP

Studies of the interaction between cAMP and protein kinase have shown that the enzyme contains two dissimilar types of subunits, and have apparently established the general mechanism by which cAMP brings about activation. The mechanism shown in Eq. (1), in which binding of cAMP to a regulatory

$$RC + cAMP \rightleftharpoons R\text{-}cAMP + C \quad (1)$$
(inactive complex) (active form)

subunit (R) causes dissociation of inactive holoenzyme to yield an active free catalytic subunit (C), was proposed by Brostrom, Reimann, Walsh, and Krebs (1970) on the basis of the following properties of cardiac muscle protein kinase preparations. (1) Cyclic AMP was not essential for activation, since aged preparations lost cAMP dependency without loss of total activity; after ageing, cAMP dependency could be restored by sulfhydryl compounds. (2) High dilution activated the enzyme; i.e., the requirement for cAMP decreased as enzyme concentration was lowered, suggesting that activation involves a dissociation process. (3) Cyclic AMP increased V_{max}, but had no effect on the affinity of ATP, casein, or histone for the enzyme, consistent with a mechanism in which cAMP exposes new catalytic sites. (4) Cyclic AMP decreased the stability of the protein kinase to heat and urea, which suggested a conformational change in the enzyme and/or its dissociation into subunits.

Essentially identical mechanisms were proposed independently by Gill and Garren (1970), Tao, Salas, and Lipmann (1970), and Kumon, Yamamura, and Nishizuka (1970) for protein kinases from adrenal cortex, rabbit reticulocytes, and rat liver, respectively, on the basis of more direct evidence indicating that cAMP´ dissociates protein kinase into a cAMP-binding protein and a cAMP-independent protein kinase. Direct evidence for the activation of skeletal muscle protein kinase by this mechanism was provided by Reimann, Brostrom, Corbin, King, and Krebs (1971). It is of interest that two of the above studies were initiated without reference to protein kinase, in attempts to characterize cAMP-binding proteins which might be involved in the action of cAMP in adrenal cortex (Gill and Garren, 1969) and reticulocyte hemolysates (Tao, Salas, and Lipmann, 1970). In each case, the cAMP-binding proteins identified were found to be associated with protein kinase.

Elucidation of the mechanism of cAMP action was aided by the fact that protein kinase preparations from adrenal cortex and liver contained varying amounts of protein with the properties of free regulatory subunits (cAMP-binding activity not associated with kinase activity), and the reticulocyte and liver preparations contained small amounts of protein kinase with the properties of free catalytic subunit (cAMP-independent protein kinase not associated with cAMP-binding activity). Upon addition of free cAMP-binding protein to partially cAMP-dependent protein kinase fractions (Gill and Garren, 1970) or to cAMP-independent fractions (Kumon, Yamamura, and Nishizuka, 1970), inhibition of protein kinase activity, which could be overcome by addition of cAMP, was observed. The reticulocyte cAMP-independent kinase and the adrenal cortex binding protein were shown to sediment at lower velocities than the cAMP-dependent protein kinase, consistent with the idea that they are subunits of the holoenzyme. Upon dissociation of adrenal cortex cAMP-dependent protein kinase with cAMP, the release of a similar, slow-sedimenting binding protein was shown (Gill and Garren, 1970). In reticulocytes, the release of binding protein plus protein kinase cosedimenting with the presumed free catalytic subunit was demonstrated (Tao, Salas, and Lipmann, 1970). In liver, binding and cAMP-independent protein kinase activity were separated by chromatography of cAMP-dependent protein kinase in the presence of cAMP (Kumon, Yamamura, and Nishizuka, 1970). The skeletal muscle cAMP-dependent enzyme was dissociated into catalytic and cAMP-binding subunits by affinity chromatography on casein-sepharose columns in the presence of cAMP, and reconstitution of cAMP-dependent enzyme by recombination of subunits originally derived from purified cAMP-dependent protein kinase was demonstrated (Reimann, Brostrom, Corbin, King, and Krebs, 1971).

Subsequent studies have amply confirmed the activation mechanism shown in Eq. (1), including extensive demonstrations of the reversibility of the reaction, for protein kinases from skeletal muscle (Brostrom, Corbin, King, and Krebs, 1971), heart (Erlichman, Hirsch, and Rosen, 1971), adrenal cortex (Gill and Garren, 1971; Garren, Gill, and Walton, 1971), liver (Kumon, Nishiyama, Yamamura, and Nishizuka, 1972), reticulocyte (Tao, 1972b), and adipose tissue (Corbin, Brostrom, Alexander, and Krebs, 1972). In addition, evidence in keeping with this mechanism has been obtained for cAMP-dependent protein kinases from anterior pituitary (Labrie, Lemarie, and Courte, 1971), testis (Reddi, Ewing, and Williams-Ashman, 1971), pineal gland (Fontana and Lovenberg, 1971), brain (Miyamoto, Petzold, Harris, and Greengard, 1971), and parotid gland (Selinger and Schramm, 1971). Operation of the mechanism *in vivo* in adipose cells is supported by observations that treatment of fat pads with epinephrine increases the relative proportion of a lower molecular weight, cAMP-independent protein kinase in extracts of the cells. In this system the cAMP-dependent and independent activities are separated by Sephadex G-100 chromatography in the presence of high salt, which prevents reassociation of the regulatory and catalytic subunits (Soderling, Corbin, and Park, 1972). It should be noted that in no case has it been established whether the regulatory subunit-catalytic subunit complex is completely inactive, or whether it has an intrinsic catalytic activity, lower than that of the free catalytic subunit.

When the regulatory and catalytic subunits are separated after dissociation of protein kinase with cAMP, the regulatory subunit is obtained in the form of a regulatory subunit-cAMP complex. Addition of this complex to isolated catalytic subunit results in the reconstitution of the cAMP-dependent holoenzyme and the release of free cAMP (Brostrom, Corbin, King, and Krebs, 1971). The binding constant for cAMP is approximately 1×10^{-8}M, as indicated by studies of the interaction of cAMP with holoenzyme (Walsh, Perkins, Brostrom, Ho, and Krebs, 1971; Garren, Gill, and Walton, 1971). However, in the absence of any catalytic subunit, binding of cAMP in the regulatory subunit-cAMP complex appears almost completely irreversible (Brostrom, Corbin, King, and Krebs, 1971; Kumon, Nishiyama, Yamamura, and Nishizuka, 1972). Of a variety of treatments tried, only prolonged dialysis of the skeletal muscle regulatory subunit-cAMP complex against Norite removed any substantial amount of ^{3}H-cAMP, and this resulted in a proportional decrease in the capacity of the preparation to recombine with catalytic subunit, indicating that removal of cAMP was accompanied by denaturation (Brostrom, Corbin, King, and Krebs, 1971). It is possible that isolated skeletal muscle regulatory subunit binds cAMP in a more irreversible manner than regulatory subunits from some other tissues. This is indicated by the fact that unlabeled cAMP does not exchange with ^{3}H-cAMP bound to the skeletal muscle subunit (Brostrom, Corbin, King, and Krebs, 1971), whereas binding of ^{3}H-cAMP to regulatory subunits isolated after dissociation with cold cAMP from liver (Kumon, Yamamura, and Nishizuka, 1970), heart (Erlichman, Hirsch, and Rosen, 1971), and pineal gland enzymes (Fontana and Lovenberg, 1971) does occur. Tao (1972b) has employed cGMP for the isolation of reticulocyte regulatory subunit free of cyclic nucleotide, since the lower affinity of cGMP may allow its removal by dialysis. However, the absence of bound cyclic nucleotide in undenatured regulatory subunit was not directly demonstrated.

An extremely tight binding of cAMP would be consistent with the suggestion of Greengard, Hayaishi, and Colowick (1969) that activation of protein kinase might involve formation of a covalent adenylyl derivative of the enzyme, with cAMP acting as the adenylyl donor by virtue of the high energy of its 3',5'-cyclic phosphoester bond. An analogous adenylylation, with ATP as the adenylyl donor, has been shown to regulate the activity of bacterial glutamine synthetase (Kingdon, Shapiro, and Stadtman, 1967). However, the fact that cAMP can be recovered unchanged from the regulatory subunit-cAMP complex after dissociation in the presence of catalytic subunit (Brostrom, Corbin, King, and Krebs, 1971) or after denaturation by a variety of agents (Tao, Salas, and Lipmann, 1970; Reimann, Walsh, and Krebs, 1971; Kumon, Nishiyama, Yamamura, and Nishizuka, 1972) indicates that adenylylation of protein kinase does not occur, unless, as suggested by Hayaishi, Greengard, and Colowick (1971), the possibility holds true that the adenylylation reaction is readily reversible and reforms cAMP when the regulatory subunit is denatured or combined with catalytic subunit.

There have been several reports of inhibition of protein kinase activity by cAMP. The acellular slime mold *Physarum polycephalum* contains, in addition to a cAMP-dependent protein kinase, a kinase catalyzing the phosphorylation of casein, but not histone, protamine, or phosvitin, which is inhibited 80 to 90% by 10^{-9}M cAMP or cGMP (Kuehn, 1971). This indicates that cAMP may promote inhibition as well as activation of protein phosphorylation in this organism. Inhibition resulting at second hand from cAMP action is, of course, well documented in the case of glycogen synthetase; in the slime mold, cAMP appears to inhibit a protein kinase directly, although mediation of this action by an intervening cAMP-stimulated phosphorylation has not been rigorously ruled out. In mammalian systems, Johnson, Maeno, and Greengard (1971) have briefly noted an inhibition of endogenous phosphorylation of liver cell plasma membranes by cAMP, and Kish and Kleinsmith (1972) have observed variable inhibitory and stimulatory effects of cAMP on protein kinase activity of liver nuclear extracts, utilizing various nuclear proteins as substrates. Finally, Kuo, Krueger, Sanes, and Greengard (1970) have found that partially purified cAMP-dependent protein kinases from a variety of bovine tissues are inhibited by cAMP when Ca^{++} replaces Mg^{++} in the assay mixture, or when flavin mononucleotide is present. The shift from cAMP activation to inhibition caused by Ca^{++} has interesting, although undemonstrated, physiological possibilities. In view of the fact that cAMP destabilizes cAMP-dependent protein kinase to heat and urea treatment (Brostrom, Reimann, Walsh, and Krebs, 1970), it seems important to show that inhibition by cAMP is reversible, since assay conditions under which the free catalytic subunit is unstable might result in expression of less protein kinase activity in the presence of cAMP than in its absence.

An effect of cAMP on the K_m of ATP, but not histone, for purified brain protein kinase has been reported by Miyamoto, Kuo, and Greengard (1969b). Although a double reciprocal plot of activity with respect to ATP concentration is nonlinear in the absence of cAMP, the K_m value is increased 17-fold in the absence of cAMP if data obtained at high ATP concentration are used to calculate K_m. In contrast, the K_m of ATP for a purified skeletal muscle protein kinase (Peak I) is unchanged by cAMP (Reimann, Walsh, and Krebs, 1971), and, in general, studies with other protein kinases have found little or no effect of cAMP on the K_m of ATP. A change in K_m of ATP is not easily accounted for by the activation mechanism discussed above, unless the brain regulatory subunit-catalytic subunit complex has intrinsic catalytic activity with a higher K_m for ATP. Another explanation might be the presence of a second, cAMP-independent histone kinase activity in the preparations, which accounts for most of the activity in the absence of cAMP. The fact that the double reciprocal plot in the absence of cAMP is nonlinear is compatible with this possibility. A purified muscle protein kinase fraction (Peak II), which showed a biphasic reciprocal plot of activity versus ATP concentration in the absence of cAMP, was found to contain a cAMP-independent protein kinase with a high K_m for ATP as a contaminant (Reimann, Walsh, and Krebs, 1971).

In addition to cAMP, the ability of other cyclic nucleotides to activate protein kinases has been tested in many studies. It has been found that the 3'5'-cyclic nucleotides of guanosine, cytidine, uridine, and inosine, as well as thymidine and deoxyadenosine will generally fully activate protein kinases if supplied in sufficiently high concentration (100 to 10,000 times the concentration of cAMP required), with cIMP being the most effective of these. However, high concentrations of cyclic nucleotides are often inhibitory to protein kinases. Cyclic 2',3' nucleotides do not activate at any concentration, nor do they block cAMP activation (Miyamoto, Kuo, and Greengard,

1969b; Schlender, Wei, and Villar-Palasi, 1969; Kuo, Krueger, Sanes, and Greengard, 1970; Gill and Garren, 1970; Reimann, Walsh, and Krebs, 1971; Labrie, Lemaire, and Courte, 1971; Majumder and Turkington, 1971a; Rubin, Erlichman, and Rosen, 1972b; Kumon, Nishiyama, Yamamura, and Nishizuka, 1972). Exceptions to the above are protein kinase fractions from certain arthropods (Kuo and Greengard, 1970a,b; Kuo, Wyatt, and Greengard, 1971) and small amounts of activity in some mammalian tissues (Greengard and Kuo, 1970) which are preferentially activated by cGMP, and are therefore designated as cGMP-dependent protein kinases. As pointed out by Kuo and Greengard (1970a), the existence of protein kinases specifically activated by low concentrations of cGMP suggests that protein phosphorylation may also play a role in the as yet undefined actions of cGMP in cells.

A large number of cAMP analogues have been studied as activators of protein kinase. These are discussed elsewhere in this volume.

V. OTHER FACTORS INFLUENCING PROTEIN KINASE ACTIVITY

Although it seems clear that activation by cAMP constitutes a major mechanism for regulation of cAMP-dependent protein kinase, the activity of the enzyme is also affected in a number of other ways, some of which may well have physiological importance.

A heat-stable protein inhibitor of phosphorylase kinase activation, which interferes with the determination of cAMP in boiled muscle extracts (Posner, Hammermeister, Bratvold, and Krebs, 1964; Posner, Stern, and Krebs, 1965), was first detected and studied by Gonzales (1962). Appleman, Birnbaumer, and Torres (1966) studied the inhibitor further and showed that it also blocked glycogen synthetase inactivation, giving an early indication that a common factor is involved in the action of cAMP on glycogen synthetase and phosphorylase kinase. The inhibitor was eventually purified extensively and shown to act on cAMP-dependent protein kinases by Walsh, Ashby, Gonzales, Calkins, Fischer, and Krebs (1971). It appears to be specific for cyclic nucleotide-dependent protein kinases, having no effect on phosphorylase kinase, phosphofructokinase, hexokinase, pyruvate kinase, and a number of other enzymes (Appleman, Birnbaumer, and Torres, 1966). Preparations of the inhibitor have been powerful tools in determining various actions of cAMP-dependent protein kinase. Evidence for the inactivation of glycogen synthetase and activation of phosphorylase kinase by the same cAMP-dependent protein kinase was provided by demonstrating parallel inhibition of both processes in the presence of varying amounts of purified inhibitor (Soderling, Hickenbottom, Reimann, Hunkeler, Walsh, and Krebs, 1970). The absence of an intervening glycogen synthetase kinase, analogous to phosphorylase kinase, in the action of cAMP-dependent protein kinase on glycogen synthetase was also demonstrated, by showing that the process of inactivation of glycogen synthetase is abruptly and completely halted by addition of the inhibitor (Soderling, Hickenbottom, Reimann, Hunkeler, Walsh, and Krebs, 1970). The role of cAMP-dependent protein kinase in activation of fat cell lipase was demonstrated in one study by using fat cell homogenates which had been titrated with inhibitor just sufficient to block endogenous protein kinase activity. Activation of the lipase by cAMP was then completely dependent on addition of exogenous protein kinase (Corbin, Reimann, Walsh, and Krebs, 1970). Use of the inhibitor also allowed the activation of phosphorylase kinase by autophosphorylation and by cAMP-dependent protein kinase to be experimentally distinguished (Walsh, Perkins, Brostrom, Ho, and Krebs, 1971).

The inhibitor purified from skeletal muscle has a molecular weight of 26,000 as determined by gel exclusion; it is stable to heating at 96°C and to precipitation by trichloroacetic acid, but is inactivated by treatment with trypsin or chymotrypsin. It acts noncompetitively with respect to ATP, protein substrate, and cAMP. It does not bind cAMP but causes a fivefold increase in the affinity of cAMP for protein kinase (Walsh, Ashby, Gonzales, Calkins, Fischer, and Krebs, 1971). The inhibitor is widely distributed in animal tissues, with the highest levels being found in brain and muscle (Ashby and Walsh, 1972a).

The inhibitor has been shown to act by forming a complex with the catalytic subunit of protein kinase. The complex is catalytically inactive and also cannot recombine with the regulatory subunit (Ashby and Walsh, 1972a). The inhibitor behaves in many respects like a regulatory subunit which is not affected by cAMP; for example, the maximum degree of inhibition of the catalytic subunit produced by regulatory subunit and inhibitor is the same. However, the two proteins differ substantially in physical properties such as sedimentation coefficient and heat stability (Ashby and Walsh, 1972a). The possibility that the effects of the inhibitor or similar factors on cyclic nucleotide-dependent protein kinases are more complicated than described by Ashby and Walsh is indicated by the studies of Kuo and Greengard (1971) and Donnelly, Kuo, Miyamoto, and Greengard (1972). These workers found that inhibitor prepared by the procedure of Appleman, Birnbaumer, and Torres (1966) from bovine muscle, bovine brain, and silkmoth larvae, as well as highly purified preparations from lobster tail muscle, stimulated the phosphorylation of certain substrates by cGMP- and cAMP-dependent protein kinases and inhibited the phosphorylation of others. In some cases, using the same substrate, stimulation of cGMP and inhibition of cAMP-dependent kinase was observed. The same effects were obtained with isolated catalytic subunits from the enzymes. These workers have adopted the designation protein kinase modulator for their preparations. They have suggested that the modulator interacts with the catalytic subunit of cyclic nucleotide-dependent protein kinases and that it may function to modify the substrate specificity of the enzymes. Although direct evidence for a physiological role for protein kinase inhibitor or modulator is lacking at present, the probability of such a role is indicated by the finding that the tissue concentration of the inhibitor is altered *in vivo* by starvation, high-protein or high-carbohydrate diet, insulin, and epinephrine (Ashby and Walsh, 1972b).

Protein kinase activity has been observed to be variously affected by magnesium and ATP, in addition to their utilization in the actual phosphorylation reaction. Huijing and Larner (1966a,b) reported that inactivation of partially purified glycogen synthetase preparations by associated glycogen synthetase kinase activity [later shown to be identical with cAMP-dependent protein kinase (Schlender, Wei, and Villar-Palasi, 1969; Soderling, Hickenbottom, Reimann, Hunkeler, Walsh, and Krebs, 1970; Villar-Palasi, Larner, and Shen, 1971)] was highly dependent on cAMP at 1 mM magnesium and essentially independent of cAMP at 10 mM magnesium and higher. The data indicated that cAMP increased the affinity of the enzyme for magnesium. Similar effects on glycogen synthetase kinase activity were obtained with more highly purified protein kinase preparations (Villar-Palasi and Schlender, 1970). A loss of cyclic AMP dependency at very high magnesium concentrations (100 mM) was observed for the activation of phosphorylase kinase (Krebs, Love, Bratvold, Trayser, Meyer, and Fischer, 1964). However, with purified protein kinase, cAMP dependence for casein phosphorylation was unchanged over the range of 1 to 10 mM magnesium (Reimann, Walsh, and Krebs, 1971). Whether interactions with magnesium affect the dependence of protein kinase on cAMP *in vivo* is not clear at present.

Haddox, Newton, Hartle, and Goldberg (1972) observed a decrease in the affinity of cAMP for skeletal muscle protein kinase caused by prior incubation with ATP and magnesium, as measured by binding of cAMP or activation of histone phosphorylation. ATP or magnesium alone had no effect. The dissociation constant of cAMP in the binding reaction was changed from 2×10^{-8} to 2×10^{-7}M and the K_m of cAMP for kinase activation from 3×10^{-9} to 5×10^{-8}M. The effect was associated with binding of ^{3}H-ATP to the enzyme, with a dissociation constant of 1.3×10^{-7}M. Unlabeled ATP, but not cAMP, exchanged with bound ^{3}H-ATP, indicating that separate binding sites exist for the cyclic nucleotide and the triphosphate. The binding site involved in this effect also appears different from the site for binding substrate ATP, since the K_m of ATP as a substrate for the enzyme is close to 10^{-5} M (Reimann, Walsh, and Krebs, 1971). These observations appear related to the finding of Brostrom, Corbin, King, and Krebs (1971) that the rate of reassociation of catalytic subunit with regulatory subunit-cAMP complex, as measured by the release of free cAMP, is increased by the presence of ATP and magnesium. The possibility that these effects are important in regulating the sensitivity of protein kinase to cAMP *in vivo* seems remote, in view of the fact that tissue concentrations of ATP are approximately 4,000 times higher than the concentration range over which differences in cAMP affinity are produced.

Observations of changes in the relative amounts of cAMP-dependent and independent glycogen synthetase kinase activity in crude extracts of rat skeletal muscle and rat diaphragm, made after treatment of rats or intact diaphragms with insulin or epinephrine, have been reported by Villar-Palasi and Wenger (1967), Shen, Villar-Palasi, and Larner (1970) and Villar-Palasi, Larner, and Shen (1971). Insulin treatment increased and epinephrine treatment decreased cAMP dependency, and these changes were found to be stable during gel filtration of the extracts to remove small molecules. In contrast, activation of purified protein kinase by cAMP could be reversed by gel filtration. Whether the hormonally induced changes in cAMP sensitivity of the glycogen synthetase kinase activity reflect some mechanism other than dissociation and association of the regulatory subunit of protein kinase, as discussed in the previous section, is not entirely clear at present. Loss of cAMP dependency of glycogen synthetase kinase activity was also observed on storage of the muscle extracts at 5°C, and this process was accelerated by fluoride. No *in vitro* conversion of cAMP-independent to dependent activity was observed (Villar-Palasi, Larner, and Shen, 1971). As noted earlier, Brostrom, Reimann, Walsh, and Krebs (1970) also observed loss of cAMP dependency of cardiac muscle protein kinase on ageing; dependency could be restored by sulfhydryl compounds.

Activation of protein kinase by preincubation with substrates has been observed in several laboratories. Reimann, Walsh, and Krebs (1971) found that preincubation of a fraction (Peak I) of skeletal muscle protein kinase with casein at pH 6.0 or lower increased activity measured in the absence of cAMP, but not in its presence. Preincubation with histone resulted in a complete loss of the requirement for cAMP at all pH values. These effects were not obtained with another skeletal muscle protein kinase fraction (Peak II). Miyamoto, Petzold, Harris, and Greengard (1971) found that brain protein kinase is partially activated (twofold as compared to eightfold by cAMP) by preincubation with histone. Protamine and poly-L-lysine were also effective. Sucrose density-gradient centrifugation following preincubation with histone revealed the probable mechanism of histone activation. Normally, the enzyme sediments in two peaks of estimated molecular weight 140,000 and 80,000, both of which are cAMP dependent. After preincubation with histone, a cAMP-independent peak of molecular weight 40,000 is also observed. Cyclic AMP, under favorable conditions, also causes the protein kinase activity to sediment in a position corresponding to molecular weight 40,000. The results strongly suggest that histone can activate protein kinase by causing dissociation of some of the regulatory subunits in the absence of cAMP. The regulatory subunits of protein kinases are somewhat acidic (Chen and Walsh, 1971; Krebs, 1972), favoring interaction with the highly basic histones. The interaction of brain protein kinase with histones is complicated by the fact that histones also cause a shift of the 140,000 molecular weight form of the enzyme to 80,000, without a change in cAMP dependency (Miyamoto, Petzold, Harris, and Greengard, 1971). Tao (1972a) subsequently has shown that preincubation of reticulocyte protein kinase with protamine results in a complete conversion of this enzyme to a slower sedimenting, cAMP-independent form corresponding to that produced by incubation with cAMP. After interaction of the enzyme with protamine, binding protein was detected sedimenting more rapidly than the holoenzyme, indicating more directly the formation of a complex or aggregate with protamine. In contrast to the brain enzyme, histones do not activate reticulocyte protein kinase, and in general the susceptibility of protein kinases from different sources to this type of activation seems to vary considerably, probably due to variations in charge and in the affinity of the regulatory subunits for catalytic subunits. Activation by this mechanism undoubtedly contributes to the variations in degree of cAMP stimulation with different substrates which have been observed in many laboratories. It remains to be determined if substrate activation of protein kinase is important *in vivo*. Complexing of histones to DNA might prevent their interaction with the regulatory subunit. The phosphorylation of liver histone *in vivo* is highly dependent on administration of cAMP or glucagon (Langan, 1969*a,b*).

In addition to regulation of the activity of protein kinase by interactions with various molecules, the possibility of regulation by control of the synthesis of protein kinase and its regulatory and catalytic subunits has been investigated in a number of studies. In synchronized Chinese hamster ovary cells (Shepherd, Noland, and Hardin, 1971) and in Chang's liver cells (Makman and Klein, 1972), no changes in the levels of protein kinase throughout the cell cycle were found. In the latter study, both cAMP-dependent and independent activities were present in the same proportion during synchronous growth, indicating that the catalytic and regulatory subunits of protein kinase are coordinately synthesized at all stages of the cycle. Protein kinase activity was also similar in cells from suspension and stationary cultures, in cells grown in two different modifications of Eagle's medium, and in cells grown in the presence of horse serum and calf plus fetal calf serum (Klein and Makman, 1972). In a line of cultured rat hepatoma cells (HTC cells), however, Granner (1972) found large differences in the relative amounts of regulatory and catalytic subunits as compared to normal rat liver. Measurements of the specific activity and cAMP dependence of partially purified protein kinase fractions showed that although cAMP-independent activity is fourfold higher in HTC cells than in liver, the stimulation produced by cAMP is less than threefold, compared to eight- to 10-fold in the liver preparations. Cyclic AMP-independent activity in the HTC cell preparation was inhibited by rat liver cAMP-binding protein, and the amount of binding protein in liver was 10-fold higher than in HTC cells, confirming the interpretation that protein kinase in these cells is partially activated due to a deficiency of regulatory subunit. Activation by this means is of interest in view of the finding that HTC cells have very low levels of adenyl cyclase and cAMP (Granner, Chase, Aurbach, and Tomkins, 1968) and respond poorly to exogenous cAMP (Stellwagen, 1972). Deficiency of regulatory subunit may depend on the clone of hepatoma cells, since a larger stimulation of HTC cell protein kinase by cAMP has been reported (Klein and Makman, 1972).

The level of cAMP-dependent protein kinase in mouse mammary epithelial cells increases seven- to ninefold during mammary gland development in pregnancy, or during insulin- and prolactin-induced cell division and differentiation in cultured explants (Majumder and Turkington, 1971b). Interestingly, prolactin alone induces only the cAMP-binding protein. Hormonal treatment of the mammary gland results in the induction of casein synthesis. Since exogenous cAMP does not induce casein synthesis even though taken up in large amounts, Majumder and Turkington suggest that hormonal induction of mammary gland differentiation may result from the induction of proteins which interact with cAMP rather than from alterations in cAMP concentration, which apparently is not the rate-limiting factor.

In developing rat brain, protein kinase activity increases approximately twofold from the fetal to the adult level between the 3rd and 21st day after birth. Cyclic AMP dependency is higher in fetal than in adult brain, with the major change occurring between 3 and 6 days after birth, suggesting differential rates of synthesis of catalytic and regulatory subunits (Gaballah, Popoff, and Sooknandan, 1971).

VI. MULTIPLE FORMS AND SUBUNIT STRUCTURE

Efforts to purify protein kinases from a variety of sources have led to the finding that activity can often be resolved into two or more components. Both cAMP-dependent and independent components have been separated. As noted earlier, in some cases the cAMP-independent component appears to be the free catalytic subunit of cAMP-dependent protein kinase (Kumon, Yamamura, and Nishizuka, 1970; Yamamura, Takeda, Kumon, and Nishizuka, 1970; Tao, Salas, and Lipmann, 1970; Chen and Walsh, 1971; Labrie, Lemaire, and Courte, 1971). In other cases, the relationship between the cAMP-independent and dependent activities is less clear (Villar-Palasi, Larner, and Shen, 1971; Majumder and Turkington, 1971a; Waddy and MacKinlay, 1971). Also, it was noted earlier that cells contain abundant protein kinase activity which acts preferentially on casein, phosvitin, and similar casein-like phosphoproteins which are generally present in tissues. These phosphoprotein kinases are distinct from cAMP-dependent protein kinases. Nevertheless, in one study (Yamamura, Kumon, Nishiyama, Takeda, and Nishizuka, 1971), a phosphoprotein kinase, acting preferentially on casein and phosvitin, was detected as a peak of cAMP-independent histone kinase activity, due perhaps to some inherent ability to phosphorylate histones or perhaps to phosphoprotein contamination in the histone substrate. The situation is further complicated by the fact that casein is a reasonably good substrate for some cAMP-dependent protein kinases (Walsh, Perkins, and Krebs, 1968) and a poor substrate for others (Langan, 1969b). Also, an apparently specific cAMP-independent histone kinase exists which is unrelated to cAMP-dependent protein kinase, as shown by its specificity for a distinct site in lysine-rich (f1) histone (Langan, 1971a, b), and by the failure of purified regulatory subunit to convert it to a cAMP-dependent form (D. A. Walsh and T. A. Langan, 1971, *unpublished results*). At present, the latter test appears to be the most useful one for establishing the relationship of a cAMP-independent protein kinase component to cAMP-dependent protein kinase.

Multiple forms of cAMP-dependent protein kinase were first demonstrated in skeletal muscle extracts by ion-exchange chromatography, which resolves two fractions (Reimann and Walsh, 1970; Villar-Palasi and Schlender, 1970). Each fraction has the same relative activity toward phosphorylase kinase, glycogen synthetase, histone, and casein; further, the heat lability of the activity toward glycogen synthetase and phosphorylase kinase is the same for the two protein kinase fractions, both in the presence and absence of cAMP (Soderling, Hickenbottom, Reimann,

Hunkeler, Walsh, and Krebs, 1970). Resolution of two cAMP-dependent protein kinase fractions by ion-exchange chromatography has subsequently been achieved with preparations from reticulocyte (Tao, Salas, and Lipmann, 1970), adrenal cortex (Gill and Garren, 1971), liver (Chen and Walsh, 1971; Eil and Wool, 1971; Yamamura, Kumon, Nishiyama, Takeda, and Nishizuka, 1971), brain (Miyamoto, Petzold, Harris, and Greengard, 1971), and lacrimal gland (Takats, Farago, and Antoni, 1972). Molecular sizing techniques such as gel filtration and gel electrophoresis resolve up to three components of cardiac muscle cAMP-dependent protein kinase (Rubin, Erlichman, and Rosen, 1972a), and sucrose density-gradient centrifugation separates two cAMP-dependent activities in anterior pituitary (Labrie, Lemaire, and Courte, 1971). Density-gradient centrifugation also further separates the activity in the first DEAE peak from skeletal muscle and brain into a heavy and a light cAMP-dependent protein kinase (Reimann, Walsh, and Krebs, 1971; Miyamoto, Petzold, Harris, and Greengard, 1971) and additional components in liver are revealed by electrofocusing (Chen and Walsh, 1971) and gel filtration (Kumon, Nishiyama, Yamamura, and Nishizuka, 1972). A total of six skeletal muscle protein kinase fractions have been obtained by Zapf and Froesch (1972) by stepwise elution from DEAE-cellulose with a series of five phosphate buffers of increasing concentration. However, the use of a stepwise elution procedure may tend to produce peaks of activity which do not contain unique components.

The relationship between the multiple forms of cAMP-dependent protein kinase which have been obtained from single tissues is not established, but a good deal of evidence suggests that many of them are produced by dissociation or degradation of protein kinases after extraction from the cell. In skeletal muscle, for example, there is evidence that very fresh extracts contain a single cAMP-dependent protein kinase, corresponding in sedimentation behavior to the heavier (6.8S) component present in purified preparations (Krebs, 1972; Walsh and Krebs, 1973). Freshly prepared cardiac muscle protein kinase behaves as a single component of molecular weight 280,000 on gel filtration, but after storage for several weeks at 4°C, two additional components of apparent molecular weight 140,000 and 90,000 are formed at the expense of the original component. Conversion to the 140,000 and 90,000 molecular weight forms is also caused by subjecting fresh preparations to centrifugation in the analytical untracentrifuge, whereas gel electrophoresis of the same preparations reveals a single component of molecular weight 140,000 (Rubin, Erlichman, and Rosen, 1972a). The brain enzyme is converted from the 140,000 to the 80,000 molecular weight form by interaction with histone, without loss of cAMP-dependency (Miyamoto, Petzold, Harris, and Greengard, 1971); in liver, production of a component of molecular weight 120,000 is prevented by the use of diisopropyl fluorophosphate. However, resolution of liver protein kinase components with molecular weights ranging from 100,000 to more than 200,000 is not affected by diisopropyl fluorophosphate, indicating that if proteases are involved in producing these components, they are not of the type inhibited by organofluorophosphate compounds (Kumon, Nishiyama, Yamamura, and Nishizuka, 1972).

Regardless of the origin of the multiple forms of cAMP-dependent protein kinase which have been detected, all have shown a close similarity with respect to properties that might be expected to affect biological function, such as affinity for cAMP and substrate specificity. In addition, by and large the cAMP-dependent protein kinases from different tissues and organisms are very similar, especially for an enzyme which is known to act on the components of processes as diverse as glycogen and triglyceride metabolism, and which is thought to mediate a very wide variety of cell responses. The similarity of protein kinases has been noted repeatedly (Kuo, Krueger, Sanes, and Greengard, 1970; Krebs, 1972; Yamamura, Inoue, Shimomura, and Nishizuka, 1972; Walsh and Krebs, 1973), and is supported by demonstrations of the action of protein kinases from one tissue on substrates from another, and by the reconstitution of cAMP-dependent protein kinases using catalytic and regulatory subunits from different tissues (Corbin, Reimann, Walsh, and Krebs, 1970; Huttunen, Steinberg, and Mayer, 1970a,b; Yamamura, Kumon, and Nishizuka, 1971; Yamamura, Inoue, Shimomura, and Nishizuka, 1972; Corbin, Brostrom, Alexander, and Krebs, 1972). Similarity among cAMP-dependent protein kinases even extends to those forms which are tightly bound to subcellular components, as far as can be told from studies in which these activities were examined in some detail (Maeno, Johnson, and Greengard, 1971; Lemaire, Pelletier and Labrie, 1971; Rubin, Erlichman, and Rosen, 1972b). Thus, although one might easily have imagined the existence of a multiplicity of cAMP-dependent protein kinases, each specific for a single protein substrate and perhaps responding to different concentrations of cAMP, the general picture that has emerged experimentally is that protein kinases from various sources do not show any large differences of obvious functional significance.

Cyclic AMP-dependent protein kinases from most tissues so far investigated exist in the form of components of molecular weight 140,000 to 160,000, as estimated from sedimentation constants of 6.8 to 7S or by other techniques; in some cases, only sedimentation constants have been reported (Tao, Salas, and Lipmann, 1970; Reimann, Walsh, and Krebs, 1971; Gill and Garren, 1971; Chen and Walsh, 1971; Miyamoto, Petzold, Harris, and Greengard, 1971; Labrie, Lemaire, and Courte, 1971; Reddi, Ewing, and Williams-Ashman, 1971; Corbin, Brostrom, Alexander, and Krebs, 1972). Fresh cardiac muscle enzyme (Rubin, Erlichman, and Rosen, 1972a), as noted above, appears to be approximately twice this size when examined by gel filtration, but shows a molecular weight of 140,000 in the gel electrophoresis system of Hedrich and Smith (1968). Kumon, Nishiyama, Yamamura, and Nishizuka (1972) estimate a value of 180,000 for a major component of liver cAMP-dependent protein kinase. There are also heterodisperse smaller and larger species in certain liver preparations, and discrete smaller components in skeletal muscle, cardiac muscle, and brain preparations, which may be due to dissociation or degradation, as discussed above. An unusually small (3.2S) protein kinase is present as the only cAMP-dependent form in mammary gland extracts (Majumder and Turkington, 1971a).

Molecular weights of catalytic subunits from brain (Miyamoto, Petzold, Harris, and Greengard, 1971), anterior pituitary (Labrie, Lemaire, and Courte, 1971), and adipose tissue (Corbin, Brostrom, Alexander, and Krebs, 1972) have been estimated at 40,000 to 45,000, whereas values of 60,000 have been obtained for reticulocyte (Tao, Salas, and Lipmann, 1970) and adrenal cortex (Gill and Garren, 1971) catalytic subunits. The regulatory subunits from the latter two tissues have also been examined and estimated to have molecular weights of 80,000 and 92,000, respectively, suggesting that the holoenzymes are composed of one regulatory and one catalytic subunit.

Detailed analyses of the subunits of the multiple forms of cAMP-dependent protein kinase found in skeletal muscle, liver, and cardiac muscle preparations have been carried out. Reimann, Walsh, and Krebs (1971) and Krebs (1972) have reported sedimentation values of 5.4, 5.2, and 5.0S for three lighter forms of skeletal muscle cAMP-dependent protein kinase. Each of these enzymes, as well as the heavier 6.8S form, contains a catalytic subunit of 4.1S. The regulatory subunits of these four forms are more variable, with sedimentation constants ranging from 3.4 to 4.9S; the 4.9S form converts into one of the lighter forms *in vitro* (Corbin and Brostrom, 1971; Krebs, 1972).

Cyclic AMP-dependent protein kinase from rat liver is resolved into two 6.8S components by DEAE-Sephadex chromatography. Isoelectric focusing of one of these in the presence of cAMP separates two catalytic subunits of isoelectric point 7.6 and 8.9, both of which have a sedimentation constant of 4.0S. The substantial difference in isoelectric points of components with the same sedimentation constant suggests that they represent true multiple forms of protein kinase rather than interconversion products. Whether both are derived from liver parenchymal cells is not known. A third catalytic subunit of isoelectric point 8.6 separated in the chromatographic step is probably not identical with either of these (Chen and Walsh, 1971). Kumon, Nishiyama, Yamamura, and Nishizuka (1972) have obtained evidence that the regulatory subunits of two forms of rat liver cAMP-dependent protein kinase may also differ. A regulatory subunit of 150,000 molecular weight, as estimated by gel filtration, is dissociated by cAMP from the 180,000 molecular weight form of the enzyme. One ageing at 4°C, this decomposes to a smaller protein of molecular weight 40,000. Regulatory subunit of the heterodisperse form of the protein kinase (molecular weight from 100,000 to greater than 200,000) was obtained only in a form of molecular weight 40,000, apparently due to instability of the dissociated subunit. The molecular weight of the catalytic subunit of all forms of the rat liver enzyme was estimated by these workers as 30,000.

All three forms of cardiac muscle protein kinase (molecular weights 280,000, 140,000, and 90,000) contain catalytic and regulatory subunits of molecular weight 42,000 and 55,000, respectively, as determined by sodium dodecylsulfate (SDS)-acrylamide gel electrophoresis. Optical density scans of stained gels suggested a 2:1 molar ratio of catalytic to regulatory subunits. Acrylamide gel electrophoresis in the absence of SDS showed two distinct catalytic subunit bands, which may have resulted from aggregation or limited proteolytic degradation (Rubin, Erlichman, and Rosen, 1972a).

Aggregation of protein kinase subunits following dissociation by cAMP appears to occur in a number of systems. Gill and Garren (1971) observed that one form of adrenal cortex catalytic subunit behaves like a dimer after dissociation by cAMP. Reddi, Ewing, and Williams-Ashman (1971) found that inclusion of cAMP in a sucrose gradient causes mammalian testis catalytic activity to sediment faster than the original cAMP-dependent enzyme. Aggregation of subunits to a size approximately the same as the original enzyme might explain a number of observations in which the sedimentation of catalytic (Gill and Garren, 1970; Tao, Salas, and Lipmann, 1970; Majumder and Turkington, 1971a) or binding (Fontana and Lovenberg, 1971) activity is unchanged by cAMP.

In rat liver extracts, Chambaut, Leray, and Nanoune (1971) have reported detecting a cAMP-binding protein, devoid of kinase activity, which sediments faster than cAMP-dependent protein kinase. It is possible that this represents a cAMP receptor unrelated to the protein kinase system, or a protein involved in regulating the intracellular concentration of free cAMP. However, the heavy cAMP-binding fraction was not tested for regulatory subunit activity in combination with catalytic subunit, and the possibility exists that it is an aggregated form of the regulatory subunit.

Sy and Richter (1972a) have separated a protein from several yeast strains which binds cAMP with a dissociation constant of 5×10^{-9}M. No protein kinase activity was detected in the preparations, using phosvitin as a substrate, either in the presence or absence of cAMP. The binding protein also has no effect on the activity of a yeast protein kinase (Rabinowitz and Lipmann, 1960), which itself is unaffected by cAMP. The function of the binding protein is not known, but it seems possible that it is involved in actions of cAMP which are not mediated by protein phosphorylation. Yeast contains adenyl cyclase and cAMP, and the concentration of the latter varies with growth conditions (Sy and Richter, 1972b).

VII. SUBCELLULAR DISTRIBUTION

Substantial quantities of cAMP-dependent protein kinase appear to be present in the cytosol of all cells in which the enzyme has been found. All highly purified preparations are evidently derived from this source, since none of the procedures employs steps which would solubilize firmly bound particulate activity. In homogenates of skeletal muscle (Walsh, Perkins, and Krebs, 1968), liver (Chen and Walsh, 1971), and mammary gland (Majumder and Turkington, 1971a), the enzyme is found almost entirely (greater than 90%) in the cytosol. In other tissues which have been examined, large amounts of protein kinase are present in both soluble and particulate forms. In frog bladder epithelial cells, 30% of the activity is present in a $100,000 \times g$ pellet (Jard and Bastide, 1970); in human erythrocytes, more than 70% of the activity is bound to the cell membrane (Rubin, Erlichman, and Rosen, 1972b). In adrenal cortex, the specific activity of protein kinase is higher in extensively washed microsomal fractions than in the soluble fraction. The highest specific activity is in the smooth membrane fraction of the endoplasmic reticulum, with little being found in free ribosomes (Walton, Gill, Abrass, and Garren, 1971).

Extensive studies of the subcellular distribution of protein kinase in brain and anterior pituitary have been carried out. In brain, 42% of the total activity was found in the cytosol. The particulate fractions, however, contained a large amount of latent activity which was expressed after treatment with Triton X-100, so that 78% of total activity was associated with particulate fractions when latent activity was accounted for. Some of this, however, may have been due to cytosol trapped in pinched-off vesicles. The microsomal and mitochondrial fractions, both of which are very complex in brain, contained the bulk of the particulate activity. The crude mitochondrial fraction was fractionated further by sucrose density-gradient procedures, and protein kinase activity was found to be high in fractions containing synaptic membranes and nerve endings and low in purified mitochondria (Maeno, Johnson, and Greengard, 1971). Association of protein kinase with nerve endings and synaptic membrane components of brain mitochondrial fractions was also reported by Gaballah and Popoff (1971). An early observation of cAMP-dependent protein kinase in brain microsome and synaptosome membrane fractions, detected by phosphorylation of endogenous protein substrates, was made by Weller and Rodnight (1970). The finding by these workers that protein kinase activity of the soluble fraction of brain is not stimulated by cAMP was probably due to the use of casein and phosvitin as substrates, since brain contains large amounts of a phosphoprotein kinase acting on these substrates (Rabinowitz and Lipmann, 1960), while the cAMP-dependent protein kinase is poorly active with casein and phosvitin (Miyamoto, Kuo, and Greengard, 1969a,b).

In anterior pituitary, 50% of the activity was found in the soluble fraction, with most of the remaining activity in a crude mitochondrial fraction. Treatment with Triton X-100, however, showed that substantial latent activity was present in the microsomal fraction. The specific activities of rough and smooth membranes and free ribosomes of this tissue were about equal, with most of the latent activity being in the smooth membranes (Lemaire, Pelletier, and Labrie, 1971). The activity present in the mitochondrial fraction may be due to contamination with adenohypophyseal secretory granules, which contain cAMP-dependent protein kinase (Labrie, Lemaire, Poirier, Pelletier, and Boucher, 1971). Lemaire, Pelletier, and Labrie (1971) also noted a low degree of activation of protein kinase activity in particulate fractions when phosphorylation of endogenous substrate was measured. They suggested that cAMP might act mainly to dissociate soluble protein kinase, allowing free catalytic subunit to bind to substrates in particulate fractions. However, the particulate fractions showed much greater cAMP-dependence when exogenous histone substrate was supplied. In view of this, the possibility should also be considered that in the absence of an excess of substrate, much of the particulate protein phosphorylation is catalyzed by enzymes unrelated to cAMP-dependent protein kinase. In line with this alternative, direct measurements of cAMP binding have shown that the highest binding activity in adrenal cortex is in a smooth membrane fraction which is also high in cAMP-dependent protein kinase activity (Walton, Gill, Abrass, and Garren, 1971). Also, in erythrocyte membranes, cAMP-binding activity is more tightly membrane bound than the catalytic subunit of cAMP-dependent protein kinase (Rubin, Erlichman, and Rosen, 1972b).

The presence of cAMP-dependent protein kinase in varying amounts in all subcellular fractions, with the possible exception of mitochondria, is in keeping with the potential role of the enzyme in mediating multiple effects of cAMP in cells. The association of large amounts of protein kinase with certain particulate components strongly suggests that protein substrates are localized in those components, and some evidence in support of this is presented in the next section. However, it is evident that large amounts of protein kinase need not be associated with all structures containing substrate, especially if the amount of protein phosphorylation involved is small relative to the total cAMP-dependent protein phosphorylation in the cell. Nuclei, for example, have been reported to contain only 6 to 8% of total cell cAMP-dependent protein kinase in brain, liver, and anterior pituitary (Maeno, Johnson, and Greengard, 1971; Chen and Walsh, 1971; Lemaire, Pelletier, and Labrie, 1971), an amount not greatly different from the nuclear contribution to total cell protein.[2] Nevertheless, histones are among the best substrates for the enzyme *in vitro,* and the phosphorylation of lysine-rich histone by cAMP-dependent protein kinase has been demonstrated *in vivo* (Langan, 1969a,b). The amount of liver lysine-rich histone which is phosphorylated in response to hormones or cAMP is quite small, however, amounting to approximately 1% of total lysine-rich histone (Langan, 1969a). Although alteration of the function of the DNA associated with this amount of histone could have very significant and selective effects on the cell, a large amount of protein kinase would not be required. The studies of Siebert and Humphrey (1965) have indicated that enzymes of the cytosol are not excluded from the nucleoplasm, so that histones may be accessible to soluble protein kinase, or a small fraction of the total protein kinase might be firmly associated with nuclear structures. Protein (histone) kinase activity has been reported in nuclei prepared in nonaqueous media (Siebert, Ord, and Stocken, 1971) and in extensively washed deoxynucleoprotein (chromatin) preparations (Burdon and Pearce, 1971), but this activity is only slightly stimulated by cAMP. The presence of cAMP-binding protein in nuclei has, however, been observed (Pierre and Loeb, 1971). In order to determine if cAMP-dependent protein kinase activity bound to nuclei might be masked by the presence of histone kinases unrelated to cAMP-dependent kinase [for example, histone kinase HK_2 (Langan, 1971a,b)], we have examined the phosphorylation of specific sites in lysine-rich histone by extensively washed chromatin preparations from liver and other tissues. Phosphorylation by these preparations is stimulated only about 20% by cAMP when total phosphate incorporation into lysine-rich

---

[2] The large difference between the specific activity of protein kinase in liver cytosol and nuclear fractions given by Chen and Walsh (1971) is in error. The correct values are 4,300 and 2.000 units per mg protein in the cytosol and nuclear fractions, respectively (L. J. Chen and D. A. Walsh, 1972, *personal communication*).

histone is measured. However, when phosphorylation at the specific site in lysine-rich histone acted upon by cAMP-dependent protein kinase (Langan, 1969a) is measured, phosphorylation is found to be stimulated two- to threefold by cAMP (Langan, 1972, unpublished results). It is apparent that cAMP-dependent protein kinase activity in washed chromatin preparations is masked by the presence of cAMP-independent histone kinase acting at other phosphorylation sites. As in any study of subcellular distribution of enzyme activity, it is difficult to rule out the possibility that activity in particulate fractions is due to binding of enzyme after homogenization of the tissue. But in view of this and other evidence that nuclear fractions contain protein kinase (Maeno, Johnson, and Greengard, 1971; Chen and Walsh, 1971, and personal communication, 1972; Lemaire, Pelletier, and Labrie, 1971), and the evidence that lysine-rich histone is phosphorylated *in vivo* (Langan, 1969a,b), it seems clear that nuclear proteins are accessible as substrates to cAMP-dependent protein kinase, either bound to nuclear structures or in soluble form in the nucleoplasm.

VIII. SUBSTRATES AND SUBSTRATE SPECIFICITY OF cAMP-DEPENDENT PROTEIN KINASE

The likelihood that a large number of hormone responses are mediated by cAMP-dependent protein kinases suggests that searches for substrates of this enzyme should be highly fruitful in elucidating pathways of hormone action. However, this approach has met with serious difficulties, due in large part to the apparent low specificity of the kinases for protein substrates. Using labeled ATP of high specific activity and sufficient purified enzyme, it is possible to detect the introduction of at least small amounts of phosphate into a large number of protein preparations. Consequently, the substrate specificity of cAMP-dependent protein kinase has been described as broad or nonspecific. However, it is the ability to phosphorylate a variety of seemingly dissimilar proteins which allows the enzyme to play a central role in transmitting hormonal signals, and it seems clear that regulatory functions cannot be carried out by an enzyme which catalyzes nonspecific phosphorylation *in vivo*. Cyclic AMP-dependent phosphorylation of lysine-rich histone *in vivo* in rat liver is quite specific for a particular serine residue (Langan, 1969a), and the same specificity is shown *in vitro* by calf liver protein kinase preparations (Langan, 1969b), as well as by the enzymes from a number of other mammalian tissues tested (Langan, 1972, *unpublished results*). The reported sequence surrounding the serine residue phosphorylated in glycogen synthetase (Larner and Sanger, 1965) is different from that found in lysine-rich histone, so the enzyme does not appear to recognize primary structure in the protein substrate. Recognition of a specific three-dimensional configuration is a more likely possibility. This might also explain the apparent lack of specificity *in vitro*, since the flexibility of proteins might allow large amounts of enzyme to force regions of a protein into a resemblance of the required configuration, especially if the substrate contains a certain fraction of partially denatured molecules.

In view of the possibility that phosphate may be introduced *in vitro* into proteins which are not physiological substrates for protein kinase, it seems important, when attempting to determine whether a particular protein is a substrate, to obtain evidence other than simple incorporation of ^{32}P from labeled ATP into the protein preparation. In particular, stoichiometry should be determined if purified substrate is available, since there seems to be no reason why values approaching one mole of phosphate per mole of protein or greater should not be obtained for *bona fide* substrates. Determining whether phosphorylation takes place at a specific site would also be valuable since, normally, specific functional modifications would be expected to require specific structural changes. However, phosphorylase kinase appears to undergo phosphorylation at a large number of sites (Riley, DeLange, Bratvold, and Krebs, 1968), so that lack of site specific phosphorylation cannot be taken as conclusive evidence that the protein is a nonphysiological substrate.

Although testing of proteins for substrate activity with cAMP-dependent protein kinase will probably provide an important approach for the identification of the components involved in various hormone responses, it is apparent that demonstration of substrate activity *in vitro* is only a tentative indication of function in hormonal regulation. More definitive evidence, such as demonstrations of altered activity *in vitro* and *in vivo*, as outlined by Krebs (1973; Table 1) is required before participation of the protein in a particular hormone response can be considered seriously.

In the case of phosphorylase kinase, phosphorylation accompanied by activation was reasonably well established as an event in the hormonal regulation of glycogen metabolism before a separate, cAMP-dependent protein kinase catalyzing this reaction was shown to exist (Walsh, Perkins, and Krebs, 1968). Discovery of cAMP-dependent protein kinase was hindered by the fact that, especially in the presence of high concentrations of ATP and magnesium, phosphorylase kinase catalyzes an autophosphorylation reaction which also results in its activation, and by the persistence of traces of protein kinase in highly purified phosphorylase kinase preparations which are sufficient to catalyze a readily observable cAMP-dependent phosphorylation and activation (DeLange, Kemp, Riley, Cooper, and Krebs, 1968; Walsh, Perkins, Brostrom, Ho, and Krebs, 1971). Phosphorylation of phosphorylase kinase in the presence of high concentrations of ATP (3 mM) and cAMP results in the incorporation of up to 3 moles of phosphate per 10^5 grams of protein. However, cAMP stimulation of this phosphorylation is poorly correlated with activation, the increase in phosphorylase kinase activity being considerably greater than the increase in phosphorylation (DeLange, Kemp, Riley, Cooper, and Krebs, 1968). In some experiments, phosphorylation continued to increase after activity reached a maximum. It is under these conditions that phosphorylation of a large number of sites in phosphorylase kinase has been demonstrated, although phosphorylation of a few of these sites is more closely correlated with activation than phosphorylation of the remainder, and the same few sites are preferentially dephosphorylated during inactivation by phosphorylase kinase phosphatase (Riley, DeLange, Bratvold, and Krebs, 1968). In more recent experiments, utilizing lower ATP concentrations (0.1 to 0.2 mM) which effectively eliminate the contribution of the phosphorylase kinase autophosphorylation reaction, good correlation between phosphorylation and activation has been observed. In the presence of added purified protein kinase and cAMP, phosphorylation parallels activation and reaches a maximum value of 0.4 moles of phosphate per 10^5 grams of protein when activation is complete (Walsh, Perkins, Brostrom, Ho, and Krebs, 1971). Although only preliminary information is available on the molecular weight and subunit composition of phosphorylase kinase (Krebs, 1972), calculation indicates that this level of phosphorylation corresponds roughly to one phosphate per molecule of either of the two larger subunits present. The site specificity of the phosphorylation which takes place under these conditions has not been reported. However, in view of the fact that much greater amounts of phosphorylation take place during the course of activation under conditions which favor autophosphorylation, it seems likely that a large fraction of the nonspecific phosphorylation that has been observed is due to the autophosphorylation reaction rather than to the action of cAMP-dependent protein kinase.

Attempts to demonstrate increased phosphorylation accompanying hormonal activation of phosphorylase kinase *in vivo* have been unsuccessful (Mayer and Krebs, 1970). It is possible that increases were obscured by the presence of large amounts of phosphate having no effect on activity, perhaps resulting from autophosphorylation *in vivo*. In any case, the technical difficulties in these experiments are great, and at present the negative result does not seem to justify any serious doubt about the role of phosphorylase kinase as a substrate for protein kinase in mediating the action of cAMP on glycogen breakdown.

The phosphorylation of glycogen synthetase was also established as a significant regulatory event before phosphorylation by a separate cAMP-dependent protein kinase preparation was demonstrated (Schlender, Wei, and Villar-Palasi, 1969; Bishop and Larner, 1969). It was further shown that the same cAMP-dependent protein kinase catalyzed the phosphorylation of both phosphorylase kinase and glycogen synthetase (Soderling, Hickenbottom, Reimann, Hunkeler, Walsh, and Krebs, 1970; Villar-Palasi, Larner and Shen, 1971). Good correlation between phosphorylation and inactivation (conversion to the D form) is observed during the course of the conversion reaction (Larner, Villar-Palasi, and Brown, 1969; Soderling, Hickenbottom, Reimann, Hunkeler, Walsh, and Krebs, 1970). Good agreement between loss of phosphate and activation (conversion to the I form) catalyzed by a crude phosphatase preparation was also found (Larner, Villar-Palasi, and Brown, 1969). However, widely different values for the stoichiometry of phosphorylation have been obtained. Soderling, Hickenbottom, Reimann, Hunkeler, Walsh, and Krebs (1970) report a maximum value of 1.0 to 1.3 moles of phosphate incorporated per 10^5 grams of glycogen synthetase during conversion, although it was not known with certainty that the enzyme preparations used in these experiments were completely in the dephospho form. Smith, Brown, and Larner (1971) found 7 moles of phosphate per 10^5 grams in glycogen synthetase preparations after complete conversion to the D form by incubation for several days in the cold (3 to 5°C) in the presence of glycogen to stabilize the enzyme (Brown and Larner, 1971). The dephospho enzyme (I form), prepared by prolonged incubation in the presence of contaminating phosphatase activity, contained virtually no phosphate. Interestingly, the subunits of glycogen synthetase were found to have a molecular weight of 90,000 by SDS gel electrophoresis following incubation with SDS and mercaptoethanol. This value was not lowered by incubation in 8 M urea and 1% SDS and inclusion of urea in the gel (Smith, Brown, and Larner, 1971). These findings imply the presence of seven phosphorylation sites per polypeptide chain, or much smaller subunits which are not dissociated by SDS-mercaptoethanol or urea. Each of these sites appears to be surrounded by the same sequence of amino acids, since only a single phosphopeptide of the sequence Glu-Ile-SerPO$_4$-Val-Arg was isolated from preparations of glycogen synthetase (Larner and Sanger, 1965). Also unusual is the fact that this sequence is identical to that found surrounding the phosphorylation sites in phosphorylase (Fischer, Graves, Crittenden, and Krebs, 1959), which is a substrate for phosphorylase kinase but not for cAMP-dependent protein kinase.

Hormonal stimulation of adipose tissue lipolysis is brought about by the activation of triglyceride lipase, as shown by measurements of increased lipase activity in homogenates of hormone-treated tissue (Hollenberg,

Raben, and Astwood, 1961; Vaughan, Berger, and Steinberg, 1964). An early indication that lipase activation might be due to cAMP-stimulated protein phosphorylation was provided by Rizack (1964), who showed that activation of the lipase in crude homogenates was dependent on the addition of cAMP and ATP. These results were confirmed by Tsai and Vaughan (1970). The mediation of this effect by protein phosphorylation was demonstrated directly in adipose tissue homogenates supplemented with the protein inhibitor of protein kinase (Corbin, Reimann, Walsh, and Krebs, 1970), and in partially purified preparations of hormone-sensitive lipase (Huttunen, Steinberg, and Mayer, 1970a,b) by showing that activation of lipase by cAMP and ATP requires the addition of cAMP-dependent protein kinase. Similar but less pronounced effects were later observed with partially purified pancreatic lipase by Santhanam and Wagle (1971). With purified adipose tissue enzyme, good correlation between lipase activation and incorporation of phosphate into protein in the lipase preparation was observed, and both processes were completely dependent on added protein kinase (Huttunen, Steinberg, and Mayer, 1970b; Huttunen and Steinberg, 1971). These studies provide convincing evidence that the cAMP-stimulated activation of triglyceride lipase is mediated by cAMP-dependent protein kinase. They are compatible with the interpretation that the lipase is a substrate for the protein kinase, but do not rule out the participation of an intervening lipase kinase with a function analogous to that of phosphorylase kinase in the activation of phosphorylase.

As noted earlier, histones are good substrates for cAMP-dependent protein kinases, and have been widely used as substrates for the assay and characterization of protein kinases from various sources. With skeletal muscle protein kinase, the rate of phosphorylation of a mixture of total thymus histones is about one-third the rate found with phosphorylase kinase (Soderling, Hickenbottom, Reimann, Hunkeler, Walsh, and Krebs, 1970), but when certain individual histone fractions are used, the rates of phosphorylation are comparable to those obtained with phosphorylase kinase and glycogen synthetase (Krebs, 1972). *In vitro,* phosphorylation of all histone fractions has been observed. In general histones f2b and f1 have been found to be phosphorylated more rapidly than fractions f2a and f3 (Langan, 1968a,b; Shepherd, Noland, and Hardin, 1971; Corbin, Brostrom, Alexander, and Krebs, 1972). Isolated catalytic subunit from cAMP-dependent liver protein kinase phosphorylates histones f1, f2a, and f3 at about half the rate observed with f2b histone (Chen and Walsh, 1971). Partially purified protein kinase preparations from a wide variety of organisms were found to be most active with histone IV (a subfraction of histone f2a) as compared to histone IIb (f2b) and Ib (f1) by Kuo and Greengard

(1970b). As pointed out by Krebs (1972), relative rates of phosphorylation of different protein substrates may be highly dependent on assay conditions, which are necessarily quite arbitrary, so that evaluation of the significance of moderate differences in the rates of phosphorylation of different substrates *in vitro* is difficult.

A phosphatase specific for enzymatically phosphorylated histones and protamine was detected and characterized in liver by Meisler and Langan (1969). The enzyme is essentially inactive with phosvitin, casein, nuclear phosphoprotein, and a wide variety of low molecular weight phosphate esters, including tryptic phosphopeptides derived from active substrates. Histone phosphatase activity was also detected in crude extracts of all rat tissues examined as well as in the number of other eukaryotic organisms, but was undetectable in extracts of several prokaryotes. Histone phosphatase activity of rat cerebral cortex has been characterized by Maeno and Greengard (1972). More than 50% of the activity in this tissue is in particulate fractions, especially in fractions containing synaptic membranes, which are also rich in cAMP-dependent protein kinase. Soluble and particulate activities exhibit different substrate specificities and metal requirements. An endogenous membrane-bound phosphorylated protein was found to be the best substrate for the particulate enzyme.

The phosphorylation of lysine-rich (f1) histone by cAMP-dependent protein kinase takes place primarily on a single specific serine residue. A tryptic peptide containing this serine residue has been isolated and found to have the partial sequence Lys-Ala-SerPO$_4$ (Thr,Ser,Glu,Pro$_2$,Gly,-Val,Ile,Leu)Lys (Langan, 1968b, 1969a). Comparison of this sequence with known sequences in lysine-rich histone (Rall and Cole, 1971) localizes the phosphorylation site at serine residue 38 (calf thymus) or 37 (rabbit thymus). Interestingly, certain components of the lysine-rich histone fraction from several tissues lack this phosphorylation site. In one component of rabbit thymus lysine-rich histone, the absence of the phosphorylation site has been shown to be due to the replacement of serine by alanine at position 37 (Langan, Rall, and Cole, 1971). This variation in primary structure, which might appear to be minor in the absence of the knowledge that it determines the presence or absence of a phosphorylation site for cAMP-dependent protein kinase, suggests that functional differences may exist between different lysine-rich histone components.

The phosphorylation of rat liver lysine-rich histone by cAMP-dependent protein kinase has been demonstrated *in vivo,* using an assay based on the isolation of the tryptic peptide containing the specific phosphorylation site. Administration of glucagon or dibutyryl cAMP stimulates phosphorylation eight- to 20-fold. Increased phosphorylation is easily detectable 15 min after administration of glucagon, and is not blocked by inhibitors of RNA or protein synthesis, indicating that simple activation of the enzyme by cAMP occurs (Langan, 1969a,b). The increased phosphorylation following insulin administration observed in intact rats (Langan, 1969b) does not occur in the isolated perfused liver (Mallette, Neblett, Exton, and Langan, 1973), indicating that the effect of insulin in the whole animal is an indirect effect, perhaps due to glucagon or epinephrine release following insulin-induced hypoglycemia. It was proposed (Langan, 1968a, 1969a, 1971a) that cAMP-stimulated histone phosphorylation might provide the primary step in the mechanism for the induction of certain enzymes whose synthesis has been shown to be increased by cAMP. In order to obtain evidence bearing on the validity of this mechanism, we have attempted to determine if phosphorylation of lysine-rich histone affects its interaction with DNA, and if this leads to increased RNA synthesis. In collaboration with Fasman and co-workers, it has been shown that phosphorylation of lysine-rich histone by cAMP-dependent protein kinase markedly alters the interaction of the histone with DNA as measured by changes in circular dichroic spectra (Adler, Schaffhausen, Langan, and Fasman, 1971). Also, phosphorylation of chromatin proteins by cAMP-dependent protein kinase results in an increased binding of actinomycin D to the chromatin, suggesting that phosphorylation alters DNA-histone interactions in the chromatin (Fontana and Lovenberg, 1972). The effect of phosphorylation on the ability of lysine-rich histone to block RNA synthesis on reconstituted chromatin templates has been studied by Watson and Langan (1973). In the most recent experiments, chromatin reconstituted with lysine-rich histone phosphorylated by cAMP-dependent protein kinase is 25 to 80% more active as a template for RNA synthesis than chromatin reconstituted with control histone. It therefore appears that, *in vitro,* phosphorylation of lysine-rich histone can affect both the interaction of the histone with DNA and its ability to inhibit RNA synthesis. These results must be interpreted with caution, however, because of the possibility that systems containing simple DNA-histone complexes or artificially reconstituted chromatin preparations may not reflect events as they occur in the cell.

As noted earlier, the cAMP-dependent phosphorylation of lysine-rich histone which occurs *in vivo* is quite limited in extent, suggesting that there may be some selectivity in the phosphorylation which takes place in the cell. Although the basis for the limited phosphorylation observed *in vivo* is not known, some mechanism providing for selectivity appears to be required if histone phosphorylation is to function in the hormonal regulation of RNA synthesis. Stimulation of RNA synthesis by cAMP in animal cells has been observed (Dokas and Kleinsmith, 1971; Varrone, Ambesi-

Impiombato, and Macchia, 1972), but histone phosphorylation was not examined in these studies, and the possibility that increased RNA synthesis was secondary to some other action of cAMP was not eliminated.

Application of the criteria proposed by Krebs (1973; Table 1) to histone phosphorylation poses difficulties because of the uncertainties concerning the cellular function to which cAMP-dependent histone phosphorylation is related. However, criteria 1, 4, and 5 appear to be satisfied. In fact histone is the only well-defined substrate for cAMP-dependent protein kinase which has been shown to be phosphorylated *in vivo* in response to hormone and cAMP. The altered interaction of phosphorylated lysine-rich histone with DNA and the increased RNA synthesis with chromatin templates containing phosphorylated histone indicate that the third criterion may also be satisfied. With respect to the second criterion, histone phosphorylation can be expected to function in enzyme induction only in cases in which cAMP acts by stimulating the transcription of mRNA. There is some evidence that this is the case in the induction of serine dehydratase (Jost, Hsie, Hughes, and Ryan, 1970; Yeung and Oliver, 1971), ornithine decarboxylase (Beck, Bellantone, and Canellakis, 1972), and enzymes for steroid synthesis in cultured interstitial cells (Shin and Sato, 1971). However, there is considerably stronger evidence that cAMP acts at the translational level in the induction of tyrosine transaminase (Chuah and Oliver, 1971; Wicks and McKibbin, 1972), phosphoenolpyruvate carboxykinase (Wicks and McKibbin, 1972), enzymes for steroid synthesis in adrenal cortex (Ferguson and Morita, 1964; Garren, Ney, and Davis, 1965) and general protein synthesis in anterior pituitary (Labrie, Beraud, Gauthier, and Lemay, 1971). At present it seems that cAMP may act at the transcriptional level in the induction of some enzymes, but better evidence is clearly needed.

Extensive phosphorylation of histones *in vivo* has also shown to occur in rapidly dividing cells, and following the addition of hormones such as insulin or prolactin to isolated cell preparations (Ord and Stocken, 1968; Turkington and Riddle, 1969; Sherod, Johnson, and Chalkley, 1970; Sung, Dixon, and Smithies, 1971; Balhorn, Rieke, and Chalkley, 1971; Balhorn, Chalkley, and Granner, 1972; Lake, 1972; Lake, Goidl, and Salzman, 1972; Neblett and Crofford, 1972). At present there is no evidence that histone phosphorylation under these conditions is related to cAMP-dependent protein kinase. Histone phosphorylation occurs during gonadotrophin-induced spermatogenesis in salmonoid fish (Marushige, Ling, and Dixon, 1969; Louie and Dixon, 1972), but the timing of the phosphorylation suggests that other factors besides simple elevation of tissue cAMP are involved in controlling this type of phosphorylation.

The evidence that cAMP stimulates protein synthesis at the level of translation has led to tests of ribosomal preparations as substrates for cAMP-dependent protein kinase. Loeb and Blat (1970) studied protein phosphorylation in a mixture of crude rat liver supernatant and ribosomes. Cyclic AMP-dependent phosphorylation was detected at a level of 0.02 nmoles of phosphate per milligram protein. The phosphorylated proteins were presumed to be ribosomal because of their solubility in 66% acetic acid. Increased phosphate incorporation into rat liver ribosomal preparations was also observed *in vivo* following glucagon administration (Blat and Loeb, 1971). The relationship of this phosphorylation reaction to that observed *in vitro* was not established. Eil and Wool (1971) examined the phosphorylation of rat liver 40S and 60S ribosomal subunits by purified preparations of cAMP-dependent protein kinase. Phosphorylation at a level of about 1 nmole per milligram of protein was observed, which was highly dependent on cAMP and on the inclusion of ribosomal subunits in the reaction mixture. Acrylamide gel electrophoresis of ^{32}P-labeled phosphorylated ribosomal proteins revealed the presence of nine radioactive bands from the 60S subunit and three from the 40S. With ribosomes from adrenal cortex, cAMP-dependent protein kinase catalyzes phosphorylation of ribosome-associated proteins which are eluted during the dissociation of 80S ribosomes into subunits by 0.88 M KCl. The phosphorylated proteins are, however, not present in a 0.5 M NH_4Cl extract of these ribosomes, which contains factors required for *in vitro* protein synthesis (Walton, Gill, Abrass, and Garren, 1971). Phosphorylation of ribosomes of rat cerebrum (Johnson, Maeno, and Greengard, 1971) and cultured mammary cells (Majumder and Turkington, 1972) by cAMP-dependent protein kinase has also been reported. *E. coli* ribosomes and ribosome-associated proteins have also been shown to serve as substrates for mammalian cAMP-dependent protein kinase (Traugh and Traut, 1972; Fakunding, Traugh, Traut, and Hershey, 1972). In view of the lack of any substantial evidence that the phosphorylation of ribosomal proteins affects their function, the fact that they can be phosphorylated to a certain extent by cAMP-dependent protein kinase should be interpreted cautiously. It is clear that the criteria of Krebs (1973; Table 1) have not been satisfied in any significant way with respect to the relationship of cAMP-dependent ribosomal protein phosphorylation to the regulation of enzyme synthesis. It should be noted that extensive cAMP-independent ribosomal protein phosphorylation has also been observed (Kabat, 1970, 1971), again indicating that only a fraction of total cell protein phosphorylation reactions are controlled by cAMP.

A variety of membrane preparations have been tested as substrates for cAMP-dependent protein kinase. Beef brain microsome and crude synaptosome membrane fractions were observed to undergo cAMP-dependent phosphorylation by bound protein kinase activity (Weller and Rodnight, 1970). The ability of various subcellular fractions of rat cerebrum to act as substrates for purified brain cAMP-dependent protein kinase has been examined and quantitated by Johnson, Maeno, and Greengard (1971). Substrate activity was expressed as pmoles of phosphate incorporated per minute per milligram of substrate protein, under conditions in which phosphorylation by a constant amount of added protein kinase was linear with amount of substrate and time. For a system with hyperbolic saturation kinetics, activity expressed in this way will vary with the apparent K_m and V_{max}, as well as with the relative proportion of substrate protein in the fraction tested. Therefore a fraction containing a relatively small proportion of substrate of low K_m and high V_{max} may show higher activity than a fraction with a larger proportion of substrate of high K_m and low V_{max}. Nevertheless, the usefulness of this type of quantitation in examining the distribution of protein kinase substrates in tissues is shown by the good recoveries (70 to 97%) of total activity in the subcellular fractions. Substrate activity was highest in cerebrum fractions containing synaptic plasma membrane fragments; fractions containing synaptic vesicles and microsomes also had high activity. Sixty-eight percent of the substrate activity in the microsomal fraction was recovered in deoxycholate-soluble membrane proteins and 4% in ribosomes and polysomes. The nuclear fraction and the soluble fraction were lower in substrate activity than other major fractions. However, their activity was still about half that of the most active synaptic membrane fractions. Protein kinase substrates as revealed by this study are therefore clearly present in all subcellular fractions of rat cerebrum. The substrate activity of the synaptic plasma membrane fraction was nearly as high as that of certain histone fractions, in spite of the fact that the membrane fraction contains a large amount of protein which is inactive as substrate. This indicates the presence of a highly active substrate for protein kinase in the membrane. The nature of this substrate has been examined by SDS-acrylamide gel electrophoresis of ^{32}P-labeled synaptic membrane fraction phosphorylated by endogenous protein kinase in the presence and absence of cAMP (Johnson, Ueda, Maeno, and Greengard, 1972). Among a large number of phosphorylated proteins resolved by the gel, cAMP stimulated the phosphorylation of a single minor protein band, with an apparent molecular weight of 100,000. The cAMP-stimulated band was found only in subcellular fractions containing synaptic plasma membrane fragments. The high substrate activity of this band suggests it may be a natural substrate for brain cAMP-dependent protein kinase. The findings are in keeping with the hypothesis (Greengard and Kuo, 1970;

Johnson, Maeno, and Greengard, 1971) that cAMP, acting via a protein kinase, may mediate some types of synaptic transmission. Further tests of this hypothesis by investigation of the phosphorylation of membrane proteins *in vivo* under conditions of varied synaptic activity are clearly of interest.

Other membrane preparations whose phosphorylation has been studied include bovine adenohypophyseal secretory granules (Labrie, Lemaire, Poirier, Pelletier, and Boucher, 1971) and rat liver plasma membranes (Shlatz and Marinetti, 1971). In both these preparations, phosphorylation by endogenous kinase is not stimulated by cAMP, and in some experiments with secretory granules inhibition was observed. Cyclic AMP inhibition of endogenous phosphorylation of rat liver plasma membranes has also been observed (Johnson, Maeno, and Greengard, 1971). However, in secretory granule preparations, phosphorylation of added histone is cAMP-stimulated (Labrie, Lemaire, Poirier, Pelletier, and Boucher, 1971), and in mixtures containing rat liver plasma membranes (60 to 70 μg) and partially purified protein kinase (300 μg), cAMP-dependent protein phosphorylation occurs (Shlatz and Marinetti, 1971). It seems possible that the endogenous phosphorylation of these membranes is another example of protein phosphorylation by kinases unrelated to cAMP-dependent protein kinase. In rat adenohypophysis, cAMP-dependent phosphorylation has also been observed in mixtures of partially purified protein kinase and soluble proteins (Howell and Montague, 1971). In *in vivo* studies, glucagon administration was found to increase protein ^{32}P-phosphate associated with rat liver microsomes and mitochondrial and lysosomal membranes 84 to 340%. There was little effect on total soluble proteins (Zahlten, Hochberg, Stratman, and Lardy, 1972).

Brain microtubule protein preparations undergo cAMP-dependent phosphorylation catalyzed by endogenous protein kinase or added partially purified brain protein kinase (Goodman, Rasmussen, DiBella, and Guthrow, 1970), although the relationship of this phosphorylation to neurotubule function has not been clarified. Phosphorylated microtubule protein was found to interact specifically with a soluble brain component, resulting in a greatly changed behavior during purification compared to nonphosphorylated protein (Murray and Froscio, 1971). This phenomenon may possibly account for the absence of reports of neurotubular protein phosphorylation *in vivo*.

The major structural proteins of a variety of RNA tumor viruses and other membrane-maturing enveloped animal viruses are phosphorylated by protein kinases present in the virions (Strand and August, 1971; Silberstein, McAuslan, and August, 1972). In the case of Rauscher murine leukemia virus, phosphorylation of the viral proteins is stimulated 25 to 50% by cAMP. At least 10 of the 30 separate proteins of the virus resolved by SDS gel electrophoresis are phosphorylated. Protamine is also phosphorylated by the viral protein kinase, but with this substrate stimulation by cAMP is less than with endogenous substrates. Calf-thymus arginine-rich and lysine-rich histones were also phosphorylated. The most effective substrate is a basic protein isolated from *E. coli* as a host factor (factor II) required for the *in vitro* replication of bacteriophage Q beta RNA. The *E. coli* protein is also an excellent substrate for beef heart cAMP-dependent protein kinase. Viral enzyme has been partially purified from frog polyhedral cytoplasmic deoxyvirus 3. Similar viral protein kinase activity has been detected by Hatanaka, Twiddy, and Gilden (1972), also in membrane-maturing viruses, but in this study no cAMP dependency for the phosphorylation of endogenous substrates was observed. Rabbit red cell protein kinase catalyzes a cAMP-stimulated phosphorylation of the proteins of adenovirus type 2 and 12 (Tao and Doerfler, 1972). Bacteriophage lambda can also be phosphorylated, but not bacteriophage T4 or tobacco mosaic virus. The adenoviruses themselves do not contain protein kinase. The possibility that virus-host interactions may be influenced by viral protein phosphorylation has been considered by each of these workers.

A homogenous basic protein purified from guinea pig seminal vesicle secretion is an excellent substrate for rat testis cAMP-dependent protein kinase (Reddi, Ewing, and Williams-Ashman, 1971). Bailey and Villar-Palasi (1971) have reported the phosphorylation of the inhibitor component of troponin (TNI) by skeletal muscle cAMP-dependent protein kinase. This finding is of interest because it suggests a potential mechanism for a link between cAMP and muscle contractility or other contractile protein functions. However, Stull, Brostrom, and Krebs (1972) observed that phosphorylation of TNI by the cAMP-dependent protein kinase is considerably slower than the phosphorylation of substrates such as glycogen synthetase or histone. These workers found that TNI was also phosphorylated by phosphorylase kinase at a rate 1/15 that obtained for the phosphorylation of phosphorylase b. However, no differences were found in the rate of phosphorylation of TNI by activated and nonactivated phosphorylase kinase over the pH range of 6 to 9. The TNI preparations, as isolated, contain protein phosphate; after maximal phosphorylation *in vitro*, the total phosphate content was estimated at 1 mole of phosphate per mole of TNI. These observations suggest that muscle Ca^{++} may exert the major control over TNI phosphorylation by virtue of its ability to activate both the active and inactive forms of phosphorylase kinase. Dephosphorylation of TNI by phosphorylase phosphatase has also been demonstrated (England, Stull, and Krebs, 1972).

The phosphorylation of *E. coli* RNA polymerase (Martelo, Woo, Reimann, and Davie, 1970) and polynucleotide phosphorylase (Thang and Meyer, 1971) by mammalian cAMP-dependent protein kinase has been reported. In both cases, an increase in enzyme activity is associated with phosphorylation. RNA polymerase phosphorylation occurs on the sigma factor. However, demonstration of phosphorylation by a protein kinase from *E. coli* is required if these phosphorylation reactions are to be considered functional.

IX. CONCLUDING REMARKS

From the large amount of work done in recent years, a reasonably coherent picture of the role of protein kinase in cAMP-mediated hormone responses appears to have emerged. This picture, which might be represented by the diagram shown in Fig. 1, has a number of unexpected aspects. The first of these is the participation of a protein phosphokinase in each of the well-documented examples of cAMP action in animal cells. It seems surprising that over the course of evolution other proteins responding to direct allosteric interaction with cAMP have not emerged, especially in view of the fact that such a protein does exist in bacteria. The second unexpected aspect is that one protein kinase, which in spite of the existence of multiple forms must be considered functionally as a single enzyme, is a common component of the chain of responses between the activation of adenyl cyclase and a variety of diverse hormone actions. One might easily have expected that distinct cAMP-dependent protein kinases specific for individual protein substrates might exist, with each being involved in mediating single or closely related responses. The absence of such alternate pathways for cAMP action seems evident not only from the lack of positive evidence for their existence, but also from the fact that cAMP binding assays, which are potentially capable of detecting any protein involved in cAMP action, have not revealed the existence of binding activity other than the protein kinase regulatory subunit. The net result of this situation is that the response of a particular cell to a rise in cAMP concentration is essentially determined by the nature of the protein kinase substrates which are present in that cell, and these in turn are determined by the regulatory processes controlling cellular differentiation and, on a shorter term basis, protein synthesis in already-differentiated cells. In other words, the response of a cell to cAMP is quite directly linked to the regulation of genetic output

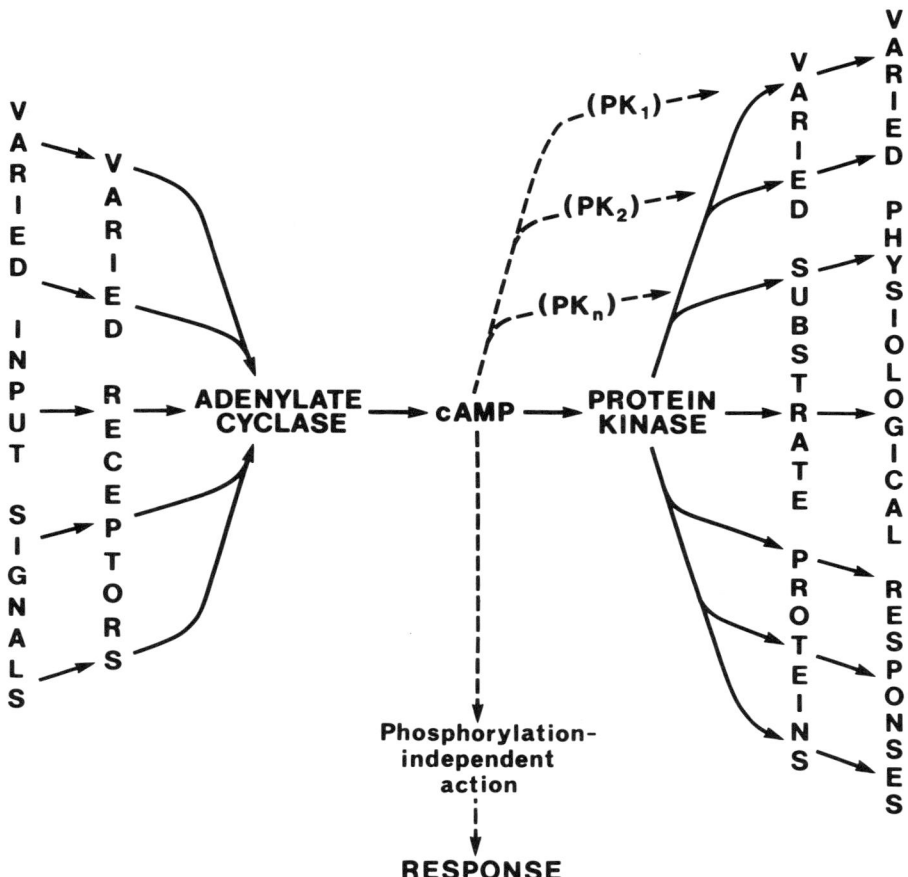

FIG. 1. Information flow in cAMP-mediated hormone responses. Solid lines show pathways which are documented by available evidence; dashed lines indicate pathways for which there are no experimental examples at present.

in that cell, providing an opportunity for a great deal of diversity in the responses to a common and essentially universally present adenyl cyclase-cAMP-protein kinase system. At the input level, diversity is again provided by gene products in the form of hormone-specific receptors associated with adenyl cyclase.

However, the simplicity with which specificity and diversity are achieved in this control scheme conceals a phenomenon which our present concepts do not handle easily, namely the unorthodox substrate specificity of protein kinase. This enzyme acts on a variety of apparently quite unrelated and dissimilar substrates. Conventional concepts of what determines enzyme specificity do not seem to apply. The ability of protein kinase to recognize a group of proteins with no obvious similarities is certainly its most unusual and striking property. The interpretation that the enzyme is truly nonspecific seems implausible, since it is hard to imagine that normal cell function, much less regulatory control, would be possible in a cell containing large amounts of an enzyme which randomly and nonspecifically phosphorylates cell proteins. As noted earlier, recognition of some type of three-dimensional configuration seems likely as an explanation for protein kinase substrate specificity. The flexibility of protein structure might then explain why this specificity is not as sharply defined as that of most enzymes acting on more rigid molecules.

As also noted, the situation depicted in the scheme suggests a framework for future investigation of the mechanism by which cAMP brings about its various effects, namely elucidation of the substrates of protein kinase. Because the results of *in vitro* experiments may be misleading when considered alone, more sophisticated methods for identifying cAMP-dependent protein phosphorylation reactions *in vivo* are clearly desirable. Unfortunately, the phosphorylation reactions of interest may be masked by the large amount of cAMP-independent protein phosphorylation which occurs in cells. Perhaps double labeling procedures using ^{32}P and ^{33}P, coupled with high-resolution protein fractionation techniques will be useful. However, a difficulty in this approach to determining mechanisms of cAMP action is that it may lead to the isolation of protein kinase substrates without providing any clues to the function, enzymatic or otherwise, of the proteins. Determination of function may be quite difficult in the absence of clues provided by conventional studies of the hormone response. Therefore, approaches originating at the point of physiological response will certainly continue to be required in order to establish the mechanisms by which cAMP and protein kinase bring about diverse actions in cells.

X. REFERENCES

Adiga, P. R., Murthy, P. V. N., and McKenzie, J. M. (1971): Stimulation by thyrotropin, long-acting thyroid stimulator, and dibutyryl 3',5'-adenosine monophosphate of protein and ribonucleic acid synthesis and ribonucleic acid polymerase activities in porcine thyroid *in vitro*. *Biochemistry*, 10:702–711.

Adler, A. J., Schaffhausen, B., Langan, T. A., and Fasman, G. D. (1971): Altered conformational effects of phosphorylated lysine-rich histone (f1) in f1-deoxyribonucleic acid complexes. Circular dichroism and immunological studies. *Biochemistry*, 10:909–913.

Agren, G., De Verdier, C.-H., and Glomset, J. (1954): A study of the phosphorus-containing proteins of cells. II. The turnover rate of P^{32}-labelled phosphoserine of the Schneider protein residues of several rat organs. *Acta Chemica Scandinavica*, 8:1570–1578.

Anderson, L., and Kelley, J. J. (1959): The dephosphorylation of casein by alkalies. *Journal of the American Chemical Society*, 81:2275–2276.
Appleman, M. M., Birnbaumer, L., and Torres, H. N. (1966): Factors affecting the activity of muscle glycogen synthetase. III. The reaction with adenosine triphosphate, Mg^{++}, and cyclic 3′5′-adenosine monophosphate. *Archives of Biochemistry and Biophysics*, 116:39–43.
Ashby, C. D., and Walsh, D. A. (1972a): Characterization of the interaction of a protein inhibitor with adenosine 3′,5′-monophosphate-dependent protein kinases. *Journal of Biological Chemistry*, 247:6637–6642.
Ashby, C. D., and Walsh, D. A. (1972b): Assessment of the role of a protein inhibitor of cyclic AMP-dependent protein kinases. *Federation Proceedings*, 31:439Abs.
Bader, H., Post, R. L., and Jean, D. H. (1967): Further characterization of a phosphorylated intermediate in ($Na^+ + K^+$)-dependent ATPase. *Biochimica et Biophysica Acta*, 143:229–238.
Baggio, B., Pinna, L. A., Moret, V., and Siliprandi, N. (1970): A simple procedure for the purification of rat liver phosvitin kinase. *Biochimica et Biophysica Acta*, 212:515–517.
Bailey, C., and Villar-Palasi, C. (1971): Cyclic AMP dependent phosphorylation of troponin. *Federation Proceedings*, 30:1147Abs.
Balhorn, R., Chalkley, R., and Granner, D. (1972): Lysine-rich histone phosphorylation. A positive correlation with cell replication. *Biochemistry*, 11:1094–1098.
Balhorn, R., Rieke, W. O., and Chalkley, R. (1971): Rapid electrophoretic analysis for histone phosphorylation. A reinvestigation of phosphorylation of lysine-rich histone during rat liver regeneration. *Biochemistry*, 10:3952–3959.
Beck, W. T., Bellantone, R. A., and Canellakis, E. S. (1972): The *in vivo* stimulation of rat liver ornithine decarboxylase activity by dibutyryl cyclic adenosine 3′,5′-monophosphate, theophylline and dexamethasone. *Biochemical and Biophysical Research Communications*, 48:1649–1655.
Belocopitow, E. (1961): The action of epinephrine on glycogen synthetase. *Archives of Biochemistry and Biophysics*, 93:457–458.
Bishop, J. S., and Larner, J. (1969): Presence in liver of a 3′:5′-cyclic AMP stimulated kinase for the I form of UDPG-glycogen glucosyltransferase. *Biochimica et Biophysica Acta*, 171:374–377.
Blat, C., and Loeb, J. E. (1971): Effect of glucagon on phosphorylation of some rat liver ribosomal proteins in vivo. *FEBS Letters*, 18:124–126.
Brostrom, C. O., Corbin, J. D., King, C. A., and Krebs, E. G. (1971): Interaction of the subunits of adenosine 3′:5′-cyclic monophosphate-dependent protein kinase of muscle. *Proceedings of the National Academy of Sciences*, 68:2444–2447.
Brostrom, C. O., Hunkeler, F. L., and Krebs, E. G. (1971): The regulation of skeletal muscle phosphorylase kinase by Ca^{2+}. *Journal of Biological Chemistry*, 246:1961–1967.
Brostrom, M. A., Reimann, E. M., Walsh, D. A., and Krebs, E. G. (1970): A cyclic 3′,5′-AMP-stimulated protein kinase from cardiac muscle. In: *Advances in Enzyme Regulation*, Vol. 8, edited by G. Weber, pp. 191–203. Pergamon Press, New York.
Brown, N. E., and Larner, J. (1971): Molecular characteristics of the totally dependent and independent forms of glycogen synthase of rabbit skeletal muscle. I. Preparation and characteristics of the totally glucose 6-phosphate dependent form. *Biochimica et Biophysica Acta*, 242:69–80.
Burdon, R. H., and Pearce, C. A. (1971): Enzymic modification of chromosomal macromolecules. III. The effect of adenine nucleotides on histone modification in chromatin and a possible influence of steroid hormones. *Biochimica et Biophysica Acta*, 246:561–571.
Burgess, R. R. (1971): RNA polymerase. *Annual Reviews of Biochemistry*, 40:711–740.
Burnett, G., and Kennedy, E. P. (1954): The enzymatic phosphorylation of proteins. *Journal of Biological Chemistry*, 211:969–980.
Chambaut, A. M., Leray, F., and Hanoune, J. (1971): Relationship between cyclic AMP dependent protein kinase(s) and cyclic AMP binding protein(s) in rat liver. *FEBS Letters*, 15:328–334.
Chambers, D. A., and Zubay, G. (1969): The stimulatory effect of cyclic adenosine 3′5′-

monophosphate on DNA-directed synthesis of β-galactosidase in cell-free system. *Proceedings of the National Academy of Sciences,* 63:118–122.

Chelala, C. A., and Torres, H. N. (1969): Interconvertible forms of muscle phosphorylase phosphatase. *Biochimica et Biophysica Acta,* 178:423–426.

Chelala, C. A., and Torres, H. N. (1970): Regulation of skeletal muscle phosphorylase phosphatase activity. *Biochimica et Biophysica Acta,* 198:504–513.

Chen, L.-J., and Walsh, D. A. (1971): Multiple forms of hepatic adenosine 3':5'-monophosphate dependent protein kinase. *Biochemistry,* 10:3614–3621.

Chuah, C.-C., and Oliver, I. T. (1971): Role of adenosine cyclic monophosphate in the synthesis of tyrosine aminotransferase in neonatal rat liver. Release of enzyme from membrane-bound polysomes *in vitro. Biochemistry,* 10:2990–3001.

Corbin, J. D., and Brostrom, C. O. (1971): Subunit analysis of skeletal muscle cyclic AMP-dependent protein kinases. *Federation Proceedings,* 30:1089Abs.

Corbin, J. D., Brostrom, C. O., Alexander, R. L., and Krebs, E. G. (1972): Adenosine 3',5'-monophosphate-dependent protein kinase from adipose tissue. *Journal of Biological Chemistry,* 247:3736–3743.

Corbin, J. D., and Krebs, E. G. (1969): A cyclic AMP-stimulated protein kinase in adipose tissue. *Biochemical and Biophysical Research Communications,* 36:328–336.

Corbin, J. D., Reimann, E. M., Walsh, D. A., and Krebs, E. G. (1970): Activation of adipose tissue lipase by skeletal muscle cyclic adenosine 3',5'-monophosphate-stimulated protein kinase. *Journal of Biological Chemistry,* 245:4849–4851.

Craig, J. W., and Larner, J. (1964): Influence of epinephrine and insulin on uridine diphosphate glucose-α-glucan transferase and phosphorylase in muscle. *Nature,* 202:971–973.

Davidson, J. N., Frazer, S. C., and Hutchison, W. C. (1951): Phosphorus compounds in the cell. I. Protein-bound phosphorus fractions studied with the aid of radioactive phosphorus. *Biochemical Journal,* 49:311–321.

de Crombrugghe, B., Chen, B., Anderson, W. B., Gottesman, M. E., Perlman, R. L., and Pastan, I. (1971): Role of cyclic adenosine 3',5'-monophosphate receptor protein in the initiation of *lac* transcription. *Journal of Biological Chemistry,* 246:7343–7348.

de Crombrugghe, B., Chen, B., Anderson, W., Nissley, P., Gottesman, M., Pastan, I., and Perlman, R. (1971): *Lac* DNA, RNA polymerase and cyclic AMP receptor protein, cyclic AMP, *lac* repressor and inducer are the essential elements for controlled *lac* transcription. *Nature New Biology,* 231:139–142.

DeLange, R. J., Kemp, R. G., Riley, W. D., Cooper, R. A., and Krebs, E. G. (1968): Activation of skeletal muscle phosphorylase kinase by adenosine triphosphate and adenosine 3',5'-monophosphate. *Journal of Biological Chemistry,* 243:2200–2208.

Dietze, G., and Hepp, K. D. (1972): Effect of 3'5'-AMP on calcium-activated ATPase in rat heart sarcolemma. *Biochemical and Biophysical Research Communications,* 46:269–278.

Dokas, L. A., and Kleinsmith, L. J. (1971): Adenosine 3',5'-monophosphate increases capacity for RNA synthesis in rat liver nuclei. *Science,* 172:1237–1238.

Donnelly, T. E., Jr., Kuo, J.-F., Miyamoto, E., and Greengard, P. (1972): Protein kinase modulator: Purification and properties. *Federation Proceedings,* 31:439Abs.

Drummond, G. I., Harwood, J. P., and Powell, C. A. (1969): Studies on the activation of phosphorylase in skeletal muscle by contraction and by epinephrine. *Journal of Biological Chemistry,* 244:4235–4240.

Eil, C., and Wool, I. G. (1971): Phosphorylation of rat liver ribosomal subunits: partial purification of two cyclic AMP activated protein kinases. *Biochemical and Biophysical Research Communications,* 43:1001–1009.

Emmer, M., de Crombrugghe, B., Pastan, I., and Perlman, R. (1970): Cyclic AMP receptor protein of *E. coli:* Its role in the synthesis of inducible enzymes. *Proceedings of the National Academy of Sciences,* 66:480–487.

England, P. J., Stull, J. T. and Krebs, E. G. (1972): Dephosphorylation of the inhibitor component of troponin by phosphorylase phosphatase. *Journal of Biological Chemistry,* 247: 5275–5277.

Erlichman, J., Hirsch, A. H., and Rosen, O. M. (1971): Interconversion of cyclic nucleotide-

activated and cyclic nucleotide-independent forms of a protein kinase from beef heart. *Proceedings of the National Academy of Sciences,* 68:731–735.
Fakunding, J., Traugh, J. A., Traut, R. R., and Hershey, J. W. B. (1972): Interaction of enzymically phosphorylated initiation factor F2 with ribosomes. *Federation Proceedings,* 31:410Abs.
Ferguson, J. J., Jr., and Morita, Y. (1964): RNA synthesis and adrenocorticotropin responsiveness. *Biochimica et Biophysica Acta,* 87:348–350.
Fischer, E. H., Graves, D. J., Crittenden, E. R. S., and Krebs, E. G. (1959): Structure of the site phosphorylated in the phosphorylase *b* to *a* reaction. *Journal of Biological Chemistry,* 234:1698–1704.
Fischer, E. H., Heilmeyer, L. M., Jr., and Haschke, R. H. (1971): Phosphorylase and the control of glycogen degradation. In: *Current Topics in Cellular Regulation,* Vol. 4, edited by B. L. Horecker and E. R. Stadtman, pp. 211–251. Academic Press, New York.
Fischer, E. H., and Krebs, E. G. (1955): Conversion of phosphorylase b to phosphorylase a in muscle extracts. *Journal of Biological Chemistry,* 216:121–132.
Fontana, J. A., and Lovenberg, W. (1971): A cyclic AMP-dependent protein kinase of the bovine pineal gland. *Proceedings of the National Academy of Sciences,* 68:2787–2790.
Fontana, J. A., and Lovenberg, W. (1972): Pineal protein kinase and the regulation of hydroxyindole metabolism. *Federation Proceedings,* 31:440Abs.
Friedman, D. L., and Larner, J. (1963): Studies on UDPG-alpha-glucan transglucosylase. Mechanism of interconversion by a phosphorylation-dephosphorylation reaction sequence. *Biochemistry,* 2:669–675.
Gaballah, S., and Popoff, C. (1971): Localization of adenosine 3',5'-monophosphate-dependent protein kinase in brain. *Journal of Neurochemistry,* 18:1795–1797.
Gaballah, S., Popoff, C., and Sooknandan, G. (1971): Changes in cyclic 3',5'-adenosine monophosphate-dependent protein kinase levels in brain development. *Brain Research,* 31:229–232.
Garren, L. D., Gill, G. N., and Walton, G. M. (1971): The isolation of a receptor for adenosine 3',5'-cyclic monophosphate (cAMP) from the adrenal cortex: the role of the receptor in the mechanism of action of cAMP. *Annals of the New York Academy of Sciences,* 185:210–226.
Garren, L. D., Ney, R. L., and Davis, W. W. (1965): Studies on the role of protein synthesis in the regulation of corticosterone production by adrenocorticotropic hormone *in vivo. Proceedings of the National Academy of Sciences,* 53:1443–1450.
Gill, G. N., and Garren, L. D. (1969): On the mechanism of action of adrenocorticotropic hormone: The binding of cyclic-3',5'-adenosine monophosphate to an adrenal cortical protein. *Proceedings of the National Academy of Sciences,* 63:512–519.
Gill, G. N., and Garren, L. D. (1970): A cyclic-3',5'-adenosine monophosphate dependent protein kinase from the adrenal cortex: comparison with a cyclic AMP binding protein. *Biochemical and Biophysical Research Communications,* 39:335–343.
Gill, G. N., and Garren, L. D. (1971): Role of the receptor in the mechanism of action of adenosine 3':5'-cyclic monophosphate. *Proceedings of the National Academy of Sciences,* 68:786–790.
Gonzalez, C. (1962): Studies on phosphorylase *b* kinase. M.S. Thesis, University of Washington, Seattle.
Goodman, D. B. P., Rasmussen, H., DiBella, F., and Guthrow, C. E., Jr. (1970): Cyclic adenosine 3':5'-monophosphate-stimulated phosphorylation of isolated neurotubule subunits. *Proceedings of the National Academy of Sciences,* 67:652–659.
Granner, D. K. (1972): Protein kinase: Altered regulation in a hepatoma cell line deficient in adenosine 3',5'-cyclic monophosphate-binding protein. *Biochemical and Biophysical Research Communications,* 46:1516–1522.
Granner, D., Chase, L. R., Aurbach, G. D. and Tomkins, G. M. (1968): Tyrosine aminotransferase: Enzyme induction independent of adenosine 3',5'-monophosphate. *Science,* 162:1018–1020.
Greengard, P., Hayaishi, O., and Colowick, S. P. (1969): Enzymatic adenylylation of pyrophosphate by 3',5'cyclic AMP: Reversal of the adenyl cyclase reaction. *Federation Proceedings,* 28:467.

Greengard, P., and Kuo, J. F. (1970): On the mechanism of action of cyclic AMP. In: *Advances in Biochemical Psychopharmacology, Vol. 3, Role of Cyclic AMP in Cell Function*, edited by P. Greengard and E. Costa, pp. 287–306, Raven Press, New York.

Haddox, M. K., Newton, N. E., Hartle, D. K., and Goldberg, N. D. (1972): ATP(Mg^{2+}) induced inhibition of cyclic AMP reactivity with a skeletal muscle protein kinase. *Biochemical and Biophysical Research Communications*, 47:653–661.

Hatanaka, M., Twiddy, E., and Gilden, R. V. (1972): Protein kinase associated with RNA tumor viruses and other budding RNA viruses. *Virology*, 47:536–538.

Hayaishi, O., Greengard, P., and Colowick, S. P. (1971): On the equilibrium of the adenylate cyclase reaction. *Journal of Biological Chemistry*, 246:5840–5843.

Hedrick, J. L., and Smith, A. J. (1968): Size and charge isomer separation and estimation of molecular weights of proteins by disc gel electrophoresis. *Archives of Biochemistry and Biophysics*, 126:155–164.

Hers, H. G., de Wulf, H., and Stalmans, W. (1970): The control of glycogen metabolism in the liver. *FEBS Letters*, 12:73–82.

Hollenberg, C. H., Raben, M. S., and Astwood, E. B. (1961): The lipolytic response to corticotropin. *Endocrinology*, 68:589–598.

Hoskins, D. D., Casillas, E. R., and Stephens, D. T. (1972): Cyclic AMP-dependent protein kinases of bovine epididymal spermatozoa. *Biochemical and Biophysical Research Communications*, 48:1331–1338.

Howell, S. L., and Montague, W. (1971): The mode of action of cyclic AMP in the rat anterior pituitary. *FEBS Letters*, 18:293.

Huijing, F., and Larner, J. (1966a): On the effect of adenosine 3′,5′ cyclophosphate on the kinase of UDPG: α-1,4-glucan α-4-glucosyl transferase. *Biochemical and Biophysical Research Communications*, 23:259–263.

Huijing, F., and Larner, J. (1966b): On the mechanism of action of adenosine 3′,5′-cyclophosphate. *Proceedings of the National Academy of Sciences*, 56:647–653.

Huttunen, J. K., and Steinberg, D. (1971): Activation and phosphorylation of purified adipose tissue hormone-sensitive lipase by cyclic AMP-dependent protein kinase. *Biochimica et Biophysica Acta*, 239:411–427.

Huttunen, J. K., Steinberg, D., and Mayer, S. E. (1970a): ATP-dependent and cyclic AMP-dependent activation of rat adipose tissue lipase by protein kinase from rabbit skeletal muscle. *Proceedings of the National Academy of Sciences*, 67:290–295.

Huttunen, J. K., Steinberg, D., and Mayer, S. E. (1970b): Protein kinase activation and phosphorylation of a purified hormone-sensitive lipase. *Biochemical and Biophysical Research Communications*, 41:1350–1356.

Jackson, J. E., Jackson, E. M., and Freeman, S. (1965): Demonstration, extraction and intracellular distribution of kidney phosphoprotein kinase activity. *Biochimica et Biophysica Acta*, 105:483–495.

Jard, S., and Bastide, F. (1970): A cyclic AMP-dependent protein kinase from frog bladder epithelial cells. *Biochemical and Biophysical Research Communications*, 39:559–566.

Jergil, B., and Dixon, G. H. (1970): Protamine kinase from rainbow trout testis. *Journal of Biological Chemistry*, 245:425–434.

Johnson, E. M., Maeno, H., and Greengard, P. (1971): Phosphorylation of endogenous protein of rat brain by cyclic adenosine 3′,5′-monophosphate-dependent protein kinase. *Journal of Biological Chemistry*, 246:7731–7739.

Johnson, E. M., Ueda, T., Maeno, H., and Greengard, P. (1972): Adenosine 3′,5′-monophosphate-dependent phosphorylation of a specific protein in synaptic membrane fractions from rat cerebrum. *Journal of Biological Chemistry*, 247:5650–5652.

Johnson, R. M., and Albert, S. (1953): Incorporation of P^{32} into the "phosphoprotein" fraction of mammalian tissue. *Journal of Biological Chemistry*, 200:335–344.

Jost, J.-P., Hsie, A., Hughes, S. D., and Ryan, L. (1970): Role of cyclic adenosine 3′,5′-monophosphate in the induction of hepatic enzymes. I. Kinetics of the induction of rat liver serine dehydratase by cyclic adenosine 3′,5′-monophosphate. *Journal of Biological Chemistry*, 245:351–357.

Jost, J.-P., and Sahib, M. K. (1971): Role of cyclic adenosine 3′,5′-monophosphate in the induction of hepatic enzymes. II. Effect of $N^6,O^{2'}$-dibutyryl cyclic adenosine 3',5'-monophosphate on the kinetics of ribonucleic acid synthesis in purified rat liver nuclei. *Journal of Biological Chemistry*, 246:1623-1629.

Kabat, D. (1970): Phosphorylation of ribosomal proteins in rabbit reticulocytes. Characterization and regulatory aspects. *Biochemistry*, 9:4160-4175.

Kabat, D. (1971): Phosphorylation of ribosomal proteins in rabbit reticulocytes. A cell-free system with ribosomal protein kinase activity. *Biochemistry*, 10:197-203.

Kemp, R. G., and Krebs, E. G. (1967): Binding of metabolites by phosphofructokinase. *Biochemistry*, 6:423-434.

Kennedy, E. P., and Smith, S. W. (1954): The isolation of radioactive phosphoserine from "phosphoprotein" of the Ehrlich ascites tumor. *Journal of Biological Chemistry*, 207:153-163.

Kingdon, H. S., Shapiro, B. M., and Stadtman, E. R. (1967): Regulation of glutamine synthetase. VIII. ATP:glutamine synthetase adenyltransferase, an enzyme that catalyzes alterations in the regulatory properties of glutamine synthetase. *Proceedings of the National Academy of Sciences*, 58:1703.

Kish, V. M., and Kleinsmith, L. J. (1972): Heterogeneity of nuclear protein kinases associated with non-histone chromatin proteins. *Journal of Cell Biology*, 55:138a.

Klein, M. I., and Makman, M. H. (1972): Cyclic AMP-dependent protein kinase in cultured cells: effect of growth conditions. *Federation Proceedings*, 31:513Abs.

Kleinsmith, L. J., and Allfrey, V. G. (1969): Nuclear phosphoproteins. I. Isolation and characterization of a phosphoprotein fraction from calf thymus nuclei. *Biochimica et Biophysica Acta*, 175:123-135.

Krane, S. M., Stone, M. J., and Glimcher, M. J. (1965): The presence of protein phosphokinase in connective tissues and the phosphorylation of enamel proteins *in vitro*. *Biochemica et Biophysica Acta*, 97:77-87.

Krebs, E. G. (1972): Protein kinases. In: *Current Topics in Cellular Regulation*, Vol. 5, edited by B. L. Horecker and E. R. Stadtman, pp. 99-133. Academic Press, New York.

Krebs, E. G. (1973): The mechanism of hormonal regulation by cyclic AMP. In: *Endocrinology, Proceedings of the 4th International Congress*, Excerpta Medica, Amsterdam (*in press*).

Krebs, E. G., DeLange, R. J., Kemp, R. G., and Riley, W. D. (1966): Activation of skeletal muscle phosphorylase. *Pharmacological Reviews*, 18:163-171.

Krebs, E. G., and Fischer, E. H. (1956): The phosphorylase *b* to *a* converting enzyme of rabbit skeletal muscle. *Biochimica et Biophysica Acta*, 20:150-157.

Krebs, E. G., and Fischer, E. H. (1962): Molecular properties and transformations of glycogen phosphorylase in animal tissues. *Advances in Enzymology*, 24:263.

Krebs, E. G., Graves, D. J., and Fischer, E. H. (1959): Factors affecting the activity of muscle phosphorylase b kinase. *Journal of Biological Chemistry*, 234:2867-2873.

Krebs, E. G., Kent, A. B., and Fischer, E. H. (1958): The muscle phosphorylase b kinase reaction. *Journal of Biological Chemistry*, 231:73-83.

Krebs, E. G., Love, D. S., Bratvold, G. E., Trayser, K. A., Meyer, W. L., and Fischer, E. H. (1964): Purification and properties of rabbit skeletal muscle phosphorylase b kinase. *Biochemistry*, 3:1022-1033.

Kuehn, G. D. (1971): An adenosine 3',5'-monophosphate-inhibited protein kinase from *Physarum polycephalum*. *Journal of Biological Chemistry*, 246:6366-6369.

Kumon, A., Nishiyama, K., Yamamura, H., and Nishizuka, Y. (1972): Multiplicity of adenosine 3',5'-monophosphate-dependent protein kinases from rat liver and mode of action of nucleoside 3',5'-monophosphate. *Journal of Biological Chemistry*, 247:3726-3735.

Kumon, A., Yamamura, H., and Nishizuka, Y. (1970): Mode of action of adenosine 3',5'-cyclic phosphate on protein kinase from rat liver. *Biochemical and Biophysical Research Communications*, 41:1290-1297.

Kundig, W., and Roseman, S. (1971): Sugar transport. I. Isolation of a phosphotransferase system from *Escherichia coli*. *Journal of Biological Chemistry*, 246:1393-1406.

Kuo, J. F., and Greengard, P. (1969a): Cyclic nucleotide-dependent protein kinases. IV. Widespread occurrence of adenosine 3',5'-monophosphate-dependent protein kinase in various tissues and phyla of the animal kingdom. *Proceedings of the National Academy of Sciences*, 64:1349–1355.

Kuo, J. F., and Greengard, P. (1969b): An adenosine 3′,5′-monophosphate-dependent protein kinase from *Escherichia coli*. *Journal of Biological Chemistry*, 244:3417–3419.

Kuo, J. F., and Greengard, P. (1970a): Cyclic nucleotide-dependent protein kinases. VI. Isolation and partial purification of a protein kinase activated by guanosine 3′,5′-monophosphate. *Journal of Biological Chemistry*, 245:2493–2498.

Kuo, J. F., and Greengard, P. (1970b): Cyclic nucleotide-dependent protein kinases. VII. Comparison of various histones as substrates for adenosine 3′,5′-monophosphate-dependent and guanosine 3′,5′-monophosphate-dependent protein kinases. *Biochimica et Biophysica Acta*, 212:434–440.

Kuo, J. F., and Greengard, P. (1971): Stimulation of cyclic GMP-dependent protein kinase by a protein fraction which inhibits cyclic AMP-dependent protein kinases. *Federation Proceedings*, 30:1089Abs.

Kuo, J. F., Krueger, B. K., Sanes, J. R., and Greengard, P. (1970): Cyclic nucleotide-dependent protein kinases. V. Preparation and properties of adenosine 3′,5′-monophosphate-dependent protein kinase from various bovine tissues. *Biochimica et Biophysica Acta*, 212:79–91.

Kuo, J. F., Wyatt, G. R., and Greengard, P. (1971): Cyclic nucleotide-dependent protein kinases. IX. Partial purification and some properties of guanosine 3′,5′-monophosphate-dependent and adenosine 3′,5′-monophosphate-dependent protein kinases from various tissues and species of arthropoda. *Journal of Biological Chemistry*, 246:7159–7167.

Labrie, F., Béraud, G., Gauthier, M., and Lemay, A. (1971): Actinomycin-insensitive stimulation of protein synthesis in rat anterior pituitary *in vitro* by dibutyryl adenosine 3′,5′-monophosphate. *Journal of Biological Chemistry*, 246:1902–1908.

Labrie, F., Lemaire, S., and Courte, C. (1971): Adenosine 3′,5′-monophosphate-dependent protein kinase from bovine anterior pituitary gland. *Journal of Biological Chemistry*, 246: 7293–7302.

Labrie, F., Lemaire, S., Poirier, G., Pelletier, G., and Boucher, R. (1971): Adenohypophyseal secretory granules. I. Their phosphorylation and association with protein kinase. *Journal of Biological Chemistry*, 246:7311–7317.

Lake, R. S. (1972): Chromatin-associated lysine-rich histone (F1) phosphokinase activity in mitotic chinese hamster cells. *Federation Proceedings*, 31:495Abs.

Lake, R. S., Goidl, J. A., and Salzman, N. P. (1972): F1-histone modification at metaphase in chinese hamster cells. *Experimental Cell Research*, 73:113–121.

Langan, T. A. (1967): A phosphoprotein preparation from liver nuclei and its effect on the inhibition of RNA synthesis by histones. In: *Regulation of Nucleic Acid and Protein Biosynthesis*, Vol. 10, edited by V. V. Koningsberger and L. Bosch, pp. 233–242. Elsevier Publishing Company, Amsterdam.

Langan, T. A. (1968a): Histone phosphorylation: Stimulation by adenosine 3′,5′-monophosphate. *Science*, 162:579–581.

Langan, T. A. (1968b): Phosphorylation of proteins of the cell nucleus. In: *Regulatory Mechanisms for Protein Synthesis in Mammalian Cells*, edited by A. San Pietro, M. R. Lamborg, and F. T. Kenny, p. 101. Academic Press, New York.

Langan, T. A. (1969a): Phosphorylation of liver histone following the administration of glucagon and insulin. *Proceedings of the National Academy of Sciences*, 64:1276–1283.

Langan, T. A. (1969b): Action of adenosine 3′,5′-monophosphate-dependent histone kinase *in vivo*. *Journal of Biological Chemistry*, 244:5763–5765.

Langan, T. A. (1971a): Cyclic AMP and histone phosphorylation. *Annals of the New York Academy of Sciences*, 185:166–180.

Langan, T. A. (1971b): Phosphorylation of separate sites in lysine-rich histone by cyclic AMP-dependent and independent protein kinases. *Federation Proceedings*, 30:1089Abs.

Langan, T. A., Rall, S. C., and Cole, R. D. (1971): Variation in primary structure at a phosphorylation site in lysine-rich histones. *Journal of Biological Chemistry*, 246:1942–1944.

Langan, T. A., and Smith, L. K. (1967): Phosphorylation of histones and protomines by a specific protein kinase from liver. *Federation Proceedings*, 26:603Abs.

Larner, J., and Sanger, F. (1965): The amino acid sequence of the phosphorylation site of

muscle uridine diphosphoglucose α-1,4-glucan α-4-glucosyl transferase. *Journal of Molecular Biology*, 11:491–500.
Larner, J., and Villar-Palasi, C. (1971): Glycogen synthetase and its control. In: *Current Topics in Cellular Regulation*, Vol. 3, edited by B. L. Horecker and E. R. Stadtman, pp. 196–236. Academic Press, New York.
Larner, J., Villar-Palasi, C., and Brown, N. E. (1969): Uridine diphosphate glucose: α-1,4-glucan α-4-glucosyltransferase in heart. Two forms of the enzyme, interconversion reactions and properties. *Biochimica et Biophysica Acta*, 178:470–479.
Lemaire, S., Pelletier, G., and Labrie, F. (1971): Adenosine 3',5'-monophosphate-dependent protein kinase from bovine anterior pituitary gland. II. Subcellular distribution. *Journal of Biological Chemistry*, 246:7303–7310.
Li, H.-C. (1972): *Personal communication*.
Lipmann, F. (1933): Über die bindung der phosphorsäure in phosphorproteinen. I. Mitteilung: Isolierung einer phosphorhaltigen aminosäure (serinphosphorsäure) aus casein. *Biochemische Zeitschrift*, 262:3.
Lipmann, F., and Levene, P. A. (1932): Serinephosphoric acid obtained on hydrolysis of vitellinic acid. *Journal of Biological Chemistry*, 98:109–114.
Loeb, J. E., and Blat, C. (1970): Phosphorylation of some rat liver ribosomal proteins and its activation by cyclic AMP. *FEBS Letters*, 10:105–108.
Lorini, M., Pinna, L. A., Moret, V., and Siliprandi, N. (1965): Localization of mitochondrial phosvitin kinase. *Biochimica et Biophysica Acta*, 110:636–639.
Louie, A. J., and Dixon, G. H. (1972): Trout testis cells. II. Synthesis and phosphorylation of histones and protamines in different cell types. *Journal of Biological Chemistry*, 247:5498–5505.
Maeno, H., and Greengard, P. (1972): Phosphoprotein phosphatases from rat cerebral cortex. *Journal of Biological Chemistry*, 247:3269–3277.
Maeno, H., Johnson, E. M., and Greengard, P. (1971): Subcellular distribution of adenosine 3',5'-monophosphate-dependent protein kinase in rat brain. *Journal of Biological Chemistry*, 246:134–142.
Majumder, G. C., and Turkington, R. W. (1971a): Adenosine 3',5'-monophosphate-dependent and -independent protein phosphokinase isoenzymes from mammary gland. *Journal of Biological Chemistry*, 246:2650–2657.
Majumder, G. C., and Turkington, R. W. (1971b): Hormonal regulation of protein kinases and adenosine 3',5'-monophosphate-binding protein in developing mammary gland. *Journal of Biological Chemistry*, 246:5545–5554.
Majumder, G. C., and Turkington, R. W. (1972): Hormonal regulation of phosphorylation of specific proteins in plasma membrane and ribosomes of cultured mammary cells. *Federation Proceedings*, 31:486Abs.
Makman, M. H., and Klein, M. I. (1972): Expression of adenylate cyclase, catecholamine receptor, and cyclic adenosine monophosphate-dependent protein kinase in synchronized culture of Chang's liver cells. *Proceedings of the National Academy of Sciences*, 69:456–458.
Makman, R. S., and Sutherland, E. W. (1965): Adenosine 3',5'-phosphate in *Escherichia coli*. *Journal of Biological Chemistry*, 240:1309–1314.
Mallette, L. E., Neblett, M., Exton, J. H., and Langan, T. A. (1973): Phosphorylation of lysine-rich histone in the isolated, perfused rat liver: Effects of glucagon, cyclic AMP and insulin. *Journal of Biological Chemistry* (*in press*).
Mansour, T. E. (1972): Phosphofructokinase. In: *Current Topics in Cellular Regulation*, Vol. 5, edited by B. L. Horecker and E. R. Stadtman, pp. 1–46. Academic Press, New York.
Martelo, O. J., Woo, S. L. C., Reimann, E. M., and Davie, E. W. (1970): Effect of protein kinase on ribonucleic acid polymerase. *Biochemistry*, 9:4807–4813.
Marushige, K., Ling, V., and Dixon, G. H. (1969): Phosphorylation of chromosomal basic proteins in maturing trout testis. *Journal of Biological Chemistry*, 244:5953–5958.
Mayer, S. E., and Krebs, E. G. (1970): Studies of the phosphorylation and activation of skeletal muscle phosphorylase and phosphorylase kinase *in vivo*. *Journal of Biological Chemistry*, 245:3153–3160.

Mayer, S. E., and Stull, J. T. (1971): Cyclic AMP in skeletal muscle. *Annals of the New York Academy of Sciences,* 185:433–448.
Meisler, M. H., and Langan, T. A. (1969): Characterization of a phosphatase specific for phosphorylated histones and protamine. *Journal of Biological Chemistry,* 244:4961–4968.
Merlevede, W., and Riley, G. A. (1966): The activation and inactivation of phosphorylase phosphatase from bovine adrenal cortex. *Journal of Biological Chemistry,* 241:3517–3524.
Miyamoto, E., Kuo, J. F., and Greengard, P. (1969a): Adenosine 3′,5′-monophosphate-dependent protein kinase from brain. *Science,* 165:63–65.
Miyamoto, E., Kuo, J. F., and Greengard, P. (1969b): Cyclic nucleotide-dependent protein kinases. III. Purification and properties of adenosine 3′,5′-monophosphate-dependent protein kinase from bovine brain. *Journal of Biological Chemistry,* 244:6395–6402.
Miyamoto, E., Petzold, G. L., Harris, J. S., and Greengard, P. (1971): Dissociation and concomitant activation of adenosine 3′,5′-monophosphate-dependent protein kinase by histone. *Biochemical and Biophysical Research Communications,* 44:305–312.
Murray, A. W., and Froscio, M. (1971): Cyclic adenosine 3′:5′-monophosphate and microtubule function: Specific interaction of the phosphorylated protein subunits with a soluble brain component. *Biochemical and Biophysical Research Communications,* 44:1089–1095.
Neblett, M. S., and Crofford, O. B. (1972): Effects of insulin and epinephrine on the phosphorylation of histones in rat adipose tissue cells. *Federation Proceedings,* 31:495Abs.
Ord, M. G., and Stocken, L. A. (1968): Variations in the phosphate content and thiol/disulphide ratio of histones during the cell cycle. Studies with regenerating rat liver and sea urchins. *Biochemical Journal,* 107:403–410.
Pastan, I. (1970): *Personal communication.*
Pastan, I., and Perlman, R. (1970): Cyclic adenosine monophosphate in bacteria. *Science,* 169:339–344.
Perlman, R. L., and Pastan, I. (1968): Regulation of β-galactosidase synthesis in *Escherichia coli* by cyclic adenosine 3′,5′-monophosphate. *Journal of Biological Chemistry,* 243:5420–5427.
Pierre, M., and Loeb, J. E. (1971): A cyclic AMP binding protein of rat liver: Interaction with an histone kinase. *Biochimie,* 53:727–734.
Pinna, L. A., Clari, G., and Moret, V. (1971): Rat liver cytosol phosphoprotein: Purification and enzymatic phosphorylation and dephosphorylation. *Biochimica et Biophysica Acta,* 236:270–278.
Posner, J. B., Hammermeister, K. E., Bratvold, G. E., and Krebs, E. G. (1964): The assay of adenosine-3′,5′-phosphate in skeletal muscle. *Biochemistry,* 3:1040–1044.
Posner, J. B., Stern, R., and Krebs, E. G. (1965): Effects of electrical stimulation and epinephrine on muscle phosphorylase, phosphorylase b kinase, and adenosine 3′,5′-phosphate. *Journal of Biological Chemistry,* 240:982–985.
Prince W. T., Berridge, M. J., and Rasmussen, H. (1972): Role of calcium and adenosine-3′:5′-cyclic monophosphate in controlling fly salivary gland secretion. *Proceedings of the National Academy of Sciences,* 69:553–557.
Rabinowitz, M. (1962): Protein kinases. In: *The Enzymes,* Vol. 6, edited by P. D. Boyer, H. Lardy, and K. Myrback, pp. 119–131. Academic Press, New York.
Rabinowitz, M., and Lipmann, F. (1960): Reversible phosphate transfer between yolk phosphoprotein and adenosine triphosphate. *Journal of Biological Chemistry,* 235:1043–1050.
Rall, S. C., and Cole, R. D. (1971): Amino acid sequence and sequence variability of the amino-terminal regions of lysine-rich histones. *Journal of Biological Chemistry,* 246:7175–7190.
Rall, T. W., and Sutherland, E. W. (1958): Formation of a cyclic adenine ribonucleotide by tissue particles. *Journal of Biological Chemistry,* 232:1065–1076.
Rall, T. W., Sutherland, E. W., and Berthet, J. (1957): The relationship of epinephrine and glucagon to liver phosphorylase. IV. Effect of epinephrine and glucagon on the reactivation of phosphorylase in liver homogenates. *Journal of Biological Chemistry,* 224:463–475.
Rall, T. W., Sutherland, E. W. and Wosilait, W. D. (1956): The relationship of epinephrine and glucagon to liver phosphorylase. III. Reactivation of liver phosphorylase in slices and in extracts. *Journal of Biological Chemistry,* 218:483–495.

Ramaley, R. F., Bridger, W. A., Moyer, R. W., and Boyer, P. D. (1967): The preparation, properties, and reactions of succinyl coenzyme A synthetase and its phosphorylated form. *Journal of Biological Chemistry*, 242:4287-4298.

Rasmussen, H. (1970): Cell communication, calcium ion, and cyclic adenosine monophosphate. *Science*, 170:404-412.

Rasmussen, H., and Tenenhouse, A. (1968): Cyclic adenosine monophosphate, Ca^{++}, and membranes. *Biochemistry*, 59:1364-1370.

Reddi, A. H., Ewing, L. L., and Williams-Ashman, H. G. (1971): Protein phosphokinase reactions in mammalian testis. Stimulatory effects of adenosine 3':5'-cyclic monophosphate on the phosphorylation of basic proteins. *Biochemical Journal*, 122:333-345.

Reimann, E. M., Brostrom, C. O., Corbin, J. D., King, C. A., and Krebs, E. G. (1971): Separation of regulatory and catalytic subunits of the cyclic 3',5'-adenosine monophosphate-dependent protein kinase(s) of rabbit skeletal muscle. *Biochemical and Biophysical Research Communications*, 42:187-194.

Reimann, E. M., and Walsh, D. A. (1970): Characterization of the adenosine 3',5'-monophosphate-stimulated protein kinase from rabbit skeletal muscle. *Federation Proceedings*, 29:601Abs.

Reimann, E. M., Walsh, D. A., and Krebs, E. G. (1971): Purification and properties of rabbit skeletal muscle adenosine 3',5'-monophosphate-dependent protein kinases. *Journal of Biological Chemistry*, 246:1986-1995.

Riley, W. D., DeLange, R. J., Bratvold, G. E., and Krebs, E. G. (1968): Reversal of phosphorylase kinase activation. *Journal of Biological Chemistry*, 243:2209-2215.

Riley, G. A., and Haynes, R. C., Jr. (1963): The effect of adenosine 3',5'-phosphate on phosphorylase activity in beef adrenal cortex. *Journal of Biological Chemistry*, 238:1563-1570.

Rizack, M. A. (1964): Activation of an epinephrine-sensitive lipolytic activity from adipose tissue by adenosine 3',5'-phosphate. *Journal of Biological Chemistry*, 239:392-395.

Robison, G. A., Butcher, R. W., and Sutherland, E. W. (1971): *Cyclic AMP*. Academic Press, New York.

Rosell-Perez, M., and Larner, J. (1964): Studies on UDPG-alpha-glucon transglucosylase. V. Two forms of the enzyme in dog skeletal muscle and their interconversion. *Biochemistry*, 3:81-88.

Rubin, C. S., Erlichman, J., and Rosen, O. M. (1972a): Molecular forms and subunit composition of a cyclic adenosine 3',5'-monophosphate-dependent protein kinase purified from bovine heart muscle. *Journal of Biological Chemistry*, 247:36-44.

Rubin, C. S., Erlichman, J., and Rosen, O. M. (1972b): Cyclic adenosine 3',5'-monophosphate-dependent protein kinase of human erythrocyte membranes. *Journal of Biological Chemistry*, 247:6135-6139.

Ruddon, R. W., and Anderson, S. L. (1972): Presence of multiple protein kinase activities in rat liver nuclei. *Biochemical and Biophysical Research Communications*, 46:1499-1508.

Santhanam, K., and Wagle, S. R. (1971): Studies on *in vitro* activation of high molecular weight pancreatic lipase. *Biochemical and Biophysical Research Communications*, 43:1369-1375.

Schlender, K. K., Wei, S. H., and Villar-Palasi, C. (1969): UDP-glucose: Glycogen α-4-glucosyltransferase I kinase activity of purified muscle protein kinase. Cyclic nucleotide specificity. *Biochimica et Biophysica Acta*, 191:272-278.

Selinger, Z., and Schramm, M. (1971): Control of reactions related to enzyme secretion in rat parotid gland. *Annals of the New York Academy of Sciences*, 185:395-402.

Sharma, S. K., and Talwar, G. P. (1970): Action of cyclic adenosine 3',5'-monophosphate *in vitro* on the uptake and incorporation of uridine into ribonucleic acid in ovariectomized rat uterus. *Journal of Biological Chemistry*, 245:1513-1518.

Shen, L. C., Villar-Palasi, C., and Larner, J. (1970): Hormonal alteration of protein kinase sensitivity to 3',5'-cyclic AMP. *Physiological Chemistry and Physics*, 2:536-544.

Shepherd, G. R., Noland, B. J., and Hardin, J. M. (1971): Histone phosphokinase levels in synchronized mammalian cells. *Experimental Cell Research*, 67:474-477.

Sherod, D. Johnson, G., and Chalkley, R. (1970): Phosphorylation of mouse ascites tumor cell lysine-rich histone. *Biochemistry*, 9:4611-4615.

Shin, S.-I., and Sato, G. H. (1971): Inhibition by actinomycin D, cycloheximide and puromycin of steroid synthesis induced by cyclic AMP in interstitial cells. *Biochemical and Biophysical Research Communications*, 45:501–507.

Shlatz, L., and Marinetti, G. V. (1971): Protein kinase mediated phosphorylation of the rat liver plasma membrane. *Biochemical and Biophysical Research Communications*, 45:51–56.

Siebert, G., and Humphrey, G. B. (1965): Enzymology of the nucleus. *Advances in Enzymology*, 27:239–288.

Siebert, G., Ord, M. G., and Stocken, L. A. (1971): Histone phosphokinase activity in nuclear and cytoplasmic cell fractions from normal and regenerating rat livers. *Biochemical Journal*, 122:721–725.

Silberstein, H., McAuslan, B. R., and August, J. T. (1972): Protein kinase and phosphate acceptor proteins of animal viruses. *Federation Proceedings*, 31:407Abs.

Smith, C. H., Brown, N. E., and Larner, J. (1971): Molecular characteristics of the totally dependent and independent forms of glycogen synthase of rabbit skeletal muscle. II. Some chemical characteristics of the enzyme protein and of its change on interconversion. *Biochimica et Biophysica Acta*, 242:81–88.

Soderling, T. R., Corbin, J. D., and Park, C. R. (1972): Hormonal effects on adipose tissue cyclic AMP-dependent protein kinase. *Federation Proceedings*, 31:440Abs.

Soderling, T. R., Hickenbottom, J. P., Reimann, E. M., Hunkeler, F. L., Walsh, D. A., and Krebs, E. G. (1970): Inactivation of glycogen synthetase and activation of phosphorylase kinase by muscle adenosine 3′,5′-monophosphate-dependent protein kinases. *Journal of Biological Chemistry*, 245:6317–6328.

Stellwagen, R. H. (1972): Induction of tyrosine aminotransferase in HTC cells by N^6, $O^{2'}$-dibutyryl adenosine 3′,5′-monophosphate. *Biochemical and Biophysical Research Communications*, 47:1144–1150.

Stone, D. B., and Mansour, T. E. (1967): Phosphofructokinase from the liver fluke *Fasciola hepatica*. I. Activation by adenosine 3′,5′-phosphate and by serotonin. *Molecular Pharmacology*, 3:161–176.

Strand, M., and August, J. T. (1971): Protein kinase and phosphate acceptor proteins in Rauscher murine leukaemia virus. *Nature New Biology*, 233:137–140.

Stull, J. T., Brostrom, C. O., and Krebs, E. G. (1972): Phosphorylation of the inhibitor component of troponin by phosphorylase kinase. *Journal of Biological Chemistry*, 247:5272–5274.

Sundararajan, T. A., Kumar, K. S. V. S., and Sarma, P. S. (1958): Some properties of protein phosphokinase from rabbit mammary gland. *Biochimica et Biophysica Acta*, 29:449–450.

Sung, M. T., Dixon, G. H., and Smithies, O. (1971): Phosphorylation and synthesis of histones in regenerating rat liver. *Journal of Biological Chemistry*, 246:1358–1364.

Sutherland, E. W., and Wosilait, W. D. (1955): Inactivation and activation of liver phosphorylase. *Nature*, 175:169–170.

Sy, J., and Richter, D. (1972a): Separation of a cyclic 3′,5′-adenosine monophosphate binding protein from yeast. *Biochemistry*, 11:2784–2787.

Sy, J., and Richter, D. (1972b): Content of cyclic 3′,5′-adenosine monophosphate and adenylyl cyclase in yeast at various growth conditions. *Biochemistry*, 11:2788–2791.

Takáts, A., Faragó, A., and Antoni, F. (1972): Adenosine 3′,5′-monophosphate dependent protein kinase in the lacrymal gland. *Biochimica et Biophysica Acta*, 268:77–80.

Takeda, M., Yamamura, H., and Ohga, Y. (1971): Phosphoprotein kinases associated with rat liver chromatin. *Biochemical and Biophysical Research Communications*, 42:103–110.

Tao, M. (1972a): Dissociation of rabbit red blood cell cyclic AMP-dependent protein kinase I by protamine. *Biochemical and Biophysical Research Communications*, 46:56–61.

Tao, M. (1972b): Rabbit red cell cyclic AMP-dependent protein kinase I: Reversible subunit interaction. *Biochemical and Biophysical Research Communications*, 47:361–364.

Tao, M., and Doerfler, W. (1972): Phosphorylation of adenovirus polypeptides. *European Journal of Biochemistry*, 27:448–452.

Tao, M., Salas, M. L., and Lipmann, F. (1970): Mechanism of activation by adenosine 3′:5′-

cyclic monophosphate of a protein phosphokinase from rabbit reticulocytes. *Proceedings of the National Academy of Sciences,* 67:408–414.
Thang, M. N., and Meyer, F. (1971): Activation of polynucleotide phosphorylase by 3′,5′-cyclic AMP, ATP-dependent protein kinase. *FEBS Letters,* 13:345–348.
Traugh, J. A., and Traut, R. R. (1972): Phosphorylation of ribosomal proteins of *E. coli* by protein kinase from rabbit skeletal muscle. *Biochemistry,* 11:2503.
Traut, R. R., and Lipmann, F. (1963): Activation of glycogen synthetase by glucose 6-phosphate. *Journal of Biological Chemistry,* 238:1213–1221.
Tsai, S.-C., and Vaughan, M. (1970): Activation of partially purified lipase from adipose tissue by ATP, $MgCl_2$ and cyclic 3′,5′-AMP. *Federation Proceedings,* 29:602Abs.
Turkington, R. W., and Riddle, M. (1969): Hormone-dependent phosphorylation of nuclear proteins during mammary gland differentiation *in vitro. Journal of Biological Chemistry,* 244:6040–6046.
Ullman, A., and Monod, J. (1968): Cyclic AMP as an antagonist of catabolite repression in *Escherichia coli. FEBS Letters,* 2:57–60.
Varrone, S., Ambesi-Impiombato, F. S., and Macchia, V. (1972): Stimulation by cyclic 3′,5′-adenosine monophosphate of RNA synthesis in a mammalian cell-free system. *FEBS Letters,* 21:99–102.
Vaughan, M., Berger, J. E., and Steinberg, D. (1964): Hormone-sensitive lipase and monoglyceride lipase activities in adipose tissue. *Journal of Biological Chemistry,* 239:401–409.
Villar-Falasi, C., and Larner, J. (1970): Glycogen metabolism and glycolytic enzymes. *Annual Reviews of Biochemistry,* 39:639–672.
Villar-Palasi, C., Larner, J., and Shen, L. C. (1971): Glycogen metabolism and the mechanism of action of cyclic AMP. *Annals of the New York Academy of Sciences,* 185:74–84.
Villar-Palasi, C., and Schlender, K. K. (1970): Purification and properties of muscle UDPG glycogen glucosyl transferase I kinase. *Federation Proceedings,* 29:938Abs.
Villar-Palasi, C., and Wenger, J. I. (1967): *In vivo* effect of insulin on muscle glycogen synthetase. Identification of the action pathway. *Federation Proceedings,* 26:563Abs.
von Euler, H., Hevesy, G., and Solodkowska, W. (1948): Turnover of ribosenucleic acid in the Jensen-sarcoma of the rat. *Arkiv für Kemi, Mineralogi och Geologi,* 26A:1–12.
Waddy, C. T., and MacKinlay, A. G. (1971): Protein kinase activity from lactating bovine mammary gland. *Biochimica et Biophysica Acta,* 250:491–500.
Walinder, O. (1968): Identification of a phosphate-incorporating protein from bovine liver as nucleoside diphosphate kinase and isolation of 1-^{32}P-phosphohistidine, 3-^{32}P-phosphohistidine, and N-Σ-^{32}P-phospholysine from erythrocytic nucleoside diphosphate kinase, incubated with adenosine triphosphate-^{32}P. *Journal of Biological Chemistry,* 243:3947–3952.
Walsh, D. A., Ashby, C. D., Gonzalez, C., Calkins, D., Fischer, E. H., and Krebs, E. G. (1971): Purification and characterization of a protein inhibitor of adenosine 3′,5′-monophosphate-dependent protein kinases. *Journal of Biological Chemistry,* 246:1977–1985.
Walsh, D. A., and Krebs, E. G. (1973): Protein kinases. In: *The Enzymes,* 3rd ed., edited by P. D. Boyer. Academic Press, New York (*in press*).
Walsh, D. A., Perkins, J. P., Brostrom, C. O., Ho, E. S., and Krebs, E. G. (1971): Catalysis of the phosphorylase kinase activation reaction. *Journal of Biological Chemistry,* 246:1968–1975.
Walsh, D. A., Perkins, J. P., and Krebs, E. G. (1968): An adenosine 3′,5′-monophosphate-dependent protein kinase from rabbit skeletal muscle. *Journal of Biological Chemistry,* 243:3763–3765.
Walton, G. M., Gill, G. N., Abrass, I. B., and Garren, L. D. (1971): Phosphorylation of ribosome-associated protein by an adenosine 3′:5′-cyclic monophosphate-dependent protein kinase: Location of the microsomal receptor and protein kinase. *Proceedings of the National Academy of Sciences,* 68:880–884.
Watson, G., and Langan, T. A. (1973): Effects of F1 histone and phosphorylated F1 histone on template activity of chromatin. *Federation Proceedings,* 32:588Abs.
Weller, M., and Rodnight, R. (1970): Stimulation by cyclic AMP of intrinsic protein kinase activity in ox brain membrane preparations. *Nature,* 225:187–188.

Wicks, W. D., and McKibbin, J. B. (1972): Evidence for translational regulation of specific enzyme synthesis by N^6, $O^{2'}$-dibutyryl cyclic AMP in hepatoma cell cultures. *Biochemical and Biophysical Research Communications*, 48:205–211.

Wilson, B. D., and Wright, R. L. (1970): Mechanisms of TSH action: Effects of dibutyryl cyclic AMP on RNA synthesis in isolated thyroid cells. *Biochemical and Biophysical Research Communications*, 41:217–224.

Yamamura, H., Inoue, Y., Shimomura, R., and Nishizuka, Y. (1972): Similarity and pleiotropic actions of adenosine 3',5'-monophosphate-dependent protein kinases from mammalian tissues. *Biochemical and Biophysical Research Communications*, 46:589–596.

Yamamura, H., Kumon, A., Nishiyama, K., Takeda, M., and Nishizuka, Y. (1971): Characterization of two adenosine 3',5'-monophosphate-dependent protein kinases from rat liver. *Biochemical and Biophysical Research Communications*, 45:1560–1566.

Yamamura, H., Kumon, A., and Nishizuka, Y. (1971): Cross-reactions of adenosine 3',5'-monophosphate-dependent protein kinase systems from rat liver and rabbit skeletal muscle. *Journal of Biological Chemistry*, 246:1544–1547.

Yamamura, H., Takeda, M., Kumon, A., and Nishizuka, Y. (1970): Adenosine 3',5'-cyclic phosphate-dependent and independent histone kinases from rat liver. *Biochemical and Biophysical Research Communications*, 40:675–682.

Yeung, D., and Oliver, I. T. (1971): The postnatal induction of serine dehydratase in rat liver. *Comparative Biochemistry and Physiology*, 40A:135–144.

Zahlten, R. N., Hochberg, A. A., Stratman, F. W., and Lardy, H. A. (1972): Glucagon-stimulated phosphorylation of mitochondrial and lysosomal membranes of rat liver in vivo. *Proceedings of the National Academy of Sciences*, 69:800–804.

Zapf, J., and Froesch, E. R. (1972): Protein kinases from rat skeletal muscle: evidence for six different fractions of the enzyme. *FEBS Letters*, 20:141–144.

Zubay, G., Schwartz, D., and Beckwith, J. (1970): Mechanism of activation of catabolite-sensitive genes: A positive control system. *Proceedings of the National Academy of Sciences*, 66:104–110.

Cyclic GMP

Nelson D. Goldberg, Robert F. O'Dea, and Mari K. Haddox

OUTLINE

I. Introduction 156
II. Occurrence and Distribution 157
III. Guanylate cyclase 161
    A. Tissue and Subcellular Distribution 161
    B. Kinetic Characteristics 164
    C. Hormonal Effects *In Vitro* 166
IV. Degradation and Excretion of Cyclic GMP 168
    A. Enzymatic Hydrolysis 168
    B. Cellular Extrusion 172
V. Effects Observed with Exogenous Cyclic GMP 173
    A. Intact Cells and Tissues 173
        1. Cyclic AMP-like effects of exogenous cyclic GMP 175
        2. Effects of exogenous cyclic GMP different from those of cyclic AMP 179
    B. Cell-free Systems 183
        1. Cyclic AMP-dependent protein kinases 183
        2. Cyclic GMP-dependent protein kinases 185
        3. Bacteria 187
VI. Agents and Conditions Which Alter the Steady-State Levels of Cyclic GMP in Biological Materials 189
    A. Mammalian Systems 189
        1. Effect of agents that promote cyclic AMP accumulation 189
        2. Acetylcholine 191
        3. Other hormonal and neurohumoral agents 196
        4. Steroids 200
        5. Nonhormonal substances 202
    B. Bacteria 206
VII. Speculation 207
VIII. Acknowledgments 213
IX. References 213

I. INTRODUCTION

Over the past dozen years center stage in the field of biological regulatory effectors has been occupied by cyclic adenosine 3',5' monophosphate (cyclic AMP). This cyclic nucleotide is now recognized as a key intracellular regulator of a number of cellular processes in a variety of living cells. Almost a decade ago the occurrence of a second cyclic nucleotide, cyclic 3',5'-guanosine monophosphate (cyclic GMP), was reported. In spite of the time elapsed and the unique structural similarity between the two nucleotides, information regarding the biological importance of cyclic GMP has until recently remained limited compared to the body of knowledge that has accumulated concerning the role of its predecessor, cyclic AMP. Testimony to the contrasting degree of attention focused on these two compounds over the past decade is the fact that in only one review (Hardman, Robison, and Sutherland, 1971) of cyclic nucleotides has the subject of cyclic GMP received any serious attention whereas during the same period dozens of reviews concerned exclusively with cyclic AMP have appeared, the most recent of which appear as the five other chapters in this volume.

There are a number of reasons for the slower progress in cyclic GMP research besides the analytical difficulties associated with its quantitative detection in biological materials. One limiting factor could perhaps be attributed to the strong allegiance to the "cyclic AMP approach" which dominates most of the investigations dealing with cyclic GMP. Although there may be a number of chemical, metabolic, and biological similarities between the two cyclic nucleotides, a sufficient number of dissimilarities have been uncovered to leave little doubt at the present time that cyclic GMP and cyclic AMP have entirely different biological roles. The properties unique to cyclic GMP may continue to go unappreciated, however, if the relativism with respect to cyclic AMP remains the only yardstick by which it is measured.

Another major reason for the slower rate of progress in defining the importance of cyclic GMP is the fact that a metabolic function for cyclic AMP (i.e., at least with respect to epinephrine- and glucagon-mediated glycogenolysis and phosphorylase activation) preceded and actually led to its discovery whereas no clue as to the biological role of cyclic GMP was uncovered until almost 7 years after its occurrence was reported. The observation, which has provided a clearer and more meaningful direction for recent investigations, was that acetylcholine-induced depression of cardiac contractility is associated with a relatively rapid accumulation of tissue cyclic GMP. This cholinergic action on the heart is opposite to that promoted by epinephrine which elevates myocardial cyclic AMP concentrations. Since this relationship between acetylcholine and cyclic GMP was established in cardiac muscle, cyclic GMP levels in other tissues have been shown to be increased by cholinergic as well as other biologically active agents which also promote cellular events that, in general, are opposite to those associated with an elevation of tissue cyclic AMP.

The picture beginning to develop, although still somewhat blurred and incomplete, represents the beginning of a story that may ultimately be explained in terms of a biological dualism between cyclic GMP and cyclic AMP. This concept, although it may represent biases exclusive to this reviewer, will serve as the basis for the organization of this review. Emphasis will, therefore, be placed on the evidence in support of the hypothesis that cyclic GMP and cyclic AMP are associated with promoting opposing biological events although experimental results to the contrary will, of course, also be presented.

II. OCCURRENCE AND DISTRIBUTION

The natural occurrence of cyclic GMP was first described by Ashman and co-workers (1963) in a report concerned with the identification of organic ^{32}P-phosphate-containing compounds present in rat urine following the administration of ^{32}P-inorganic phosphate. In that study the identification of cyclic GMP was made possible by comparing the physical, chemical, and biochemical properties of a ^{32}P-containing compound from urine that fluoresced under ultraviolet light with authentic cyclic GMP chemically synthesized and contributed by Dr. George Drummond. Semi-quantitative, isotope-dilution procedures were employed in the first report and subsequently (Price et al., 1967) to establish that the rate of cyclic GMP excretion in the urine of rats and humans was approximately $\frac{1}{2}$ to $\frac{1}{5}$ the rate found for cyclic AMP. It was postulated by this group of investigators that urinary cyclic GMP probably derives from body tissues where it may be generated from GTP by an enzymic reaction analogous to the one catalyzed by adenylate cyclase in which cyclic AMP is generated from ATP. This hypothesis has been borne out, and the supporting evidence will be dealt with in detail in the sections that follow. It is also of considerable interest that in the early studies concerned with urinary cyclic GMP excretion it was determined that cyclic GMP and cyclic AMP were the two major and probably the only naturally occurring organophosphate compounds present in urine as well as the only detectable nucleotides excreted with carbon-nitrogen skeleta and phosphate moieties intact. Attempts by Price, Ashman, and Melicow (1967), Hardman, Davis, and Sutherland (1966), Goldberg (*unpublished data*), and Steiner, Parker, and Kipnis (1970) to detect other cyclic nucleotides in urine or animal tissues have been unsuccessful. Cyclic AMP and cyclic GMP therefore appear to be the only naturally occurring 3′,5′-cyclic nucleotides and effects obtained with other, synthetic cyclic nucleotides, although of value in establishing structure-activity relationships, probably have little if any physiological relevance.

The occurrence of cyclic GMP as a natural constituent of mammalian

TABLE 1. *Tissue levels of cyclic GMP*

Tissue	Source	Concentration (reference in parentheses) (moles $\times$ 10^{-8}/kg, wet wt.)
Heart ventricle	Rat	11 (B); 4 (D); 6, 12 (E)^{1,2}; 12 to 24 (F)
	Calf	4 (E)^{1,2}
Heart atrium	Calf	8 (E)^{1,2}
Cerebellum	Mouse	63 (G,L); 61 (H); 31 (O)
	Rat	86 (E)^{1,2}; 52 to 92 (N)^{1,2}
	Rabbit	13 to 30 (E)^{1,2}
Cerebral cortex	Mouse	3.6 (L)
	Rabbit	8 to 14 (E)^{1,2}
Forebrain	Mouse	11 (H)
	Rat	20 to 35 (A); 4 (E)^{1,2}; 6 (R)
Striatum, thalamus, hippocampus, hypothalamus, brainstem spinal cord	Mouse	3 to 7 (G)
Brain (whole)	Rat	7 (D); 5 (E)^{1,2}
Lung	Rat	116 (E)^{1,2}; 48 to 62 (N)^{1,2}; 37 (R)
Liver	Rat	4 to 7 (A); 0.9 (D); 3 (E)^{1,2}; 2 (F); 1 to 5 (N)^{1,2}
Kidney	Rat	3 to 5 (A,D)
Stomach	Rat	3 to 4 (D)
Intestine, proximal	Rat	10 (D)
Intestine, distal	Rat	4 (D)
Intestinal, mucosal	Dog	2 to 3 (C)
	Rat	5 to 16 (F)
Intestine (whole)	Rat	6 (D); 12 (H)
Spleen	Rat	5 (D); 22 (R)
Thymus	Rat	56 (R)
Pituitary	Rat	0.8 (F); 0.7 (M)[1]
Testes	Rat	13 (R)
Ductus deferens	Rat	5 (I)[1]
Uterus	Rat	4 (D); 2 to 4 (S)[1]
Sperm	Bovine	12 (J)[3]
	Human	62 (K)
	Trout	23 to 30 (K)
	Sea urchin	10 to 90 (K)
Egg	Trout	5 (K)
	Sea urchin	n.d.[4] (K,V)[1]
Egg yolk	Chicken	5 (K)
Adipocytes	Rat	2 to 5 (F)[2]; 9 (BB)^{1,2}

Adipose tissue	Rat	6 (E)[1,2]
Lymphocytes	Human	100 to 150 (F)[3]; 90 to 250 (CC)[3]; 10 to 20 (GG)[3]
Macrophages	Guinea pig	50 to 100 (T)[3]
Platelets	Human	1 to 20 (DD)
Skeletal muscle	Rat	2 to 4 (D,E[1,2], F)
	Frog	2 (D)
	Lobster	5 (E)[1,2]
Skin	Human	1 to 4 (U)
	Mouse	10 (U)
	Rat	3 to 4 (D)
	Frog	4 (D)
Whole body	Cricket	250 to 380(D)
	Minnow	14 (D)
	Earthworm	9 (D)
Bacteria	E. coli	4 (Q)[2,5]; 78 (Q)[2,6]
	B. licheniformis	8 (Q)[2,5]; 70 (Q)[2,6]
Body Fluids		
CSF	Human	n.d.[4] to 7 nmoles/L (P)
Plasma	Human	1.8 to 6 nmoles/L (F,Y); 4 to 10 nmoles/L (Z)
Plasma, mesenteric	Dog	24.5 nmoles/L (AA)
Plasma, aortic	Dog	13.3 nmoles/L (AA)
Semen	Human	10 to 40 nmoles/L (K)
Urine	Human	1.4 μmoles/L (A); 0.94 μmoles/L (E); 0.3 to 1.8 μmoles/gm creatinine (P,W,X)
	Rat	0.5 μmoles/L (EE); 0.15 μmoles/L (FF)[7]
	Dog	0.3 to 0.6 μmoles/L (A)

(A) Goldberg et al., 1969; (B) George et al., 1970; (C) Schafer et al., 1971; (D) Ishikawa et al., 1969; (E) Kuo et al., 1972; (F) Steiner et al., 1972b; (G) Steiner et al., 1972c; (H) W. D. Lust and N. D. Goldberg, *unpublished data;* (I) Schultz et al., 1972a; (J) Garbers et al., 1971; (K) Gray, 1970; (L) Ferrendelli et al., 1970; (M) Peake et al., 1972; (N) Murad et al., 1971; (O) Ferrendelli et al., 1972; (P) Broadus et al., 1970; (Q) Goldberg et al., 1973; (R) W. J. George and N. D. Goldberg, *unpublished data;* (S) N. D. Goldberg, C. Sanford, D. K. Hartle, and M. K. Haddox, *unpublished data;* (T) J. W. Hadden, M. K. Haddox, and N. D. Goldberg, *unpublished data;* (U) N. D. Goldberg, M. K. Haddox, and J. Voorhees, *unpublished data;* (V) N. D. Goldberg, M. K. Haddox, D. K. Hartle, C. Blomquist, and J. W. Hadden, *unpublished data;* (W) Kaminsky, et al., 1970b; (X) Murad et al., 1972; (Y) Steiner et al., 1970; (Z) Ball et al., 1970, 1972; (AA) Blonde et al., 1972; (BB) Jarett et al., 1972; (CC) Hadden et al., 1972; (DD) N. D. Goldberg, M. K. Haddox, D. K. Hartle, and J. White, *unpublished data;* (EE) Hardman et al., 1969; (FF) M. K. Haddox, A. Delong, H. Rassmussen, and N. D. Goldberg, *unpublished data;* (GG) M. K. Haddox, J. W. Hadden, and N. D. Goldberg, *unpublished data.*

[1] Perfused or incubated *in vitro.*
[2] Converted to moles/kg wet weight from pmoles/mg protein assuming protein content 20% of wet weight.
[3] Expressed in terms 10^{-8} moles/kg upon converting pmoles/10^6 cells assuming that 10^{13} cells is equivalent to 1 kg (wet weight).
[4] n.d. — not detectable.
[5] Grown on succinate.
[6] Grown on glucose.
[7] Continuously perfused animal.

tissue (i.e., rat kidney, liver, and brain) was first reported by Goldberg et al. (1969). Ishikawa et al. (1969) established that cyclic GMP is widely distributed in nature by demonstrating its presence in a variety of tissues in the rat and other phylogenetic forms. Since these disclosures, a number of other laboratories have contributed information regarding the distribution of cyclic GMP in biological materials employing innovative analytical procedures for its quantitation (Murad et al., 1971; Steiner et al., 1972a,b; Kuo et al., 1972; Schultz et al., 1972a).

Table 1 is a summary of the information now available regarding the distribution and tissue levels of cyclic GMP. Cyclic GMP has been detected in all phyla of the animal kingdom examined including prokaryotes. No report has been made of cyclic GMP in plants, and one attempt in our laboratory several years ago to detect it in bean seedlings was unsuccessful. With our present knowledge of its ubiquitous distribution, it would be surprising if cyclic GMP were not present in plants. Cyclic AMP was not detected in plants until 10 years after its discovery in animal tissues (Galsky and Lippincott, 1969).

A second consideration is that the absolute tissue levels of cyclic GMP can vary from tissue to tissue over a considerably broad range of 10^{-8} to almost 10^{-6} moles/kg (wet weight). Although it is generally considered that tissue cyclic GMP levels range between 10^{-8} and 10^{-7} moles/kg, and in a given tissue are usually 1/10 to 1/100 the levels of cyclic AMP (10^{-7} to 10^{-6} moles/kg), the fact that concentrations of the two cyclic nucleotides are (or under certain conditions may become) quite comparable should also be appreciated. Examples of the latter situation occur in mouse cerebellum, rat thymus gland and lung, sperm from different sources, guinea pig macrophages, and some insects. There is no way at this time to assess the importance of the greater or lesser basal concentrations of cyclic GMP in different tissues or cells. The fact that cyclic GMP levels are in some instances similar to those of cyclic AMP has, however, served as a confidence factor for those in the field struggling against the tide of attention focused on cyclic AMP. In a more serious vein, these recent disclosures of relatively high levels of cyclic GMP in certain tissues may aid in resolving the disparity between the relative levels of the two cyclic nucleotides in body fluids (i.e., urine and plasma) where cyclic GMP concentrations are $\frac{1}{2}$ to $\frac{1}{5}$ those of cyclic AMP as compared to the 10- to 100-fold differences between the two cyclic nucleotide concentrations in some tissues.

The data in Table 1 also show that the absolute tissue levels of cyclic GMP vary from two- to threefold and in one case fivefold (i.e., rat liver) in a given tissue analyzed in a given laboratory and three- to fivefold and in one case eightfold (i.e., rat forebrain) for a given tissue when analyzed in different laboratories. There are, however, more instances of fairly close agreement than disagreement among laboratories, and the magnitude of the differences falls within the range often encountered among animals or from experiment to experiment. Considering the diversity of analytical procedures employed and the high degree of sensitivity and specificity demanded of the analytical approach, the data represent a tribute to the analytical skills of these investigators and especially to their technical help.

Along with the expected development of the field and a greater understanding of the factors which regulate cyclic GMP generation and removal, the degree of variation will undoubtedly be reduced. Assuming for a moment that the bulk of these differences are not due to any inherent error in analytical procedures, it might be concluded that they reflect a reasonably great sensitivity of the components involved in cyclic GMP metabolism to subtle cellular alterations. It also seems appropriate to emphasize that the values shown in Table 1 represent all of the data presently available from laboratory findings reported within a 3-year period prior to the writing of this chapter. They are, therefore, just the rudiments of what one might predict will develop into a large body of knowledge.

III. GUANYLATE CYCLASE

The first descriptions of an enzyme that catalyzed the generation of cyclic GMP from GTP (i.e., guanylate cyclase) by a reaction that appeared to be analogous to the one in which cyclic AMP is formed from ATP were reported independently by White, Aurbach, and Carlson (1969), White and Aurbach (1969), Hardman and Sutherland (1969), and Schultz, Bohme, and Munske (1969) within a very short period of time. A number of properties of guanylate cyclase which distinguish it from adenylate cyclase have been uncovered. Differences have been found with regard to apparent subcellular distribution, sensitivity to activating agents, effect of detergents, and substrate and cation requirements.

A. Tissue and Subcellular Distribution

Guanylate cyclase activity has been detected in all mammalian tissues examined to date, and the ability of cells in other animals and bacteria (Clark and Bernlohr, 1972a) to generate cyclic GMP from GTP has also been reported. Attempts to demonstrate guanylate cyclase activity in sea urchin and chicken eggs and mammalian sperm have been unsuccessful (Gray et al., 1970; Gray, 1970). On the other hand, the activity of guanylate cyclase in sea urchin sperm is two or three orders of magnitude greater than that measured in any mammalian cell (Gray, 1970). In mammalian tissues the assayable activity, *in vitro,* of guanylate cyclase varies over a 10-fold range in different organs with the greatest apparent activity in lung and the least in skeletal muscle (Hardman and Sutherland, 1969).

One unique property of guanylate cyclase that appears to distinguish it from adenylate cyclase is its apparent subcellular distribution. Unlike mammalian adenylate cyclase which is considered to be totally particulate, the major portion of guanylate cyclase activity appears to reside in the soluble fraction after homogenization of most tissues (White et al., 1969; White and Aurbach, 1969; Hardman and Sutherland, 1969; Schultz et al., 1969). Of the total enzyme activity found in rat lung, spleen, and liver homogenates, 80 to 90% was found to be present in the high-speed supernatant fraction (Hardman and Sutherland, 1969). A soluble designation for guanylate cyclase is, however, far from the general rule. Even in tissues in which a majority of the activity in homogenates appeared to be cytoplasmic, there was detectable activity (10 to 20%) in the particulate fraction. In some tissues (i.e., heart) the distribution between soluble and particulate fractions was intermediate whereas in others the major portion of the activity was found in the particulate fraction (i.e., rat small intestine). In addition, the enzyme activity present in certain lower phylogenetic forms, notably sea urchin sperm (Gray, 1970) and the bacterium *Bacillus licheniformis* (Clark and Bernlohr, 1972a), appears to be entirely particulate. One interpretation of these varying patterns of distribution is that different tissues and species may have marked differences in the subcellular locale of the enzyme; another is that guanylate cyclase may be associated with cellular membranes *in situ* but in some tissues may easily undergo dissociation during the homogenization or fractionation procedures. Indirect evidence for the latter was provided by Ishikawa et al. (1969) who demonstrated that Triton X-100 could produce a 2.5-fold increase in the activity of guanylate cyclase from a particulate fraction of rat small intestine. It was also of interest, from the standpoint of establishing a separate identity for guanylate cyclase, that Triton markedly depressed adenylate cyclase activity from the same tissue. More recently Hardman et al. (1971, 1972) have reported that non-ionic detergents could produce a seven- to 10-fold stimulation of particulate guanylate cyclase activity from rat lung, liver or spleen, but only a two- to threefold enhancement of the assayable activity in the $100,000 \times g$ supernatant fraction. The guanylate cyclase activity unmasked by solubilizing agents, therefore, can be equal to or greater than the activity found in tissues in which 80 to 90% of the total enzyme activity appeared to be soluble without addition of detergents. These findings strongly suggest that the soluble nature of guanylate cyclase in at least some tissues may be more apparent than real, arising from an *in vitro* artifact related to the sensitivity of the tissue to disruptive fractionation procedures, and that a major portion of the enzyme activity may be membrane bound.

Establishing the precise subcellular distribution of guanylate cyclase would aid in clarifying a number of important but still puzzling features of the possible involvement of cyclic GMP as an intracellular effector or "messenger" for extracellular biological signals. Evidence has been obtained establishing a relationship between cellular cyclic GMP accumulation and the action of certain neurohormones and polypeptide hormones (see Section VI) which are at this time believed to find their final action at the cell membrane. Compartmentation of guanylate cyclase in the cell membrane would support the possibility that stimulation of cellular cyclic GMP generation may result from an interaction of these agents with a membranous component linked to and intimately involved with the regulation of guanylate cyclase activity. On the other hand, a cytoplasmic localization of guanylate cyclase would suggest the internalization of the biological signal itself (e.g., steroids or prostaglandins) or make obligatory the participation of a third component. Interaction of the hormone with a receptor on the cell membrane would then result in the generation or more probably the transport of such a component (i.e., either extracellular or membranous) which could serve as an activator of cytoplasmic guanylate cyclase activity. Although an association of guanylate cyclase with the cell membrane might suggest a more direct link between hormone receptor and catalytic portion of the enzyme the participation even in this case of an additional obligatory component is not necessarily excluded. It appears that calcium may be a good candidate for an intermediary acting between hormone and soluble guanylate cyclase or an obligatory component for a membranous enzyme. This statement is based upon the recent report by Schultz et al. (1972b)[1] that cholinergically induced accumulation of cyclic GMP in smooth muscle is dependent upon the presence of calcium in the medium (see Section VI).

The information on subcellular distribution may also provide some insight into the problem encountered in demonstrating hormonal activation of guanylate cyclase *in vitro*. Most of these studies have been carried out with the soluble fraction of tissue homogenates or with homogenates from tissues in which the bulk of the activity after homogenization appears to be soluble. If guanylate cyclase activity in intact cells is particulate but becomes "solubilized" during homogenization or other preparatory procedures, it is conceivable that the activity examined may represent a com-

---

[1] From a presentation, based on the abstract cited, by Dr. G. Schultz at the Fifth International Congress on Pharmacology, San Francisco, Calif., July 23, 1972.

pletely activated form of the enzyme that has been dissociated from a regulatory or repressor component in the membrane and is, therefore, not further activable. Such an explanation might also aid in accounting for the greater than expected apparent activity of guanylate cyclase which appears to be comparable to that of adenylate cyclase in most instances. The latter is inconsistent with the fact that the tissue steady-state levels of cyclic AMP are one to two orders of magnitude greater than those of cyclic GMP.

Reports that guanylate cyclase appeared to be a cytoplasmic enzyme in some tissues lent support to the idea that hormones such as the steroids and thyroxine which enter the interior of the cell may be the agents responsible for regulating its activity. This concept was, perhaps, reinforced by the early reports (Hardman et al., 1966, 1969) that the reduced urinary excretion of cyclic GMP following hypophysectomy or adrenalectomy could be restored to near normal by the administration of cortisol alone in the latter instance or in combination with large doses of thyroxine or a mixture of hormones from the pituitary. The possibility that steroid hormones may indeed influence guanylate cyclase activity, but in a direction opposite to that implied by the studies just cited involving urinary excretion, is indicated by the observations that tissue steady-state levels of cyclic GMP may be elevated following adrenalectomy and diminished after steroid treatment (Steiner et al., 1972b; Goldberg et al., 1973) (see Section VI). The possibility that this action of steroids to lower tissue cyclic GMP levels may derive from a reduction or suppression of guanylate cyclase activity is supported by the recent observations of Thompson, Williams, and Kompton[2] that the measurable activity of guanylate cyclase in target tissues is elevated after adrenalectomy or orchidectomy. These investigators also found that treatment of intact animals with cortisol or aldosterone in some cases lowered the assayable enzyme activity in homogenates.

B. Kinetic Characteristics

A majority of the kinetic investigations of guanylate cyclase have been conducted with the enzyme activity recovered from the soluble fraction of mammalian cell homogenates (White and Aurbach, 1969; Hardman and Sutherland, 1969). Apparent K_m values for GTP of 0.02 to 0.1 mM for the activity from rat lung and 0.3 mM for a guanylate cyclase from beef lung have been determined. Kinetic constants for particulate guanylate cyclase from mammalian tissues have not yet appeared but activity in the particulate form from bacteria (*B. licheniformis*) (Clark and Bernlohr, 1972a) and from sea

---

[2] W. J. Thompson, *personal communication*.

urchin sperm (Gray, 1970) have been reported to have apparent K_m values with respect to GTP of 5 and 0.3 mM, respectively.

The requirement of guanylate cyclase activity for manganese is another of its properties which appears to set it apart from adenylate cyclase. The cation requirement of the latter can be satisfied almost equally as well by either Mg^{2+} or Mn^{2+}, whereas guanylate cyclase was shown to be 10 times more active in the presence of Mn^{2+} (apparent $K_m = 0.5$ mM) than with equimolar concentrations of Mg^{2+} (White and Aurbach, 1969; Hardman and Sutherland, 1969). Evidence that the reaction rate of guanylate cyclase was critically dependent upon an optimal ratio of GTP to Mn^{2+} was provided by White and Aurbach (1969) and Hardman and Sutherland (1969) who demonstrated that at fixed low concentrations of Mn^{2+} (1 to 3 mM), increasing concentrations of GTP (> 1 mM) produced a marked inhibition of enzyme activity. This inhibition could be overcome with higher concentrations of Mn^{2+} (10 mM). These findings are somewhat analogous to a situation encountered with adenylate cyclase (Birnbaumer et al., 1970) where Mg^{2+} has been postulated to interact with the enzyme protein at two sites: a substrate site for Mg^{2+} and ATP (or the ATP·Mg^{2+} chelate) and another which may represent an activator site for the cation alone. More recently, Hardman et al. (1972) have reported that Ca^{2+} (0.1 to 3 mM), although even less effective than Mg^{2+} in satisfying the cation requirement alone, could enhance guanylate cyclase activity several fold when present together with Mn^{2+} and concentrations of GTP equal to or greater than those of Mn^{2+}. One interpretation offered for these results was that Ca^{2+} competitively displaced Mn^{2+} from complex with GTP, resulting in an increased availability of free Mn^{2+} for interaction at the presumed activator site. This stimulatory effect of Ca^{2+} was not seen when Mn^{2+} was replaced by Mg^{2+}. Whether physiological levels of Ca^{2+} can exert a similar effect in intact tissue is unknown; however, the finding is an intriguing one in view of the demonstrated inhibitory effect of Ca^{2+} on hormone-induced activation of adenylate cyclase activity in different tissues (Birnbaumer et al., 1969; Drummond and Duncan, 1970). These findings again raise the possibility that there may be an intimate relationship between cholinergic action, calcium translocation, and tissue cyclic GMP production.

Although attempts to uncover a biologically active substance that serves as an activator of guanylate cyclase activity *in vitro* have been unsuccessful, a number of cellular metabolites have been found to be inhibitory to the enzyme. The list includes various nucleoside tri-, di-, and monophosphates as well as oxaloacetate and phosphoenolpyruvate (White and Aurbach, 1969; Hardman and Sutherland, 1969). However, the concentrations of these substances found to be effective were one to three orders of magnitude greater than the concentrations actually present in tissue except in the case of ATP. It would seem reasonable to conclude at this time that of these inhibitory metabolites only ATP deserves serious consideration as a possible physiologically important effector of guanylate cyclase activity. From the work of White and Aurbach (1969), it would seem that the inhibitory effect of ATP does not derive merely from a chelation of Mn^{2+} since 0.4 mM ATP was shown to produce a 50% inhibition of guanylate cyclase activity with high concentrations (10 mM) of Mn^{2+}. The possibility that ATP may serve as a suppressor of guanylate cyclase activity is of potential interest from the viewpoint of regulation since guanine nucleotides (i.e., GTP or GDP) have been reported to be relatively specific modifiers (i.e. enhancers) of adenylate cyclase activity (Rodbell et al., 1971). Therefore, high steady-state levels of both ATP and GTP (a situation which would be expected because changes in the levels of one triphosphate are usually reflected by the same relative changes in the others) could conceivably maximize the cellular expression of a cyclic AMP-linked signal while suppressing the generation of cyclic GMP, perhaps the mediator of a competing signal.

C. Hormonal Effects in vitro

Another characteristic of guanylate cyclase that distinguishes it from the enzyme that promotes cyclic AMP generation is its unresponsiveness, in cell-free systems, to agents that stimulate adenylate cyclase activity. Fluoride, which enhances the activity of adenylate cyclase in almost all mammalian tissues, did not stimulate the rate of cyclic GMP generation in any mammalian tissue preparation tested *in vitro* (White et al., 1969; White and Aurbach, 1969; Hardman and Sutherland, 1969; Hardman et al., 1971; Steiner et al., 1972*b*). In addition there was no effect of the halogen on the guanylate cyclase activity in sea urchin sperm (Gray, 1970; Hardman et al., 1971) or in *B. licheniformis* (Clark and Bernlohr, 1972*a*). However, adenylate cyclase activity in sea urchin sperm was also unaffected by fluoride (Hardman et al., 1971). The results of a number of the studies cited indicate that epinephrine, glucagon, adrenocorticotrophic hormone, parathyroid hormone, or antidiuretic hormone had no effect on guanylate cyclase activity in different preparations of mammalian tissue which contained adenylate cyclase activity responsive to one or more of these hormonal agents. The lack of an effect of these hormones *in vitro* is consistent with their ineffectiveness to promote tissue cyclic GMP accumulation *in vivo* (see Section VI). Dexamethasone, which can be shown to diminish cyclic GMP concentrations in some tissues, was reported to have no effect on guanylate cyclase activity in cell-free systems. Thyroxine was also ineffective under similar conditions (Steiner et al., 1972b).

Perhaps somewhat more surprising is the inability to demonstrate guanylate cyclase activation *in vitro* even by an agent such as acetylcholine, which has been found to stimulate the accumulation of cyclic GMP in a variety of tissues (see Section VI). More thorough investigations of possible hormone effects on the particulate rather than the soluble form of the enzyme may aid in uncovering the much sought after hormonal sensitivity in cell-free systems. It is also possible that hormone activation may derive from a reduction in the requirement of guanylate cyclase for GTP which may be present in very low concentrations in the vicinity of the enzyme, especially if the enzyme associated with the cell membrane is assumed to be representative of the species under control of membrane-active agents. The GTP concentration in the cell membrane or other particulate compartments becomes critical in this case. At present there is no information available regarding the concentrations of GTP in cell membranes. The possibility that they may be extremely low, relative to the cytoplasmic concentrations which are in the range of 0.3 to 0.5 mM[3] is suggested by the observations of Rodbell et al. (1971) that the GTP (or GDP) requirement for restoring full hormonal sensitivity to adenylate cyclase in isolated liver membrane preparations is in the range of 0.1 μM. This may be a requirement peculiar to only this membrane component or the experimental conditions, but if it were representative of GTP levels in this compartment, it is conceivable that hormone stimulation of guanylate cyclase and/or the influx of calcium resulting from interaction of the hormone with the membrane could result in a lowering of the enzyme requirement for GTP from the reported values of 0.02 to 0.3 mM to a much lower concentration which may approximate the cell membrane concentration. The millimolar (i.e., saturating) substrate concentrations of GTP usually used in cell-free systems could, therefore, obscure the stimulatory effect of a hormone acting through such a mechanism.

It is tempting to extend this line of reasoning and speculate further. If adenylate and guanylate cyclases were architecturally arranged in the membrane in such a way that the two enzymes were in close proximity, an activation of guanylate cyclase activity leading to a greater utilization of GTP for cyclic GMP synthesis could deprive adenylate cyclase of GTP, a component which apparently can impart hormonal sensitivity to this enzyme (Rodbell et al., 1971). A decrease in the sensitivity of cardiac muscle adenylate cyclase to adrenergic agents (Murad et al., 1962; Kuo et al.,

---

[3] N. D. Goldberg, *unpublished data.*

1972) and glucagon (Kuo et al., 1972) after exposure to acetylcholine (which promotes cyclic GMP accumulation) has in fact been observed. The consequence of the foregoing would be a decreased sensitivity of the cell to a cyclic AMP-linked signal during the expression of an incoming, opposing cyclic GMP-linked signal. The preceding is, of course, highly speculative and will be difficult but not impossible to test in the future.

IV. DEGRADATION AND EXCRETION OF CYCLIC GMP

A. Enzymatic Hydrolysis

Since another chapter in this volume is devoted to a complete coverage of the enzymic hydrolysis of cyclic nucleotides, the subject will be dealt with here only as it relates to the action of phosphodiesterase on cyclic GMP and the effect of cyclic GMP on the activity of the enzyme.

Prior to the discovery by Ashman et al. (1963) that cyclic GMP was a naturally occurring compound, it was noted by various investigators that cyclic GMP could serve as substrate for the phosphodiesterase believed to be responsible for the degradation of cyclic AMP. Drummond and Perrott-Yee (1961) reported that a soluble phosphodiesterase preparation from rabbit brain hydrolyzed cyclic GMP at 33% of the rate observed with cyclic AMP. A similar order of substrate preference was also reported for the activity present in dog heart (Nair, 1966) and rat brain (Cheung, 1967). Although these early studies were performed with concentrations of cyclic nucleotide substrates far greater (0.1 to 1 mM) than those encountered in mammalian tissues, they provided the first insight into the complex nature of the enzymic system catalyzing the hydrolysis of these compounds.

Since the early investigations with the enzyme, evidence has accumulated indicating that different phosphodiesterases either specific or more selective for cyclic GMP or cyclic AMP may exist. This was first suggested by the study of Brooker et al. (1968) which showed that cyclic GMP did not affect the rate of cyclic AMP hydrolysis by a soluble rat brain preparation. If a single enzyme were responsible for the hydrolysis of both cyclic nucleotides, cyclic GMP would have been expected to serve as a competitive substrate inhibitor of cyclic AMP degradation. By conducting their kinetic studies with substrate concentrations in the micromolar range, these investigators also established K_m values of 2 and 5 μM for cyclic AMP and cyclic GMP, respectively. Alternate substrate inhibition studies reported by O'Dea et al. (1970) and by Goldberg et al. (1970) with a soluble rat brain enzyme preparation also revealed that cyclic GMP did not alter the degradation of cyclic AMP whereas cyclic AMP could inhibit the hydrolysis of cyclic GMP competitively. The K_i value determined for cyclic AMP (10 μM) as an inhibitor of cyclic GMP hydrolysis was, however, almost an order of magnitude greater than the determined K_m values for either cyclic AMP or cyclic GMP (2 μM) in this system. These results suggested that different enzyme sites or, more probably, different species of the enzyme were responsible for the hydrolysis of the two cyclic nucleotides—one highly specific for cyclic AMP that did not degrade cyclic GMP and another more selective for cyclic GMP than cyclic AMP that could, however, degrade both. It was also shown in these kinetic studies that theophylline, puromycin, and papaverine, known modifiers of phosphodiesterase activity with millimolar substrate levels of cyclic AMP, also inhibited competitively the hydrolysis of both cyclic nucleotides when they were present in micromolar concentrations. Imidazole, a compound shown to produce an apparent activation of phosphodiesterase with millimolar substrate levels of cyclic AMP (Butcher and Sutherland, 1962), was, however, found to inhibit competitively cyclic GMP hydrolysis without altering the rate of degradation of cyclic AMP when this cyclic nucleotide was present in micromolar concentrations. The inhibitory effect of imidazole on cyclic GMP hydrolysis raised the interesting possibility that the variety of effects this compound produces (such as its cholinergic-like action on heart function) may be related to its inhibitory effect on cyclic GMP degradation and a resulting elevation of the tissue levels of this cyclic nucleotide. In the past these effects of imidazole have been attributed to its potential to lower cyclic AMP concentrations through the apparent stimulatory effect on cyclic AMP hydrolysis. The newer concept would be consistent with the observed association of cholinergic action with elevated tissue cyclic GMP concentration in heart and other tissues (see Section VI), although this has not yet been demonstrated to occur after imidazole treatment.

Beavo et al. (1970) and O'Dea et al. (1971) provided evidence that a partially purified phosphodiesterase from beef heart hydrolyzes both cyclic nucleotides but degrades cyclic GMP preferentially, a characteristic exhibited by one of the activities in crude brain extracts. With the beef heart enzyme, each cyclic nucleotide could interfere with the hydrolysis of the other but a preference was exhibited for cyclic GMP by K_m and K_i values for cyclic guanylate (ca. 1 μM) that were approximately an order of magnitude lower than the comparable values for cyclic AMP as substrate and as inhibitor.

Evidence that there are distinct molecular forms of phosphodiesterase that are highly selective or specific for the two cyclic nucleotides was first reported by Thompson and Appleman (1971a,b) and Kakiuchi et al. (1971). Separation of the activities from rat brain by gel filtration disclosed a high molecular weight protein (ca. 400,000) more selective on the basis of K_m for cyclic GMP than cyclic AMP. A low molecular weight enzyme (ca. 200,000) present in several rat tissues except liver appeared to degrade cyclic AMP exclusively, and in liver a high molecular weight protein was found which degraded cyclic GMP more selectively. Clark and Bernlohr (1972b) have recently reported the occurrence of a phosphodiesterase in *B. licheniformis* which hydrolyzes cyclic GMP but not cyclic AMP. It appears, therefore, that the degradation of the two cyclic nucleotides may indeed be handled by distinctly different proteins in some if not all tissues and that a selective means for controlling the tissue steady-state levels of cyclic GMP and cyclic AMP through different regulatory influences on these enzyme activities may be possible. This concept would be consistent with the observations that changes in tissue steady-state levels of cyclic GMP and cyclic AMP usually occur independently or reciprocally (see Section VI).

A comment should be made here regarding the numerous observations that cyclic GMP can serve as an apparent inhibitor of cyclic AMP hydrolysis in intact or broken cell systems and especially the conclusion that such an effect of cyclic GMP may account for alterations in the tissue steady-state levels of cyclic AMP (Rosen, 1970; Murad et al., 1970; Whitfield et al., 1971; Goren and Rosen, 1971). Although the conclusion seems to be a reasonable one from the observations made under the experimental conditions employed, it may not be representative of the situation regarding the hydrolysis of these two cyclic nucleotides under physiological conditions. The conclusion implies that increases in tissue cyclic GMP levels would be accompanied by increases in cyclic AMP concentration, a situation that is hardly ever encountered when changes in endogenous cyclic nucleotide levels have been examined in a reasonably homogeneous tissue. The question of physiological relevance arises primarily because concentrations of exogenous cyclic GMP used were far greater than those found in tissues even following treatment with agents which promote its accumulation. Even when micromolar concentrations of cyclic GMP are employed to inhibit cyclic AMP hydrolysis (Beavo et al., 1970; O'Dea et al., 1971), the physiological significance of the inhibitory effect could be questioned because the species of phosphodiesterase involved was more selective for cyclic GMP than for cyclic AMP (on the basis of apparent K_m values). It is possible that the species of phosphodiesterase that exhibits selectivity for cyclic GMP may not serve to promote cyclic AMP hydrolysis in intact cells because, as discussed earlier, in most tissues phosphodiesterases exist which exhibit lower K_m values for cyclic AMP and appear to hydrolyze cyclic AMP exclusively.

Further support for independent or reciprocal control of tissue concentrations of the two cyclic nucleotides that might derive from regulatory influences at the level of their enzymic hydrolysis has appeared. Beavo et al. (1971) made the intriguing observation that micromolar concentrations of cyclic GMP can stimulate two- to threefold the rate of phosphodiesterase catalyzed hydrolysis of cyclic AMP in a number of mammalian tissues. Manganiello and Vaughan (1972) have recently reported a similar phenomenon with a phosphodiesterase activity in fibroblasts. The stimulatory effect of cyclic GMP was seen by Beavo et al. (1971) with particulate preparations of the enzyme in most tissues and with a soluble form from liver and thymus. The kinetic parameter altered by cyclic GMP was not clearly defined in these studies; the effect occurred in a substrate concentration range of cyclic AMP between 0.5 and 40 μM in the presence of 2 μM cyclic GMP although much lower concentrations (0.08 μM) could be shown to enhance activity. If this stimulatory effect of cyclic GMP represents a physiological regulatory mechanism for controlling cyclic AMP levels, it would be manifest as a decline in cyclic AMP concentration with increasing intracellular cyclic GMP levels within a discrete concentration range of the two cyclic nucleotides.

Franks and MacManus (1971) have also reported a stimulatory effect of cyclic GMP on the rate of phosphodiesterase-catalyzed degradation of cyclic AMP by rat thymocytes. This effect occurred at low concentrations (0.1 to 10 μM) of cyclic GMP when the cyclic AMP level was 10 μM. At lower levels of cyclic AMP (1 μM), a similar range of cyclic GMP concentrations was inhibitory. High levels of cyclic GMP (> 10 μM) inhibited cyclic AMP hydrolysis under all conditions.

Another regulatory influence on the activity of a phosphodiesterase that could lead to a somewhat selective effect on the degradation of the cyclic nucleotides has been uncovered by Kakiuchi and his co-workers (1970a,b, 1971, 1972). They have found that a fraction of phosphodiesterase activity from rat cerebral cortex, when supplemented with a nondialyzable, heat-stable, phosphodiesterase-activating factor (PAF) (Cheung, 1969, 1971) could be stimulated by Ca^{2+} in such a way that the enzyme becomes more selective for degrading cyclic GMP. This effect of calcium (in conjunction with PAF) appeared to be brought about by increasing the rate and decreasing (by an order of magnitude) the K_m for cyclic GMP when it was the substrate while increasing the V_{max} but not affecting the K_m when cyclic AMP was the substrate (Kakiuchi et al., 1972).

The situation regarding the regulation of cyclic nucleotide phosphodiesterases, therefore, appears at this time to be a complex one involving different forms of the enzyme with differing selectivities for the two cyclic nucleotides, activator proteins, modulating effects of cations, and effects of the cyclic nucleotides themselves that may be of a regulatory nature.

B. Cellular Extrusion

In addition to the well-recognized role of phosphodiesterase in the removal of cyclic AMP and cyclic GMP, there is also evidence that the release of these cyclic nucleotides by the cell could represent another mechanism that may aid in providing for minute-to-minute control or for the general regulation of intracellular cyclic nucleotide levels.

The evidence at present suggests that cyclic GMP may be transferred to the extracellular space in some tissues, then transported via the plasma to other organs where a portion may be metabolized or excreted. Consistent with the above is the finding by Broadus et al. (1970a) that the cyclic GMP normally found in human urine arises almost entirely from the plasma by a process of simple glomerular filtration. It appears that the rate at which cyclic GMP is delivered to the kidney is rather constant over a given time interval and independent of urine volume changes. Greater than 50-fold fluctuations in urinary excretion rate resulting from the infusion of hypotonic or hypertonic solutions in dogs (Goldberg et al., 1969) or humans (Broadus et al., 1971) did not affect the excretion rate of cyclic GMP. Murad and Pak (1972) have, however, uncovered a diurnal variation in the urinary excretion of cyclic GMP and cyclic AMP in humans with peak excretion rates for both occurring in the afternoon. More complete details of these studies are presented in the chapter, "Clinical Studies and Applications of Cyclic Nucleotides" which appears in this volume.

Blonde, Wehmann, and Steiner (1972) have recently conducted studies in which cyclic nucleotide levels were measured in venous and arterial blood obtained from catheters implanted in selected vessels in the dog. They found that the plasma levels of cyclic GMP in superior mesenteric vein were double those from the aorta or hepatic vein whereas the plasma levels of cyclic AMP in all three vessels were equivalent. It was concluded that cyclic GMP released from the small intestine may account for a major fraction of this cyclic nucleotide in plasma and that the liver removes a significant fraction of circulating cyclic GMP. As did Broadus et al. (1970a), Blonde, Wehmann, and Steiner concluded that no more than 20% of the plasma cyclic GMP is excreted by the kidney. An additional 20% was calculated to be removed and metabolized by renal tissue. Taken collectively these results indicate that about 60% of the plasma cyclic GMP is removed from plasma by extrarenal tissue. Whether the circulating plasma cyclic nucleotides which are in the 10^{-8}M range contribute significantly to the intracellular cyclic nucleotide pools of the tissues which appear to be responsible for their removal from plasma would depend upon the rate at which they are hydrolyzed or excreted by these tissues. One point made by the distribution studies is that cyclic GMP may cross the cell membrane of some tissues from the extra- to the intracellular compartments and in the opposite direction as well. The fact that the volume of distribution of injected ^{3}H-cyclic GMP was found to exceed the extracellular space by 38% (Broadus et al., 1970a) supports this conclusion but does not aid in answering the question of why the addition of exogenous cyclic GMP is often without effect in intact cell preparations. It may be that hepatic cells are unique in their ability to permit the transport of cyclic GMP and that this organ accounts for the removal of a majority of the circulating cyclic GMP. This possibility is reinforced by the results of Exton et al. (1971) showing a distribution of exogenous cyclic GMP in liver that exceeded the sucrose (extracellular) space by almost 2.5-fold.

The observation by Gray (1970) that cyclic GMP is present in seminal fluid of normal and castrate males indicates that the cells of some of the tissues in the urogenital system exclusive of the testes may release cyclic GMP with or into this body fluid.

From the evidence just presented it would seem that a significant fraction of the cyclic GMP generated intracellularly may indeed be released from the cell. Just what part such a route of cyclic GMP removal may play in the overall scheme of controlling the levels of this cyclic nucleotide in a particular tissue or in the total organism remains to be established.

V. EFFECTS OBSERVED WITH EXOGENOUS CYCLIC GMP

A. Intact Cells and Tissues

In order to define a biological role for cyclic GMP it is of fundamental importance to demonstrate that the cyclic nucleotide itself can produce a definable effect on a specific cellular process. However, the effects of exogenous cyclic GMP should be consistent with the cellular expression of biological signals that can be demonstrated to be associated with an elevation of endogenous cyclic GMP concentrations. A majority of the attempts to uncover an influence of cyclic GMP on cellular function by introducing it or its dibutyryl derivative into intact cell systems have, with only a few but important exceptions, produced results which portray cyclic GMP as a poor substitute for cyclic AMP. Most of these effects produced by exogenous cyclic GMP are opposite to those produced by agents that promote increases in the tissue concentrations of endogenous cyclic GMP.

In experiments designed to uncover effects of exogenous cyclic GMP, concentrations of the cyclic nucleotide used (0.1 to 10 mM) represent as much as 10^6 times the concentration known to occur in most tissues. The justification for the use of these excessive concentrations of cyclic GMP apparently derives from a precedent established with cyclic AMP which has been shown to mimic certain hormone actions in intact cell preparations when introduced in millimolar concentrations. There is, however, no *a priori* reason to expect that if excessive concentrations of one compound produce a desirable effect that even greater excesses of another introduced extracellularly will similarly promote effects representative of the physiological actions it may produce intracellularly. There are good reasons to suspect that high artificially produced concentrations of cyclic GMP could produce effects in some tissues different than those brought about intracellularly by concentrations in a much lower range. One obvious reason (see Section V, B) is that cyclic GMP at concentrations approximately 100-fold greater than those of cyclic AMP can activate a protein kinase which is unquestionably more specific for cyclic adenylate. This protein kinase is now conceded to be the intracellular component involved in promoting at least some of the cellular events attributable to agents that stimulate cyclic AMP generation. Assuming that intracellular components may exist [e.g., the cyclic GMP-specific protein kinase discovered by Kuo and Greengard (1970a)] with dissociation (or activation) constants for cyclic GMP more closely approximating the tissue concentrations of this cyclic nucleotide (i.e., 10^{-8} M range) it would seem that cyclic GMP, in the submicromolar range normally present intracellularly, would support an interaction with such a specific cellular component. Much higher concentrations produced artificially, but never normally achieved intracellularly, might then affect, nonspecifically, other components such as the cyclic AMP-dependent protein kinase. An assumption inherent to this line of reasoning (which is expanded upon in Section VI, A, 5) is that events promoted by cyclic AMP-linked components are dominant over those normally linked to cyclic GMP.

Another possibility (also described in greater detail in Section VI, A, 2 and 5) is that some cyclic GMP-mediated events may require the presence of another component (e.g., extracellular calcium). An artificially produced increase in cyclic GMP alone, unaccompanied by an increase in the second obligatory component, might not provide for the interaction of cyclic GMP with its intended intracellular receptor; instead, at the relatively high concentrations of cyclic GMP usually used and in the absence of the companion component, the cyclic nucleotide may then act as a poor effector of a cyclic AMP-regulated system.

Since a majority of the reports describing effects of exogenous cyclic

GMP are at odds with the findings regarding changes in endogenous levels and considering that explanations may be forthcoming to aid in resolving the inconsistency, the section that follows is presented with reservation and a suggestion to the reader that it be viewed with an abundance of caution. Some of the observations that have been pursued beyond the point of merely adding the cyclic nucleotide and measuring a response may, nevertheless, be of potential value in helping to elucidate the biological role of cyclic GMP. The highlights of the work in this phase of cyclic GMP investigation are probably to be found in the few experiments, referred to earlier as the important exceptions, in which effects of this cyclic nucleotide or its dibutyryl derivative have been uncovered that mimic the actions of agents which have been found to promote the elevation of endogenous cyclic GMP concentrations.

1. *Cyclic AMP-Like Effects of Exogenous Cyclic GMP*

In one of the earliest studies in which effects of exogenous cyclic GMP were examined, Glinsmann and Hern (1969) found that at concentrations of 0.1 to 1 mM, cyclic GMP was nearly as effective as cyclic AMP in promoting glucose output, glycogenolysis, phosphorylase activation, and the conversion of glycogen synthetase from the *I* to the less active *D*-form in the isolated perfused rat liver. These effects of cyclic GMP occurred in the absence of any increase in tissue cyclic AMP levels, a finding which supports the idea that the glucagon-like action of cyclic GMP is not brought about indirectly through an elevation of tissue cyclic AMP levels. Similar results were obtained with a comparable *in vitro* system by Conn and Kipnis (1969), Conn et al. (1971), and Exton et al. (1971), whereas Guder and Wieland (1970) observed some of the same cyclic AMP-mimicking effects of cyclic GMP in isolated rat kidney tubules. In the more inquiring study of Exton et al. (1971), it was found that cyclic GMP was $\frac{1}{3}$ to $\frac{1}{2}$ as potent as cyclic AMP in activating glycogen phosphorylase, stimulating glucose production from ^{14}C-lactate, and promoting potassium release from liver cells. By studying the partition of cyclic AMP and cyclic GMP between perfusion fluid and liver, these investigators also attempted to explain the discrepancy between the apparent equipotencies of the two cyclic nucleotides in the isolated perfused liver and the much weaker activity of cyclic GMP (ca. $\frac{1}{100}$) compared to cyclic AMP in cell-free systems. Several interesting observations were made: first, cyclic AMP and cyclic GMP disappeared from the perfusion fluid at almost identical rates. The disappearance was attributed to removal by hepatic tissue. After perfusion with the cyclic nucleotides, the hepatic concentration of cyclic AMP was approximately 20% of the perfusate concentration while the tissue level of cyclic GMP was almost 70% of that in the circulating fluid. The concentration of ^{14}C-sucrose measured in liver following perfusion was also 20% of the perfusion fluid concentration and the water space of hepatic tissue 74%. It was concluded that cyclic AMP accumulated in the extracellular space while cyclic GMP accumulated in both intra- and extracellular hepatic fluid compartments and that the levels of the two cyclic nucleotides attained in their respective compartments after perfusion approached those of the recirculating fluid. These results led to the conclusion that the much greater intracellular accumulation of cyclic GMP than of cyclic AMP could probably account for the unexpected effectiveness of cyclic GMP in the perfused liver in contrast to its relative ineffectiveness on the cyclic AMP-dependent protein kinase. A similar difference in apparent distribution of the two cyclic nucleotides has been reported in the perfused rat heart (Hardman et al., 1971) although no effect of cyclic GMP in this tissue preparation has been reported. These results do not necessarily support the concept that cyclic GMP is transported into liver better than cyclic AMP but may indicate that cyclic AMP is metabolized more rapidly than cyclic GMP upon entering the hepatocyte. The results are difficult to reconcile with the findings of Thompson and Appleman (1971) that the species of phosphodiesterase in rat liver is more selective for cyclic GMP than cyclic AMP.

It was also of interest in the studies cited above to find that a small decrease in hepatic cyclic AMP concentration rather than an increase resulted after perfusion with cyclic GMP. This observation motivated other investigators in this group (Beavo et al., 1971) to examine the effects of cyclic GMP on phosphodiesterase-catalyzed degradation of cyclic AMP which ultimately led to the disclosure that micromolar concentrations of the former could indeed enhance the rate of cyclic AMP hydrolysis (see Section IV).

Some insight into the possible biological role of cyclic GMP may be extracted from another observation made by Exton et al. (1971) that insulin did not interfere with the effects produced by perfusion with cyclic GMP but could reverse those induced by submaximal concentrations of cyclic AMP. Since the metabolic effects of insulin on hepatic tissue (Kreutner and Goldberg, 1967; Bishop and Larner, 1967) are antagonistic to those of epinephrine and glucagon whose actions are believed to be mediated by an elevation of hepatic cyclic AMP, these results would be consistent with the proposal (see Section VI) that the action of such an agent may be associated with the generation of cellular cyclic GMP. In this situation the small concentration of endogenous cyclic GMP which would presumably be generated by an action of insulin would merely add to the larger millimolar pool already produced artificially that mimics the effects of micromolar cyclic AMP in this system (see introduction to this Section).

On the other hand, the possibility must still be considered that insulin does not affect cyclic GMP levels in liver (Goldberg et al., 1969) or in adipose tissue (Jarett et al., 1972) but can, as proposed earlier, bring about a lowering of tissue cyclic AMP concentration (Jefferson et al., 1967) perhaps through a direct or indirect action on phosphodiesterase activity (Senft et al., 1968). Metabolic effects of insulin in liver (Bishop et al., 1971; Nichols and Goldberg, 1972); muscle (Goldberg et al., 1967; Craig et al., 1968), and adipose tissue (Jarett et al., 1972) that are not attributable to an obligatory lowering of tissue cyclic AMP levels have, however, been reported and serve as a basis for the continuing effort to uncover an intracellular mediator of insulin action.

Cyclic GMP has been found to produce other cyclic AMP-like effects in liver. Friedman et al. (1971) reported that cyclic GMP (0.5 mM), like cyclic AMP or glucagon, produced a hyperpolarization of rat liver cells which was associated with an efflux of both potassium and calcium from the liver. It is not known what the relation is between the recognized intracellular actions of cyclic AMP and the cell membrane effects described here.

Effects of exogenous cyclic GMP on lipid metabolism in intact cells have in several instances been shown to be similar to those induced by cyclic AMP. Braun et al. (1969) reported that cyclic GMP (1 to 10 mM) was 20% as effective as cyclic AMP in promoting lipolysis in rat epididymal fat cells. Similarly, Murad et al. (1971) observed that cyclic GMP in the range of 0.1 to 5 mM, although less effective than cyclic AMP, could also inhibit theophylline-induced glycerol production in an unmodified Krebs-Ringer phosphate medium or, as does cyclic AMP, stimulate lipolysis in a phosphate-saline medium in which sodium was substituted for calcium and other cations. The proposal (see Section VI, A, 5) that a second component such as calcium may be required for some actions of cyclic GMP and may be inhibitory to some of those mediated by cyclic AMP would not be inconsistent with these observations. An additional speculation that could be offered from these observations is that cyclic AMP may produce cyclic GMP-like effects when a particular concentration of calcium is achieved intracellularly. In the studies of Murad et al. (1971), it was found that cyclic GMP in either of the two incubation media used caused an accumulation of tissue and medium cyclic AMP which was progressive with increasing concentrations of exogenous cyclic GMP. The latter taken together with the observation that cyclic GMP could be shown to inhibit the hydrolysis of cyclic AMP by fat cells as well as by kidney and liver homogenates led to the conclusion that the apparent effects of cyclic GMP may have resulted from an accumulation of cyclic AMP. In contrast to the above, Kitabchi et al. (1970) reported that cyclic AMP (2.5 to 5 mM) and, particularly, cyclic TMP produced an insulin-like effect, stimulating the conversion of glucose to carbon dioxide and lipid in isolated fat cells, whereas cyclic GMP and dibutyryl cyclic AMP exhibited an inhibitory effect on glucose oxidation. This group postulated that the action of insulin on fat cell metabolism was probably mediated by a cyclic nucleotide structurally similar to cyclic TMP. Bron and Rous (1971) found that both cyclic AMP and cyclic GMP when administered intraperitoneally could stimulate gluconeogenesis *in vivo* and inhibit the incorporation of ^{14}C-pyruvate and ^{3}H-acetate into fatty acids of mouse liver. When these investigators measured the incorporation of radioactive precursors into fatty acids extracted from the whole body, cyclic GMP was found to stimulate slightly the lipogenic process whereas under the same conditions cyclic AMP did not produce any significant effect.

The effects of exogenous cyclic GMP on adrenal steroidogenesis have also been examined by numerous laboratories. Glinsmann et al. (1969) and Mahafee et al. (1970) reported that cyclic GMP and cyclic AMP were almost equipotent in promoting steroidogenesis in rat adrenal quarters *in vitro*. Rivkin and Chasin (1971) observed that cyclic GMP (3.0 mM) enhanced steroidogenesis twofold in rat adrenal cell suspensions prepared by collagenase digestion of the glandular tissue, a response significantly enhanced by the inclusion of 5 mM theophylline in the incubation medium. A 10-fold increase in steroidogenesis in the presence of 8mM cyclic GMP was reported by Kitabchi et al. (1971) in trypsin-digested rat adrenal cells. These workers also observed that this response to cyclic GMP was partially blocked by caffeine (10 mM), an effect apparently resulting from an independent action of the methylxanthine because no phosphodiesterase activity was detectable in this cell preparation.

The growth of several malignant cell lines in culture was found by Heidrick and Ryan (1971) to be inhibited equally as well by cyclic GMP as by cyclic AMP, whereas the former appeared to be more effective in inhibiting the growth of nonmalignant cultured cells. The specificity of this effect remains unresolved since a qualitatively similar response was produced by 2′,3′-cyclic nucleotides as well.

The unique concentration-dependent effects of cyclic GMP to mimic an apparent stimulatory effect of cyclic AMP on the proliferation of rat thymocytes reported by Whitfield et al. (1971) are difficult to assess in view of the recent observations (see Section VI) that the process of cell division in a number of other cells is inhibited by cyclic AMP and that in human lymphocytes mitogenic action is linked to an accumulation of cellular cyclic

GMP (Hadden et al., 1972; Goldberg et al., 1973). However, it is noteworthy that the effects of cyclic GMP to stimulate DNA synthesis in thymocytes were brought about by concentrations as low as 10^{-11}M in an intact cell system. It remains to be proven whether the effects observed with cyclic AMP and/or cyclic GMP all represent the expression of mitogenic action or an acceleration of the mitotic process in a small population of cells in which the initial events in cell division have already been triggered.

2. Effects of Exogenous Cyclic GMP Different From Those of Cyclic AMP

Krause, Halle, and Wollenberger (1972) have recently demonstrated that dibutyryl (db)-cyclic GMP when added to pulsating embryonic rat heart cells in culture can mimic the cholinergic action of carbamylcholine; db-cyclic GMP or the cholinergic agent decreased the rate of spontaneous cardiac cell contractions 15%. There was no effect of 5'GMP, GTP, cyclic GMP, or butyrate on the pulsations of these cells. In the same system, db-cyclic AMP and epinephrine were shown to produce a positive chronotropic effect (i.e., 25% increase in rate of contraction). These observations are consistent with the discovery by George et al. (1970) that elevated myocardial cyclic GMP levels are associated with cholinergically induced depression of cardiac function and with the findings of others (Robison et al., 1965; Cheung and Williamson, 1965) that the positive chronotropic and inotropic effects of catecholamines are associated with increases in myocardial cyclic AMP concentrations. Experiments in our laboratory to demonstrate a chronotropic or inotropic effect of cyclic GMP or db-cyclic GMP with either embryonic chicken hearts or adult rat hearts under a variety of different experimental conditions have been unsuccessful. The failure on the part of one group and success on the part of another to demonstrate such effects are not too surprising considering that a similar situation existed and a number of years elapsed before rather consistent effects of cyclic AMP or its dibutyryl derivative could be demonstrated on heart function (Kukovetz and Pöch, 1970). The basis for the success with the embryonic heart cells in culture may stem from a developmental and/or species difference in the relative permeability to db-cyclic GMP or some as yet unknown factor.

Another example of an effect obtainable with db-cyclic GMP in intact tissue which, in a sense, points to an association between cyclic guanylate and cholinergic action as well as an antagonism between cyclic AMP and cyclic GMP mediated actions was reported by Puglisi, Berti, and Paoletti (1971). They found that db-cyclic AMP antagonized acetylcholine-induced contraction of isolated rat fundic strips whereas db-cyclic GMP (8 mM) mimicked the effect of the cholinergic agent. Still unexplained and yet to be confirmed was the observation by this group that atropine could block the db-cyclic GMP-induced contraction of this smooth muscle preparation. The conclusion drawn from these studies was that cyclic GMP may be involved in promoting acetylcholine release in this organ [4] No other evidence in support of such a possibility has appeared. It is possible that the blocking effect of atropine may be related to an inhibitory action of this drug on calcium transport which appears to be required for the expression of some cholinergic muscarinic actions.[1]

An effect recently obtained with exogenous cyclic GMP that may ultimately prove to be representative of another role played by this cyclic nucleotide is that it can stimulate the secretion of growth hormone. Cehovic, Posternak, and Charollais (1972a) and Cehovic, Robison, and Bass (1972b) reported that cyclic GMP (3 to 6 mM) increased, *in vitro,* the rate of growth hormone synthesis and release from anterior pituitaries of female rats and the synthesis of the hormone in glands from male rats. Cyclic AMP had no effect on either function in pituitaries from female rats but increased the release of growth hormone from those of male rats. A further involvement of cyclic GMP as a possible mediator of growth hormone release was uncovered by Peake, Steiner, and Daughaday (1972) who showed that cyclic GMP (0.1 to 10 mM) produced a significant increase in the release of immunoassayable growth hormone from pituitary explants of male rats and that aminophylline potentiated this effect of the cyclic nucleotide. Cyclic AMP had no effect on growth hormone release with or without aminophylline, but db-cyclic AMP was effective. The possibility that cyclic GMP may be involved as an intracellular regulator of growth hormone secretion is consistent with the conclusion drawn by Cehovic, Dettbarn, and Welsch (1971) that the process may be controlled cholinergically since they could demonstrate that paraoxon treatment increased plasma growth hormone levels significantly in both male and female rats after short (3 day) or long (14 day) term treatment with the cholinesterase inhibitor.

Further investigations will be required to clarify the situation with regard to possible sex differences in controlling growth hormone secretion, the different hormonal or neurohumoral factors which may serve to regulate it, and the possible roles of cyclic GMP and cyclic AMP as intracellular mediators of their actions. Evidence that cyclic AMP may, under certain conditions, stimulate growth hormone release has appeared (Cehovic

---

[4] From the presentation by Dr. R. Paoletti at the Fifth International Congress on Pharmacology, San Francisco, Calif., July 23, 1972.

et al., 1970; Steiner et al., 1970; Ewart and Taylor, 1971) and may indicate that growth hormone secretion is controlled by different signals which are expressed through the actions of either cyclic nucleotide. It may be worthwhile noting, however, that the concept that cyclic AMP serves as an effector of growth hormone secretion (Peake et al., 1972) or protein synthesis in the anterior pituitary (Labrie et al., 1971) derives from effects obtained with the dibutyryl derivative of this cyclic nucleotide. Dibutyryl cyclic AMP has been observed to mimic the effects of cyclic GMP in systems where cyclic AMP is either totally ineffective [i.e., growth hormone release (Peake et al., 1972)] or less effective than either of the former [i.e., stimulation of sperm motility and respiration (Garbers et al., 1971)]. In both of these cases, inhibitors of phosphodiesterase appear to induce greater increases in the levels of endogenous cyclic GMP than in those of cyclic AMP. It is possible, therefore, that (a) db-cyclic AMP may substitute for cyclic GMP rather than cyclic AMP in some biological systems or (b) db-cyclic AMP may inhibit phosphodiesterase-promoted hydrolysis of cyclic GMP more selectively than the degradation of cyclic AMP in some tissues and that the effects seen upon addition of db-cyclic AMP are those resulting from an accumulation of endogenous cyclic GMP. The different effects seen in a number of intact cell systems with cyclic AMP as compared to those obtained with its commonly used dibutyryl derivative (Bdolah and Schramm, 1965; Wilber et al., 1968; Solomon et al., 1970) certainly raise the question as to whether all of the effects seen with the latter are truly representative of those normally mediated by cyclic AMP.

Other effects of exogenous cyclic GMP in intact cell systems also suggest that its role may be distinctly different from that of cyclic AMP. Pagliara and Goodman (1970) found that cyclic GMP (0.5 mM) but not 5'GMP decreased the production of glucose and ammonia from glutamate and glutamine in rat renal cortical slices. In contrast, cyclic AMP (0.5 mM) was found in these studies to exert a positive effect on both gluconeogenesis and ammonia production. Although no unequivocal conclusions can be drawn from these results, it is noteworthy that opposite effects were produced by the two cyclic nucleotides on the gluconeogenic process. These effects are analogous to those expected of hormonal agents such as insulin which suppresses and epinephrine which facilitates gluconeogenesis. Bourgoignie et al. (1969) also claim to have noted a difference in the effects of the two cyclic nucleotides on toad bladder function. They reported that cyclic GMP (2 mM), unlike cyclic AMP, had no effect on toad bladder permeability to water, and yet both cyclic nucleotides induced an increase in short-circuit current of this organ. Along the same lines, it was noted in some early studies (Goldberg et al., 1969) that intra-arterial (renal) injection of cyclic

GMP in the dog increased urinary excretion of sodium with no apparent effect on potassium excretion.

Differences in the effects of the two cyclic nucleotides were also reported by Yeung and Oliver (1968) who found that parenteral administration of cyclic GMP, unlike cyclic AMP, did not induce hepatic serine dehydratase synthesis and by Jost et al. (1970) who found that cyclic AMP but not cyclic GMP injected into rats promoted phosphoenolpyruvate carboxykinase synthesis by the liver. Mitznegg, Hach, and Hein (1971) reported that the dibutyryl derivative of cyclic GMP or monobutyryl cyclic IMP or UMP did not inhibit oxytocin-induced contraction of the isolated rat uterus whereas the response to the hormone could be shown to be suppressed by cyclic AMP. Similarly, in studies with functional mouse adrenal tumor cells in culture, Dorval (1970) and Masui and Garren (1971) found that the effect of cyclic AMP to stimulate steroidogenesis could not be reproduced by cyclic GMP. In various other adrenal gland preparations (see above), effects on steroidogenesis by the two cyclic nucleotides have been reported to be similar. However, in the experiments of Masui and Garren with cultured adrenal tumor cells it was also observed that ^{3}H-thymidine incorporation into DNA and morphological differentiation of the cells promoted by either ACTH or cyclic AMP was not affected by cyclic GMP. Gericke et al. (1970) found that cyclic AMP and cyclic CMP but not cyclic GMP could inhibit the synthesis of antibodies *in vitro*.

An effect of exogenous cyclic GMP that is clearly opposite to the effect produced by exogenous cyclic AMP in an intact cell system and may provide a very special insight into the mechanism of cyclic GMP action has been reported by Whitfield and MacManus (1972). These investigators have found that intermediate concentrations of cyclic GMP (10^{-9} to 10^{-7}M), unlike low (10^{-11} to 10^{-10}M) or high (10^{-6}M) levels which raise cellular cyclic AMP concentrations and appear to stimulate thymocyte proliferation, have neither of the latter effects. However, the intermediate concentrations of cyclic GMP can, in the presence of increasing concentrations of extracellular calcium and at appropriate cellular cyclic AMP levels, inhibit the mitogenesis that appears to be induced by PGE_1 or cyclic AMP. Although these studies clearly point out that exogenous cyclic GMP and cyclic AMP can promote opposite effects in this system under certain conditions and that the effects produced by cyclic GMP are dependent upon a given concentration of extracellular calcium, a number of questions are also raised by the observations. First, is the puzzling situation with regard to the effect of cyclic AMP (and agents promoting its accumulation) stimulating rather than inhibiting mitogenesis (see Section VI, A, 5), and, second, does cyclic GMP produce its effects extra- or intracellularly, when exogenous concentrations in the range of 10^{-11} to 10^{-8}M are employed?

B. Cell-free Systems

Certain problems inherent to intact cell systems, such as the barrier imposed by the cell membrane, are eliminated in experiments conducted with cell-free systems. However, the interpretation of experimental results with the latter may be equally as difficult or misleading for a number of other reasons. In the introduction to a recently published monograph on cyclic AMP by Robison, Butcher, and Sutherland (1971), Sutherland pointed out the dilemma which can arise when attempting to relate the results obtained with a hormone in a cell-free system to the physiological action produced by the agent. Numerous effects of hormones on soluble enzymes have been uncovered over the years but more often than not the alterations induced or the components affected could not be associated with any alterations in cell function known to be produced by the hormone. The problem usually arises because concentrations of the agents tested far exceed those ever achieved intracellularly and the effect on the activity of the enzyme sooner or later falls out of favor as an unphysiological, nonspecific action. A similar view may ultimately be taken of the effects of cyclic GMP in cell-free systems observed in some of the early studies conducted. In these initial explorations, the effectiveness of cyclic GMP as a substitute for cyclic AMP was tested in a number of cell-free systems. The results showed that cyclic GMP could indeed replace cyclic AMP as an effector of the system or enzyme but that in general about 100 times more cyclic GMP than cyclic AMP was required to produce comparable effects. Although these results served to indicate that cyclic GMP probably does not serve as the natural effector of these systems, some investigators mistakenly interpreted them to indicate that cyclic GMP and cyclic AMP promote the same events but that cyclic GMP is a much less potent effector. This interpretation is difficult to reconcile with not only the much greater requirement for cyclic GMP but also the fact that the tissue concentrations of cyclic GMP are very often more than an order of magnitude lower than those of its adenosine counterpart.

1. *Cyclic AMP-Dependent Protein Kinases*

As early as 1962, before the natural occurrence of cyclic GMP was reported, Rall and Sutherland (1962) showed that cyclic GMP was 1/500 as potent as cyclic AMP in bringing about the activation of dog liver glycogen phosphorylase. Schlender et al. (1969) found that both cyclic nucleotides could bring about the same degree of activation of rabbit muscle glycogen synthetase-*I* kinase (later to be shown to be identical to the cyclic AMP-activatable protein kinase) but that the values of K_a were 0.07 and 9.9 mM for cyclic AMP and cyclic GMP, respectively. The same relative ineffectiveness of cyclic GMP compared to cyclic AMP was shown in the activation of glycogen synthetase-I kinase from rat skeletal muscle (Walaas et al., 1969) and rat liver (Glinsman and Hern, 1969). In the series of experiments conducted by Schlender et al. (1969), it was also demonstrated that the cyclic AMP-promoted activation of glycogen synthetase-I kinase was not antagonized by cyclic GMP but that the activation produced by 2 mM cyclic GMP and increasing concentrations of cyclic AMP were additive. When a combination of the two cyclic nucleotides was tested in the liver phosphorylase system (Murad et al., 1969), there was no antagonism by cyclic GMP.

Since the discovery by Walsh et al. (1968) that a protein kinase is the intracellular component with which cyclic AMP interacts in the initiation of the sequence of events leading to glycogen phosphorylase activation in skeletal muscle, it has been shown that the same or a similar cyclic AMP-activable protein kinase is also responsible for the phosphorylation and inactivation of glycogen synthetase (i.e., glycogen synthetase-I kinase) (Schlender et al., 1969; Soderling et al., 1970). Cyclic AMP is also probably responsible for the activation of tissue hormone-sensitive triglyceride lipase (Corbin et al., 1970) as well as the phosphorylation of other cellular proteins which may be involved in the expression of other hormone actions. With few exceptions cyclic GMP has been shown to be a relatively poor substitute for cyclic AMP as an activator of protein kinases from a variety of tissues in a number of different phyla (Kuo and Greengard, 1969). This relative ineffectiveness of cyclic GMP compared to cyclic AMP has been shown with protein kinases from rabbit reticulocytes (Tao, 1970), adipose tissue (Corbin and Krebs, 1969; Kuo and Greengard, 1969), brain and skeletal muscle (Kuo and Greengard, 1969; Kuo et al., 1970a; Miyamoto et al., 1969), adrenal cortex (Gill and Garren, 1970), frog bladder (Jard and Bastide, 1970), and rat liver (Kumon et al., 1972). More recently, Majumder and Turkington (1971a,b) have reported that a hormonally inducible protein kinase from mouse mammary gland can be stimulated about 200% by high concentrations of cyclic GMP and other cyclic nucleotides; apparent K_a values for cyclic AMP and cyclic GMP were determined to be 0.03 and 1.0 mM, respectively.

It is also difficult to demonstrate complexing of cyclic GMP to partially purified preparations of this kinase as determined by retention of cyclic nucleotide-protein complexes on cellulose ester filters (O'Dea et al., 1971). This is undoubtedly a reflection of a relatively high value of K_s for cyclic GMP as compared to K_s values of 10^{-9} to 10^{-8} M for cyclic AMP (Gilman,

1970; Haddox et al., 1972). Tao (1972) used the poor characteristics of cyclic GMP binding to the cyclic AMP-dependent protein kinase to advantage in purifying the regulatory subunit of this enzyme. Both cyclic GMP and cyclic AMP protected the regulatory subunit against heat inactivation, but only cyclic GMP was easily removed by dialysis from complex with the subunit.

There have been reports that interference with cyclic AMP binding to the protein kinase (i.e., the cyclic AMP-selective holoenzyme) by cyclic GMP only occurs at relatively high concentrations of the latter (Majumder and Turkington, 1971b) or when the ratio of cyclic GMP to cyclic AMP is quite large (Gilman, 1970). However, more recently it has been observed (Goldberg, 1972) that concentrations of cyclic GMP in the 10^{-9} to 10^{-8} M range significantly inhibit (20 to 30%) cyclic AMP binding to a skeletal muscle protein kinase. Also, the release of ^{3}H-cyclic AMP from complex with the regulatory subunit of the kinase freed of catalytic subunit is promoted by 10^{-10} to 10^{-8} M cyclic GMP. One intriguing aspect of cyclic GMP-induced interference with cyclic AMP binding is that with both the holoenzyme or isolated regulatory subunit the inhibitory effect of cyclic GMP occurs only at the extremely low concentrations of the cyclic nucleotide, disappearing at concentrations above 10^{-8} M (Goldberg, 1972).

2. Cyclic GMP-Dependent Protein Kinases

During the course of a systematic study of the distribution and characteristics of cyclic AMP-dependent protein kinase activity in a number of vertebrate and invertebrate species, Kuo and Greengard (1969) reported that for any given enzyme preparation, high concentrations (0.25 mM) of cyclic IMP, cyclic GMP, cyclic UMP, and cyclic CMP could stimulate protein kinase activity as effectively as optimal concentrations (5 μM) of cyclic AMP. However, it was also noted that with a partially purified protein kinase from lobster tail muscle cyclic GMP was at least as effective as cyclic AMP in the micromolar range in facilitating kinase activity. Upon further purification of the activity in lobster muscle, protein kinases more selective for cyclic GMP or cyclic AMP were separated (Kuo and Greengard, 1970a). The protein kinase more sensitive to activation by cyclic GMP, referred to as the cyclic GMP-dependent kinase, exhibited a K_a value of 0.08 μM for cyclic guanylate and a value 50-fold higher for cyclic AMP. A second kinase had a reverse order of sensitivity for the activators with apparent K_a values of 0.02 and 1.2 μM for cyclic AMP and cyclic GMP, respectively. A cyclic GMP-dependent protein kinase may also have been seen in brain, bladder, and uterine tissue of mammals (Greengard and Kuo, 1970). A protein kinase in the 50,000 × g supernatant fraction of rat cerebellar homogenates was found by Hoffman (1972) to have an apparent K_a for cyclic GMP of about 0.04 μM and a 25-fold greater requirement for cyclic AMP. An interesting effect of inorganic phosphate to enhance cyclic GMP stimulability and to suppress cyclic AMP activability was also reported for the kinase activity in cerebellar tissue.

Cyclic GMP-dependent protein kinase activity in arthropods has been the most extensively studied because animal tissues from this phylum appear to be a rich source of the enzyme. Like cyclic AMP-dependent kinase, the cyclic GMP-activable enzyme also appears to be cytoplasmic and inhibited by tissue components in crude extracts (Kuo and Greengard, 1970a; Kuo et al., 1971). However, a heat-stable protein factor prepared from tissues of mammals or arthropods and believed to be similar or identical to the inhibitor of cyclic AMP-dependent protein kinases (Appleman et al., 1966) was found to stimulate (twofold) cyclic GMP-dependent protein kinase activity from lobster muscle under certain assay conditions (Kuo and Greengard, 1971). The stimulatory effect was observed only when cyclic GMP concentrations were optimal and when a mixture of histone or arginine-rich histone was used as the protein substrate. The protein factor was inhibitory at low concentrations of cyclic GMP or with other histone fractions (Kuo and Greengard, 1971). There appears to be a wide variation in the relative distribution of cyclic GMP- and cyclic AMP-dependent protein kinases in different tissues and among similar tissues in different species (Kuo et al., 1971). In one tissue, the silk moth fat body, there was a preponderance of the cyclic GMP-dependent species, whereas in a number of other arthropod tissues the activity of this form of the kinase was equivalent to or less than that of the cyclic AMP-dependent form. Relative differences with respect to divalent cation requirements for cyclic AMP- and cyclic GMP-dependent activities have also been uncovered (Kuo and Greengard, 1970a; Kuo et al., 1971). In a study of the effects of various 3′,5′-cyclic nucleotide analogues as activators of protein kinase activities, it was found (Kuo and Greengard, 1970b) that the 3′-methylene cyclic phosphate analogue of cyclic AMP which is partially active in stimulating cyclic AMP-dependent kinase is inactive with the cyclic GMP-dependent enzyme. Although no entirely uniform pattern of differences among the kinases from different species and tissues has yet emerged, it seems that in arthropods the cyclic GMP- and cyclic AMP-dependent kinases are distinctly different biochemically and that they may, therefore, serve different physiological roles as suggested by Kuo et al. (1971). This conclusion will, of course, require that a correlation be made between an elevation of tissue cyclic GMP levels and the phosphorylation of an endogenous protein that can be identified as a component intimately involved in the regulation of the cellular process affected. The existence of a cyclic GMP-dependent protein kinase does serve to strengthen the position of cyclic GMP as a biological regulator independent of cyclic AMP. However, the hypothesis that all of the effects of cyclic GMP may be mediated through the activation of a protein kinase may require considerable modification since it has already been reported in at least two instances that an effect of cyclic GMP can be demonstrated in cell free systems [i.e. activation of phosphodiesterase (Beavo et al., 1971) and inhibition of transcription in bacteria (Riggs et al., 1971; Nissley et al., 1972)] where there is no apparent requirement for ATP.

The discovery of protein kinases with a much lower activation constant for cyclic GMP than cyclic AMP has led to the development of analytical procedures for the measurement of tissue and body fluid cyclic GMP based upon activation by the cyclic nucleotide of phosphotransferase activity (Kuo et al., 1972) and competition with ^{3}H-cyclic GMP for binding to the enzyme (Murad et al., 1971; Murad and Gilman, 1971).

3. Bacteria

In *Escherichia coli*, and other microorganisms not studied as extensively, cyclic AMP has been shown to function in a regulatory capacity which, under at least some conditions, involves controlling the expression of certain loci of the bacterial chromosome and in turn enhancing biosynthesis of inducible enzymes required for the metabolism of nutrients other than glucose and certain other carbohydrates. This regulatory role of cyclic AMP appears to involve the formation of a binary complex between the cyclic nucleotide and a specific bacterial protein designated as the cyclic AMP receptor protein (CRP) (Pastan and Perlman, 1970) or catabolite gene activator protein (CAP) (Riggs et al., 1971). The cyclic AMP–CRP complex is believed to interact with the promoter site of the *lac* and *gal* operons to facilitate RNA polymerase interaction with a region on the promoter site which, in turn, would initiate mRNA transcription for *beta*-galactosidase and galactokinase synthesis. (See Pastan and Perlman, 1970, for a review of this subject.) It was shown by Zubay et al. (1970) and by Emmer et al. (1970) that cyclic GMP at concentrations of 0.1 to 1 mM can inhibit the binding of cyclic AMP to CRP and also antagonize cyclic AMP-induced stimulation of *beta*-galactosidase synthesis in *in vitro* systems. In the experiments reported by Zubay et al. (1970) cyclic GMP at a concentration of 0.1 mM inhibited 70% of the binding of ^{3}H-cyclic AMP (5 μM) to CRP while 3'GMP or 5'GMP (0.1 mM) had little if any effect. A K_i value of approximately 10 μM for cyclic GMP as an inhibitor of cyclic AMP binding to CRP was determined by Emmer et al. (1970), and, in more recent studies conducted under equilibrium conditions with highly purified CRP from *E. coli*, a K_i value of 7 μM was determined. Purified CRP was also used by Anderson et al. (1972) to show that cyclic GMP and several analogues of cyclic AMP inhibited both the binding of ^{3}H-cyclic AMP to the receptor protein and cyclic AMP-induced transcription of *gal* operon DNA.

In the process of defining the interactions that take place between the various components involved in stimulating transcription of the *lac* operon, Riggs et al. (1971) found that purified CRP can complex with DNA in the absence of cyclic AMP, as determined by retention of labeled complexes on nitrocellulose filters, but that the binding of CRP to DNA is greatly enhanced by cyclic AMP. CRP complexing with DNA in the absence of cyclic AMP was shown to be completely inhibited by cyclic GMP at a concentration of 0.37 mM, the only concentration tested. If proven to be correct, the effect of cyclic GMP to dissociate CRP-DNA complex would represent the first manifestation in a functional system of an effect of this cyclic nucleotide that is not merely the reversal of a cyclic AMP-induced event. Nissley et al. (1972) have, however, presented evidence indicating that the binding of more highly purified CPR to DNA from bacterial, avian, fish, and mammalian sources is totally dependent upon cyclic AMP and inhibitable by cyclic GMP as a result of a competition between the two cyclic nucleotides for CRP. Whether the purification procedure used by Nissley et al. (1972) removed cyclic AMP or another component associated with certain forms of native CRP remains to be established.

Finally, the synthesis of *gal* mRNA and galactokinase has also been shown to be dependent on cyclic AMP and this stimulatory effect of cyclic AMP to be inhibitable by cyclic GMP (0.1 to 1 mM) in *in vitro* systems (Parks et al., 1971; Nissley et al., 1971). In the study by Nissley et al. (1971) it was shown that incubation of cyclic AMP, CRP, *gal* operon DNA, and RNA polymerase led to the formation of a rifampicin-resistant preinitiation complex which upon the addition of nucleoside triphosphates supported *gal* mRNA synthesis. Cyclic GMP (2 mM) prevented the formation of the preinitiation complex and caused its dissociation after it had formed.

One difference between the *E. coli lac* and *gal* operons is that guanosine 3'-diphosphate, 5'-diphosphate (ppGpp) (Cashel, 1969) is the only other naturally occurring nucleotide that, like cyclic GMP, can inhibit cyclic AMP-induced synthesis of galactokinase *in vitro* but that, unlike cyclic GMP, ppGpp stimulates the synthesis of *beta-* galactosidase in reconstituted bacterial systems. What relationship may exist, if any, between ppGpp and cyclic GMP other than that both derive from GTP and are phosphorylated at 3' and 5' positions of the ribosyl moiety remains to be determined.

Although the results of these investigations indicate that cyclic GMP can produce effects in bacterial systems, *in vitro*, which oppose those induced by cyclic AMP, a question that again arises concerns the relatively high concentrations (0.1 to 1 mM) of cyclic GMP that have been employed to produce these effects. The levels of cyclic guanylate in bacteria have never been found to exceed 1 μmole/kg, wet weight (see Section III). It should be noted, however, that the effects of cyclic GMP, with only one exception, have been observed in *in vitro* systems fortified with rather high levels of cyclic AMP to induce transcription. It is conceivable that lower levels of cyclic GMP may be more effective on components of the transcription mechanism or other systems in the intact organism *in vivo* when the concentrations of cyclic AMP (i.e., endogenous) are not as high as those created in the cell-free systems.

VI. AGENTS AND CONDITIONS WHICH ALTER THE STEADY-STATE LEVELS OF CYCLIC GMP IN BIOLOGICAL MATERIALS

At the present time the greatest progress in defining a biological role for cyclic GMP has been achieved with experiments in which a relationship has been established between changes that occur in steady-state levels of this cyclic nucleotide in tissues *in vivo* or in intact cells *in vitro* and alterations in cell function brought about by biologically active agents. This approach, although not an end in itself, is one that was of fundamental importance in elucidating the biological role of cyclic AMP. All aspects of the criterion established by Sutherland and Robison (1966) regarding changes in the tissue levels of cyclic AMP in relation to hormone action (i.e., that interaction of a hormone with the cell should lead to an increase in the concentration of the intracellular mediator preceding observable effects of the hormone on cell function or metabolism) have not yet been satisfied in the case of cyclic GMP. However, the identification of some agents which promote cellular cyclic GMP accumulation, which is probably the basic first step, has been accomplished.

A. Mammalian Systems

1. *Effect of Agents that Promote Cyclic AMP Accumulation*

The search for an agent whose action might be associated with an alteration in tissue cyclic GMP concentration was initiated by determining what influence, if any, agents known to promote cyclic AMP generation in tissue had on steady-state levels of cyclic guanylate. This information was of fundamental importance at the outset in helping to establish whether the biological roles of the two cyclic nucleotides might be similar or different. In the first study of this type (Goldberg et al., 1969), it was found that the administration of epinephrine or glucagon to rats had no effect on the concentration of hepatic cyclic GMP in spite of the fact that cyclic AMP levels in the liver were greatly elevated by these agents. It was also shown that the alloxan-induced diabetic state, which leads to increases in the concentrations of cyclic AMP in rat liver (Jefferson et al., 1967; Nichols and Goldberg, 1972), did not influence hepatic cyclic GMP levels. The conclusions drawn from these early studies (Goldberg et al., 1969) were that cyclic GMP generation is probably not induced by hormones which promote the generation of cyclic AMP and that the steady-state levels of the two cyclic nucleotides are probably regulated by different biological factors. Further support for this concept can be found in the results obtained more recently by several other investigators. Steiner et al. (1970) also found that glucagon treatment of rats had no effect on hepatic cyclic GMP levels and that incubation of fat cells with ACTH or epinephrine which caused the expected increases in cellular cyclic AMP levels did not alter the steady-state concentration of cyclic GMP. The ineffectiveness of cyclic AMP-linked hormones to promote increases in tissue cyclic GMP was also demonstrated by Kuo et al. (1972) with epinephrine and glucagon on heart ventricular slices and with norepinephrine on rabbit cerebellar slices. In the latter instance cyclic GMP levels in cerebellar tissue were diminished after exposure to norepinephrine, a response also noted by George et al. (1970) in isolated rat hearts perfused with isoproterenol. A separation of effects on the two cyclic nucleotides was also demonstrated in mouse forebrain shortly after electroconvulsive shock treatment (Goldberg et al., 1970) and within the first 60 sec following decapitation (Goldberg et al., 1970; Steiner et al., 1970, 1972c). Under these conditions dramatic increases in tissue cyclic AMP levels were associated with no change or a decrease in cyclic GMP levels. At later times after decapitation, cyclic GMP and cyclic AMP concentrations were both found to drop far below control levels; a more complicated relationship between the relative levels of the two cyclic nucleotides occurred at later times following ECS (Lust et al., 1972; Goldberg et al., 1973; see also Section VI, A, 5). The effect of cholera toxin to induce increases in cyclic AMP levels of dog intestinal mucosa was also shown to be unaccompanied by any changes in the levels of cyclic GMP (Schafer et al., 1970).

Hardman, Davis, and Sutherland (1966, 1969) were the first to demonstrate that the levels of the two cyclic nucleotides may be controlled by different biological factors when they reported that removal of the pituitary gland produced a rather selective lowering of urinary cyclic GMP excretion with little influence on cyclic AMP excretion. In other studies dealing with the urinary excretion of cyclic nucleotides, it has been found that the administration of glucagon (Hardman et al., 1969; Broadus et al., 1970b; Murad and Pak, 1972; Steiner et al., 1972b) or *beta* adrenergic agents (Kaminsky et al., 1970a; Ball et al., 1970, 1972; Murad and Pak, 1972) which promote increases in the rate of cyclic AMP excretion had little or no effect on the rate of excretion of cyclic GMP.

Using as a guideline the knowledge that tissue cyclic GMP and cyclic AMP levels are controlled independently and probably do not share identical biological roles, a search was initiated to uncover a physiologically important substance which could promote the accumulation of tissue cyclic GMP. The first substance found to produce such an effect was acetylcholine.

2. *Acetylcholine*

George, Polson, O'Toole, and Goldberg (1970) found that perfusion of isolated rat hearts with acetylcholine (0.37 μM) produced an elevation of myocardial cyclic GMP concentration coincident with the first signs of cholinergically induced depression of cardiac function. The increase of about 250% occurred within 10 sec after introduction of the cholinergic agent and declined by 30 sec to a value 40% above the control level. The levels of cyclic AMP were found to be unaffected initially but by 20 sec declined about 40% before returning to the control level at 30 sec. An approximately linear relationship was established between the increases detected in cardiac tissue cyclic GMP concentration and the decreases in the amplitude of cardiac contractility produced by acetylcholine. However, no significant correlation was found between cardiac muscle cyclic GMP changes and the decrease in cardiac rate produced by acetylcholine. It was reasoned at the time that heart rate was probably determined by only a small group of cells in the sinoatrial node discharging pulses at the highest rate, and it would have been necessary but not feasible to examine only that small number of cells in which the response was controlled to determine whether changes in cyclic GMP levels are in any way related to changes in the chronotropic response. It was also determined at that time and subsequently (Goldberg et al., 1973) that treatment of the isolated perfused hearts with atropine blocked the cholinergically induced effects on both cardiac function and elevation of myocardial cyclic GMP. The concentration of cardiac cyclic GMP was also found to be diminished (50%) after perfusion of hearts with isoproterenol alone or in the presence of theophylline.

George, Kadowitz, and Polson (1972) have contributed to our understanding of cholinergic action on heart function by demonstrating with electrically paced, isometrically contracting rat hearts that acetylcholine decreases the peak tension developed as well as the rate of peak tension development but not the time to peak tension. These results support the concept (which has often been disputed) that acetylcholine does indeed have a direct action on rat ventricular myocardium. In these studies it was again established that the cardiac action of acetylcholine is associated with an elevation of myocardial cyclic GMP concentrations and a decrease in cyclic AMP levels. However, it was reported by these investigators (George et al., 1972)[5] that a statistically significant correlation between cholinergic depression of contractility was obtained with respect to the increases in cyclic GMP concentrations but not to the decreases in myocardial cyclic AMP levels.

Since the relationship between acetylcholine action and tissue cyclic GMP accumulation was made in rat myocardium, a similar relationship has been reported in a number of mammalian tissues and cells both *in vivo* and *in vitro* (Table 2). Kuo et al. (1972) examined the effects of acetylcholine (1 and 0.3 μM) on rat ventricular slices in incubation media containing 3 mM theophylline (to minimize the loss of cyclic nucleotides by enzymic hydrolysis) and found that five- to 10-fold increases in cyclic GMP levels were produced within 30 sec after exposure to the cholinergic agent. The tissue levels of cyclic GMP declined to intermediate levels at 2 min and to control values at 5 min. In this report it was also shown that the acetylcholine-induced increases were partially prevented (ca. 50%) when either glucagon or isoproterenol were included in the incubation media along with acetylcholine. In the ventricular slices, acetylcholine alone did not affect the basal levels of cyclic AMP but partially blocked the increases usually seen after exposure of the tissue to either glucagon or epinephrine. In another series of experiments, these investigators showed that cyclic GMP levels in slices of rabbit cerebral and cerebellar tissues (without theophylline in the incubation media) were also elevated (70 to 300%) 30 and 60 sec after exposure to acetylcholine. In all experiments with brain tissue, as in the case with cardiac tissue, peak increases in cyclic GMP appeared to occur within 1 to 2 min and to return to control levels by 6 min. The transient nature of the increase in cyclic GMP concentration may

---

[5] M. K. Haddox and N. D. Goldberg, *unpublished data*.

TABLE 2. *Agents capable of elevating cyclic GMP levels*

Tissue	Source	Agents	Reference
Heart ventricle[2]	Rat	Ach	George et al., 1970
Heart ventricle[1]	Calf	Ach	Kuo et al., 1972
Cerebellum	Mouse	Maaloxone[5]	Goldberg, 1972
Cerebellum	Mouse	Oxotremorine	Ferrendelli et al., 1970
Cerebellum	Mouse	d-Amphetamine	Ferrendelli et al., 1972
Cerebellum[1]	Rabbit	Ach, histamine	Kuo et al., 1972
Cerebellum	Mouse	ECS (postictal-depression)	Lust et al., 1972
Cerebral cortex	Mouse	Oxotremorine, atropine	Ferrendelli et al., 1970
Cerebral cortex[1]	Rabbit	Ach	Kuo et al., 1972
Forebrain	Mouse	ECS (postictal-depression)	Lust et al., 1972
Lung[1]	Rabbit	Methacholine	Goldberg et al., 1973
Uterus[2]	Rat[3]	Oxytocin, 5-HT, methacholine	Goldberg et al., 1973
Uterus[2]	Rat[3]	Prostaglandin $F_{2\alpha}$	Goldberg et al.,[9]
Ductus deferens[2]	Rat	Carbachol	Schultz et al., 1972a,b
Ductus deferens[2]	Rat	Norepi	Schultz and Hardman[11]
Intes. sm. muscle[1]	Rat	Carbachol, histamine, 5-HT, K^+	Schultz et al., 1972b[12]
Submaxillary gland[1]	Rat	Carbachol, K^+	Schultz et al., 1972b[12]
Anterior pituitary[1]	Rat	Hypothalamic Extract	Peake et al., 1972
Ova[2]	Sea Urchin	Sperm	Goldberg et al.[13]
Platelets[2]	Human	Epi, Collagen	Goldberg et al.[10]
Lymphocytes[4]	Human	PHA, Con A	Hadden et al., 1972
Lymphocytes[4]	Human	Ach	Goldberg, 1972
Plasma	Human	Norepi, Epi[14]	Ball et al., 1970, 1972
Urine	Human	Norepi, Epi[14]	Ball et al., 1972
Urine	Human	PTH	Kaminsky et al., 1970
Urine	Human	Ca^{2+}	Kaminsky et al., 1970; Murad et al., 1971
Urine	Rat	Calcitonin, PTH[6]	Goldberg et al., 1973
Urine	Dog	Methacholine	Goldberg et al., 1973
Bacteria	*E. coli*	Glucose[7]	Goldberg et al., 1973

Abbreviations used are: acetylcholine, Ach; electroconvulsive shock, ECS; phytohemagglutinin, PHA; concanavalin A, Con A; serotonin, 5-HT; epinephrine, Epi; norepinephrine, Norepi; parathyroid hormone, PTH; intestinal smooth muscle, Intes. sm. muscle.

[1] Slices, *in vitro*.
[2] Whole organ or cells, *in vitro*.
[3] Diethylstilbesterol treated.
[4] From peripheral blood; believed to be the thymic-dependent population.
[5] A centrally active acetylcholinesterase inhibitor.
[6] Four hr after hormone administration.
[7] In cultures grown on succinate; the increase followed a transient (15 min) depletion.
[8] Appeared to be blocked by atropine.
[9] N. D. Goldberg, C. Sanford, D. K. Hartle, F. A. Kuehl, and M. K. Haddox, *unpublished data*.
[10] N. D. Goldberg, M. K. Haddox, D. K. Hartle, and J. White, *unpublished data*.
[11] G. Schultz and J. G. Hardman, *personal communication*.
[12] In the report by Dr. Schultz, at the Fifth International Congress on Pharmacology, San Francisco, Calif., July 23, 1972, based on the abstract cited.
[13] N. D. Goldberg, M. K. Haddox, D. K. Hartle, C. Blomquist, and J. W. Hadden, *unpublished data*.
[14] In the presence of a *beta*-adrenergic blocking agent.

have reflected the hydrolysis of acetylcholine by tissue cholinesterase or a characteristic unique to alterations produced in tissue cyclic GMP.

Ferrendelli and co-workers (1970) tested the possibility of a relationship between acetylcholine action and cyclic GMP accumulation by monitoring the changes in brain tissue cyclic nucleotide levels following the administration of oxotremorine, a drug whose effects are believed to derive from a release of acetylcholine (Cox and Potkonjak, 1969). Treatment of mice with oxotremorine resulted in increases in cyclic GMP levels in tissue from both the cerebral cortex and cerebellum which were greatest (70%) at 3 to 5 min when drug symptoms were also reported to be maximal, but by 20 min had declined to control (cerebral cortex) or below control (cerebellum) levels. There was also a lowering of cyclic AMP concentration in these two brain areas which was most apparent (25 to 50%) 30 min after injection of the drug. Pretreatment of the animals with atropine prevented oxotremorine-induced tremors and increases in tissue cyclic GMP levels. It was also reported that increasing concentrations (30 to 240 mg/kg) of this cholinergic blocking agent itself produced a progressive lowering of cerebellar cyclic GMP levels but elevated the concentration of cyclic GMP in the cerebral cortex. Relatively high concentrations of atropine have been found to increase cyclic GMP levels in rat uterine tissue also.[6] The decreases in cerebellar cyclic GMP resulting from atropine administration would be expected from the cholinergic receptor blocking action of the drug if cyclic GMP generation is indeed linked in some way to this receptor. However, no satisfactory explanation can be offered at this time regarding the apparent atropine-induced increases in cerebral cortical or uterine cyclic GMP levels. The fact that higher doses of atropine produced sleep in these animals is, however, of considerable interest and consistent with the observed association between CNS depression and the elevation of cyclic GMP levels in certain areas of brain tissue following electroconvulsive shock treatment (Lust et al., 1972; Goldberg et al., 1973). Another possibility to consider is that high levels of atropine in some tissues may act as a partial agonist or, when occupying the receptor, may stimulate cyclic GMP production but prevent the transport of extracellular calcium which along with cyclic GMP may be required for the full expression of some cholinergic signals.

The increase in cyclic GMP concentration in mouse cerebellum quick frozen *in situ* following the administration of a centrally active cholinesterase

---

[6] W. J. George, in a presentation, based on the abstract cited, at the 56th Annual Meeting of the Federation of American Societies for Experimental Biology, Atlantic City, N.J., April 12, 1972.

inhibitor, maaloxone (Goldberg et al., 1973), provides more direct supporting evidence for the contention by Ferrendelli et al. (1970) that a relationship between cholinergic action and cyclic GMP accumulation also exists in brain tissue.

More recently, Schultz and co-workers (1972a) reported that incubation of ductus deferens with carbachol (0.1 mM) led to a two- to fivefold elevation in the tissue levels of cyclic GMP within 1 min. No change was detected in cyclic AMP concentration. Carbachol-induced increases in ductus deferens cyclic GMP have also been shown (Schultz et al., 1972b) to be dose dependent with half maximal increases occurring at 0.01 mM. Schultz and his co-workers (1972b)[1] reported that increases in cyclic GMP levels of ductus deferens and rat submaxillary gland promoted by carbachol, or potassium-induced increases in intestinal tissue cyclic GMP, did not occur when calcium was absent from the incubation media. Eliminating calcium from the media also led to a lowering of the endogenous levels of cyclic GMP in smooth muscle; the readdition of this cation brought about a restoration of endogenous tissue cyclic GMP concentration and a return of the response on the part of the tissues to accumulate cyclic GMP and to contract (i.e., ductus deferens and intestines). These observations are of great potential importance since they indicate that cholinergically induced accumulation of tissue cyclic GMP, like many known cholinergic effects on cell function, is dependent upon the presence of calcium. The stimulation of guanylate cyclase activity by calcium (Hardman et al., 1971, 1972) (see Section IV, B) is consistent with the above, and taken together these observations suggest that a neurohumorally induced increase in cellular or membrane calcium might lead to an enhanced generation of cyclic GMP and that full expression of some hormone effects are brought about perhaps through an action of cyclic GMP and calcium in the interior of the cell.

As shown in Table 2 increases of 100 to 300% in the tissue levels of cyclic GMP following exposure to methacholine (1 μM) have also been demonstrated, *in vitro*, in uteri from diethylstilbesterol-treated rats, in rabbit lung tissue, and in human peripheral blood lymphocytes (Goldberg et al., 1973). In each case the elevation of cyclic GMP occurred promptly and was not accompanied by any significant alteration in the tissue cyclic AMP levels. The increase in uterine cyclic GMP was associated with a cholinergically induced contraction of this organ. Bronchiolar smooth muscle contraction is known to be stimulated by cholinergic agents and relaxation promoted by *beta*-adrenergic agents which enhance cyclic AMP accumulation in this tissue as well as in the uterus. The cellular process affected in lymphocytes by cholinergic stimulation is not yet defined but it has been found in studies with these cells that acetylcholine (5×10^{-7} M), like phytohemagglutinin, can stimulate cellular uptake of isotopically labeled calcium.[7] The uptake of calcium is greater when stimulated by phytohemagglutinin but with both agents reaches a maximum at 10 min; then, after a return to control levels, the uptake increases gradually over a prolonged period of time.[7] Acetylcholine treatment of lymphocytes also promotes significant increases (30%) in ^{3}H-uridine incorporation into RNA and ^{3}H-leucine incorporation into protein (40%), but the cholinergic agent alone does not appear to stimulate DNA synthesis.[7] The increase in ^{3}H-thymidine incorporation into DNA promoted by phytohemagglutinin is, however, augmented in the presence of acetylcholine.[7] Another effect of acetylcholine that may be associated with the observed increased generation of cyclic GMP (and, perhaps, increased calcium uptake) is that it enhances the ability of antigen-stimulated lymphocytes to inhibit the migration of macrophages.[7]

The infusion of dogs with methacholine (0.1 to 1.0 µg/kg/min) has also been found (Goldberg et al., 1973) to promote a rapid doubling in urinary excretion of cyclic GMP. In this case the excretion of cyclic AMP was also elevated although the latter event occurred minutes after the increased excretion of cyclic GMP. The increase in cyclic AMP excretion may derive indirectly from the effects of catecholamines discharged from the adrenal medulla upon cholinergic stimulation. The source of the greater amounts of cyclic GMP appearing in the urine has not been determined.

It should be pointed out that stimulation of rat sciatic nerve, which would be expected to stimulate the release of acetylcholine from nerve terminals associated with the neuromuscular junction, was not found to elevate cyclic GMP levels in rat gastrocnemeus muscle (Goldberg et al., 1973). This would indicate that cholinergic actions of the muscarinic but not the nicotinic type are associated with cyclic GMP accumulation.

3. *Other Hormonal and Neurohormonal Agents*

One characteristic common to cholinergic stimulation of the muscarinic type is that the effects produced on cellular function and metabolism are usually opposite to those promoted by agents such as *beta* adrenergic catecholamines which stimulate the generation of cyclic AMP. Several other agents which share this cyclic AMP-opposing characteristic have also been examined for their ability to enhance tissue cyclic GMP accumulation and have in most instances been found to do so (Table 2).

---

[7] J. W. Hadden, E. M. Hadden, and N. D. Goldberg, *unpublished data.*

The administration of oxytocin, serotonin (Goldberg et al., 1973) or prostaglandin $F_{2\alpha}$[8] has been shown to promote cyclic GMP accumulation (2.5- to threefold) in uteri (*in vitro*) from diethylstilbestrol-treated rats. No accompanying changes in the levels of cyclic AMP were seen. These three chemically distinct substances (a polypeptide hormone, a biogenic amine and a prostaglandin), like cholinergic agents, stimulate uterine smooth muscle contractility, an action opposed by agents that stimulate cyclic AMP generation in this tissue (Triner et al., 1971). The increases in uterine cyclic GMP were found at the earliest time examined (45 sec) after addition of each of the hormonal agents and a measurable contraction of smooth muscle was also apparent at the time the tissue cyclic GMP levels were increased. The very earliest time that a hormonally induced increase in the concentration of uterine cyclic guanylate can be detected is not yet known, nor has it been established that the increase actually precedes the hormone-induced initiation of contraction.

Schultz et al. (1972*b* [1] also reported that both serotonin and histamine, like acetylcholine, increased cyclic GMP concentrations in intestinal smooth muscle without changing the levels of tissue cyclic AMP. All three agents stimulate intestinal contractility. Compounds which enhance cyclic AMP generation in this tissue and cyclic AMP itself (Kawasaki et al., 1969) have been shown to inhibit intestinal motility. Kuo et al. (1972) were the first to report that histamine could induce increases (two- to fourfold) in tissue cyclic GMP concentrations; this effect of histamine was uncovered in slices of rabbit cerebral cortex but could not be demonstrated with slices from the cerebellum. A marked (two- to 10-fold) elevation of cerebral cortical cyclic AMP concentration accompanied the elevation of cerebral cortical cyclic GMP in response to histamine. These results suggest that the activation of certain histaminergic and serotonergic receptors may bring about an increased generation of cyclic GMP while other receptors for these neurohormones in the same (i.e., brain) or different tissues may be linked to adenylate cyclase and promote an increased synthesis of cyclic AMP.

Similarly in the case of catecholamines, some evidence has appeared indicating that *alpha*-adrenergic agonist action may be associated with the accumulation of tissue cyclic GMP while effects of the *beta*-adrenergic variety, as suggested by Robison, Butcher, and Sutherland (1967), may be linked to adenylate cyclase and increased production of cyclic AMP. It was also proposed (Robison et al., 1967) that the actions of *alpha*-adrenergic

---

[8] C. Sanford, D. K. Hartle, M. K. Haddox, F. A. Kuehl, and N. D. Goldberg, *unpublished data*.

agents which are generally opposite in a given tissue to those of *beta*-adrenergic agents might derive from a lowering of tissue cyclic AMP. Ball and co-workers (1970, 1972) and Kaminsky et al. (1970*a*) monitored the changes in plasma cyclic nucleotide levels in humans after the administration of *alpha*- or *beta*-adrenergic agents alone or in combination with adrenergic blocking agents and could demonstrate rather clearly that plasma cyclic GMP levels were elevated as a result of *alpha*-adrenergic stimulation and that cyclic AMP concentrations increased following *beta*-adrenergic stimulation. Schultz et al. (1972*a*) in a preliminary communication reported an inability to detect any changes in tissue cyclic GMP (or cyclic AMP) concentrations upon the addition of norepinephrine to ductus deferens (*in vitro*), a tissue with only *alpha*-adrenergic receptors. More recently, however, Schultz and Hardman[9] have found norepinephrine to induce small increases (ca. 50%) in cyclic GMP concentrations of ductus deferens. This effect of *alpha*-adrenergic stimulation appeared in some experiments to be blocked by atropine. The latter may indicate that a portion of the apparent *alpha*-adrenergic response may be mediated indirectly through a cholinergic mechanism. Several naturally occurring smooth muscle stimulants exhibit a combination of direct as well as indirect actions which are apparently mediated in some tissues through a release of acetylcholine from nerve terminals (Paton, 1968).

Evidence that *alpha*-adrenergic stimulation may indeed be associated with rapid cellular accumulation of cyclic GMP can be found in the recent demonstration that platelet aggregation induced by epinephrine was accompanied by an increase of over threefold in the level of this cyclic nucleotide in human platelets.[10] The increase occurred within 30 sec after exposure to the catecholamine when the very first obvious sign of aggregation was detectable and when no alteration in the level of cyclic AMP was measurable. Epinephrine-induced platelet aggregation is believed to be brought about through an *alpha*-adrenergic mechanism (Cole et al., 1971). Collagen, an agent somewhat less potent than epinephrine as an inducer of this process, was also found to increase (70 to 100%) the concentration of platelet cyclic GMP.[10]

Another polypeptide hormone for which some evidence has been obtained to indicate that it may enhance the generation of cyclic GMP is calcitonin. This hormone, which promotes bone calcification and lowers plasma calcium levels, could, in a sense, be viewed as an antagonist of parathyroid

---

[9] G. Schultz and J. G. Hardman, *personal communication*.
[10] N. D. Goldberg, M. K. Haddox, D. K. Hartle, and J. White, *unpublished data*.

hormone which stimulates bone resorption, renal and intestinal uptake of calcium, and the elevation of plasma and urinary calcium levels. Parathyroid hormone has also been shown to produce striking increases in urinary cyclic AMP excretion (Chase and Aurbach, 1967; Aurbach et al., 1969). All of the studies in which a possible relationship between calcitonin and cyclic GMP have been tested involve the monitoring of changes that occur in the urinary excretion of cyclic nucleotides. The interpretation of the results obtained are, therefore, subject to the uncertainties inherent to any effects that represent the sum of those occurring in all the tissues of the body. From the demonstration by Kaminsky et al. (1970b) with human subjects that large hypercalcemic doses of parathyroid hormone or the infusion of calcium itself led to a doubling of urinary cyclic GMP excretion, it seemed possible to consider an involvement of calcitonin since increased circulating calcium levels stimulate its release. However, this possibility was discounted (Kaminsky et al., 1970a,b) when it was found that the administration of calcitonin did not alter the rate of cyclic GMP excretion whereas the infusion of calcium did increase cyclic GMP excretion. Murad and Pak (1972) also claim to have observed no significant effect on cyclic GMP excretion after infusion of calcitonin in humans. In experiments[11] conducted with conscious thyroparathyroidectomized rats, with permanently implanted bladder cannulae to allow a continuous collection of urine, it was found that the infusion of calcitonin produced an immediate elevation (two- to threefold) in the excretion of cyclic GMP (Goldberg et al., 1973) and a decrease, no change, or an increase (twofold) in urinary cyclic AMP excretion. When an increase in cyclic AMP excretion did occur, it followed the increase in cyclic GMP excretion by at least 30 min, and the levels of both cyclic nucleotides under these circumstances fluctuated in an oscillatory fashion. It was also found in other experiments carried out with rats that the infusion of parathyroid hormone could stimulate cyclic GMP excretion (Goldberg et al., 1973), in agreement with the results obtained in humans (Kaminsky et al., 1970b). However, the increase in urinary cyclic GMP excretion after parathyroid hormone infusion was delayed for 3 to 5 hr in contrast to the almost immediate increase seen upon infusion of calcitonin. The increase in cyclic GMP excretion after parathyroid hormone infusion seemed to be closely related to the elevation in urinary calcium excretion which also occurred 4 hr after infusion of the hormone was started. It is possible, therefore, that the increased urinary excretion of cyclic GMP in rats following the infusion

---

[11] These studies were conducted in collaboration with Drs. A. DeLong and H. Rassmussen of the University of Pennsylvania.

of calcitonin may stem from a direct action of the polypeptide hormone to stimulate cyclic GMP generation in target tissues whereas the delayed increase following parathyroid hormone administration, which is associated with a delayed hormone-induced hypercalcemia, may arise secondarily from an action of calcium. The latter would be consistent with the observed effect of calcium to stimulate cyclic GMP excretion (Kaminsky et al., 1970*a,b*; Gilman and Murad, 1972) as well as the recently reported[1] dependence on calcium for maintenance of endogenous tissue cyclic GMP levels and the responsiveness of tissues to hormones that stimulate the accumulation of this cyclic nucleotide. It might be concluded from the results obtained thus far that hormones affecting calcium metabolism and/or calcium itself may influence cyclic GMP metabolism and/or excretion.

4. *Steroids*

The first evidence that cyclic GMP may be directly or indirectly associated with endocrine or neuroendocrine function derived from reports by Hardman et al. (1966, 1969) that removal of the pituitary gland did not affect the whole body concentration of cyclic GMP but diminished (50%) the urinary excretion of cyclic GMP in rats. The administration of thyroxine combined with glucocorticoid but not the glucocorticoid alone restored the excretion rate to normal. Adrenalectomy also caused a lowering of cyclic GMP excretion which could be reversed by glucocorticoid administration while injections of thyroxine were found to restore the diminished rate of urinary cyclic GMP excretion resulting from thyroparathyroidectomy (Hardman et al., 1966, 1969). In contrast to these experiments in which effects on the urinary excretion of cyclic GMP were examined, it was found by Goldberg et al. (1973) that after removal of the adrenal gland there was a detectable elevation in the concentration of cyclic GMP in both lung (25%) and renal (20%) tissue. Treatment of adrenalectomized rats with cortisol decreased the levels of cyclic GMP (50 to 60%) and a single injection (4 hr) of either cortisol or aldosterone lowered the concentrations of this cyclic nucleotide in renal tissue over 50% in adrenalectomized rats previously treated with theophylline (Goldberg et al., 1973). It has also been observed[12] that preincubation of human peripheral blood lymphocytes with cortisol is inhibitory to phytohemagglutinin-induced mitogenic action and that cortisol prevents the elevation of lymphocyte cyclic GMP concentration usually induced by this mitogen (see below). Steiner et al. (1972*b*) have

---

[12] N. D. Goldberg, M. K. Haddox, E. M. Hadden, and J. W. Hadden, *unpublished data*.

implicated glucocorticoids as agents which can suppress cyclic GMP levels in some, but not all, tissues. They observed that hypophysectomy doubled and adrenalectomy tripled the concentrations of cyclic guanylate in skeletal muscle whereas treatment with dexamethasone decreased the elevated levels in adrenalectomized animals about 40%. Adrenalectomy and dexamethasone treatment had qualitatively similar effects on skeletal muscle cyclic AMP levels but the percentage changes were considerably smaller than those observed with cyclic GMP. These investigators did not detect any changes in the levels of renal cortical or hepatic tissue cyclic GMP under the same experimental conditions. Recent observations by Thompson, Williams, and Kompton[2] that adrenalectomy leads to an elevation and dexamethasone a suppression of assayable guanylate cyclase activity in skeletal muscle and lung are consistent with the observations cited above in regard to the changes in steady-state levels of cyclic GMP that occur in these tissues with similar alterations in endocrine status. The results of the studies in tissue therefore suggest that steroid hormones may indeed influence tissue cyclic GMP concentration through a direct or indirect effect on guanylate cyclase activity. Although the results reported thus far favor an influence that would lead to a lowering of tissue cyclic GMP levels, the data of Hardman et al. (1966, 1969), demonstrating an opposite effect on urinary excretion, indicate that steroid effects may represent a positive or negative influence depending, perhaps, upon the target organ.

In contrast to the report of Hardman et al. (1969) that hypophysectomy did not alter whole body cyclic GMP concentration of the rat is the finding[13] that removal of the pituitary caused a reduction of about 50% in the levels of cyclic GMP in thymus, lung, liver, spleen, and testes. Little or no change was seen in the levels of this cyclic nucleotide in heart, intestine, or brain. Growth hormone injections over a period of 8 to 10 days which produced a significant increase in body weight of the hypophysectomized animals had no significant effect to restore the cyclic GMP concentration of the tissues in which it had been lowered. Although growth hormone itself appears to have no effect on tissue cyclic GMP levels it seems from the disclosures by Peake et al. (1972), Cehovic and Posternak (1971), and Cehovic et al. (1971) that cyclic GMP may have an effect on the biosynthesis and/or secretion of the hormone. The work of Peake and his colleagues, although described in greater detail in Section V, might also be considered here briefly from the standpoint of the increases inducible in endogenous anterior pituitary cyclic GMP concentrations that may be related to the secretion of growth hormone. The finding that hypothalamic extracts containing growth hormone

---

[13] W. J. George and N. D. Goldberg, *unpublished data*.

releasing factor could stimulate the release of this hormone from anterior pituitary explants and at the same time increase the endogenous levels of both cyclic AMP (300 to 400%) and cyclic GMP (100%) is in itself less than convincing evidence that the latter is intimately involved in the process of growth hormone secretion. However, the argument in favor of such an involvement of cyclic GMP is strengthened by the findings that exogenous cyclic GMP but not cyclic AMP (at 50-fold higher concentrations) could stimulate release of the hormone and that this effect of cyclic GMP could be potentiated by aminophylline. In addition, higher concentrations of aminophylline which also stimulated secretion of the hormone increased tissue cyclic GMP levels 15-fold and cyclic AMP concentrations only threefold. Additional supporting evidence for the latter, referred to earlier in Section V (i.e., Cehovic and Posternak, 1971; Cehovic et al., 1971), is the demonstration that exogenous cyclic GMP but not cyclic AMP is effective in stimulating the synthesis of growth hormone and that secretion of the hormone may be stimulated cholinergically.

One of the questions posed by the observations of Peake et al. (1972) that will warrant clarification in the future is whether the effects produced by phosphodiesterase inhibitors, which in the past have been viewed as cyclic AMP-induced, might in some instances derive from an elevation in the levels of tissue cyclic GMP.

5. Nonhormonal Substances

The elevation of cyclic GMP levels following the administration of phosphodiesterase inhibitors has been observed in a number of tissues (Table 2) including brain and kidney (Goldberg et al., 1969), small intestine (Ishikawa et al., 1969), pituitary explants (Peake et al., 1972), sperm (Garbers et al., 1971), and ductus deferens (Schultz et al., 1972a). In view of the fact that agents commonly used as phosphodiesterase inhibitors in intact cell systems have also been shown to be equally effective in inhibiting the enzymic hydrolysis of both cyclic GMP and cyclic AMP by tissue extracts (Goldberg et al., 1970; O'Dea et al., 1970), the increases seen in tissue cyclic GMP levels after treatment with these agents are not too surprising. Furthermore, from the experiments reported to date it would appear that there is a proportionately greater increase in cyclic GMP than in cyclic AMP tissue levels upon treatment with an agent such as theophylline. The contrast is especially marked in the case of the pituitary gland cited above. In bull sperm (Garbers et al., 1971), caffeine promoted over a twofold elevation of cyclic GMP, and a much higher concentration of the methylxanthine increased cyclic AMP levels less than twofold. Incubation of ductus deferens with a phosphodiesterase inhibitor was found by Schultz et al. (1972a) to increase the levels of cyclic GMP in this tissue five- to eightfold compared to increases of three- to fivefold in cyclic AMP concentrations. In view of the preceding, it would seem that the criterion often used to aid in establishing cyclic AMP as the mediator of a hormonal action, which requires that inhibitors of phosphodiesterase reproduce the effect presumably mediated by cyclic AMP, should be revised to accommodate the possible involvement of cyclic GMP which also increases as a result of such treatment. Except for the case in the anterior pituitary, however, it would seem fair to say that the effects produced in intact cell systems by phosphodiesterase inhibitors are usually characteristic of those expected from increases in the concentration of cyclic AMP. If the tissue concentrations of both cyclic nucleotides are indeed increased by phosphodiesterase inhibitor treatment, one explanation for the cyclic AMP-like response that occurs is that the intracellular influence of cyclic AMP may be dominant over that of cyclic GMP. The dominance of cyclic AMP could conceivably be offset by increases in cyclic GMP concentration that are far greater than those occurring in cyclic AMP as, perhaps, in the case of the anterior pituitary after aminophylline treatment. Another possibility is that cyclic GMP-promoted events may in some instances require an obligatory second component such as calcium. The increases in tissue cyclic GMP levels resulting from phosphodiesterase inhibitor treatment, such as the increases that are apparently achieved upon addition of exogenous cyclic GMP (Exton et al., 1971) (or derivatives), may therefore only partially satisfy the requirement of a cyclic GMP-sensitive system.

Increases in brain tissue cyclic GMP levels brought about by agents and conditions that apparently act indirectly through the release of naturally occurring substances in this tissue which are as yet unidentified have also been described. Although it was found initially (Goldberg et al., 1971) that the tonic and clonic stages of convulsive behavior following electroconvulsive shock treatment of mice are associated with increases only in brain cyclic AMP levels, a more extensive study revealed that increases of three- to fourfold occur in the levels of cyclic GMP in both cerebellar and cerebral tissue during the postictal depressive phase (Lust et al., 1972; Goldberg et al., 1973). The independent nature of the changes in the two cyclic nucleotides was most striking in cerebellar tissue where it was shown that the elevated levels of cyclic AMP which arose during the tonic and clonic convulsive periods declined toward control levels during postictal depression when the concentration of cyclic GMP in this area of the brain increased maximally. The reciprocal changes that can occur in the levels of the two cyclic nucleotides is well illustrated under these conditions but the basis for the changes seen is not known. Although it is possible that the increases in cyclic AMP may result in part from the anoxia resulting from electroconvulsive shock treatment (Sattin, 1971), it does not seem probable that the delayed increases in cyclic GMP could be attributable to such a condition since cyclic GMP levels in mouse brain have been shown to undergo no change in the early (Goldberg et al., 1971) or a decrease (Steiner et al., 1972c) during the later stages of anoxia.

Ferrendelli, Kinscherf, and Kipnis (1972) have presented evidence indicating that drugs affecting the release and/or metabolism of monoaminergic transmitters in brain tissue also influence cerebellar cyclic GMP levels. The administration of D-amphetamine in a dose that caused hyperactivity in mice increased the levels of cerebellar cyclic GMP more than twofold 10 min after injection. Both chlorpromazine and reserpine pretreatment, which diminished motor activity of the animals and blocked the hyperactivity elicited by D-amphetamine, also diminished cerebellar cyclic GMP levels 62 and 73%, respectively. D-Amphetamine had a greatly reduced effect to raise the levels of cyclic GMP that had been diminished by previous treatment with chlorpromazine. However, atropine which was shown to prevent oxotremorine-induced increases in cerebellar cyclic GMP (Ferrendelli et al., 1970) was ineffective in blocking the increases promoted by D-amphetamine. These observations are of potentially great interest but are difficult to interpret because of the complex nature of both the drug actions and the central nervous system itself. One conclusion that may be drawn, however, is that a neurotransmitter(s) other than acetylcholine may stimulate cyclic GMP generation in certain areas of the brain. From the recent disclosures that substances present in brain such as serotonin, histamine, prostaglandin $F_{2\alpha}$, and possibly norepinephrine (i.e., *alpha*-adrenergic stimulation) can induce tissue accumulation of cyclic GMP in other tissues (see Section VI, A, 3), it seems possible that the apparent effect of D-amphetamine noted here may derive from the release of one of these substances, although a direct effect of the drug is also possible.

The process of cell division has not yet been proven to be under hormonal control, but in a number of recent experiments it has been shown that cell proliferation can be influenced dramatically by cyclic AMP (Otten et al., 1971; Sheppard, 1971). Cyclic AMP or agents which stimulate its production can inhibit the proliferation of rapidly growing "malignant" cells grown in culture. The transformation of "normal" cells, which exhibit contact inhibition, to cells with growth characteristics of "malignant" cells is now believed to occur as a result of a lowering of cellular cyclic AMP concentration. The possibility that there is an active signal for the initiation of cell division represented by an elevation of cyclic GMP levels, rather than just a passive one brought about by a lowering of cyclic AMP concentrations, has recently been proposed (Hadden et al., 1972; Goldberg et al., 1973). This hypothesis was developed from the discovery that phytohemagglutinin and concanavalin A, mitogenic agents commonly used to induce lymphocyte proliferation, increase the levels of cyclic GMP in these cells five- to 50-fold within 5 to 20 min. The cellular concentrations of cyclic AMP are not altered by purified preparations of either mitogenic agent, but a less purified preparation of phytohemagglutinin which exhibited a considerable degree of agglutinating activity increased the levels of cyclic adenylate about 70%. Taking into consideration the fact that phytohemagglutinin, like most polypeptide and neurohormones, finds its final action at the surface of the cell membrane and that events leading to cell division take place in the nucleus, it was postulated that cyclic GMP may represent a membrane to nuclear signal. Preliminary evidence[14] that cyclic GMP at concentrations of 10^{-9} M can stimulate both RNA synthesis and histone acetylation in isolated lymphocyte nuclei, two events that occur early in the process of cell division, supports this concept.

An important biological event tantamount to an initiation of cell proliferation is the process of fertilization. It was reasoned that ova, like lymphocytes, are cells whose normal functions include cell proliferation, the difference lying in the biological substance which normally serves as the stimulus for initiating the process in the two cell types: lymphocytes are poised to proliferate upon interaction with antigen whereas an ovum normally undergoes cell division when fertilized by sperm. An earlier report by Gray et al. (1970) that unfertilized sea urchin ova had no detectable level of cyclic GMP has recently been confirmed.[15] However, within 5 to 10 min after fertilization with sea urchin sperm, levels of cyclic GMP are detectable in ova (washed free of unattached sperm).[15] The earliest time that a finite amount of cyclic GMP may appear is not known because of the limitation in sensitivity of the analytical procedure. The possibility that the cyclic GMP detected was generated by guanylate cyclase deriving from the sperm rather than the ova cannot be ruled out but was considered to be highly unlikely after determining that the contribution by sperm [which do indeed contain relatively high levels of cyclic GMP and guanylate cyclase activity (Gray et al., 1970)] to the total mass of the washed sample of fertilized ova is negligible. The results of these experiments combined with those obtained with lymphocytes in which a relationship was established between cyclic

---

[14] J. W. Hadden and J. Meetz, *unpublished data*.
[15] N. D. Goldberg, M. K. Haddox, C. Blomquist, and J. W. Hadden, *unpublished data*.

GMP and the mitogenic process suggest that an elevation of cyclic GMP concentration in ova may represent an active intracellular signal, generated upon fertilization, that promotes the process of cell division.

B. Bacteria

In the preceding section it was pointed out that exogenous cyclic AMP and cyclic GMP can produce opposite effects on both *lac* and *gal* mRNA synthesis *in vitro;* cyclic AMP seems to serve as an obligatory stimulator and cyclic GMP can inhibit or reverse the cyclic AMP-induced effect, presumably through a competition with cyclic AMP for the cyclic AMP receptor protein. In order to establish that cyclic GMP does indeed represent the physiological antagonist of cyclic AMP action in bacteria, implied by these studies, it will be necessary to demonstrate: (1) that the endogenous, steady-state levels of cyclic GMP and cyclic AMP change in a reciprocal fashion (elevated levels of cyclic GMP arising under conditions when inducible enzyme synthesis is repressed and diminished levels when derepression and elevated cyclic AMP concentrations are evident), and (2) that the endogenous levels of cyclic GMP in bacteria *in vivo* are similar to those with which effects can be produced *in vitro*. Neither of the two criteria has been satisfied completely but some recent results indicate that reciprocal changes in the levels of cyclic GMP and cyclic AMP can be demonstrated in *E. coli* (K-12 strain) grown under certain conditions (Goldberg et al., 1973).[16] In a growth medium containing a relatively high concentration of glucose, which lowers the levels of bacterial cyclic AMP (Makman and Sutherland, 1965) and represses the induction of *beta*-galactosidase (Pastan and Perlman, 1970) but supports maximal bacterial growth, the levels of cyclic GMP were found to be maximal. As growth proceeded and glucose in the medium was consumed, there was a marked lowering in the concentration of cyclic GMP and a progressive increase in the level of bacterial cyclic AMP. When growth ceased, presumably due to a depletion of the carbon source, cyclic AMP levels became maximal and those of cyclic GMP minimal. A reciprocal relationship between the changes in the two cyclic nucleotides could also be demonstrated when succinate was used as the carbon source (Goldberg et al., 1973). Succinate does not serve as a repressor of *beta*-galactosidase synthesis. This was reflected by a considerably higher basal level of cyclic AMP in *E. coli* when they were grown in its presence. However, as growth progressed, small increases in cyclic

---

[16] These studies were conducted in collaboration with Dr. R. W. Bernlohr of the Department of Microbiology at the University of Minnesota.

AMP concentration with corresponding decreases in the levels of cyclic GMP were detectable. At the first sign of a slowing of the growth rate (i.e., departure from log phase growth), the rate at which the concentrations of cyclic AMP increased and those of cyclic GMP decreased became greater. However, the relationship between the two cyclic nucleotides in bacteria may be more complicated than the reciprocal one shown by the two experiments just described. When a high concentration of glucose was added to cultures growing on succinate, there was an immediate lowering of cyclic AMP concentrations as well as a transient (15 min) depletion of cyclic GMP to almost undetectable levels before the latter increased beyond the level existing prior to the addition of glucose. It would appear, therefore, that the overall changes in the levels of the two cyclic nucleotides in bacteria are clearly reciprocal under some but not all conditions. It might also be pointed out that under most conditions tested the highest ratios of cyclic GMP to cyclic AMP were found in cultures undergoing the greatest rate of growth whereas the lowest ratios were associated with a slowing or cessation of bacterial proliferation. It is possible, therefore, that these two cyclic nucleotides may, under at least some conditions, have an influence on bacterial cell division in addition to affecting the induction of certain bacterial enzymes. It remains to be demonstrated that concentrations of cyclic GMP in the micromolar range have any cyclic AMP-opposing effects or other effects in *in vitro* systems. As suggested earlier (Section V, B, 3), the conditions under which cyclic GMP may be most effective *in vivo* may not yet have been examined *in vitro*. It would seem from the changes found to occur *in vivo* that the influence of cyclic GMP is manifest when the levels of cyclic AMP are diminished and its influence minimal. In *in vitro* systems the effects of cyclic GMP have been examined in systems maximally stimulated by relatively high concentrations of cyclic AMP.

VII. SPECULATION

From the investigations completed thus far, enough evidence has accumulated to warrant serious consideration of the hypothesis that cyclic GMP is involved in promoting cellular events that are antagonistic to those mediated through an elevation of cellular cyclic AMP. The evidence most clearly in support of a "dualism" between the two cyclic nucleotides derives from the observations that the cellular events induced by agents or conditions which promote cyclic AMP generation are opposed to those that can be shown to be associated with an accumulation of endogenous cyclic GMP.

The only results that could be considered incompatible with the "dualism" theory, or at least not supporting it, derive from some of the experiments in which the effects of very high concentrations of exogenous cyclic GMP have been examined. Until these cyclic AMP-like effects can be reproduced with concentrations of cyclic GMP that approach the levels normally occurring intracellularly, it would seem from the information now available that they result from nonspecific interactions with either the cyclic AMP-dependent protein kinase or other cellular components.

In contrast to the "dualism" theory of biological control through opposing actions of cyclic GMP and cyclic AMP is the concept that bidirectional changes (i.e., increases and decreases) in only intracellular cyclic AMP concentrations can provide for both the positive and negative regulatory influences imposed upon some cellular processes. The latter could be described as a "unitary" concept of regulation imposed through cyclic AMP alone. The "unitary" concept has been used to explain, for example, the following: regulation of cardiac contractility [i.e., epinephrine stimulating by raising and acetylcholine suppressing by lowering cyclic AMP levels (Murad et al., 1962)]; the control of lipolysis [i.e., epinephrine promoting by increasing and insulin inhibiting by decreasing cyclic AMP concentrations (Butcher et al., 1968)]; the control of platelet aggregation [i.e., PGE_1 inhibiting by elevating and epinephrine stimulating by lowering cyclic AMP levels (Salzman and Levine, 1971)]; the regulation of hepatic glycogen metabolism [i.e., glucagon stimulating glycogenolysis and inhibiting glycogenesis by elevating tissue cyclic AMP levels and insulin producing the opposite effects by lowering the levels of cyclic AMP (Jefferson et al., 1967)]; the control of cell proliferation [i.e., increased cellular cyclic AMP levels inhibiting and lowered levels either permitting or promoting the process (Otten et al., 1971; Sheppard, 1972). Although, in at least some of the examples cited, decreases in cellular cyclic AMP levels can be demonstrated with the agent or condition promoting the cyclic AMP-opposing event, there has yet to be a clear demonstration that the decrease actually precedes the effect induced on cellular function or that the decreases are proportional to the magnitude of the effect produced. Evidence to the contrary has in fact appeared. Agents promoting cyclic AMP-opposing events can be shown in some cases to have no cyclic AMP-lowering effect (Goldberg et al., 1967, 1973; Craig et al., 1969; Cole et al., 1971; Hadden et al., 1972), and in other cases small decreases that can be detected appear to occur after the first signs of the hormonally induced cellular modification (George et al., 1970; Nichols and Goldberg, 1972). The "dualism" concept is by no means incompatible with a lowering of cellular cyclic AMP levels. However, according to the hypothesis, the cyclic AMP-opposing event would be initiated by the generation of an "active signal," represented by an elevation of cellular cyclic GMP concentration rather than a passive one, represented by the lowering of cyclic AMP concentration. A decrease in the level of the latter could be viewed as a secondary event that might permit a greater cyclic GMP-linked response. A lowering of cyclic AMP levels, secondary to an increase in the concentration of cyclic GMP, could conceivably be brought about through the facilitatory effect cyclic GMP has been shown to have on the phosphodiesterase-promoted hydrolysis of cyclic AMP (Beavo et al., 1971).

What then can be said about the mechanism(s) by which cyclic GMP may act? The discussion of such a subject would, of course, be a highly speculative one, since the biochemical or molecular basis of cellular events that may be affected by cyclic GMP has not yet been elucidated. From the information now available suggesting an involvement of cyclic GMP with the mediation of certain cellular processes and from our understanding of cyclic AMP action in at least one defined metabolic system, some rather general concepts can be put forth that may aid in arriving at the underlying mechanism(s) for the proposed antagonistic actions of the two cyclic nucleotide effectors. Of all the cellular events known to be affected by agents that stimulate cyclic AMP generation in mammalian systems, the control of glycogen metabolism in skeletal muscle is the only one in which the regulatory components involved have been characterized and a mechanism for cyclic AMP action elucidated. Although all of the components of this metabolic system in liver and adipose tissue have not been characterized to the extent that they have in skeletal muscle, it is likely that in these tissues similar or identical components are influenced in the same way.

In these tissues, agents that stimulate cyclic AMP production stimulate glycogenolysis and inhibit glycogenesis. This is thought to be accomplished as a result of cyclic AMP interaction with and activation of a protein kinase. The activated protein kinase promotes the activation of phosphorylase kinase which in turn catalyzes the activation of glycogen phosphorylase; each activating reaction involves the phosphorylation of enzyme protein. The same cyclic AMP-dependent protein kinase that initiates the events leading to phosphorylase activation also appears to promote the phosphorylation of glycogen synthetase. The phospho-form of glycogen synthetase, however, is the inactive form of the enzyme in most tissues. An increase in cellular cyclic AMP can, therefore, through activation of a protein kinase, bring about the activation of glycogen phosphorylase and inactivation of glycogen synthetase. Cholinergic stimulation has been shown to promote the inactivation of phosphorylase [e.g., heart (Hess et al., 1962)] and activation of glycogen synthetase [e.g., liver (Schimazu, 1967)]. Acetylcholine actions of this type appear to be associated with increases in cellular cyclic

GMP. If, according to the "dualism" hypothesis, cyclic GMP were involved in promoting the metabolic events opposed by cyclic AMP in this system (i.e., inactivation of phosphorylase and activation of glycogen synthetase), it would be expected that cyclic GMP would bring about an activation of the phosphatase(s) that opposes or reverses the action of the kinase. There is evidence that glycogen synthetase-D phosphatase can undergo a relatively rapid activation (Nichols and Goldberg, 1972), and some evidence that it may indeed exist in two (active and inactive) forms (Bishop, 1970). Kuo et al. (1970b) have suggested that all of the effects produced by cyclic GMP may be mediated through the activation of protein kinases, a proposal also made to account for all the effects mediated by cyclic AMP (Kuo and Greengard, 1969). If activation of the phosphatase was brought about by a phosphorylation of the enzyme (promoted by a cyclic GMP-dependent kinase), it would be necessary to postulate that the reversal of this effect would require the removal of phosphate from the enzyme protein by a second phosphatase that should be activatable by the opposing intracellular effector cyclic AMP, again through an action of a cyclic AMP-dependent kinase. Such a sequence of events based on the activation of only protein kinases by the two cyclic nucleotides would require an infinite number of interconversion repetitions; it would, therefore, seem unavoidable to expect that at some point in the activation-inactivation scheme cyclic GMP must interact directly with a phosphoprotein phosphatase if the action of cyclic AMP-opposing agents are indeed mediated through this cyclic nucleotide.

A property characteristic of allosteric effectors is that they usually have reciprocal influences on opposing enzyme activities (Stadtman, 1966). If this were true in the case of cyclic GMP and cyclic AMP in this particular system, it would be expected that the former should also have an inhibitory influence on the activity of the cyclic AMP-dependent protein kinase and that cyclic AMP should suppress the activity of the phosphatase(s). Some evidence that cyclic AMP has an inhibitory influence on the activity of phosphorylase phosphatase has appeared (Merlevede and Riley, 1966; Chelala and Torres, 1969). Cyclic GMP (at submicromolar concentrations) can cause the release of a significant fraction of cyclic AMP bound to the regulatory subunit of a cyclic AMP-dependent protein kinase from skeletal muscle (Goldberg, 1972). Of the regulatory influences of cyclic GMP and cyclic AMP that have just been postulated (for the glycogen metabolic system in these particular tissues), only the stimulatory effect of cyclic GMP on phosphoprotein phosphatase activity remains a purely speculative one.

What significance can then be attributed to the cyclic GMP-dependent protein kinase in lobster muscle (Kuo and Greengard, 1970a) and to a similar activity reported to occur in uterus, urinary bladder, and brain (Greengard and Kuo, 1970). The cyclic GMP-dependent protein kinase from lobster muscle exhibits a 50-fold greater requirement for cyclic AMP than cyclic GMP. The problem encountered in considering the possible significance of a kinase with these kinetic properties in the mammalian tissues cited above is that concentrations of cyclic AMP in these tissues have been shown to be 20- to 100-fold higher than those of cyclic GMP. The fact that the K_a value of cyclic GMP is lower by approximately the same magnitude as the tissue concentration minimizes or eliminates the specificity that might derive from the apparent differences in affinity of the two cyclic nucleotides for this enzyme. On the other hand, in cerebellar tissue (Greengard and Kuo, 1970; Hoffman, 1972) where the concentrations of cyclic GMP and cyclic AMP are comparable (ca. 10^{-6} moles/kg wet weight), a protein kinase more highly selective for cyclic GMP (Hoffman, 1972) could represent an enzyme that may be activatable rather specifically by this cyclic nucleotide *in vivo*. However, even in the case of the cerebellar enzyme, specificity for the activating cyclic nucleotide *in vivo* may fall far short of being adequate since the cerebellar concentration of cyclic AMP and the K_a value for this effector with the more cyclic GMP-dependent kinase [9×10^{-7}M (Hoffman, 1972)] are about equal. The fact that all of the regulatory influences of cyclic GMP and cyclic AMP may not be restricted to only the activation of protein kinases is underscored by the observation that cyclic GMP can oppose the action of cyclic AMP upon interaction with a receptor protein in bacteria that appears to promote the initiation of transcription (Emmer et al., 1970; Zubay et al., 1970) and that it may also interact with certain cyclic nucleotide phosphodiesterases to stimulate the hydrolysis of cyclic AMP in some tissues (Beavo et al., 1971; Manganiello and Vaughan, 1972). Neither of these two cyclic GMP-promoted effects appears to require ATP.

The possible multiplicity of mechanisms by which cyclic GMP could affect cell functions may be related to the different and often contrasting effects produced by biologically active agents that promote cyclic GMP accumulation. Another question that arises, therefore, is whether all of the effects produced by cyclic GMP are mediated as a result of an interaction with the same cellular components. For example, in the case of cardiac muscle where acetylcholine (presumably through an action of cyclic GMP) suppresses contractility and promotes an inactivation of phosphorylase, it is conceivable, as pointed out above, that the activity of a phosphatase is facilitated and the activity of a kinase suppressed. However, in uterine smooth muscle agents such as acetylcholine stimulate contractility and bring about an activation of phosphorylase (Diamond and Brody, 1966). Two possibilities to consider are: (1) that cyclic GMP may interact with and affect the same cellular components in all tissues but that the sequence of biochemical events initiated by cyclic GMP does not produce the same functional response in all tissues, or (2) that the sequence of biochemical events that produce a given response is the same in all tissues, and, therefore, that cyclic GMP interacts with different cellular components to produce different responses.

In the first case it would be expected that activation of a phosphoprotein phosphatase, if such as enzyme were indeed activatable by cyclic GMP, would lead to the different alterations in function associated with an elevation of cyclic GMP in both cardiac and smooth muscle (i.e., inhibition and stimulation, respectively). In the second case it would be necessary to postulate that at least two different components were activatable by cyclic GMP, perhaps a phosphoprotein phosphatase and a protein kinase.

The work performed to date which favors the concept of a dualism between cyclic GMP and cyclic AMP derives from experiments conducted in systems that appear to be modulated by opposing biological signals (e.g., epinephrine vs. acetylcholine on heart function, smooth muscle contractility, etc.). These could be considered to be bidirectionally controlled systems. It is also possible that other biological systems exist that are controlled only monodirectionally in a manner analogous to the "turning on" of a spring loaded water spigot. The basal flow or activity of such a system would be zero and the only regulatory influence a positive or stimulatory one. The extent to which it may be "turned on" would depend, of course, upon the strength of the signal. An example of such a biological system may be the steroidogenic apparatus of the adrenal cortex and its control through ACTH. Steroidogenesis is stimulated by ACTH but there does not appear to be any naturally occurring antagonist of ACTH action. It is possible, however, that a monodirectionally controlled system may be stimulated by different biological signals and that these signals may be expressed through the generation of different intracellular effectors. It is conceivable that both cyclic GMP and cyclic AMP may represent intracellular mediators of different biological signals in a monodirectionally controlled system. In this case, the two cyclic nucleotides might promote similar or the same events rather than opposing ones.

One final comment can be made regarding the enzyme(s) that cyclic GMP is presumed to activate and the specificity of such an enzyme in respect to the substrate that it may act upon. It is possible that one enzyme may be shared by both cyclic GMP and cyclic AMP but serve to promote different, perhaps opposing effects by catalyzing similar reactions with different protein substrates. The substrate specificity of the enzyme may be determined by the cyclic nucleotide, perhaps, in combination with a specific cofactor or effector metabolite that is generated during a particular biological state. It is also conceivable that the metabolite generated under some conditions may serve as a modifier of the protein substrate and render it more or less susceptible to modification by the cyclic nucleotide-activatable enzyme.

Whatever the mechanism by which cyclic GMP may modulate cellular events, one prediction that can be offered is that a number of misconceptions will probably arise before the unique nature of its action(s) becomes fully appreciated. Perhaps the single most important concept that can be recommended as a guideline for future investigations is that the similarities between cyclic GMP and cyclic AMP should be kept in mind but that the differences, no matter how subtle they may seem or where they may occur, should be explored to the fullest.

VIII. ACKNOWLEDGMENTS

The senior author would like to express his deep appreciation to Dr. Joseph Larner for introducing him to the subject of cyclic nucleotides, buoying his interest during the course of the early work, and, through his critical assessment of how the metabolic effects of insulin may be brought about, setting the stage for a search for a cyclic AMP antagonist.

IX. REFERENCES

Anderson, W. B., Perlman, R. L., and Pastan, I. (1972): Effect of adenosine 3',5'-monophosphate analogues on the activity of the cyclic adenosine 3',5'-monophosphate receptor in *Escherichia coli*. *Journal of Biological Chemistry*, 247:2717–2722.

Appleman, M. M., Birnbaumer, L., and Torres, H. N. (1966): Factors affecting the activity of muscle glycogen synthetase. III. The reaction with adenosine triphosphate, Mg^{2+}, and cyclic 3',5'-adenosine monophosphate. *Archives of Biochemistry and Biophysics*, 116:39–43.

Ashman, D. F., Lipton, R., Melicow, M. M., and Price, T. D. (1963): Isolation of adenosine 3',5'-monophosphate and guanosine 3',5'-monophosphate from rat urine. *Biochemical and Biophysical Research Communications*, 11:330–334.

Aurbach, G. D., Potts, J. T., Chase, L. R., and Melson, G. L. (1969): Polypeptide hormones and calcium metabolism. *Annals of Internal Medicine*, 70:1243–1265.

Ball, J. H., Kaminsky, N. I., Broadus, A. E., Hardman, J. G., Sutherland, E. W., and Liddle, G. W. (1970): Effects of catecholamines and adrenergic blocking agents on cyclic nucleotides in human plasma. *Clinical Research*, 18:336.

Ball, J. H., Kaminsky, N. I., Hardman, J. G.; Broadus, A. E., Sutherland, E. W., and Liddle, G. W. (1972): Effects of catecholamines and adrenergic-blocking agents on plasma and urinary cyclic nucleotides in man. *Journal of Clinical Investigation*, 51:2124–2129.

Bdolah, A., and Schramm, M. (1965): The function of 3',5' cyclic AMP in enzyme secretion. *Biochemical and Biophysical Research Communications*, 18:452–454.

Beavo, J. A., Hardman, J. G., and Sutherland, E. W. (1970): Hydrolysis of cyclic guanosine and adenosine 3',5'-monophosphates by rat and bovine tissues. *Journal of Biological Chemistry*, 245:5649–5655.

Beavo, J. A., Hardman, J. G., and Sutherland, E. W. (1971): Stimulation of adenosine 3',5'-

monophosphate hydrolysis by guanosine 3′,5′-monophosphate. *Journal of Biological Chemistry*, 246:3841–3846.

Birnbaumer, L., Pohl, S. L., Michiel, H., Krans, J., and Rodbell, M. (1970): The actions of hormones on the adenyl cyclase system. In: *Advances in Biochemical Psychopharmacology, Vol. 3, Role of Cyclic AMP in Cell Function*, edited by P. Greengard and E. Costa, pp. 185–208. Raven Press, New York.

Birnbaumer, L., Pohl, S. L., and Rodbell, M. (1969): Adenyl cyclase in fat cells. I. Properties and the effects of adrenocorticotropin and fluoride. *Journal of Biological Chemistry*, 244: 3468–3476.

Bishop, J. S. (1970): Inability of insulin to activate liver glycogen transferase D phosphatase in the diabetic pancreatectomized dog. *Biochimica et Biophysica Acta*, 208:208–216.

Bishop, J. S., Goldberg, N. D., and Larner, J. (1971): Insulin regulation of hepatic glycogen metabolism in the dog. *American Journal of Physiology*, 220:499–506.

Bishop, J. S., and Larner, J. (1967): Rapid activation-inactivation of liver uridine diphosphate glucose-glycogen transferase and phosphorylase by insulin and glucagon *in vivo*. *Journal of Biological Chemistry*, 242:1355–1356.

Blonde, L., Wehman, R. E., and Steiner, A. L. (1972): Regulation of cyclic nucleotide levels in canine plasma. *Clinical Research*, 20:541.

Bourgoignie, J., Guggenheim, S., Kipnis, D. M., and Klahr, S. (1969): Cyclic guanosine monophosphate: Effects on short-circuit current and water permeability. *Science*, 165:1362–1363.

Braun, T., Hechter, O., and Bar, H. P. (1969): Lipolytic activity of ribonucleotide and deoxyribonucleotide-3′,5′-cyclic monophosphates in isolated rat fat cells. *Proceedings of the Society for Experimental Biology and Medicine*, 132:233–236.

Broadus, A. E., Hardman, J. G., Kaminsky, N. I., Ball, J. H., Sutherland, E. W., and Liddle, G. W. (1971): Extracellular cyclic nucleotides. *Annals of the New York Academy of Sciences*, 185:50–66.

Broadus, A. E., Kaminsky, N. I., Hardman, J. G., Sutherland, E. W., and Liddle, G. W. (1970a): Kinetic parameters and renal clearances of plasma adenosine 3′,5′-monophosphate and guanosine 3′,5′-monophosphate in man. *Journal of Clinical Investigation*, 49:2222–2236.

Broadus, A. E., Kaminsky, N. I., Northcutt, R. C., Hardman, J. G., Sutherland, E. W., and Liddle, G. W. (1970b): Effects of glucagon on adenosine 3′,5′-monophosphate and guanosine 3′,5′-monophosphate in human plasma and urine. *Journal of Clinical Investigation*, 49: 2237–2245.

Bron, T., and Rous, S. (1971): Comparison de quelques effets du 3′,5′-GMP cyclique et du 3′,5′-AMP cyclique sur le métabolisme glucolipidique de la souris vivante. *Biochimica et Biophysica Acta*, 237:156–166.

Brooker, G., Thomas, L. J., Jr., and Appleman, M. M. (1968): The assay of adenosine 3′,5′-cyclic monophosphate and guanosine 3′,5′-cyclic monophosphate in biological materials by enzymatic radioisotopic displacement. *Biochemistry*, 7:4177–4181.

Butcher, R. W., Baird, C. E., and Sutherland, E. W. (1968): Effects of lipolytic and antilipolytic substances on adenosine 3′,5′-monophosphate levels in isolated fat cells. *Journal of Biological Chemistry*, 243:1705–1712.

Butcher, R. W., and Sutherland, E. W. (1962): Adenosine 3′,5′-phosphate in biological materials. I. Purification and properties of cyclic 3′,5′-nucleotide phosphodiesterase and use of the enzyme to characterize adenosine 3′,5′-phosphate in human urine. *Journal of Biological Chemistry*, 237:1244–1250.

Cashel, M. (1969): The control of ribonucleic acid synthesis in *E. coli*. IV. Relevance of an unusual phosphorylated compound from amino acid-starved stringent strains. *Journal of Biological Chemistry*, 244:3133–3141.

Cehovic, G., Dettbarn, W.-D., and Welsch, F. (1971): Paraoxon: Effects on rat brain cholinesterase and on growth hormone and prolactin of pituitary. *Science*, 175:1256–1258.

Cehovic, G., Lewis, U. J., and VanderLaan, W. P. (1970): Release of growth hormone (GH) *in vitro* with cyclic 3′,5′-adenosine monophosphate. *Comptes Rendus Academy of Sciences*, 270:3119.

Cehovic, G., Posternak, T., and E. Charollais (1972a): A study of the biological activity and

resistance to phosphodiesterase of some derivatives and analogues of cyclic AMP. In: *Advances in Cyclic Nucleotide Research, Vol. 1, Physiology and Pharmacology of Cyclic AMP,* edited by P. Greengard and G. A. Robison, pp. 521–540. Raven Press, New York.
Cehovic, G. D., Robison, G. A., and Bass, A. D. (1972b): Sex dependent response in the release of growth hormone *in vitro* with different cyclic nucleotides. *Excerpta Medica International Congress,* Series No. 256, p. 83.
Chase, L. R., and Aurbach, G. D. (1967): Parathyroid function and the renal excretion of 3′,5′-adenylic acid. *Proceedings of the National Academy of Sciences,* 58:518–525.
Chelala, C. A., and Torres, H. N. (1969): Interconvertible forms of muscle phosphorylase phosphatase. *Biochimica et Biophysica Acta,* 178:423–426.
Cheung, W. Y. (1967): Properties of cyclic 3′,5′-nucleotide phosphodiesterase from rat brain. *Biochemistry,* 6:1079–1087.
Cheung, W. Y. (1969): Cyclic 3′,5′-nucleotide phosphodiesterase. Preparation of a partially inactive enzyme and its subsequent stimulation by snake venom. *Biochimica et Biophysica Acta,* 191:303–315.
Cheung, W. Y. (1971): Cyclic 3′,5′-nucleotide phosphodiesterase. Evidence for and properties of a protein activator. *Journal of Biological Chemistry,* 246:2859–2869.
Cheung, W. Y., and Williamson, J. R. (1965): Kinetics of cyclic adenosine monophosphate changes in rat heart following epinephrine administration. *Nature,* 207:979–981.
Clark, V. L., and Bernlohr, R. W. (1972a): Guanyl cyclase of *Bacillus licheniformis. Biochemical and Biophysical Research Communications,* 46:1570–1576.
Clark, V. L., and Bernlohr, R. W. (1972b): Cyclic GMP phosphodiesterase activity of *Bacillus licheniformis. Biochemical and Biophysical Research Communications (submitted).*
Cole, B., Robison, G. A., and Hartmann, R. C. (1971): Studies on the role of cyclic AMP in platelet function. *Annals of the New York Academy of Sciences.* 185:477–487.
Conn, H. O., Karl, I. S., Steiner, A., and Kipnis, D. M. (1971): Studies of the mechanism of action of 3′,5′-cyclic nucleotides on hepatic glucose production. *Biochemical and Biophysical Research Communications,* 45:436–443.
Conn, H. O., and Kipnis, D. M. (1969): The effect of various 3′,5′-nucleotides on gluconeogenesis and glycogenolysis in the perfused rat liver. *Biochemical and Biophysical Research Communications,* 37:319–326.
Corbin, J. D., and Krebs, E. G. (1969): A cyclic AMP-stimulated protein kinase in adipose tissue. *Biochemical and Biophysical Research Communications,* 36:328–336.
Corbin, J. D., Reimann, E. M., Walsh, D. A., and Krebs, E. G. (1970): Activation of adipose tissue lipase by skeletal muscle cyclic adenosine 3′,5′-monophosphate-stimulated protein kinase. *Journal of Biological Chemistry,* 245:4849–4851.
Cox, B., and Potkonjak, D. (1969): The relationship between tremor and change in brain acetylcholine concentration produced by injection of tremorine or oxotremorine in the rat. *British Journal of Pharmacology,* 35:295–303.
Craig, J. W., Rall, T. W., and Larner, J. (1969): The influence of insulin and epinephrine on adenosine 3′,5′-phosphate and glycogen transferase in muscle. *Biochimica et Biophysica Acta,* 177:213–219.
Diamond, J., and Brody, T. M. (1966): Relationship between smooth muscle contraction and phosphorylase activation. *Journal of Pharmacology and Experimental Therapeutics,* 152:212–220.
Drummond, G. I., and Duncan, L. (1970): Adenyl cyclase in cardiac tissue. *Journal of Biological Chemistry,* 245:976–983.
Drummond, G. I., and Perrott-Yee, S. (1961): Enzymatic hydrolysis of adenosine 3′,5′-phosphoric acid. *Journal of Biological Chemistry,* 236:1126–1129.
Emmer, M., de Crombrugghe, B., Pastan, I., and Perlman, R. (1970): Cyclic AMP receptor protein of *E. coli:* Its role in the synthesis of inducible enzymes. *Proceedings of the National Academy of Sciences, U.S.A.,* 66:480–487.
Ewart, R. B. L., and Taylor, K. W. (1971): The regulation of growth hormone secretion from the isolated rat anterior pituitary *in vitro. Biochemical Journal,* 124:815–826.
Exton, J. H., Hardman, J. G., Williams, T. F., Sutherland, E. W., and Park, C. R. (1971):

Effects of guanosine 3',5'-monophosphate on the perfused rat liver. *Journal of Biological Chemistry,* 246:2658–2664.

Exton, J. H., Jefferson, L. S., Jr., Butcher, R. W., and Park, C. R. (1966): Gluconeogenesis in the perfused liver. *American Journal of Medicine,* 40:709–715.

Ferrendelli, J. A., Kinscherf, D. A., and Kipnis, D. M. (1972): Effects of amphetamine, chlorpromazine and reserpine on cyclic GMP and cyclic AMP levels in mouse cerebellum. *Biochemical and Biophysical Research Communications,* 46:2114–2120.

Ferrendelli, J. A., Steiner, A. L., McDougal, D. B., Jr., and Kipnis, D. M. (1970): The effect of oxotremorine and atropine on cGMP and cAMP levels in mouse cerebral cortex and cerebellum. *Biochemical and Biophysical Research Communications,* 41:1061–1067.

Friedman, N., Somlyo, A. V., and Somlyo, A. P. (1971): Cyclic adenosine and guanosine monophosphates and glucagon: Effect on liver membrane potentials. *Science,* 171:400–402.

Galsky, A. G., and Lippincott, J. A. (1969): Production and inhibition of alpha-amylase production in barley endosperm by cyclic 3',5'-adenosine monophosphate and adenosine diphosphate. *Plant Cell Physiology,* 10:607–620.

Garbers, D. L., Lust, W. D., First, N. L., and Lardy, H. A. (1971): Effects of phosphodiesterase inhibitors and cyclic nucleotides on sperm respiration and motility. *Biochemistry,* 10:1825–1831.

George, W. J., Polson, J. B., O'Toole, A. G., and Goldberg, N. D. (1970): Elevation of guanosine 3',5'-cyclic phosphate in rat heart after perfusion with acetylcholine. *Proceedings of the National Academy of Sciences,* 66:398–403.

George, W. J., Kadowitz, P. J., and Polson, J. B. (1972): Influence of acetylcholine on contractility and cyclic AMP levels in the perfused rat heart. *Federation Proceedings,* 31:556 Abs.

Gericke, D., Chandra, P., Haenzel, I., and Wacker, A. (1970): Studies on the effect of nucleoside cyclic 3',5'-monophosphates on antibody synthesis by spleen cells. *Hoppe-Seyler's Zeitschrift für Physiologische Chemie,* 351:305–308.

Gill, G. N., and Garren, L. D. (1970): A cyclic-3',5'-adenosine monophosphate dependent protein kinase from the adrenal cortex: Comparison with a cyclic AMP binding protein. *Biochemical and Biophysical Research Communications,* 39:335–343.

Gilman, A. G. (1970): A protein binding assay for adenosine 3',5'-cyclic monophosphate. *Proceedings of the National Academy of Sciences,* 67:305–312.

Glinsmann, W. H., and Hern, E. P. (1969): Inactivation of rat liver glycogen synthetase by 3',5'-cyclic nucleotides. *Biochemical and Biophysical Research Communications,* 36:931–936.

Glinsmann, W. H., Hern, E. P., Linarelli, L. G., and Farese, R. V. (1969): Similarities between effects of adenosine 3',5'-monophosphate and guanosine 3',5'-monophosphate on liver and adrenal metabolism. *Endocrinology,* 85:711–719.

Goldberg, N. D., Villar-Palasi, C., Sasko, H., and Larner, J. (1967): Effects of insulin-treatment on muscle 3',5'-cyclic adenylate levels *in vivo* and *in vitro*. *Biochimica et Biophysica Acta,* 148:665–672.

Goldberg, N. D., Dietz, S. B., and O'Toole, A. G. (1969): Cyclic guanosine 3',5'-monophosphate in mammalian tissues and urine. *Journal of Biological Chemistry,* 244:4458–4466.

Goldberg, N. D., Lust, W. D., O'Dea, R. F., Wei, S., and O'Toole, A. G. (1970): A role of cyclic nucleotides in brain metabolism. In: *Advances in Biochemical Psychopharmacology, Vol. 3, The Role of Cyclic AMP in Cell Function,* edited by P. Greengard and E. Costa, pp. 67–87. Raven Press, New York.

Goldberg, N. D. (1972): Possible biological role(s) of cyclic 3',5'-guanosine monophosphate (cyclic GMP). *Fifth International Congress on Pharmacology,* Abstracts, pp. 229–230.

Goldberg, N. D., Haddox, M. K., Hartle, D. K., and Hadden, J. W. (1973): The biological role of cyclic 3',5'-guanosine monophosphate. *Fifth International Congress on Pharmacology (in press).*

Goren, E. N., and Rosen, O. M. (1971): The effect of nucleotides and a nondialyzable factor on the hydrolysis of cyclic AMP by a cyclic nucleotide phosphodiesterase from beef heart. *Archives of Biochemistry and Biophysics,* 142:720–723.

Gray, J. P. (1970): Adenosine 3',5'-monophosphate and guanosine 3',5'-monophosphate: Formation and occurrence in gametes and embryos. Ph.D. dissertation. Vanderbilt University, Nashville, Tennessee.
Gray, J. P., Hardman, J. G., Bubring, T., and Sutherland, E. W. (1970): High guanyl cyclase activity in sea urchin spermatozoa. *Federation Proceedings*, 29:608 Abs.
Greengard, P., and Kuo, J. F. (1970): On the mechanism of action of cyclic AMP. In: *Advances in Biochemical Psychopharmacology, Vol. 3, Role of Cyclic AMP in Cell Function*, edited by P. Greengard and E. Costa, pp. 287–306. Raven Press, New York.
Hadden, J. W., Hadden, E. M., Haddox, M. K. and Goldberg, N. D. (1972): Guanosine 3',5'-cyclic monophosphate: A possible intracellular mediator of mitogenic influences in lymphocytes. *Proceedings of the National Academy of Sciences*, 69:3024–3027.
Haddox, M. K., Newton, N. E., Hartle, D. K., and Goldberg, N. D. (1972): ATP (Mg^{2+})-induced inhibition of cyclic AMP reactivity with a skeletal muscle protein kinase. *Biochemical and Biophysical Research Communications*, 47:653–661.
Hardman, J. G., Davis, J. W., and Sutherland, E. W. (1966): Measurement of guanosine 3',5'-monophosphate and other cyclic nucleotides. *Journal of Biological Chemistry*, 241:4812–4815.
Hardman, J. G., Davis, J. W., and Sutherland, E. W. (1969): Effects of some hormonal and other factors on the excretion of guanosine 3',5'-monophosphate and adenosine 3',5'-monophosphate in rat urine. *Journal of Biological Chemistry*, 244:6354–6362.
Hardman, J. G., Robinson, G. A., and Sutherland, E. W. (1971): Cyclic nucleotides. *Annual Review of Physiology*, 33:311–336.
Hardman, J. G., and Sutherland, E. W. (1969): Guanyl cyclase, an enzyme catalyzing the formation of guanosine 3',5'-monophosphate from guanosine triphosphate. *Journal of Biological Chemistry*, 244:6363–6370.
Hardman, J. G., Beavo, J. A., Gray, J. P., Chrisman, T. D., Patterson, W. D., and Sutherland, E. W. (1971): The formation and metabolism of cyclic GMP. *Annals of the New York Academy of Sciences*, 185:27–35.
Hardman, J. G., Chrisman, T. D., Gray, J. P., Suddath, J. L., and Sutherland, E. W. (1972): Guanylate cyclase: Alteration of apparent subcellular distribution and activity by detergents and cations. *Fifth International Congress on Pharmacology*, Abstracts of invited papers, pp. 227–228.
Heidrick, M. L., and Ryan, W. L. (1971): Metabolism of 3',5'-cyclic AMP by strain L cells. *Biochimica et Biophysica Acta*, 237:301–309.
Hess, M. E., Shanfield, J., and Haugaard, N. (1962): The role of the autonomic nervous system in the regulation of heart phosphorylase in the open-chest rat. *Journal of Pharmacology*, 135:191–196.
Hoffman, F. (1972): Stimulation of protein kinase preparations from rat cerebellum by guanosine-3':5'-monophosphate and adenosine-3':5'-monophosphate. *Fifth International Congress on Pharmacology*, Abstracts of volunteer papers, p. 104.
Ishikawa, E., Ishikawa, S., Davis, J. W., and Sutherland, E. W. (1969): Determination of guanosine 3',5'-monophosphate in tissues and of guanyl cyclase in rat intestine. *Journal of Biological Chemistry*, 244:6371–6376.
Jard, S., and Bastide, F. (1970): A cyclic AMP-dependent protein kinase from frog bladder epithelial cells. *Biochemical and Biophysical Research Communications*, 39:559–566.
Jarett, L., Steiner, A. L., Smith, R. M., and Kipnis, D. M. (1972): The involvement of cyclic AMP in the hormonal regulation of protein synthesis in rat adipocytes. *Endocrinology*, 90:1277–1284.
Jefferson, L. S., Exton, J. H., Butcher, R. W., Sutherland, E. W., and Park, C. R. (1967): Role of adenosine 3',5'-monophosphate in the effects of insulin and anti-insulin serum on liver metabolism. *Journal of Biological Chemistry*, 243:1031–1038.
Jost, J. P., Hsie, A. W., Hughes, S. D., and Ryan, L. (1970): Role of cyclic adenosine 3',5'-monophosphate in the induction of hepatic enzymes. I. Kinetics of the induction of rat liver serine dehydratase by cyclic adenosine 3',5'-monophosphate. *Journal of Biological Chemistry*, 245:351–357.

Kakiuchi, S., and Yamazaki, R. (1970*a*): Stimulation of the activity of cyclic 3',5'-nucleotide phosphodiesterase by calcium ion. *Proceedings of the Japanese Academy*, 46:387–392.
Kakiuchi, S., and Yamazaki, R. (1970*b*): Calcium dependent phosphodiesterase activity and its activating factor (PAF) from brain. III. Studies on cyclic 3',5'-nucleotide phosphodiesterase. *Biochemical and Biophysical Research Communications*, 41:1104–1110.
Kakiuchi, S., Yamazaki, R., and Teshima, Y. (1971): Cyclic 3',5'-nucleotide phosphodiesterase. IV. Two enzymes with different properties from brain. *Biochemical and Biophysical Research Communications*, 42:968–974.
Kakiuchi, S., Yamazaki, R., and Teshima, Y. (1972): Cyclic 3',5'-nucleotide phosphodiesterase of brain and regulation of its activity by Ca^{2+}. *Fifth International Congress on Pharmacology*, Abstracts of invited papers, pp. 234–235.
Kaminsky, N. I., Ball, J. H., Broadus, A. E., Hardman, J. G., Sutherland, E. W., and Liddle, G. W. (1970*a*): Hormonal effects on extracellular cyclic nucleotides in man. *Transactions of the Association of American Physicians*, 83:235–243.
Kaminsky, N. I., Broadus, A. E., Hardman, J. G., Jones, D. J., Jr., Ball, J. H., Sutherland, E. W., and Liddle, G. W. (1970*b*): Effects of parathyroid hormone on plasma and urinary adenosine 3',5'-monophosphate in man. *Journal of Clinical Investigation*, 49:2387–2395.
Kawasaki, A., Kashimoto, T., and Yoshida, H. (1969): Effects of 3',5'-cyclic adenosine monophosphate and its dibutyryl derivative on the motility of isolated rat ileum. *Japanese Journal of Pharmacology*, 19:494–501.
Kitabchi, A. E., Solomon, S. S., and Brush, J. S. (1970): The insulin-like activity of cyclic nucleotides and their inhibition by caffeine on the isolated fat cells. *Biochemical and Biophysical Research Communications*, 39:1065–1072.
Kitabchi, A. E., Wilson, D. B., and Sharma, R. K. (1971): Steroidogenesis in isolated adrenal cells of rat. II. Effect of caffeine on ACTH and cyclic nucleotide-induced steroidogenesis and its relation to cyclic nucleotide phosphodiesterase (PDE). *Biochemical and Biophysical Research Communications*, 44:898–1004.
Kowal, J. (1970): Adrenal cells in tissue culture. VIII. Dissociation of the steroidogenic and glycolytic effects of cyclic nucleotides and ACTH. *In Vitro*, 6:174–179.
Krause, E.-G., Halle, W. and Wollenberger, A. (1972): Effect of dibutyryl cyclic GMP on cultured beating rat heart cells. In: *Advances in Cyclic Nucleotide Research*, Vol. 1, *Physiology and Pharmacology of Cyclic AMP*, edited by P. Greengard, R. Paoletti, and G. A. Robison, pp. 301–305. Raven Press, New York.
Kreutner, W., and Goldberg, N. D. (1967): Dependence on insulin of the apparent hydrocortisone activation of hepatic glycogen synthetase. *Proceedings of the National Academy of Sciences*, 58:1515–1519.
Kukovetz, W. R., and Pöch, G. (1970): Cardiostimulatory effects of cyclic 3',5'-adenosine monophosphate and its acylated derivatives. *Archives of Pharmacology and Pathology*, 266:236–254.
Kumon, A., Nishiyama, K., Yamamura, H., and Nishizuka, Y. (1972): Multiplicity of adenosine 3',5'-monophosphate-dependent protein kinases from rat liver and mode of action of nucleotide 3',5'-monophosphate. *Journal of Biological Chemistry*, 247:3726–3734.
Kuo, J. F., and Greengard, P. (1969): Cyclic nucleotide-dependent protein kinases, IV. Widespread occurrence of adenosine 3',5'-monophosphate-dependent protein kinase in various tissues and phyla of the animal kingdom. *Proceedings of the National Academy of Sciences*, 64:1349–1355.
Kuo, J. F., and Greengard, P. (1970*a*): Cyclic nucleotide–dependent protein kinases. VI. Isolation and partial purification of a protein kinase activated by guanosine 3',5'-monophosphate. *Journal of Biological Chemistry*, 245:2493–2498.
Kuo, J. F., and Greengard, P. (1970*b*): Stimulation of adenosine 3',5'-monophosphate-dependent and guanosine 3',5'-monophosphate-dependent protein kinases by some analogs of adenosine 3',5'-monophosphate. *Biochemical and Biophysical Research Communications*, 40:1032–1038.
Kuo, J. F., and Greengard, P. (1970*c*): Cyclic nucleotide-dependent protein kinases. VII.

Comparison of various histones as substrates for adenosine 3',5'-monophosphate-dependent and guanosine 3',5'-monophosphate-dependent protein kinases. *Biochimica et Biophysica Acta,* 212:434–440.

Kuo, J. F., Krueger, B. K., Sanes, J. R., and Greengard, P. (1970*a*): Cyclic nucleotide-dependent protein kinases. V. Preparation and properties of adenosine 3',5'-monophosphate-dependent protein kinase from various bovine tissues. *Biochimica et Biophysica Acta,* 212: 79–91.

Kuo, J. F., and Greengard, P. (1971): Stimulation of cyclic GMP-dependent protein kinase by a protein fraction which inhibits cyclic AMP-dependent protein kinases. *Federation Proceedings,* 30:1089.

Kuo, J. F., Lee, T. P., Reyes, P. L., Walton, K. G., Donnelly, T. E., Jr., and Greengard, P. (1972): Cyclic nucleotide-dependent protein kinases. X. An assay method for the measurement of guanosine 3',5'-monophosphate in various biological materials and a study of agents regulating its levels in heart and brain. *Journal of Biological Chemistry,* 247:16–22.

Kuo, J. F., Sanes, J., and Greengard, P. (1970*b*): Guanosine 3',5'-monophosphate-dependent protein kinases. *Federation Proceedings,* 29:601.

Kuo, J. F., Wyatt, G. R., and Greengard, P. (1971): Cyclic nucleotide-dependent protein kinases. IX. Partial purification and some properties of guanosine 3',5'-monophosphate-dependent and adenosine 3',5'-monophosphate-dependent protein kinases from various tissues and species of arthropoda. *Journal of Biological Chemistry,* 246:7159–7167.

Labrie, F., Béraud, G., Gauthier, M., and Lemay, A. (1971): Actinomycin-insensitive stimulation of protein synthesis in rat anterior pituitary *in vitro* by dibutyryl adenosine 3',5'-monophosphate. *Journal of Biological Chemistry,* 246:1902–1908.

Lust, W. D., Passonneau, J. V., and Goldberg, N. D. (1972): Reciprocal changes in 3',5'-cyclic AMP and 3',5'-cyclic GMP following electroconvulsive shock. *Federation Proceedings,* 31:555.

Mahaffee, D., Watson, B., and Ney, R. L. (1970): The relationship between nucleotide structure and the stimulation of adrenal steroidogenesis. *Clinical Research,* 18:73.

Majumder, G. C., and Turkington, R. W. (1971*a*): Adenosine 3',5'-monophosphate–dependent and –independent protein phosphokinase isoenzymes from mammary gland. *Journal of Biological Chemistry,* 246:2650–2657.

Majumder, G. C., and Turkington, R. W. (1971*b*): Hormonal regulation of protein kinases and adenosine 3',5'-monophosphate-binding protein in developing mammary gland. *Journal of Biological Chemistry,* 246:5545–5554.

Makman, R. S., and Sutherland, E. W. (1965): Adenosine 3',5'-phosphate in *Escherichia coli. Journal of Biological Chemistry,* 240:1309–1314.

Manganiello, V., and Vaughan, M. (1972): Prostaglandin E_1 effects on adenosine 3',5'-cyclic monophosphate concentration and phosphodiesterase activity in fibroblasts. *Proceedings of the National Academy of Sciences,* 69:269–273.

Masui, H., and Garren, L. D. (1971): Inhibition of replication in functional mouse adrenal tumor cells by adrenocorticotropic hormone mediated by adenosine 3':5'-cyclic monophosphate. *Proceedings of the National Academy of Sciences,* 68:3206–3210.

Merlevede, W., and Riley, G. A. (1966): The activation and inactivation of phosphorylase phosphatase from bovine adrenal cortex. *Journal of Biological Chemistry,* 241:3517–3524.

Mitznegg, P., Hach, B., and Heim, F. (1971): The influence of guanosine 3',5'-monophosphate and other cyclic nucleotides on contractile responses induced by oxytocin in isolated rat uterus. *Life Sciences,* 10:1285–1289.

Miyamoto, E., Kuo, J. F., and Greengard, P. (1969): Cyclic nucleotide-dependent protein kinases. III. Purification and properties of adenosine 3',5'-monophosphate-dependent protein kinase from bovine brain. *Journal of Biological Chemistry,* 244:6395–6402.

Murad, F., Chi, Y.-M., Rall, T. W., and Sutherland, E. W. (1962): Adenyl cyclase. III. The effect of catecholamines and choline esters on the formation of adenosine 3',5'-phosphate by preparations from cardiac muscle and liver. *Journal of Biological Chemistry,* 237:1233–1238.

Murad, F., and Gilman, A. G. (1971): Adenosine 3',5'-monophosphate and guanosine 3',5'-monophosphate: A simultaneous protein binding assay. *Biochimica et Biophysica Acta,* 252:397-400.
Murad, F., Manganiello, V., and Vaughan, M. (1970): Effects of guanosine 3',5'-monophosphate on glycerol production and accumulation of adenosine 3',5'-monophosphate by fat cells. *Journal of Biological Chemistry,* 245:3352-3360.
Murad, F., Manganiello, V. and Vaughan, M. (1971): A simple, sensitive protein-binding assay for guanosine 3',5'-monophosphate. *Proceedings of the National Academy of Sciences,* 68:736-739.
Murad, F. and Pak, C. Y. (1972): Clinical application of cyclic AMP levels. *Fifth International Congress on Pharmacology,* Abstracts of invited papers, pp. 240-241.
Murad, F., Rall, T. W., and Vaughan, M. (1969): Conditions for the formation, partial purification and assay of an inhibitor of adenosine 3',5'-monophosphate. *Biochimica et Biophysica Acta,* 192:430-435.
Nair, K. G. (1966): Purification and properties of 3',5'-cyclic nucleotide phosphodiesterase from dog heart. *Biochemistry,* 5:150-157.
Nichols, W. K., and Goldberg, N. D. (1972): The relationship between insulin and apparent glucocorticoid promoted activation of hepatic glycogen synthetase. *Biochimica et Biophysica Acta,* 279:245-258.
Nissley, S. P., Anderson, W. B., Gottesman, M. E., Perlman, R. L., and Pastan, I. (1971): *In vitro* transcription of the *gal* operon requires cyclic adenosine monophosphate receptor protein. *Journal of Biological Chemistry,* 246:4671-4678.
Nissley, P., Anderson, W. B., Gallo, M., and Pastan, I. (1972): The binding of cyclic adenosine monophosphate receptor to deoxyribonucleic acid. *Journal of Biological Chemistry,* 247: 4264-4269.
O'Dea, R. F., Haddox, M. K., and Goldberg, N. D. (1970): Kinetic analysis of a soluble rat brain cyclic nucleotide phosphodiesterase. *Federation Proceedings,* 29:473Abs.
O'Dea, R. F., Haddox, M. K., and Goldberg, N. D. (1971): Interaction with phosphodiesterase of free and kinase-complexed cyclic adenosine 3',5'-monophosphate. *Journal of Biological Chemistry,* 246:6183-6190.
Otten, J., Johnson, G. S., and Pastan, I. (1971): Cyclic AMP levels in fibroblasts: Relationship to growth rate and contact inhibition of growth. *Biochemical and Biophysical Research Communications,* 44:1192-1198.
Pagliara, A. S., and Goodman, A. D. (1970): Effect of 3',5'-GMP and 3',5'-IMP on production of glucose and ammonia by renal cortex. *American Journal of Physiology,* 218: 1301-1306.
Parks, J. S., Gottesman, M., Perlman, R. L., and Pastan, I. (1971): Regulation of galactokinase synthesis by cyclic adenosine 3',5'-monophosphate in cell-free extracts of *Escherichia coli. Journal of Biological Chemistry,* 246:2419-2424.
Pastan, I., and Perlman, R. (1970): Cyclic adenosine monophosphate in bacteria. *Science,* 169:339-344.
Paton, W. D. M., and AbooZar, M. (1968): The origin of acetylcholine released from guineapig intestine and longitudinal muscle strips. *Journal of Physiology,* 194:13-33.
Peake, G. T., Steiner, A. S., and Daughaday, D. H. (1972): Guanosine 3',5'-cyclic monophosphate is a potent pituitary growth hormone secretogogue. *Endocrinology,* 90:212-216.
Price, T. D., Ashman, D. F., and Melicow, M. M. (1967): Organophosphates of urine including adenosine 3',5'-monophosphate and guanosine 3',5'-monophosphate. *Biochimica et Biophysica Acta,* 138:452-465.
Puglisi, L., Berti, R., and Paoletti, R. (1971): Antagonism of dibutyryl-guo-3':5'-P and atropine on stomach smooth muscle contraction. *Experientia,* 27:1187-1188.
Rall, T. W., and Sutherland, E. W. (1962): Adenyl cyclase. II. The enzymatically catalyzed formation of adenosine 3',5'-phosphate and **inorganic pyrophosphate** from adenosine triphosphate. *Journal of Biological Chemistry,* 237:1228-1232.
Riggs, A. D., Reiness, G., and Zubay, G. (1971): **Purification and DNA-binding** properties of

the catabolite gene activator protein. *Proceedings of the National Academy of Sciences,* 68:1222–1225.
Rivkin, I., and Chasin, M. (1971): Nucleotide specificity of the steroidogenic response of rat adrenal cell suspensions prepared by collagenase digestion. *Endocrinology,* 88:664–670.
Robison, G. A., Butcher, R. W., Øye, I., Morgan, H. E., and Sutherland, E. W. (1965): The effect of epinephrine on adenosine 3′,5′-phosphate levels in the isolated perfused rat heart. *Molecular Pharmacology,* 1:168–177.
Robison, G. A., Butcher, R. W., and Sutherland, E. W. (1967): Adenyl cyclase as an adrenergic receptor. *Annals of the New York Academy of Sciences,* 139:703–723.
Robison, G. A., Butcher, R. W., and Sutherland, E. W. (1971): *Cyclic AMP.* Academic Press, New York.
Rodbell, M., Birnbaumer, L., Pohl, S. L., and Krans, H. M. J. (1971): The glucagon-sensitive adenyl cyclase system in plasma membranes of rat liver. *Journal of Biological Chemistry,* 246:1877–1882.
Rosen, O. M. (1970): Interaction of cyclic GMP and cyclic AMP with a cyclic nucleotide phosphodiesterase of the frog erythrocyte. *Archives of Biochemistry and Biophysics,* 139: 447–449.
Salzman, E. W., and Levine, L. (1971): Cyclic 3′,5′-adenosine monophosphate in human blood platelets. II. Effect of N^6-2′-0-dibutyryl cyclic 3′,5′-adenosine monophosphate on platelet function. *Journal of Clinical Investigation,* 50:131–141.
Sattin, A. (1971): Increase in the content of adenosine 3′,5′-monophosphate in mouse forebrain during seizures and prevention of the increase by methylxanthines. *Journal of Neurochemistry,* 18:1087–1096.
Schafer, D. E., Lust, W. D., Sircar, B., and Goldberg, N. D. (1970): Elevated concentration of adenosine 3′:5′-cyclic monophosphate in intestinal mucosa after treatment with cholera toxin. *Proceedings of the National Academy of Sciences,* 67:851–856.
Schlender, K. K., Wei, S. H., and Villar-Palasi, C. (1969): UDP-glucose: glycogen α-4-glucosyl-transferase I kinase activity of purified muscle protein kinase. Cyclic nucleotide specificity. *Biochemica et Biophysica Acta,* 191:272–278.
Schultz, G., Böhme, E., and Munske, K. (1969): Guanyl cyclase. Determination of enzyme activity. *Life Sciences,* 8:1323–1332.
Schultz, G., Hardman, J. G., Davis, J. W., Schultz, K., and Sutherland, E. W. (1972a): Determination of cyclic GMP by a new enzymatic method. *Federation Proceedings,* 31:440Abs.
Schultz, G., Hardman, J. G., Schultz, K., Baird, C. E., Parks, M. A., Davis, J. W., and Sutherland, E. W. (1972b): Cyclic GMP and cyclic AMP in ductus deferens and submaxillary gland of the rat. *Fifth International Congress on Pharmacology,* Abstracts of volunteer papers, p. 206.
Senft, G., Schultz, G., Munske, K., and Hoffman, M. (1968): Influence of insulin on cyclic 3′,5′-AMP phosphodiesterase activity in liver skeletal muscle, adipose tissue and kidney. *Diabetologia,* 4:322–329.
Sheppard, J. R. (1971): Restoration of contact-inhibited growth to transformed cells by dibutyryl adenosine 3′:5′-cyclic monophosphate. *Proceedings of the National Academy of Sciences,* 68:1316–1320.
Sheppard, J. R. (1972): Difference in the cyclic adenosine 3′,5′-monophosphate levels in normal and transformed cells. *Nature New Biology,* 236:14–16.
Shimazu, T. (1967): Glycogen synthetase activity in liver: Regulation by the autonomic nerves. *Science,* 156:1256–1257.
Smith, J. W., Steiner, A. L., and Parker, C. W. (1971): Human lymphocyte metabolism. Effects of cyclic and noncyclic nucleotides on stimulation by phytohemagglutinin. *Journal of Clinical Investigations,* 50:442–448.
Soderling, T. R., Hickenbottom, J. P., Reimann, E. M., Hunkeler, F. L., Walsh, D. A., and Krebs, E. G. (1970): Inactivation of glycogen synthetase and activation of phosphorylase kinase by muscle adenosine 3′,5′-monophosphate-dependent protein kinases. *Journal of Biological Chemistry,* 245:6317–6328.

Solomon, S. S., Brush, J. S., and Kitabchi, A. E. (1970): Divergent biological effects of adenosine and dibutyryl adenosine 3',5'-monophosphate on the isolated fat cell. *Science*, 169:387–388.
Stadtman, E. R. (1966): Allosteric regulation of enzyme activity. *Advances in Enzymology*, 28:41–154.
Steiner, A. L., Parker, C. W., and Kipnis, D. M. (1972a): Radioimmunoassay for cyclic nucleotides. I. Preparation of antibodies and iodinated cyclic nucleotides. *Journal of Biological Chemistry*, 247:1106–1113.
Steiner, A. L., Pagliara, A. W., Chase, L. R., and Kipnis, D. M. (1972b): Radioimmunoassay for cyclic nucleotides. II. Adenosine 3',5'-monophosphate and guanosine 3',5'-monophosphate in mammalian tissues and body fluids. *Journal of Biological Chemistry*, 247:1114–1120.
Steiner, A. L., Ferrendelli, J. A., and Kipnis, D. M. (1972c): Radioimmunoassay for cyclic nucleotides. III. Effect of ischemia, changes during development and regional distribution of adenosine 3',5'-monophosphate and guanosine 3',5'-monophosphate in mouse brain. *Journal of Biological Chemistry*, 247:1121–1124.
Steiner, A. L., Parker, C. W., and Kipnis, D. M. (1970): The measurement of cyclic nucleotides by radioimmunoassay. In: *Advances in Biochemical Psychopharmacology*, Vol. 3, *Role of Cyclic AMP in Cell Function*, edited by P. Greengard and E. Costa, pp. 89–111. Raven Press, New York.
Sutherland, E. W., and Robison, G. A. (1966): The role of cyclic 3',5'-AMP in responses to catecholamines and other hormones. *Pharmacological Reviews*, 18:145–161.
Tao, M. (1972): Dissociation of rabbit red blood cell cyclic AMP-dependent protein kinase I by protamine. *Biochemical and Biophysical Research Communications*, 46:56–61.
Tao, M., Salas, M. L., and Lipmann, F. (1970): Mechanism of activation by adenosine 3':5'-cyclic monophosphate of a protein phosphokinase from rabbit reticulocytes. *Proceedings of the National Academy of Sciences*, 67:408–414.
Thompson, W. J., and Appleman, M. M. (1971a): Multiple cyclic nucleotide phosphodiesterase activities from rat brain. *Biochemistry*, 10:311–316.
Thompson, W. J., and Appleman, M. M. (1971b): Characterization of cyclic nucleotide phosphodiesterases of rat tissues. *Journal of Biological Chemistry*, 246:3145–3150.
Triner, L., Nahas, G. G., Vulliemoz, Y., Overweg, N. I. A., Verosky, M., Habif, D. V., and Ngri, S. H. (1971): Cyclic AMP and smooth muscle function. *Annals of the New York Academy of Sciences*, 185:458–476.
Villar-Palasi, C., and Larner, J. (1961): Insulin treatment and increased UDPG-glycogen transglucosylase activity in muscle. *Archives of Biochemistry and Biophysics*, 94:436–442.
Walaas, O., Walaas, E., and Osaki, S. (1968): The effect of nucleoside 2',3'-cyclophosphates and nucleoside 3',5'-cyclic phosphates on UDP-glucose: α-1,4-glucan α-4-glucosyl transferase. In: *Control of Glycogen Metabolism*, edited by W. J. Whelan. Academic Press, New York.
Walsh, D. A., Perkins, J. P., and Krebs, E. G. (1968): An adenosine 3',5'-monophosphate-dependent protein kinase from rabbit skeletal muscle. *Journal of Biological Chemistry*, 243:3763–3774.
White, A. A., and Aurbach, G. D. (1969): Detection of guanyl cyclase in mammalian tissues. *Biochimica et Biophysica Acta*, 191:686–697.
White, A. A., Aurbach, G. D., and Carlson, S. F. (1969): Identification of guanyl cyclase in mammalian tissues. *Federation Proceedings*, 28:473.
Whitfield, J. F., and MacManus, J. P. (1972): Calcium-mediated effects of cyclic GMP on the stimulation of thymocyte proliferation by prostaglandin E_1. *Proceedings of the Society for Experimental Biology and Medicine*, 139:818–824.
Whitfield, J. F., MacManus, J. P., Franks, D. J., Gillan, D. J., and Youdale, T. (1971): The possible mediation by cyclic AMP of the stimulation of thymocyte proliferation by cyclic GMP. *Proceedings of the Society for Experimental Biology and Medicine*, 137:453–457.
Wilber, J. F., Peake, G. T., and Utiger, R. D. (1968): Cyclic 3',5'-adenosine monophosphate

(cyclic-AMP) stimulation of pituitary hormone secretion *in vitro. Journal of Laboratory and Clinical Medicine,* 72:1025–1026.

Yeung, D., and Oliver, I. T. (1968): Induction of phosphopyruvate carboxylase in neonatal rat liver by adenosine 3′,5′-cyclic monophosphate. *Biochemistry,* 7:3231–3239.

Zubay, G., Schwartz, D., and Beckwith, J. (1970): Mechanism of activation of catabolite-sensitive genes: A positive control system. *Proceedings of the National Academy of Sciences,* 66:104–110.

Advances in Cyclic Nucleotide Research, Vol. 3.
P. Greengard and G. A. Robison, Editors.
Raven Press, New York © 1973.

The Chemistry and Biological Properties of Nucleotides Related to Nucleoside 3′,5′-Cyclic Phosphates

Lionel N. Simon, Dennis A. Shuman, and Roland K. Robins

OUTLINE

I. Introduction to the Chemistry of Cyclic Phosphates 226
II. Physical and Chemical Properties of Nucleoside Cyclic Phosphates 238
   A. Spectral Measurements..238
   B. Hydrolysis Studies...241
III. Synthesis of Nucleoside Cyclic Phosphates Related to Adenosine
   3′,5′-Cyclic Phosphate..247
   A. Structural Modification of the Heterocyclic Base........................249
      1. Substitution on purine..249
      2. Heterocyclic ring systems related to purine 260
      3. Pyrimidine and other heterocyclic systems.........................262
   B. Structural Modification of the Carbohydrate Moiety 264
      1. Arabinofuranosyl cyclic nucleotides......................................264
      2. Xylofuranosyl cyclic nucleotides 269
      3. Lyxofuranosyl cyclic nucleotides...270
      4. 2′-Derivatives of ribofuranosyl cyclic nucleotides....................272
      5. Hexopyranosyl cyclic nucleotides 277
      6. Alditol cyclic nucleotides 280
   C. Structural Modification of the Cyclic Phosphate Moiety 281
      1. Nucleoside 3′,5′-cyclic phosphorothioates...............................281
      2. Nucleoside 3′,5′-cyclic phosphoramidates 283
      3. Nucleoside 3′,5′-cyclic thiophosphoramidates.........................287
      4. Alkyl nucleoside 3′,5′-cyclic phosphates................................287
      5. Nucleoside cyclic phosphonates 288
      6. Nucleoside cyclic pyrophosphates 289
   D. Addenda..290

IV. Introduction to the Biological Activity of Analogues of Nucleoside 3',5'-Cyclic Phosphates 291
V. Biological Activity (*In Vitro*) 294
   A. Effects on Isolated Enzyme Systems 294
      1. Cyclic nucleotide-dependent protein kinases 294
      2. cAMP receptor protein (*E. coli*) 306
      3. Cyclic nucleotide phosphodiesterases 308
   B. Crude Tissue Homogenates or Subcellular Fractions 318
      1. Glycogenolysis, activation of phosphorylase 318
      2. Lipolysis, activation of lipases 320
   C. Whole-Cell Suspensions or Tissue Slices 323
      1. Glycogenolysis 324
      2. Lipolysis, adrenal steroidogenesis 325
      3. Regulation of tumor cell growth, antiviral activity, enzyme induction in tissue culture 328
   D. Whole Tissue, Organ Perfusion, Tissue Fragment 332
      1. Anterior pituitary hormone release 332
      2. Mediator release from sensitized lung fragments 334
      3. Arterial or tracheal chain smooth muscle relaxation 334
      4. Glycogenolysis in perfused liver 336
VI. Biological Effects *In Vivo* 336
VII. Acknowledgments 339
VIII. List of Abbreviations Used 339
IX. References 340

I. INTRODUCTION TO THE CHEMISTRY OF CYCLIC PHOSPHATES

The phosphorus in phosphates is attached to oxygen atoms in an approximately tetrahedral arrangement (inorganic phosphate is exactly tetrahedral) reminiscent of sp³ tetrahedral carbon. The 3s and 3p orbitals on phosphorus are energetically available for sp³ hybridization making possible the formation of four covalent bonds. This does not describe the entire bonding, however, since the phosphoryl oxygen bond lengths are experimentally found to be shorter than those calculated for ordinary single covalent bonds (Collin, 1969). The incorporation of further $d_\pi-p_\pi$ bonding has been generally accepted to account for the observed shorter bond lengths (for a critical review of the importance of $d_\pi-p_\pi$ bonding, see Bartell et al. 1970). This ability of phosphorus to utilize d-orbitals of fairly low energy for $d_\pi-p_\pi$ bonding may result in multiple bond character when the atom to which the phosphorus donates an electron pair possesses electrons in orbitals of the same symmetry as the empty d-orbitals. The resonance hybrids **1, 2, 3** are then possible.

$$(R_3P \to O) \; R_3\overset{+}{P}\!\!-\!\!\overset{-}{O} \leftrightarrow R_3P\!=\!O \leftrightarrow R_3\overset{-}{P}\!\!\equiv\!\!\overset{+}{O}$$
$$ 1 \phantom{R_3\overset{+}{P}\!\!-\!\!\overset{-}{O}\;} 2 3$$

The degree of P—O bond character has been determined by molecular orbital calculations (Wagner, 1963; Serafini et al., 1971) and experimentally verified to be directly related to the electronegativity of the substituent R (Mauret et al., 1968, 1969). The P—O bond character was described by Wagner (1963) to vary from a nearly pure coordinate single bond (1) in $(CH_3)_3PO$ to almost a triple bond in F_3PO. Phosphodiesters seem to be best represented by 2 and will be written as such throughout this chapter.

The chemical reactivity of phosphate esters towards nucleophiles has been attributed, in part, to the difference in electronegativities of phosphorus and oxygen. The difference results in a partial positive charge on phosphorus thereby facilitating nucleophilic attack on the electrophilic phosphorus atom. The molecular conformations and their relation to chemical reactivity of phosphate esters have been correlated, by molecular orbital calculations, with the amount of d_π–p_π bonding in a phosphate ester. Of particular interest is the finding by Collin (1966) and Boyd (1969) that ring strain in cyclic phosphate esters results in lowered occupation of phosphorus 3d orbitals and provides a greater positive charge on phosphorus than in the acyclic esters. The positive charge on phosphorus decreases in the relative order, five-membered cyclic triester > acyclic triester > five-membered cyclic diester > six-membered cyclic diester > acyclic diester, which is in general agreement with observed nucleophilic chemical reactivity on phosphorus.

The observed increased reactivity of phosphomonoester dianions (presumably less electrophilic relative to positive charge) compared to phosphodiester anions has been explained by a change in the nucleophilic reaction mechanism. Whereas the reactivity of phosphodiesters is characterized by an $S_N2(P)$ mechanism, the monoester dianion is characterized by an $S_N1(P)$ mechanism in which the bulk of the nucleophilic 'push' is supplied by the two negative charges on the three oxygen atoms bound to phosphorus, resulting in a very reactive metaphosphate ion (PO_3^-). In general then, the nucleophilic reactivity of phosphoesters is triester > monoester > diester where the respective cyclic ester has increased activity relative to the acyclic ester. The detailed reaction mechanisms of phosphate esters other than those pertaining to the chemistry of cyclic phosphates will not be discussed further since these reactions have been adequately reviewed elsewhere (Cox and Ramsay, 1964; Bunton, 1970; Kirby and Younos, 1970a,b; Khan and Kirby, 1970; Khan et al., 1970; Bromilow and Kirby, 1972).

The disubstituted phosphate esters are the least reactive toward nucleophilic attack on phosphorus. The presence, however, of a sterically suited nucleophile can produce a very substantial nucleophilic rate enhancement on hydrolysis or transesterification. Acid-catalyzed phosphoryl migration with a mono- or diester (**4 ⇌ 5**) or acid-catalyzed hydrolysis of a diester (**4 → 9**) can occur readily with *cis*-diol systems as shown in Scheme I (Bruice and Benkovic, 1966). Base catalysis (**6 → 8, 7 → 8**) can occur by an $S_N2(P)$ type mechanism (**10 → 11**) (Cox and Ramsey, 1964; Usher et al., 1970).

The degree of participation of a neighboring hydroxyl group in the formation of a cyclic phosphate is dependent on the stereochemical position of the hydroxyl group. Hydrolysis studies of various phosphates have provided some general information on the participation of neighboring hydroxyl groups which are useful in understanding the chemical formation of cyclic phosphates (Khorana, 1961). Stereochemically a β-hydroxyl group is very favorable for neighboring group participation and when the possibility exists for the formation of a five- or six-membered cyclic phosphate, the five-membered cyclic phosphate is favored. Participation of an acyclic hydroxyl group is greatest for formation of five- to six-membered rings [e.g., no participation was observed in the hydrolysis of P^1,P^2-bis(4-hydroxybutyl) pyrophosphate]. In a six-membered ring (such as cyclohexane and pyranoses), participation of a neighboring hydroxyl can occur from both cis (axial-equatorial in chair conformation) and trans (equatorial-equatorial) configurations resulting in the five-membered cyclic phosphates 12 and 13. A trans (axial-axial) configuration for a five-membered cyclic phosphate is stereochemically unfavorable. The formation of six-membered cis-1,3 (14) and pyranose-trans-4,6 (15) cyclic phosphates is sterically

12 cis (a,e)   **13** trans (e,e)   **14** cis-1,3   **15**

favorable, and the degree of neighboring hydroxyl participation in their formation is, of course, dependent on the conformation of the pyranose or cyclohexane (Cawley and Letters, 1971).

In five-membered furanose ring systems, the approximately planar ring sterically favors the cis configuration of a cyclic phosphate. Trans-fused six-membered cyclic phosphates are sterically possible, with adenosine 3′,5′-cyclic phosphate as the best known example. Less efficient participation of a cis-hydroxyl is found in the formation of six- or seven-membered cyclic phosphates relative to the formation of a five-membered cyclic phosphate. This is shown by the sole formation of the 2′,3′-cyclic phosphate **16** by the cyclization of a mixture of the 2′(3′) phosphites of 1-(β-D-lyxofuranosyl)uracil (**17**) (Holý and Šorm, 1969).

The presence of a cis-hydroxyl in furanose phosphates can produce dramatic hydrolysis rate enhancements of phosphate esters; the classical example is the facile alkaline and RNAse hydrolysis of RNA. Intramolecular attack by the 2'-hydroxyl on the phosphodiester linkage gives a cyclic phosphate **18** which on further hydrolysis gives a mixture of 2' and 3' mononucleotides **19** (Usher et al., 1970).

The absence in DNA of the 2'-hydroxyl requisite for cyclic phosphate formation explains the relative stability of DNA to alkaline hydrolysis (Markham and Smith, 1952).

The synthesis of a cyclic phosphate is dependent not only on the stereochemistry of the parent compound but also on the nature of the phosphorylating agent. The direct formation of cyclic phosphates can occur during the phosphorylation of a diol; cis-diols are particularly suited for this type of reaction. Pantetheine 2,4-cyclic phosphate (**20**) (Baddiley and Thain, 1953), chloromycetin cyclic phosphate (**21**) (Mosher et al., 1953), cis-1,3-cyclohexylidine phosphate (**22**) (Brown and Higson, 1957), and riboflavin 4',5'-cyclic phosphate (**23**) (Forrest and Todd, 1950) are examples prepared by direct reaction of the diols with phosphorus oxychloride ($POCl_3$).

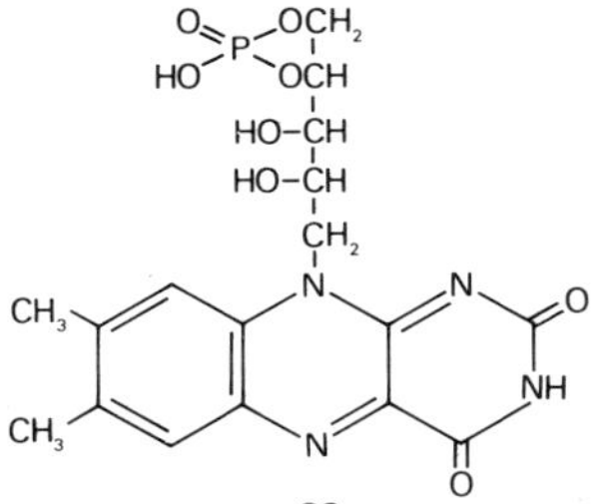

23

Stable *cis*- and *trans*-cyclic phosphate triesters are isolated by treatment of 1,2-*O*-isopropylidenexylose (**24**) (Moffatt and Khorana, 1957), methyl (or phenyl) glucoside (Baddiley et al., 1954), and 9-β-D-glucopyranosyladenine (**25**) (Barker and Foll, 1957) with phenylphosphorochloridates. The cyclic phosphate diesters **26** and **27** are obtained by hydrogenation or hydrolysis (Moffatt and Khorana, 1957; Barker and Foll, 1957). A variation of this method is used in the formation of ribo- (**28**) and 2'-deoxyribonucleoside 3', 5'-cyclic phosphates. Whereas *cis*-2', 3'-cyclic nucleotides

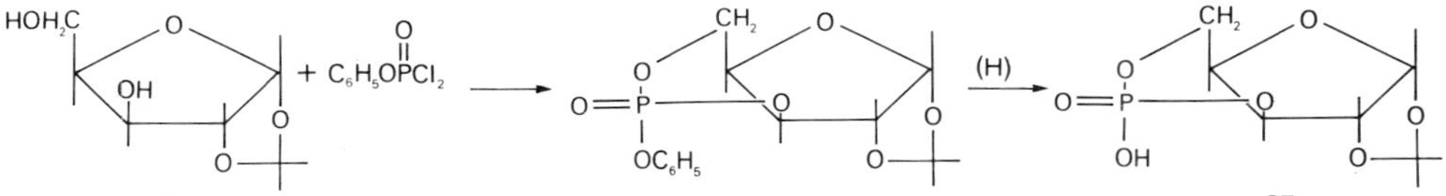

Ad = 9 − adenyl

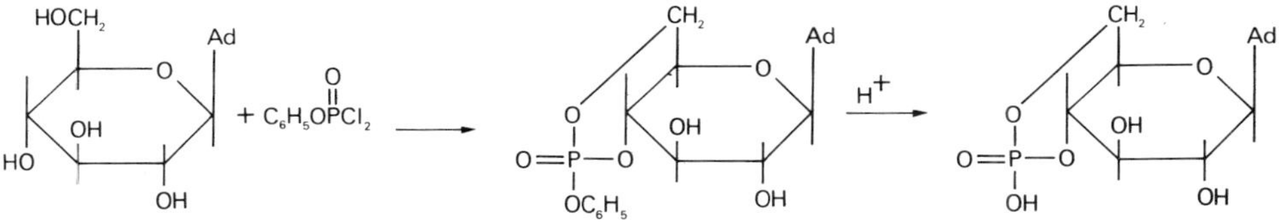

are easily formed by aqueous acid or base catalyzed transesterifications, the formation of *trans*-3',5'-cyclic nucleotides is more difficult (Smith et al., 1961). Nucleoside 3',5'-cyclic phosphates (**28**) are formed in high yield by transesterification of "active" phosphates; i.e., phosphate esters with a good cleaving group (e.g., *p*-nitrophenoxide, 2,4-dinitrophenoxide, diphenylphosphate, fluoride) under strongly basic conditions in an aprotic solvent (Borden and Smith, 1966*a*). The "active" phosphate esters, e.g., the nucleoside 5'-*p*-nitrophenyl phosphate (**30**), are obtained by phosphorylation of an appropriately blocked nucleoside **29** with a variety of reagents (e.g., tetra-*p*-nitrophenyl pyrophosphate, *bis*-*p*-nitrophenylphosphorochloridate, or *p*-nitrophenyl phosphate and dicyclohexylcarbodiimide). The "active" phosphate can also be synthesized from an unblocked nucleoside 5'-phosphate (**31**) by reaction with *p*-nitrophenol and carbodiimide, with diphenylphosphorochloridate, or with 2,4-dinitrofluorobenzene; the latter reagent yields the nucleoside 5'-phosphorofluoridate (Borden and Smith, 1966*b*).

Nucleosides can undergo acid-catalyzed transesterification with tri-substituted phosphites to give initially a cyclic phosphorus triester **32** which rapidly hydrolyzes to a mixture of monophosphites **33** (Holý and Sŏrm, 1966*a*). Oxidation of the nucleoside phosphite **33** gives an "active" phosphate **34** initiating a rapid intramolecular transesterification to the nucleoside 2',3'-cyclic phosphate (Holý et al., 1965; Holý and Smrt, 1966; Holý, 1970; Bald and Holý, 1971; Holý and Bald, 1971). The selectivity of the transesterification is a function of the stereochemistry of the diol and the phosphite. Unprotected ribonucleosides react selectively and quantitatively with triethylphosphite at the *cis*-2',3'-position, resulting ultimately in high yields of the 2',3'-cyclic nucleotides. The selectivity is lost, however, with triphenyl phosphite which can phosphitylate any sugar hydroxyl

(Holý and Sörm, 1966b). The selectivity of transesterification and oxidative cyclization is dependent on the stereochemistry of the carbohydrate and is in agreement with the previous observations concerning the formation of cyclic phosphates. In the cases of 1-aldopentafuranosyl derivatives where cis-five- or six-membered cyclic phosphorus triesters can be formed (e.g., ribo, lyxo, xylo), transesterification and then oxidative cyclization to a cyclic phosphate occurred in high yields. In arabinofuranosyl derivatives where seven-membered ring formation is necessary, the yield was considerably lower. The 1-β-D-2'-deoxyribofuranosides of uracil and thymine did not react at all, showing that the formation of a trans-cyclic phosphorus triester was much less favorable (Holý and Sörm, 1969).

The reactivity of the "active phosphate" 34 formed during the oxidation of the phosphites is sufficiently high to form six-membered trans-3',5' as well as seven-membered cyclic phosphates, as shown by the formation of 6-azauridine 3',5'-cyclic phosphate (35) and 1-β-D-lyxofuranosyluracil 2',5'-cyclic phosphate (36) from the respective 5'-monophosphite (Holý et al., 1965).

The selectivity of transesterification for pyranosides is demonstrated by the reactions of triethyl phosphite with 1-(2-deoxy-β-D-ribopyranosyl) thymine (**37**) and 1-(2-deoxy-β-D-xylopyranosyl)thymine (**38**) (Holý and Šorm, 1969). Under conditions where the *cis*-diol **37** reacted quantitatively with triethyl phosphite, no reaction was observed with the *trans*-diol **38**. A 17% yield of the monophosphites **39** was obtained by excess phosphite and a longer reaction time. Quantitative oxidative-cyclization of **39** with hexachloroacetone did occur, however, to give the *trans*-3′,4′-cyclic phosphate **40**.

Several procedures have been used for the cyclization of phosphate monoesters (for reviews on the synthesis of nucleoside phosphates, see Ueda and Fox, 1967; Hutchinson, 1970, 1971). As expected, cyclization of a phosphate leading to *cis*-fused cyclic phosphates is relatively easy. Quantitative yields of nucleoside 2′,3′-cyclic phosphates (**41**) can be obtained under mild conditions with acid chlorides (Michelson, 1958, 1959), acid anhydrides (Brown et al., 1952), dimethylformamide acetals (Holý et al., 1969), or carbodiimides (Dekker and Khorana, 1954; Smith et al., 1958; Naylor and Gilham, 1966). The first two reagents give a reactive mixed anhydride **42** which is readily attacked by a *cis* neighboring hydroxyl, giving the cyclic phosphate **41** and the respective anion **43**. The cyclization of cytidine, adenosine, and guanosine 2′(3′) phosphates with dimethylformamide acetal (**44**) at 80 to 100° occurred quantitatively, presumably through an intermediate such as **45** or **46**. This procedure was preferred to the acid chloride or dicyclohexycarbodiimide methods for the preparation of cytidine 2',3'-cyclic phosphate. The presence of a 5'-phosphate group [i.e., cytidine 2'(3'),5'-diphosphate] was found not to interfere with the dimethylformamide acetal cyclization to the 2',3'-cyclic nucleotide (Holý et al., 1969).

The cyclization of nucleoside 2'(3') phosphates (**47**) to the nucleoside 2',3'-cyclic phosphate **48** also occurred with dimethylformamide at elevated temperatures (Yurkevich et al., 1969). However, when tri-*n*-butylammonium adenosine 5'-phosphate (**49**) was refluxed in dimethylformamide, no adenosine 3',5'-cyclic phosphate (**50**) was detected: instead, dephosphorylation of the nucleotide with formation of an "activated phosphate" occurred giving adenosine 2',3'-cyclic phosphate (**48**), a complex mixture of adenosine nucleotides, adenosine, and adenine (Ueda and Kawai, 1970).

Carbodiimides (commonly dicyclohexylcarbodiimide) have been extensively used for the cyclization of monophosphates. The initial reaction involves the formation of the isourea 51 (Smith et al., 1958) which can undergo attack by a stereochemically favorable hydroxyl group giving intramolecular cyclization, e.g., the formation of a cis-2′,3′-cyclic nucleotide 52 from a nucleoside 2′(3′)-monophosphate (Dekker and Khorana, 1954). If intramolecular cyclization is less favorable, e.g., in the 3′,5′-*trans*esterification of a ribofuranosyl nucleoside, more vigorous reaction conditions are required utilizing anhydrous media and elevated temperatures (Smith et al., 1961). Attack by another phosphate on the intermediate 51 initially produces the dinucleoside pyrophosphate 53. Further treatment of the pyrophosphate 53 with dicyclohexylcarbodiimide at elevated temperatures (usually refluxing pyridine) gives presumably the isourea adduct 54 (Weimann and Khorana, 1962) or a more reactive intermediate therefrom, such as the trimetaphosphate 55 (Jacob and Khorana, 1964; Blackburn et al., 1969), which then yields *via* intramolecular cyclization the nucleoside 3′,5′-cyclic phosphate (56).

Ribonucleoside 3′,5′-cyclic phosphates are therefore most conveniently prepared from nucleoside 5′-phosphates since nucleoside 3′-phosphates would require prior blocking of the 2′-hydroxyl to prevent predominantly nucleoside 2′,3′-cyclic phosphate formation.

The formation of nucleoside 3′,5′-cyclic phosphates by the DCC method has the disadvantage of the requirement of high temperatures and of side reactions with other functional groups, such as sulfhydryl (Mikolajczyk,

1966; Holý, 1968). In such cases, intramolecular cyclization *via* the intermediate **53** or a more reactive species thereof has been achieved by treatment of the requisite nucleotide with an arylsulfonyl chloride. For example, whereas adenosine 5'-phosphorothioate **210** was desulfurized by dicyclohexycarbodiimide, treatment of the nucleotide with tri-*iso*-propylbenzenesulfonyl chloride gave the desired adenosine 3',5'-*O,O*-cyclic phosphorothioate (**209**) (Eckstein, 1970).

One of the earliest reported chemical methods for the formation of cyclic phosphates was by intramolecular nucleophilic attack of a phosphate anion on carbon (Bailly, 1921a,b). Glycerol 1,3-cyclic phosphate (**57**) (Bailly, 1921a,b, 1922), ethylene glycol cyclic phosphate (**58**) (Lecoq, 1956; Kumamoto et al., 1956), and recently 2',5'-dideoxy-5'-thio-nucleoside 3′,5′-cyclic phosphorothioates (Chladek and Nagyvary, 1972) and 1-β-D-xylofuranosylthymine 3′,5′-cyclic phosphate (Russell and Moffatt, 1969) have been synthesized by this method.

ClCH₂—CH(OH)—CH₂—OP(=O)(O⁻)—O⁻ ⟶ CH₂O—CHOH—CH₂O\P(=O)(OH) **57**

XCH₂—[...]—CH₂—OP(=O)(O⁻)—O⁻ ⟶ CH₂O—[...]—CH₂O\P(=O)(OH) **58**

X = Cl or Br

II. PHYSICAL AND CHEMICAL PROPERTIES OF NUCLEOSIDE CYCLIC PHOSPHATES

A. Spectral Measurements

The molecular structure of crystalline cytidine 2′,3′-cyclic phosphate has been studied by X-ray crystallography (Coulter and Greaves, 1970). Two nucleotides were found per asymmetric unit, both in the *syn* conformation about the glycosidic linkage. The conformation of the ribose ring of one nucleotide is planar, whereas the other has oxygen-1′ puckered toward carbon-5′. The strain in both these conformations (reflected in their chemical instability) was shown by the O(2′)–P–O-(3′) angles of 96.2° and 96.1° which are smaller than the 103° found in less strained esters. Further, the C(2′)–C(3′)–C(4′) angles are larger than usually observed and the C(3′)–O(3′) distances are longer than expected. (Fig. 1.)

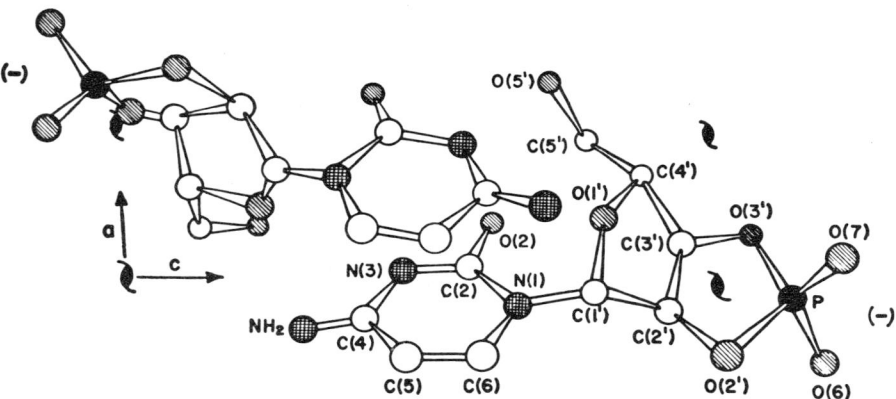

FIG. 1. A schematic view of cytidine 2′,3′-cyclic phosphate anions down the *b* axis (Coulter and Greaves, 1970). Copyright 1970 by the American Association for the Advancement of Science.

The ^{31}P, ^{13}C, and ^{1}H chemical shifts and coupling constants have been measured for several nucleoside 2',3'-cyclic phosphates (Jardetzky, 1962; Smith and Jardetzky, 1968; Smith et al., 1972; Lapper et al., 1972). As expected from molecular models and previous X-ray studies, conformational mobility of the ribose ring is possible and is observed by comparison of the ^{1}H–^{31}P couplings in the ribose rings. A rapid interconversion between puckered forms occurs with little preference for any form (Smith et al., 1973; Lapper et al., 1972).

The molecular conformation of adenosine 3',5'-cyclic phosphate (Watenpaugh et al., 1968) and uridine 3',5'-cyclic phosphate (Coulter, 1968) in the crystalline state is quite different from the corresponding 2',3'-cyclic nucleotides. In contrast to the conformational mobility of ribonucleoside 2',3'-cyclic phosphates, the 3',5'-*trans*-fused phosphate of ribonucleoside 3',5'-cyclic phosphates restricts the conformation of the furanose ring to twist C(3')-*endo*-C(4')-*exo* with normal bond lengths and angles. Adenosine 3',5'-cyclic phosphate crystallizes with two molecules per unit cell differing only in their *syn-anti* conformation about the glycosidic bond. The phosphorus atoms in both molecules are in a chair conformation. Uridine 3',5'-cyclic phosphate also crystallizes with two molecules per unit cell; however, both molecules exhibit glycosyl angles in the *anti* range. A structural analogue of adenosine 3',5'-cyclic phosphate, adenosine 3',6'-cyclic phosphonate, where the 5'-O has been substituted by a methylene, also has a *syn*, twist C(3')-*endo*-C(4')-*exo* conformational structure in the solid state (Sundaralingam and Abola, 1972). (Fig. 2.)

The ^{13}C, ^{31}P, and ^{1}H chemical shifts and coupling constants have been reported on some *ribo* (Jardetzky, 1962; Smith et al., 1973; Lapper et al., 1972) and *arabino* 3',5'-cyclic nucleotides and 8-substituted adenosine 3',5'-cyclic phosphates (Schweizer and Robins, 1973). The ^{1}H and ^{13}C

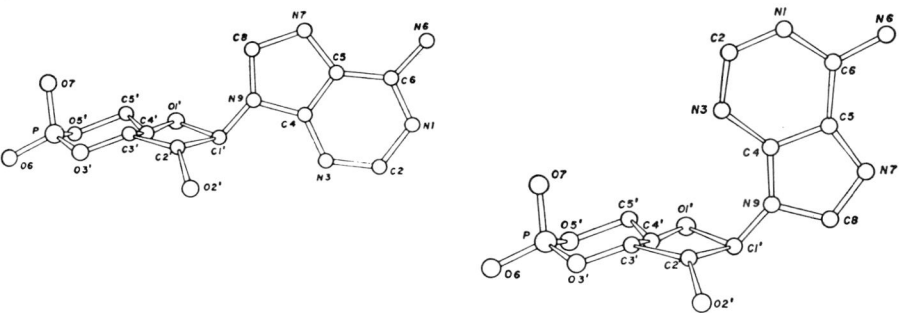

FIG. 2. A schematic view of cAMP showing different *syn-anti* conformations about the glycosidic bond (Watenpaugh et al., 1968). Copyright 1968 by the American Association for the Advancement of Science.

spectra are consistent with twist C(3')-*endo*-C(4')-*exo* conformations. This conformation of the ribose ring causes the phosphorus atom to be too far removed in ribonucleoside 3',5'-cyclic phosphates for deshielding of base protons as seen in the pmr of 5'-nucleotides (Schweizer et al., 1968) and *arabino*-nucleoside 2',5'-cyclic phosphates (Lee et al., 1971). Additionally, the coupling $J_{1'-2'}$ is observed to be less than 1 Hz for ribonucleoside 3',5'-cyclic phosphates. These two observations have a diagnostic importance in structural determinations.

The ^{31}P spectra of adenosine 3',5'-cyclic phosphate and several derivatives in solution at room temperature may be interpreted in terms of either the chair or twist conformation for the phosphate ring (Schweizer and Robins, 1973). A comparison of the chemical shifts of H-2' and H-3' of various *arabino* and *ribo* nucleoside 3',5'-cyclic phosphates suggested that adenosine 3',5'-cyclic phosphate, inosine 3',5'-cyclic phosphate, and uridine 3',5'-cyclic phosphate are probably weighted toward the *anti* conformation in solution at room temperature, whereas guanosine 3',5'-cyclic phosphate and 8-substituted purine cyclic nucleotides are probably *syn*. Similarly, 9-β-D-arabinofuranosyladenine 3',5'-cyclic phosphate probably exists preferentially as the *anti* conformer, whereas the analogues with bulky 8-substituents are mainly *syn* (Schweizer and Robins, 1973). Large differences in optical rotatory dispersion amplitudes between adenosine and adenosine 3',5'-cyclic phosphate, however, have been interpreted as the latter having predominantly *syn* conformation (Klee and Mudd, 1967).

Magnesium, calcium, and manganous salts of gold (III) · adenosine 3',5'-cyclic phosphate have been prepared and are very water soluble, whereas the corresponding salts of gold (III) · adenosine 5'-phosphate, gold (III) · adenosine 5'-diphosphate, and gold (III) · adenosine 5'-triphosphate are relatively insoluble. The relative solubility of these heavy metal complexes has been suggested as potentially useful for cytochemical localizations of enzymes by electron microscopy (Gibson et al., 1971).

The proton magnetic resonance of copper (II) complexes of adenosine (2')(3')(5') monophosphates have been compared with the spectra of adenosine 2',3' and 3',5'-cyclic phosphates. A comparison of the chemical shifts and broadening effects with varying copper (II) concentrations suggested that greater base stacking and no phosphate-base deshielding occurred in the cyclic nucleotides relative to the nucleoside phosphomonoesters (Berger and Eichhorn, 1971).

The mass spectra of trimethylsilyl derivatives of nucleotides including adenosine 3',5'-cyclic phosphate have been determined. The spectra of adenosine 3',5'-cyclic phosphate retained the important peaks observed with 5'-nucleotides and also exhibited two additional peaks apparently unique to the 3′,5′-cyclic structure. An increased stability of the cyclic phosphate was shown in the greater abundance of M ion (molecular ion) and M-15 ion (Lawson et al., 1971).

B. Hydrolysis Studies

The stability of cyclic phosphates is a function of ring size and phosphate ring configuration. The stability of simple cyclic phosphates has been correlated well with the amount of ring strain and the degree of $d_\pi-p_\pi$ bonding in the phosphate bonds (Collin, 1966; Boyd, 1969). The hydrolysis of methyl ethylene cyclic phosphate (**59**, five-membered triester), ethylene glycol cyclic phosphate (**60**, five-membered diester), propane-1,3-diol cyclic phosphate (**61**, six-membered diester), and butane-1,4-diol cyclic phosphate (**62**, seven-membered diester) is typical of each class of cyclic phosphate. The five-membered triester **59** hydrolyzed approximately 10^6 times faster than the acyclic triester trimethyl phosphate (Westheimer, 1968). Similarly, the five-membered cyclic diester **60** hydrolyzed approximately 10^7 times faster (half-life of 50 min in 0.5 N sodium hydroxide at 25°C) than dimethyl phosphate (Khorana, 1961). The six-membered cyclic diester **61** hydrolyzed only approximately 10 times faster, and the seven-membered diester **62** approximately the same as dimethyl phosphate. Different hydrolysis rates

exist within each group where additional ring strain is imposed by *trans* and *cis*-fused rings. In such cases, *cis*-fused rings were generally found to be more stable to hydrolysis than *trans*-fused ring systems (Khorana, 1961). Differences in fused ring size (e.g., in the six-membered cyclic phosphates pyranose 4,6-cyclic phosphate and furanose 3,5-cyclic phosphate), ring conformation, or the presence of other substituents, would be expected to influence the stability of cyclic phosphates and make general conclusions more difficult.

The acid or base hydrolysis of a nucleoside 2′,3′-cyclic phosphate is a simple case since facile ring opening of the cyclic phosphate ring occurs before glycosyl cleavage at approximately the same rate of hydrolyses of other five-membered cyclic phosphates (Kochetkov et al., 1972). In 0.1 N HCl

TABLE 1. *Acidic hydrolysis of ribo(deoxyribo)nucleoside 3',5'-cyclic phosphates and ribo(deoxyribo)nucleoside (2')(3')(5') phosphates in 1 N HCl at 100°C and 50°C*

Compound	Half-life (min)	Temperature (°C)	Products
Adenosine 3',5'-cyclic phosphate	30[a]	100	Adenine, transient adenosine, orthophosphoric acid, ribose, trace of ribose phosphates
Adenosine 5' or 2'(3') phosphate	2–4[a]	100	Adenine
2'-Deoxyadenosine 3',5'-cyclic phosphate	Unaffected after 2 hr[b]	50	
2'-Deoxyadenosine 3',5'-cyclic phosphate	3[b]	100	Adenine
2'-Deoxyadenosine 5'-phosphate	5[b]	50	Adenine
Guanosine 3',5'-cyclic phosphate	28[a]	100	Guanine, transient guanosine phosphates
Guanosine 2'(3')-phosphate	1.5[a]	100	Guanine
2'-Deoxyguanosine 3',5'-cyclic phosphate	Unaffected after 2 hr[b]	50	
2'-Deoxyguanosine 5'-phosphate	5[b]	50	Guanine
Cytidine 3',5'-cyclic phosphate	26[a]	100	Cytosine 9%, cytidine 5'-phosphate 11%, cytidine 2'(3') phosphate 70%
Cytidine 2'(3')-phosphate	[a]	100	9% Cytosine after 2 hr
2'-Deoxycytidine 3',5'-cyclic phosphate	Unaffected after 2 hr[b]	50	
2'-Deoxycytidine 3',5'-cyclic phosphate	22[b]	100	Cytosine
2'-Deoxycytidine 5'-phosphate	Unaffected after 2 hr[b]	50	
2'-Deoxycytidine 5'-phosphate	110[b]	100	Cytosine
Thymidine 3',5'-cyclic phosphate	13[b]	50	Thymine
Thymidine 5'-phosphate	Unaffected after 2 hr[b]	50	
Uridine 3',5'-cyclic phosphate	8[a]	100	Uracil (67%), uridine 5'-phosphate (6%), uridine 2'(3')-phosphate (27%), ribose, orthophosphate
Uridine 5'-phosphate	Unaffected after 2 hr[a]	100	

[a] Smith et al., 1961
[b] Drummond et al., 1964

at 25°C, the hydrolysis of nucleoside 2',3'-cyclic phosphates is complete in 4 hrs (Markham, 1957).

The hydrolysis of nucleoside 3',5'-cyclic phosphates is considerably more complicated. Whereas nucleosides and their 5'-phosphates generally hydrolyze in acid in the order purine > pyrimidine, the presence of the respective 3',5'-cyclic phosphates reversed this order. In fact, the D-ribo- and 2'-deoxy-D-ribo-3',5'-cyclic phosphates of adenosine and guanosine were more stable to acidic hydrolysis than their respective nucleotides, whereas the reverse was found for uridine and thymidine 3',5'-cyclic phosphates (Table 1) (Smith et al., 1961; Drummond et al., 1964). Recently, kinetic data (Zoltewicz et al., 1970; Panzica et al., 1972; Zoltewicz and Clark, 1972; Hevesi et al., 1972) on the hydrolysis of purine and pyrimidine nucleosides suggest that hydrolysis proceeds *via* the mechanism outlined below (Mechanism *A*). The rate-determining step is the cleavage of the nitrogen-glycosyl bond to give the heterocycle and a carboxonium ion **63**. If hydrolysis of a

purine 3',5'-cyclic phosphate occurs *via* a carboxonium ion, then the formation of the carboxonium ion could be hindered in the transition state by the rigid nonplanar *trans*-fused ring system, resulting in an increased stability to hydrolysis (indeed, hydrolysis of the phosphate ring may be prerequisite for glycosidic cleavage). This does not explain the labilization of the glycosidic bond of the respective 3',5'-cyclic phosphate of uridine and thymidine, as compared with cytidine. An alternate mechanism *B* of hydrolysis *via* the anhydro intermediates **64** and/or **65** is, however, consistent with these results (Ueda and Fox, 1967). Thymidine 2,3' or 2,5'-anhydronucleosides are also known to be rapidly hydrolyzed under similar acidic conditions (Ueda and Fox, 1967).

Examination of the pK_a's of adenosine, guanosine, uridine, thymidine, and cytidine lend further support to these mechanisms for hydrolysis. Protonation of adenosine, guanosine, and cytidine occurs in 1 N HCl (pH 0.1) (Sober, 1970). The protonation of the heterocyclic base would reduce the electron density at N-3 (purine) and O-2 (cytidine), thereby decreasing the potential of anhydroformation by nucleophilic displacement of phosphate. Uridine ($pK_a = -3.38$) (Shapiro and Danzig, 1972) is not fully protonated under these conditions, and anhydroformation is still possible. The different rates of hydrolysis of uridine 3',5'-cyclic phosphate and cytidine 3',5'-cyclic phosphate can then be explained by the protonation of the cytidine base where Mechanism *A* is favored over Mechanism *B*. The fact that cytidine 3',5'-cyclic phosphate hydrolyzed without cleavage of the glycosyl bond and at a rate similar to adenosine 3',5'-cyclic phosphate and guanosine 3',5'-cyclic phosphate (transient nucleotides are detected) suggests that all three cyclic nucleotides hydrolyze by a similar mechanism, with the rate-determining step most probably the hydrolysis of the phosphate ring. Proof of the validity of these mechanisms awaits more detailed kinetic studies.

The alkaline hydrolysis of nucleoside 3',5'-cyclic phosphates is shown in Table 2. The nucleoside 3',5'-cyclic phosphates are less labile to hydrolysis than the corresponding 5'-nucleotides with glycosyl cleavage observed with adenosine 3',5'-cyclic phosphate and uridine 3',5'-cyclic phosphate. The catalytic effect of polyvalent cations on the rate of hydrolysis of nucleotides is well known (Kochetkov and Budovski, 1972). The catalytic effect of barium ions on hydrolysis of the phosphate ring of nucleoside 3',5'-cyclic phosphates is shown by their complete hydrolysis in 30 min with 0.2 M barium hydroxide at 100°C to the respective 5'- and 3'-nucleotides in the ratio 1:5 (Smith et al., 1961). It is of interest that careful examination of

TABLE 2. *Basic hydrolysis of ribonucleoside 3',5'-cyclic phosphates and ribonucleoside 5'-phosphates in 1 N NaOH at 100°C*

Compound	Half-life (min)	Products
Adenosine 3',5'-cyclic phosphate	36[a]	Adenine, transient formation of adenosine, gradual loss of chromophore
Adenosine 5'-phosphate	10% Hydrolyzed after 2 hr[b]	Adenine and two minor unknown compounds
Uridine 3',5'-cyclic phosphate	90[b]	Uridine 5'-phosphate (trace), uridine 3'-phosphate (major product), uracil (about $1/4$ the amount of uridine 3'-phosphate and unidentified product (trace), destruction of chromophore
Uridine 5'-phosphate	Unaffected after 2 hr[a]	
Thymidine 3',5'-cyclic phosphate	120[c]	80% Thymidine, 3'-phosphate and 20% thymidine 5'-phosphate

[a] Lipkin et al., 1959
[b] Smith et al., 1961
[c] Tener et al., 1958

the barium hydroxide hydrolysis of the pyrimidine cyclic nucleotides revealed new nucleotides with unchanged ultraviolet spectrum with similar, but different, chromatographic mobilities to the respective 5'- and 3'-nucleotides. These nucleotides were postulated to form *via* an anhydronucleoside **64** which hydrolyzed by hydroxide attack at the pyrimidine base giving a *xylo*-nucleotide (Smith et al., 1961). The need for more in-depth studies is apparent for a better understanding of the role that the cyclic phosphate ring plays in the rate of hydrolysis of nucleoside cyclic phosphates.

The free energy of hydrolysis of the 3'-bond of adenosine 3',5'-cyclic phosphate [see Eq. (C)] has been determined by Greengard and his colleagues (Greengard, 1971) to be 3,000 calories per mole greater than the hydrolysis of adenosine 5'-triphosphate [see Eq. (D)]. "The high free-

$$c\text{-AMP}^{-1} + H_2O \rightleftharpoons 5'\text{-AMP}^{-2} + H^+ \quad \Delta G° = -11.9 \text{ Kcal/mole} \quad (C)$$

$$ATP^{-4} + H_2O \rightleftharpoons ADP^{-3} + \text{inorganic phosphate} + H^+$$
$$\Delta G° = -8.9 \text{ kcal/mole} \quad (D)$$

energy change associated with the hydrolysis of adenosine 3',5'-cyclic phosphate to adenosine 5'-phosphate indicates the thermodynamic possibility that adenosine 3',5'-phosphate might interact with its receptor by formation of a covalent bond" (Greengard, 1971). The possibility of reversible phosphorylation of enzymes by nucleoside 3',5'-cyclic phosphates is supported by the reversible phosphorylation of α-chymotrypsin with catechol cyclic phosphate (**66**) (Kaiser et al., 1971). The enthalpy of hydrolysis of nucleoside 3',5'- and 2',3'-cyclic phosphates as determined calorimetrically by enzymatic hydrolysis is listed in Table 3 (Greengard,

TABLE 3. *Enthalpy of hydrolysis of cyclic nucleotides*

3' Bond of 3',5'-cyclic nucleotide	ΔH, kcal mole^{-1}
cAMP	−14.1
cGMP	−10.5
cIMP	−13.4
cdAMP	−13.0
cUMP	−12.0
2' Bond of 2',3'-cyclic nucleotide	ΔH, kcal mole^{-1}
cAMP	− 9.4
cGMP	− 9.5
cCMP	− 8.1
cUMP	− 7.8

1971). The large change in free energy in the hydrolysis of adenosine 3',5'-cyclic phosphate is shown to be of an enthalpic nature. The results of Greengard and his colleagues (Greengard, 1971) further reveal that

> "for all of the cyclic nucleotides studied, the data indicated that the bond hydrolyzed was of a high-energy nature. Also, the enthalpy of hydrolysis of each cyclic 3',5'-nucleotide was greater than that of the corresponding cyclic 2',3'-nucleotide. Third, changing the nature of the base moiety of the cyclic 2',3'-nucleotides had only a slight effect on the measured enthalpy of hydrolysis; in contrast, the enthalpy of hydrolysis of the 3',5'-nucleotides was markedly affected by the nature of the purine or pyrimidine base. Examination of the data shows that the heat of hydrolysis of cyclic 3',5'-GMP is from 2.5 to 3.6 kcal mole^{-1} lower than the heats observed for the other cyclic 3',5'-purine nucleotides. It seems significant in this connection that of the purine derivatives used in this study, only guanosine was substituted at position 2. The building of scale models of the cyclic nucleotides used in this study

shows that the 2-amino group of the guanine ring may be within 2 Å of one of the unesterified oxygen atoms of the cyclic 3',5'-phosphate group. This would allow the possibility of hydrogen bonding or some other weak interaction that could result in charge delocalization and stabilization of the cyclic phosphate group of cyclic 3',5'-GMP. There is essentially no difference between the enthalpies of hydrolysis of cyclic 2',3'-AMP and cyclic 2',3'-GMP, and model building shows little possibility for interaction between the cyclic 2',3'-phosphate group and the purine ring.

Cyclic 3',5'-AMP has the highest enthalpy of hydrolysis of any of the cyclic nucleotides investigated to date. In addition, as discussed above, the free energy of hydrolysis of the 3'-bond of cyclic 3',5'-AMP was found to be -11.9 kcal mole^{-1}. It is thus possible that the energetic properties of the cyclic phosphate moiety of cyclic 3',5'-AMP were an important aspect of the central role that this compound came to assume in the course of the evolution of mechanisms for the regulation of metabolism."

III. SYNTHESIS OF NUCLEOSIDE CYCLIC PHOSPHATES RELATED TO ADENOSINE 3',5'-CYCLIC PHOSPHATE

The synthesis of nucleoside 3',5'-cyclic phosphates historically begins with the discovery by Sutherland's group (Rall et al., 1957; Sutherland and Rall, 1957) of the formation of a heat-stable factor by tissue particles which activated glycogen phosphorylase (Sutherland and Rall, 1958; Rall and Sutherland, 1958). The heat-stable factor was subsequently shown to be identical to a nucleotide isolated in low yield from the barium hydroxide hydrolysis of adenosine 5'-triphosphate (**67**) (Cook et al., 1957; Sutherland and Rall, 1957; Lipkin et al., 1959a,b). The nucleotide was first assigned a cyclic dinucleotide structure (Cook et al., 1957; Sutherland and Rall, 1957) which was later corrected to the monomeric cyclic phosphate, adenosine 3',5'-cyclic phosphate (**68**) (Lipkin et al., 1959a). The enzymatic formation of adenosine-, 2'-deoxyadenosine- and guanosine 3',5'-cyclic phosphates from the respective nucleoside 5'-triphosphates is now well known (Hardman et al., 1971). The synthesis of adenosine 3',5'-cyclic phosphate and 2'-deoxyadenosine 3',5'-cyclic phosphate by bacterial fermentation is of interest as a low cost source of nucleoside 3',5'-cyclic phosphates (Kikkoman Shoyu Co., 1971a,b).

Inosine 3',5'-cyclic phosphate (**69**) has been also prepared by barium hydroxide hydrolysis of inosine 5'-triphosphate or by deamination of adenosine 3',5'-cyclic phosphate (Lipkin et al., 1959a; Meyer et al., 1972b).

A considerable improvement in the synthesis of adenosine 3′,5′-cyclic phosphate was achieved by the treatment of adenosine 5′-phosphate with dicyclohexylcarbodiimide at high dilution (Smith et al., 1961). Cyclohexyl isocyanate has also been used as a dehydrating agent for the preparation of adenosine 3′,5′-cyclic phosphate; however, prior blocking of the N^6-amino function was required to prevent formation of a N^6-cyclohexylcarbamoyl derivative (Naito and Sano, 1965). Ynamines **70** have also been used as dehydrating agents for the synthesis of adenosine 3′,5′-cyclic phosphate (Fujimoto and Naruse, 1968) and, interestingly, at a lower solvent dilution than that used in the cyclohexyl isocyanate or dicyclohexylcarbodiimide procedures.

Base-catalyzed transesterification of "active" phosphates of adenosine 5′-phosphate, e.g., adenosine 5′-*p*-nitrophenyl phosphate (**71**), readily occurred in anhydrous dimethylsulfoxide at room temperature in high yields to give adenosine 3′,5′-cyclic phosphate (Borden and Smith, 1966). Both the carbodiimide and the "activated" ester method have been used extensively in the general synthesis of nucleoside 3′,5′-cyclic phosphates.

A. Structural Modification of the Heterocyclic Base

1. *Substitutions on Purine*

The synthesis of nucleoside 3',5'-cyclic phosphates modified at the heterocycle has been achieved starting from appropriately substituted nucleosides or nucleotides and by transformations of nucleoside 3',5'-cyclic phosphates. Both methods have been used to obtain the 6-substituted-9-β-D-ribofuranosylpurine 3',5'-cyclic phosphates **72** and **78**. In the latter method, inosine 3',5'-cyclic phosphate (**69**) was used (Meyer et al., 1972b; Boehringer Mannheim GmbH, 1970a, 1971) as a precursor for the synthesis of **72** in the following manner. Acetylation of inosine 3',5'-cyclic phosphate (**69**) gave 2'-O-acetylinosine 3',5'-cyclic phosphate which on treatment with refluxing phosphorus oxychloride gave the key intermediate 6-chloro-9-β-D-ribofuranosylpurine 3',5'-cyclic phosphate (**73**). Nucleophilic displacement of the 6-chloro of **73** occurred with various sulfur, nitrogen, and oxygen nucleophiles to give the respective 6-substituted 9-β-D-ribofuranosylpurine 3',5'-cyclic phosphates (**72**). 9-β-D-Ribofuranosylpurine-6-thione 3',5'-cyclic phosphate (**74**), prepared in good yield from **73** (Meyer et al., 1972b), has also been synthesized from the respective 5'-nucleotide by the dicyclohexylcarbodiimide procedure in low yield (Thomas and Montgomery, 1968).

6-Alkylamino-9-β-D-ribofuranosylpurine 3',5'-cyclic phosphates (**78**) have also been synthesized (Posternak et al., 1969) from 6-chloro-9-(2,3,

O-isopropylidene-β-D-ribofuranosyl) purine (**75**). Nucleophilic substitution on the nucleoside **75** gave the 6-alkylamino nucleoside **76** which was phosphorylated with phosphorus oxychloride and deblocked to give the respective 6-alkylaminonucleotide **77**. The 6-alkylamino cyclic nucleotides **78** were then obtained by the base-catalyzed transesterification of their *p*-nitrophenyl esters **79**.

Acylation of adenosine 3′,5′-cyclic phosphate with an acyl anhydride or an acyl halide gives first the 2′-O-acyl derivative **80** which can be further acylated to N^6,2′-O-diacyladenosine 3′,5′-cyclic phosphate (**81**) (Falbriard et al., 1967). In this manner the N^6,2′-O-diacetyl, butyryl, hexanoyl, octanoyl, lauroyl, stearoyl, benzoyl, adamantoyl (Falbriard et al., 1967), palmitoyl (Shuman et al., 1972c; Weinryb et al., 1972), diazomalonyl (Brunswick and Cooperman, 1971), and aminocaproyl (Wilchek et al., 1971) derivatives **81** have been synthesized from adenosine 3′,5′-cyclic phosphate. The N^6-acyladenosine 3′,5′-cyclic phosphates (**82**) were obtained

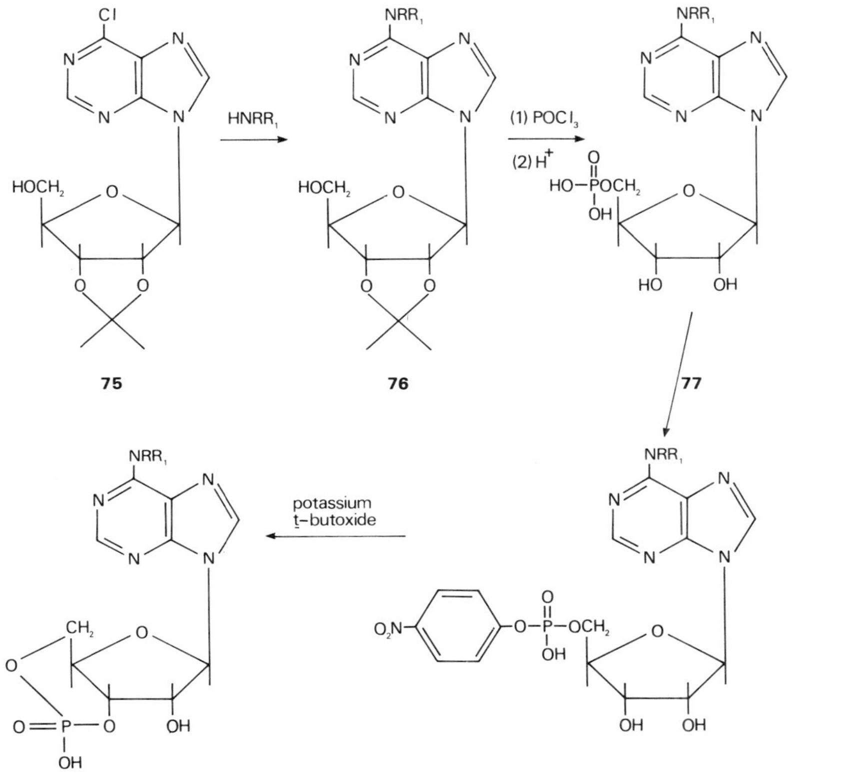

R = CH₃, R₁ = H
R = R₁ = CH₃
R = H, R₁ = n-C₄H₉
R = H, R₁ = t-C₄H₉

by selective 2'-O-deacylation (Falbriard et al., 1967) of the $N^6,2'$-O-diacyl derivatives **81**. N^6-ε-Aminocaproyladenosine 3',5'-cyclic phosphate has been covalently bound to a Sepharose column to give an affinity column for the chromatography of protein kinases (Wilcheck et al., 1971).

The stability of $N^6,2'$-O-dibutyryladenosine 3',5'-cyclic phosphate has been investigated by Swislocki (1970) and Blecher et al. (1970). Dry $N^6,2'$-O-dibutyryladenosine 3',5'-cyclic phosphate was stable for longer than 4 months at −15°C, whereas aqueous solutions slowly decomposed to N^6-monobutyryladenosine 3',5'-cyclic phosphate. Significant 2'-O-deacyla-

R = acetyl, butyryl, hexanoyl, octanoyl, lauroyl, stearoyl, benzoyl, adamantoyl, diazomalonyl, aminocaproyl

tion of $N^6,2'$-O-dibutyryladenosine 3',5'-cyclic phosphate to predominantly N^6-monobutyryladenosine 3',5'-cyclic phosphate occurred at 37°C after 1 hr in basic media (bicarbonate, 0.154 M, pH 9.6) as well as in Krebs-Ringer bicarbonate (pH 7.4).

Few 2,6-disubstituted purine 3',5'-cyclic phosphates have been reported. Guanosine 3',5'-cyclic phosphate (**83**) occurs naturally and is formed from guanosine 5'-triphosphate by guanylate cyclase (Hardman et al., 1971). Guanosine 3',5'-cyclic phosphate (**83**) has been chemically synthesized from p-nitrophenyl guanosine 5'-phosphate (Borden and Smith, 1966) and also by the cyclization of N^2-benzoylguanosine 5'-phosphate (Strauss and Fresco, 1965; Smith et al., 1961; Posternak and Falbriard, 1971; Fujimoto and Naruse, 1968) and guanosine 5'-phosphate (Strauss and Fresco, 1965).

Deamination of guanosine 3′,5′-cyclic phosphate (**83**) occurred readily to give xanthosine 3′,5′-cyclic phosphate (**84**) (Shuman et al., 1973a). Acetylation of guanosine 3′,5′-cyclic phosphate to the 2′-O-acetyl derivative **85** followed by treatment with POCl$_3$ gave 2-amino-6-chloro-9-β-D-ribofuranosylpurine 3′,5′-cyclic phosphate (**86**) (Meyer et al., 1972a; Boehringer Mannheim GmbH, 1970a, 1971). Diazotization of **86** in the presence of concentrated hydrochloric acid gave 2,6-dichloro-9-β-D-ribofuranosylpurine 3′,5′-cyclic phosphate (**87**) which by treatment with ammonium hydroxide gave 2-chloroadenosine 3′,5′-cyclic phosphate (**88**) (Meyer et al., 1972a). 2-Chloroadenosine 3′,5′-cyclic phosphate (**88**) and 2-bromoadenosine 3′,5′-cyclic phosphate have also been prepared from the respective 2-haloadenosine by phosphorylation and then dicyclohexylcarbodiimide cyclization (Jastorff and Freist, 1972).

Ac = Acetyl

2-Amino-6-chloro-9-(2,3,5-tri-O-acetyl-β-D-ribofuranosyl)purine (**89**) (Gerster and Robins, 1966) has been used for the synthesis of 2,6-diamino-9-β-D-ribofuranosylpurine 3′,5′-cyclic phosphate (**92**) (Posternak et al., 1971). Deacetylation of **89** with methanolic ammonia, then treatment with concentrated ammonium hydroxide, gave 2,6-diamino-9-β-D-ribofuranosylpurine (**90**). The isopropylidene derivative of **90** was then phosphorylated with phosphorus oxychloride in the presence of formic anhydride and pyridine to give, after acidic hydrolysis, 2,6-diamino-9-β-D-ribofuranosylpurine 5′-phosphate (**91**). Cyclization of **91** by dicyclohexycarbodiimide in pyridine-dimethylformamide gave 2,6-diamino-9-β-D-ribofuranosylpurine 3′,5′-cyclic phosphate (**92**).

The synthesis of 6,8-disubstituted nucleoside 3′,5′-cyclic phosphates has been extensively explored with 6-amino(or 6-keto) 8-substituted-9-β-D-ribofuranosylpurine 3′,5′-cyclic phosphates. The key intermediates were

8-bromoadenosine 3′,5′-cyclic phosphate (**93**) and 8-bromoinosine 3′,5′-cyclic phosphate (**105**). Bromination of adenosine 3′,5′-cyclic phosphate (Ikehara and Uesugi, 1969; Muneyama et al., 1971; Boehringer Mannheim GmbH, 1970b, 1971; Posternak et al., 1971) gave good yields of 8-bromoadenosine 3′,5′-cyclic phosphate (**93**), which was also obtained by cyclization of 8-bromoadenosine 5′-phosphate (**94**) with dicyclohexylcarbodiimide (Posternak et al., 1971). Facile nucleophilic displacement of the 8-bromo group of **93** occurred with thiourea (Muneyama et al., 1971; Boehringer Mannheim GmbH, 1970b, 1971) or with alkylthiols (Muneyama et al., 1971; Posternak et al., 1971) to give 8-thioadenosine 3′,5′-cyclic phosphate (**95**) or the 8-alkylthioadenosine 3′,5′-cyclic phosphates (**96**), respectively. The 8-(β-aminoethylthio) derivative (**97**) has been used for the covalent binding of adenosine 3′.5′-cyclic phosphate to Sepharose (Tesser et al., 1972).

8-Chloroadenosine 3′,5′-cyclic phosphate (**98**) and 8-iodoadenosine 3′,5′-cyclic phosphate (**99**), which can not be obtained readily by direct halogenation of adenosine 3′,5′-cyclic phosphate, have been synthesized from 8-thioadenosine 3′,5′-cyclic phosphate by molecular halogen and excess halide ion (Shuman et al., 1972c). 8-Aminoadenosine 3′,5′-cyclic phosphate (**100**) was obtained by hydrogenation of 8-azidoadenosine 3′,5′-cyclic phosphate (**101**) (Muneyama et al., 1971) and by cyclization of the respective nucleotide (**102**) (Posternak et al., 1971). Nucleophilic substitution of **93** with alkoxides gave the 8-alkoxy cyclic nucleotides **103**a (Muneyama et al., 1971; Boehringer Mannheim GmbH, 1970b, 1971). Amines and **93** gave the 8-substituted aminopurine 3′,5′-cyclic phosphate **103**b (Muneyama et al., 1971; Boehringer Mannheim GmbH, 1970b, 1971). 8-Oxoadenosine 3′,5′-cyclic phosphate (**104**) was obtained by hydrogenation of 8-benzyloxyadenosine 3′,5′-cyclic phosphate (Boehringer Mannheim GmbH, 1970b, 1971) or by treatment of **93** with a mixture of sodium acetate, acetic anhydride, and glacial acetic acid to give N^6,2′-O-diacetyl-8-oxoadenosine 3′,5′-cyclic phosphate which was deacetylated to the 8-oxo derivative **104** (Muneyama et al., 1971; Posternak et al., 1971).

Bromination of inosine 3′,5′-cyclic phosphate **69** gave 8-bromoinosine 3′,5′-cyclic phosphate (**105**) (Boehringer Mannheim GmbH, 1970b, 1971) which was also synthesized by deamination of 8-bromoadenosine 3′,5′-cyclic phosphate (**93**) (Shuman et al., 1973a; Boehringer Mannheim GmbH, 1970b, 1971). 8-Substituted inosine 3′,5′-cyclic phosphates were obtained by direct nucleophilic displacement of the 8-bromo such as in the preparation of 8-thioinosine 3′,5′-cyclic phosphate (**106**) (Shuman et al., 1973a) and 8-benzylaminoinosine 3′,5′-cyclic phosphate (**107**) (Boehringer Mannheim GmbH, 1970b, 1971), or alternatively by deamination of the 8-substituted adenosine 3′,5′-cyclic phosphates (**108**) (Shuman et al., 1973a).

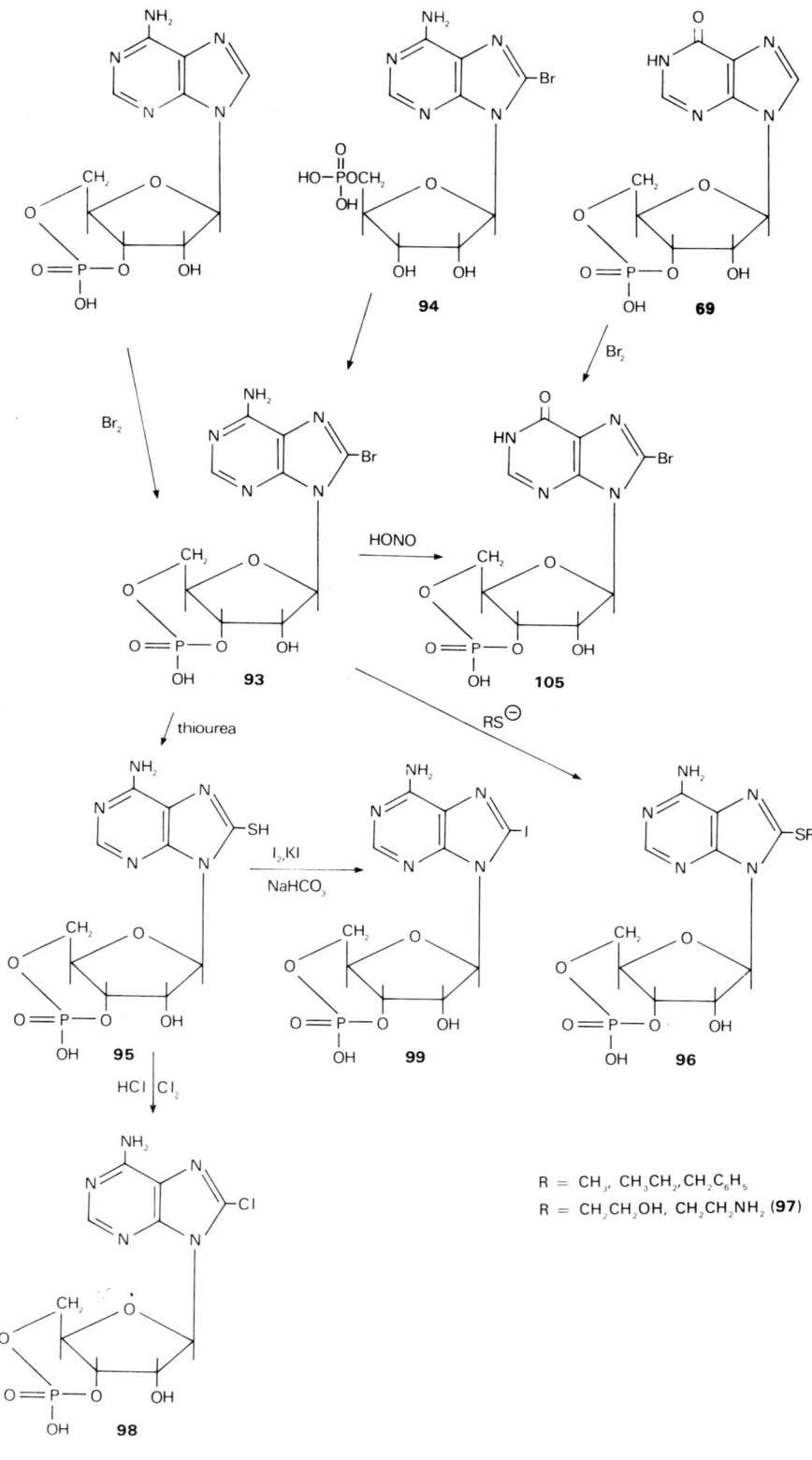

R = CH$_3$, CH$_3$CH$_2$, CH$_2$C$_6$H$_5$
R = CH$_2$CH$_2$OH, CH$_2$CH$_2$NH$_2$ (**97**)

103a, R = OCH$_3$, OCH$_2$C$_6$H$_5$, etc.
103b, R = NHCH$_3$, NHCH$_2$C$_6$H$_5$, NHCH$_2$CH$_2$OH, etc.

Hydrogenation of 8-azidoinosine 3′,5′-cyclic phosphate gave 8-aminoinosine 3′,5′-cyclic phosphate (Shuman et al., 1973a).

A few 2,6,8-trisubstituted purine 3′,5′-cyclic phosphates have been

X = Br, N₃, CH₃S, CH₃CH₂S, C₆H₅CH₂S

synthesized from guanosine 3',5'-cyclic phosphate (**83**). Bromination of guanosine 3',5'-cyclic phosphate (**83**) gave 8-bromoguanosine 3',5'-cyclic phosphate (**109**) (Shuman et al., 1973*a;* Boehringer Mannheim GmbH, 1970*b*, 1971) which on deamination with sodium nitrite in acetic acid gave 8-bromoxanthosine 3',5'-cyclic phosphate (**110**) (Shuman et al., 1973*a*). Treatment of 8-bromoguanosine 3',5'-cyclic phosphate (**109**) with methylamine, dimethylamine, or benzylamine gave the respective 8-alkylaminoguanosine 3',5'-cyclic phosphates (**111**a) (Shuman et al., 1973*a*). 8-Thioguanosine 3',5'-cyclic phosphate, obtained from 8-bromo-cGMP and thiourea, was alkylated with methyl iodide or benzyl bromide to give the respective 8-alkylthioguanosine 3',5'-cyclic phosphate (**111**b) (Shuman et al., 1973*a*).

111 a, X = N(CH$_3$)$_2$, NHCH$_3$, NHCH$_2$C$_6$H$_5$

b, X = SCH$_3$, SCH$_2$C$_6$H$_5$

2. Heterocyclic Ring Systems Related to Purine

A few nucleoside 3',5'-cyclic phosphates with heterocyclic ring systems related to purine have been synthesized. Tubercidin (7-deazaadenosine) 5'-phosphate has been cyclized to tubercidin 3',5'-cyclic phosphate (112) (Hanze, 1968) by the dicyclohexylcarbodiimide procedure. Toyocamycin 3',5'-phosphate (113) (Shuman et al., 1972c) and 5,6-dimethyl-1-α-D-ribofuranosylbenzimidazole 3',5'-cyclic phosphate (114) (Friedrich, 1963) were also obtained by cyclization of the respective 5'-phosphates.

The interesting structural isomers of adenosine 3',5'-cyclic phosphate, 3-β-D-ribofuranosyl-adenine 3',5'-cyclic phosphate (115) (Cehovic et al., 1968) and 9-β-L-ribofuranosyladenine 3',5'-cyclic phosphate (116) (Holý and Fruhaufova, 1972) have been synthesized.

Adenosine 3',5'-cyclic phosphate-1-N-oxide (117) (Falbriard et al., 1967) has been prepared from adenosine 3',5'-cyclic phosphate and perphthalic acid. Alkylation of **117** with alkyl halides in dimethylsulfoxide gave the 1-*N*-alkoxides (**118**) (Meyer et al., 1972c).

Inosine 3',5'-cyclic phosphate has been methylated to give 7-methylinosine 3',5'-cyclic phosphate (**119**) (Anderson et al., 1972).

3. Pyrimidine and Other Heterocyclic Systems

1-β-D-Ribofuranosylpyrimidine 3′,5′-cyclic phosphates [e.g., uridine 3′,5′-cyclic phosphate (120) and cytidine 3′,5′-cyclic phosphate (121)] have been prepared by procedures similar to those used for synthesis of purine 3′,5′-cyclic nucleotides. Cyclization of the requisite 5′-nucleotide by the dicyclohexylcarbodiimide (Smith et al., 1961; Drummond et al., 1964) or by the ynamine procedure (Fujimoto and Naruse, 1968), or, preferably, base-catalyzed cyclization (Borden and Smith, 1966a) of an "active" phosphate derivative gave good yields of the pyrimidine 3′,5′-cyclic phosphates 120 and 121. In the case of (121) the N^4-benzoyl derivative of cytidine 5′-phosphate was required for solubilization in pyridine.

120   121

The oxidation of 6-azauridine 5′-phosphite (122) with trichloroacetic anhydride gave the expected 6-azauridine 5′-phosphate (123) as well as 6-azauridine 3′,5′-cyclic phosphate (35) and P^1,P^2-di(6-azauridine-5′yl) pyrophosphate presumably via the intermediate 122a. The cyclic phosphate 35 was also prepared in high yield from the 5′-phosphate 123 by the dicyclohexylcarbodiimide procedure (Holý et al., 1965).

The alkylation of inosine 3′,5′-cyclic phosphate (69) with lactones of the general formula 124 gave N^1-alkylated nucleotides (125) which ring opened by heating in alkali to 5-amino-N-substituted-1-β-D-ribofuranosylimidazole-4-carboxamide 3′,5′-cyclic phosphates (126a) (Begmeyer et al., 1971). Cyclization of 5-amino-1-β-D-ribofuranosylimidazole-4-carboxamide 5′-phosphate by the dicyclohexylcarbodiimide procedure gave 5-amino-1-β-D-ribofuranosylimidazole-4-carboxamide 3′,5′-cyclic phosphate (126b) (Begmeyer et al., 1971). A structurally related compound to 126b, 1-β-D-

CYCLIC NUCLEOTIDE DERIVATIVES

126a, R = (CH$_2$)$_n$-A, where A = A = CO$_2$H or SO$_3$H
126b, R = H

ribofuranosyl-1,2,4-triazole-3-carboxamide 3',5'-cyclic phosphate (**127**), has been synthesized from the respective 5'-nucleotide (Shuman et al., 1972c).

127

B. Structural Modification of the Carbohydrate Moiety

1. *Arabinofuranosyl Cyclic Nucleotides*

The synthesis of 2',5'- and 3',5'-arabinofuranosyl cyclic nucleotides has been pursued in several ways. The phosphorylation of 9-β-D-arabinofuranosyladenine (**128**) in trimethylphosphate gave 9-β-D-arabinofuranosyladenine 5'-phosphate (**129**) with traces of 9-β-D-arabinofuranosyladenine 2',5'-cyclic phosphate (**130**) (Hubert-Habart and Goodman, 1969). Direct cyclization of **129** with dicyclohexylcarbodiimide gave high yields of the 2',5'-cyclic phosphate **130** (Hubert-Habart and Goodman, 1969). The 2',5'-cyclic phosphate **130** was also obtained in low yield during the basic hydrolysis of the 5'-β-cyanoethylphosphate derivative of **128** (Furth and Cohen, 1967).

Since the cyclization of 9-β-D-arabinofuranosyladenine 5'-phosphate (**129**) gave the 2',5'-cyclic phosphate (**130**), other methods for the synthesis of the 3',5'-cyclic phosphate **131** were studied. In addition to the possible enzymatic conversion of 9-β-D-arabinofuranosyladenine 5'-triphosphate to the *arabino*-3',5'-cyclic nucleotide (**131**) (Ortiz, 1972), two chemical methods for the synthesis of **131** have been developed. One method (Lee et al., 1971) involved the selective blocking of the 2'-O-position of 9-β-D-arabinofuranosyladenine (**128**) by the synthesis of 2'-O-benzyl-9-β-D-arabinofuranosyladenine (**132**) from 2-O-benzyl-3,5-di-O-p-nitrobenzoyl-α-D-arabinofuranosyl chloride and 6-benzamidopurine. The 2'-O-benzyl nucleoside **132** was then selectively phosphorylated to the 2'-O-benzyl-5'-nucleotide **133**. Cyclization of **133** with dicyclohexylcarbodiimide followed by hydrogenolysis of the benzyl group gave 9-β-D-arabinofuranosyladenine 3',5'-cyclic phosphate (**131**). As with the ribonucleoside 3',5'-phosphates, the cyclic phosphate **131** was more stable to acidic hydrolysis than the 5'-nucleotide **129**.

9-β-D-Arabinofuranosyladenine 3',5'-cyclic phosphate (**131**) has also been prepared from adenosine 3',5'-cyclic phosphate by inversion of configuration of the 2'-hydroxyl (Khwaja et al., 1973). The ability of nucleosides to undergo intramolecular cyclonucleoside formation was used as a general route to *arabino*-3',5'-cyclic nucleotides from the more readily available *ribo*-3',5'-cyclic nucleotides. 8-Bromo-2'-O-tosyladenosine 3',5'-cyclic phosphate (**134**), obtained by tosylation of 8-bromoadenosine 3',5'-cyclic phosphate (**93**), was converted to 8-hydroxy-2'-O-tosyladenosine 3',5'-cyclic phosphate (**135**) by treatment with acetic acid, acetic anhydride and sodium acetate followed by methanolic ammonia at room temperature. The anhydro derivative **136** was formed by heating 8-hydroxy-2'-O-tosyladenosine 3',5'-cyclic phosphate (**135**) in a solution of methanolic

ammonia or sodium acetate in dimethylformamide. Ring opening of the anhydronucleoside **136** by nucleophilic attack at the 8-purinyl position with hydroxide, methoxide, ammonia or hydrosulfide gave the 8-oxo (**137**), 8-methoxy (**138**), 8-amino (**139**), and 8-thio (**140**) ara-3',5'-cyclic nucleotides, respectively. Desulfurization of 8-thio-9-β-D-arabinofuranosyladenine 3',5'-cyclic phosphate (**140**) with Raney-nickel gave 9-β-D-arabinofuranosyladenine 3',5'-cyclic phosphate (**131**) (Khwaja et al., 1972, 1973).

Treatment of 8-bromo-2'-O-tosyladenosine 3',5'-cyclic phosphate (**136**) with thiourea gave the 8,2'-anhydro-8-thionucleotide **141** which with

Raney-nickel readily gave 2'-deoxyadenosine 3',5'-cyclic phosphate (**142**) (Khwaja et al., 1972).

Facile 2′,5′-cyclic phosphate formation by the dicyclohexylcarbodiimide procedure occurred with the 5′-phosphate of N^4-benzoyl-1-β-D-arabinofuranosylcytosine (**143**) to give 1-β-D-arabinofuranosylcytosine 2′,5′-cyclic phosphate (**144**) in high yield (Wechter, 1969). Typical of seven-membered cyclic phosphates the 2′,5′-cyclic phosphate ring of **144** was stable to acid and base and to conditions which completely hydrolyzed the D-*ribo*-3′,5′-cyclic nucleotides.

To obtain 1-β-D-arabinofuranosylcytosine 3′,5′-cyclic phosphate (**145**), the 1-β-D-arabinofuranosylcytosine 3′-phosphate (**146**) (Wechter, 1967) was synthesized from cytosine 2′,3′-cyclic phosphate (**147**) (Nagyvary, 1969). Cyclization of 1-β-D-arabinofuranosylcytosine 3′-phosphate (**146**) then gave 1-β-D-arabinofuranosylcytosine 3′,5′-cyclic phosphate (**145**) (Long et al., 1972; Kreis and Wechter, 1972).

Treatment of 1-β-D-arabinofuranosylcytosine 3′,5′-cyclic phosphate (**145**) with hydrogen sulfide in pyridine-water gave 1-β-D-arabinofuranosyl-4-thiouracil 3′,5′-cyclic phosphate (**148**) which was methylated with methyl iodide to 1-β-D-arabinofuranosyl-4-methylthio-2-pyrimidone 3′,5′-cyclic phosphate (**149**). Deamination of 1-β-D-arabinofuranosylcytosine 3′,5′-cyclic phosphate (**145**) gave 1-β-D-arabinofuranosyluracil 3′,5′-cyclic phosphate (**150**) (Long et al., 1972).

The reaction between 1-β-D-arabinofuranosyluracil (**151**) and triethyl phosphite occurred slowly to give a low yield of nucleoside phosphites which, upon treatment with hexachloroacetone, gave a single cyclic nucleotide. The structure of this nucleotide was suggested to be 1-β-D-arabinofuranosyluracil 3′,5′-cyclic phosphate (**150**) based on its ease of hydrolysis in 0.1 N hydrochloric acid and 0.2 M sodium hydroxide at 50°C (Holý and Sŏrm, 1969). The formation of the 3′,5′-cyclic phosphate **150** from an unprotected arabinonucleoside is in contrast to formation of the 2′,5′-cyclic phosphates **130** and **143** by the phosphorus oxychloride and dicyclohexylcarbodiimide procedures and deserves further investigation.

2. *Xylofuranosyl Cyclic Nucleotides*

The significance of the stereochemical position of hydroxyl groups in the ease of cyclic phosphate formation is illustrated by the direct formation of 9-β-D-xylofuranosyladenine 3′,5′-cyclic phosphate (**152**) from 9-β-D-xylofuranosyladenine (**153**) with phosphorus oxychloride in triethylphosphate (Hubert-Habart and Goodman, 1969). Similar phosphorylation conditions gave only traces of a 2′,5′-cyclic phosphate from 9-β-D-arabinofuranosyladenine (**131**) (Hubert-Habart and Goodman, 1969) and no reported formation (Yoshikawa et al., 1969) of a 3′,5′-cyclic phosphate from adenosine.

Similar results were found in the phosphorylation of adenosine and 9-β-D-xylofuranosyladenine (**153**) with pyrophosphoryl chloride in acetonitrile (Imai et al., 1969). Basic hydrolysis of 2′,3′-di-O-acetyl-9-β-D-xylofuranosyladenine 5′-phosphoromorpholidate has also been reported to give, exclusively, the 3′,5′-cyclic phosphate **152** (Ikehara and Ohtsuka, 1963a). Deamination of **152** with nitrous acid gave 9-β-D-xylofuranosylhypoxanthine 3′,5′-cyclic phosphate (**154**) (Imai et al., 1969).

Triethyl phosphite undergoes transesterification with 1-β-D-xylofuranosyluracil (**155**) to give initially the cyclic triester **156**. Hydrolysis of the cyclic triester **156** and oxidative cyclization of the resulting nucleoside phosphites gave 1-β-D-xylofuranosyluracil 3′,5′-cyclic phosphate (**157**) (Holý and Šorm, 1969).

U = Uracil

3. *Lyxofuranosyl Cyclic Nucleotides*

The formation of three lyxofuranosyl cyclic nucleotides is stereochemically possible from lyxofuranosyl nucleosides. The dicyclohexycarbodiimide cyclization of the 2′(3′) phosphate **158** or the oxidative cyclization of the 2′(3′) phosphite **159** of 1-β-D-lyxofuranosyluracil gave only 1-β-D-lyxofuranosyluracil 2',3'-cyclic phosphate (160) (Ukita and Hayatsu, 1961; Holý and Sŏrm, 1969). The cyclization of the 5'-phosphate 161 or the 5'-phosphite 162, however, gave a mixture of 1-β-D-lyxofuranosyluracil 3',5'-cyclic phosphate (163) and 1-β-D-lyxofuranosyluracil 2',5'-cyclic phosphate (164). The acidic and basic stability of the 3',5'-cyclic phosphate 163 and the 2',5'-cyclic phosphate 164 are consistent with other six- and seven-membered cyclic phosphates, in that both 163 and 164 were more stable to hydrolysis than the five-membered cyclic phosphate 160 (Ukita and Hayatsu, 1961). Intramolecular alcoholysis of the 2',5'-cyclic phosphate 164 occurred in acid or base to give the 3',5'-cyclic phosphate 163 (or vice versa, since product assignments were unknown) (Ukita and Hayatsu, 1961). The 2',5'- and 3',5'- and 2',3'-cyclic phosphates of 1-α-L- and α-D-lyxofuranosylthymine were similarly prepared from the respective 2',3'- and 5'-phosphites (Holý and Sŏrm, 1969).

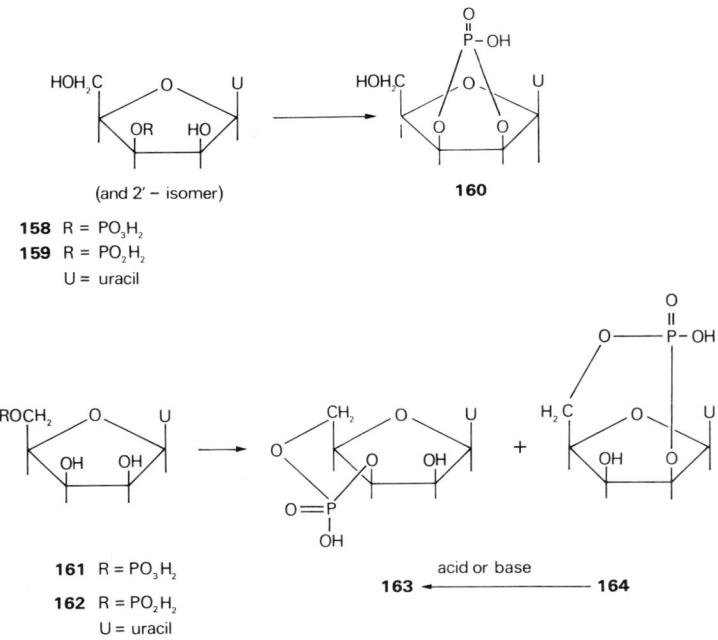

1-(2-Deoxy-β-D-*threo*-pentofuranosyl)thymine 3',5'-cyclic phosphate (165) has been synthesized by intramolecular nucleophilic displacement of iodide from 3'-deoxy-3'-iodothymidine 5'-phosphate (166) by the phosphate anion (Russell and Moffatt, 1969). Treatment of 1-(2-deoxy-β-D-*threo*-pentofuranosyl)thymine 5'-diphenylphosphate (169) with 1,5-diazabicyclo-[4.3.0]nonene-5 in dioxane gave the separable diastereomers 167 and 168.

Alkaline hydrolysis of the diastereomers **167** and **168** or the diphenylphosphate **169** also gave 1-(2-deoxy-β-D-*threo*-pentofuranosyl)thymine 3′,5′-cyclic phosphate (**165**). The 1-(2-deoxy-β-D-*threo*-pentofuranosyl)thymine 3′,5′-cyclic phosphate (**165**) was considerably more stable to basic hydrolysis than thymidine 3′,5′-cyclic phosphate and was not hydrolyzed by barium hydroxide under conditions known to hydrolyze thymidine 3′,5′-cyclic phosphate (Russell and Moffatt, 1969).

4. 2′-Derivations of Ribofuranosyl Cyclic Nucleotides

Other modifications of the carbohydrate moiety of 3′,5′-cyclic nucleotides include the synthesis of 2′-deoxyribofuranosyl 3′,5′-cyclic nucleotides. The 2′-deoxy-3′,5′-cyclic phosphates of adenosine, guanosine, uridine, cytidine, inosine, and thymidine have been conveniently synthesized as in the case of the *ribo* derivatives by the dicyclohexylcarbodiimide (Drummond et al., 1964) and ynamine procedures (Fujimoto and Naruse, 1968) or by the base-catalyzed "active phosphate" method (Borden and Smith, 1966a). 2′-Deoxyadenosine 3′,5′-cyclic phosphate (**142**) has also been prepared enzymatically (Hirata and Hayaishi, 1966; Kikkoman Shoyu, 1971) and by desulfurization of 8,2′-anhydro-8-thio-9-β-D-arabinofuranosyladenine 3′,5′-cyclic phosphate (**141**) (Khwaja et al., 1972).

Bromination of 2′-deoxyadenosine 3′,5′-cyclic phosphate gave 8-bromo-2′-deoxyadenosine 3′,5′-cyclic phosphate (Boerhinger Mannheim GmbH, 1970b, 1971); however, no other related 8-substituted derivatives were reported.

6-Chloro-9-(2-deoxy-β-D-ribofuranosyl)purine   3′,5′-cyclic   phosphate

(170) has been reported (Honjo et al., 1968) by treatment of 2'-deoxyinosine 3',5'-cyclic phosphate (171) with phosphorus oxychloride. Nucleophilic displacement of the 6-chloro function of 170 with sodium hydrosulfide gave 9-(2-deoxy-β-D-ribofuranosyl)purine-6-thione 3',5'-cyclic phosphate (172) which was methylated to 6-methylthio-9-(2-deoxy-β-D-ribofuranosyl)purine 3',5'-cyclic phosphate (173).

The reaction of acid chlorides and anhydrides with *ribo*nucleoside 3',5'-cyclic phosphates (Falbriard et al., 1967) has provided many 2'-O-acyl*ribo*nucleoside 3',5'-cyclic phosphates (see structures 80). The acylation of tritiated adenosine 3',5'-cyclic phosphate with diazomalonyl chloride has provided photo-affinity labeled adenosine 3',5'-cyclic phosphates (174) (Brunswick and Cooperman, 1971). These derivatives 174 have been used to study enzyme binding sites by generation of a reactive carbene upon photolysis leading to a covalent bond between the enzyme and adenosine 3',5'-cyclic phosphate.

$R_1 = COCN_2CO_2C_2H_5$, $R_2 = H$
$R_1 = R_2 = COCN_2CO_2C_2H_5$
$R_1 = H$, $R_2 = COCN_2CO_2C_2H_5$

The radioactive iodinated tyrosyl methyl ester of 2′-O-succinyl nucleoside 3′,5′-cyclic phosphates are used in a radioimmunoassay for the measurement of adenosine, guanosine, inosine, and uridine 3′,5′-cyclic phosphates (Steiner et al., 1972a,b). The 2′-O-succinyl derivatives (**175**) were prepared (Steiner et al., 1972c) from the respective cyclic nucleotides and succinic anhydride. The 2′-O-succinyl derivatives **175** were coupled with tyrosine methyl ester *via* mixed carboxylic-carbonic acid reaction to give the tyrosine methyl ester derivative **176**. The succinyl-cyclic nucleotide tyrosine methyl ester was then iodinated with $^{125}I_2$ to give the ^{125}I-succinyl-cyclic nucleotide tyrosine methyl esters (**177**).

B = Adenine, Guanine, Hypoxanthine and Uracil

The 3',5'-cyclic phosphate moiety has been used for selective introduction of a blocking group at the 2'-position of nucleoside 3'- or 5'-phosphates. Reaction of dihydropyran with guanosine 3',5'-cyclic phosphate and uridine 3',5'-cyclic phosphate gave the respective 2'-O-tetrahydropyranyl derivatives **178** and **179** (Straus and Fresco, 1965; Smith and Khorana, 1959). Basic hydrolysis of 2'-O-tetrahydropyranyluridine 3',5'-cyclic phosphate (**179**) gave presumably a mixture of 2'-O-tetrahydropyranyluridine 5'- and 3'-phosphates, **180** and **181**. The possible mechanism of hydrolysis of uridine 3',5'-cyclic phosphate, as discussed in Section II A, creates some doubt, however, as to the correct configuration of the 3'-hydroxyl of **180**. Enzymatic hydrolysis of 2'-O-tetrahydropyranylguanosine 3',5'-cyclic phosphate (**178**) gave only the 5'-phosphate **182**.

2'-O-Alkylation of nucleoside 3',5'-cyclic phosphates with alkyl halides has provided a general route to 2'-O-alkylated adenine, hypoxanthine, cytosine, and uracil 5'-nucleotides. Adenosine 3',5'-cyclic phosphate has been alkylated with methyl (or ethyl) iodide under basic conditions in dimethylformamide to give 2'-O-methyladenosine 3',5'-cyclic phosphate (**183**) (Tazawa et al., 1972). 2'-O-Methyladenosine 3',5'-cyclic phosphate (**183**) was also prepared by cyclization of 2'-O-methyladenosine 5'-phosphate (**184**) with dicyclohexylcarbodiimide (Miller et al., 1973).

Deamination of **183** gave 2'-O-methylinosine 3',5'-cyclic phosphate (**185**) (Tazawa et al., 1972). The 2'-O-methyl-5'-nucleotides **184** and **186** were then formed from the cyclic nucleotides by phosphodiesterase (PDE) hydrolysis (Tazawa et al., 1972).

Adenosine 5'-phosphorofluoridate has been prepared by treatment of adenosine 5'-phosphate with 2,4-dinitrofluorobenzene (**187**) (Borden and Smith, 1966). A similar attempt by Khwaja and Robins (1972) to prepare adenosine 3',5'-cyclic phosphorofluoridate gave, however, as the only isolated product 2'-*O*-(2,4-dinitrophenyl)adenosine 3',5'-cyclic phosphate (**188**) which was also prepared in a similar manner by Jastorff and Freist (1972).

5. Hexopyranosyl Cyclic Nucleotides

Stereochemically, the formation of 4',6'-cyclic phosphates of hexopyranosyl derivatives can be favorable as illustrated by the cyclization of 9-β-D-glucopyranosyladenine 6'-phosphate (**189**) with trifluoroacetic anhydride, at room temperature, to 9-β-D-glucopyranosyladenine 4',6'-cyclic phosphate (**190**) (Barker and Foll, 1957). The 4',6'-cyclic phosphate **190** was also obtained *via* direct cyclic phosphate formation of 9-β-D-glucopyranosyladenine (**191**) with phenyl phosphodichloridate (Barker and Foll, 1957) or by basic hydrolysis of 6-benzamido-9-(2,3,4-tri-*O*-acetyl-6-diphenylphosphoryl-β-D-glucopyranosyl)purine (**192**) (Nohara et al., 1966) and possibly from 9-β-D-glucopyranosyladenine 6'-phosphoromorpholidate (**193**) (Ikehara and Tazawa, 1966).

The ease of cyclic phosphate formation from glucosyl phosphates is further exemplified by the ease of formation of cyclic phosphates from 1-β-D-glucopyranosyluracil 2',3',4', or 6'-monophosphate under the mild conditions of dicyclohexylcarbodiimide and aqueous pyridine at room temperature (Zmudzka et al., 1962). The direct formation of a 4',6'-cyclic phosphate of 7-β-D-glucopyranosyltheophylline has been suggested (Baddiley et al., 1954) in the reaction of the latter with phosphorus oxychloride in pyridine (Fisher, 1914).

CYCLIC NUCLEOTIDE DERIVATIVES

Ac = Acetyl

The synthesis of cyclic nucleotides with pyranose sugars has been studied extensively (Holý, 1969) in the area of uracil and thymine 1-pyranosyl derivatives. Transesterification of pyranosyl derivatives possessing cis-hydroxyls [e.g., α- and β-anomers of 1-(2-deoxy-D-ribopyranosyl) (**37**) and 1-(2-deoxy-L-ribopyranosyl)thymine or 1-(D-ribopyranosyl)thymine α- and β-anomers (**194**)] with triethylphosphite afforded high yields of the nucleoside phosphites which easily formed cyclic phosphates by oxidative cyclization. Transesterification with triethyl phosphite was found to occur slower when *trans*-hydroxyls were present such as in 1-(2-deoxy-β-D-glucopyranosyl)thymine (**195**). The importance of carbohydrate conformation (1*C* or *C*1) in the formation of cyclic phosphates is suggested by the fact that only the *trans*-3',4'- and *trans*-4',6'-cyclic phosphates (**196** and **197**) and no *cis*-3',6'-cyclic phosphate were detected after oxidative cyclization of a mixture of the nucleoside phosphites of **195**.

The transesterification of 1-(α-D-mannopyranosyl)thymine (**198**) was

T = Thymine

complicated and gave mono- and diphosphite derivatives (Hóly, 1969). Oxidative cyclization of these phosphites gave the 2′,3′-cyclic phosphate **199** and the *bis*(cyclic phosphate) **200**. Acetic acid hydrolysis of the *bis*(cyclic phosphate) **200** gave the 2′(3′)-phosphate 4′,6′-cyclic phosphate which was hydrolyzed to the 4′,6′-cyclic phosphate **201** with *E. coli* alkaline phosphatase.

6. Alditol Cyclic Nucleotides

Riboflavin 4′,5′-cyclic phosphate (**23**) has been prepared directly from riboflavin and phosphorus oxychloride (Forrest and Todd, 1950). 2′,3′-di-*O*-Butyrylriboflavin 4′5′-cyclic phosphate was obtained, interestingly, by treatment of riboflavin 5′-phosphate with butyric anhydride or a butyryl halide (Tokyo Tanabe Seiyaku Co., 1972).

9-D-Erythrityladenine 3′,4′-cyclic phosphate (**202**) has been reported to be formed by basic hydrolysis of 9-D-erythrityladenine 5′-triphosphate (Ikehara and Ohtsuka, 1963b) and by basic hydrolysis of erythrityladenine monophosphoromorpholidate (Ikehara and Ohtsuka, 1963c).

C. Structural Modification of the Cyclic Phosphate Moiety

1. *Nucleoside 3′,5′-Phosphorothioates*

Substitution of the phosphate oxygens by sulfur in nucleoside 3′,5′-cyclic phosphates has been achieved by various methods. The greater nucleophilic character toward carbon by sulfur as compared to oxygen with thiophosphates (Akerfeldt, 1962) has been utilized for the synthesis of nucleoside 3′,5′-cyclic phosphorothioates where sulfur has been incorporated within the cyclic phosphate ring. The acid-labile 5′-deoxy-5′-thioinosine 5′-phosphorothioate (**204**a) (Hampton et al., 1969; Haga et al., 1971) and 5′-deoxy-5′-thioadenosine 5′-phosphorothioate (**204**b) (Shuman et al., 1973b) were prepared by nucleophilic displacement of the 5′-iodo of **203** by thiophosphate anion. 5′-Deoxy-5′-thioinosine- and 5′-deoxy-5′-thioadenosine 3′,5′-cyclic phosphorothioates (**205**) were then prepared from the corresponding 5′-deoxy-5′-thionucleoside 5′-phosphorothioates (**204**) by the dicyclohexylcarbodiimide procedure (Shuman et al., 1973b). Inspection of molecular models of 5′-deoxy-5′-thionucleoside 3′,5′-cyclic phosphorothioates reveals that there is less strain in the 3′,5′-cyclic phosphorothioate ring than in a corresponding 3′,5′-cyclic phosphate ring. This observation is reflected in the ease of formation of the 2′,5′-dideoxy-5′-thionucleoside 3′,5′-cyclic phosphorothioates **206** by aqueous base-catalyzed intramolecular displacement of the 5′-O-tosyl of the nucleotides **207** (Chladek and Nagyvary, 1972). The 2′,5′-dideoxy-5′-thionucleoside 3′,5′-cyclic phosphorthioates (**206**)

CYCLIC NUCLEOTIDE DERIVATIVES

were also easily formed at room temperature from the cyanoethyl esters **208**. The nucleoside 3',5'-cyclic phosphorothioates **206** were stable at 0.1 N KOH at 20°C for 24 hr (Chladek and Nagyvary, 1972). As expected from the known greater stability of acyclic and cyclic phosphodiesters as compared to phosphomonesters (see sections IA and IIB) the 5'-deoxy-5'-thionucleoside 3',5'-cyclic phosphorothioates (**205**) were more stable to acidic hydrolysis than the 5'-deoxy-5'-thionucleoside 5'-phosphorothioates (**204**). The cyclic phosphorothioates **205** were stable for 15 hr at 50°C to pH 3, 5, and 11 buffers.

Adenosine 3',5'-cyclic phosphorothioate (**209**) was synthesized (Eckstein, 1970) from N^6-dimethylaminomethyleneadenosine 5-phosphorothioate (**210**) by activation of the phosphorothioate moiety with triisopropylbenzenesulfonyl chloride. Unexpectedly, there was no formation of adenosine 3',5'-cyclic phosphate detected. Dicyclohexylcarbodiimide was not used to cyclize **210** since desulfurization to adenosine 5'-phosphate occurred. The separation of the two possible diastereomers of **209** has not been achieved.

2. Nucleoside 3',5'-Cyclic Phosphoramidates

The phosphoryl oxygen and phosphate oxygens of nucleoside 3',5'-phosphates have also been replaced by nitrogens. Nucleoside 3',5'-cyclic phosphoramidates with an exocyclic amino group have been synthesized from nucleoside 3',5'-cyclic phosphates (Meyer et al., 1973) and by cyclization of a nucleoside 5'-phosphoramide by nucleophilic attack on carbon (Russell and Moffatt, 1969). In the former method (Meyer et al , 1973), the synthesis of nucleoside 3',5'-phosphoramidates was achieved by activation of nucleoside 3',5'-phosphates to nucleophilic attack by reaction with phosphorus oxychloride (presumably by formation of a 3',5'-cyclic phosphorochloridate). Heating of 2'-O-acetylinosine 3',5'-cyclic phosphate (**211**), N^6-2'-O-diacetyladenosine 3',5'-cyclic phosphate (**212**), and N^2-2'-O-diacetylguanosine 3',5'-cyclic phosphate (**213**) with phosphorus oxychloride followed by treatment with an alkylamine gave the intermediate acyl nucleoside 3',5'-cyclic phosphoramidates **214** and **215**. Further basic hydrolysis gave the nucleoside 3',5'-phosphoramidates **216** and **217**. Acidic hydrolysis of the nucleoside 3',5'-cyclic phosphoramidates **216** and **217** in pH 1 buffer at 37°C for 3 hr gave approximately 50% hydrolysis to mainly adenosine 3',5'-cyclic phosphate and guanosine 3',5'-cyclic phosphate, respectively.

Activation of the phosphate of uridine 3',5'-cyclic phosphate (**218**) with diphenylphosphorochloridate gave the mixed anhydride **219**, which was used for the synthesis of the uridine 3',5'-cyclic (methyl phenylalanyl)-phosphoramidate (**220**) (Preobrazhenskaya et al., 1967). Acid and base hydrolysis, with cleavage of the phosphate ester bonds, of the phosphoramidate **220** was faster than for uridine 3',5'-cyclic phosphate.

Intramolecular nucleophilic displacement of the 3'-iodo function by the phosphate anion of 3'-deoxy-3'-iodothymidine 5'-phosphoromorpholidate (**221**) gave 1-(2-deoxy-β-D-*threo*-pentofuranosylthymine 3',5'-cyclic phosphoromorpholidate (**222**) (Russell and Moffatt, 1969). In contrast to the nucleoside 3',5'-cyclic phosphoramidates **214, 216,** and **217**, the P—N bond of **222** could not be selectively hydrolyzed under the conditions tried.

Ac = Acetyl

218 → **219** (U=Uracil) → **220**

221 (T=Thymine) → **222**

Because of the instability of 5'-amino-5'-deoxyadenosine 5'-phosphoramidate (**223**), the direct synthesis of 5'-amino-5'-deoxyadenosine 3',5'-cyclic phosphoramidate (**224a**) from **223** was unsuccessful (Shen, 1970). The 5'-amino-5'-deoxyadenosine 3',5'-cyclic phosphoramidates **224** were synthesized, however, by transesterification of the "activated" *bis*-(*p*-nitrophenyl)phosphoramidate esters **225** (Murayama et al., 1971). The isolation of the phosphoramidate diester **226** from the reaction suggested that the direct synthesis of the cyclic phosphoramidates **224** occurred *via* **226**. The isolation of the diastereomers of **226** represents one of the few reported nucleoside 3',5'-cyclic phosphate diastereomers to be characterized. The cyclic phosphoramidates **224** are much less stabile to acidic hydrolysis than adenosine 3',5'-cyclic phosphate with significant P—N hydrolysis occurring after 5 hr at 37°C in buffer solutions with pH < 7.

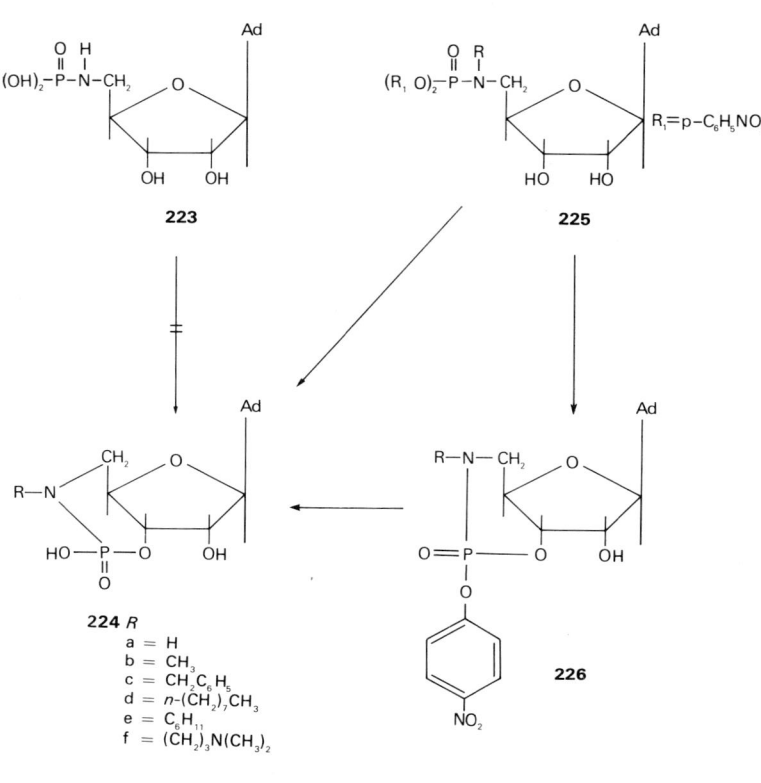

3. Nucleoside 3',5'-Cyclic Thiophosphoramidates

The synthesis (Jastorff and Krebs, 1972) of the 5'-deoxy-5'-aminoadenosine 3',5'-thiophosphoramidates (**227**) required stronger alkaline conditions and higher temperatures than for analogous 5'-deoxy-5'-aminonucleoside 3',5'-phosphoramidates (**224**) but, interestingly, gave compounds which were more resistant to P—N acidic hydrolysis. The higher relative stabilities found for the N-octyl derivatives **224d** and **227b** to acidic hydrolysis were attributed to a steric effect. In contrast to the P=O phosphoramidates **224**, diastereomers of the P=S diesters **228** could not be detected by chromatography or ^{31}P spectral determinations. Treatment of the p-nitrophenyl diester **228** with sodium methoxide gave quantitative formation of the methyl thiophosphoramidate **229**.

4. Alkyl Nucleoside 3',5'-Cyclic Phosphates

The formation of the methyl and ethyl esters of adenosine 3',5'-cyclic phosphate **230** from the silver salt of adenosine 3',5'-cyclic phosphate (**231**) and methyl (ethyl) iodide has been reported by Falbriard et al. (1967). In view of the known instability of cyclic phosphorus triesters (Westheimer, 1968), the cyclic phosphate triester moiety of alkyl adenosine 3',5'-cyclic phosphates (**230**) would be expected to be considerably more labile to hydrolysis than the phosphate ring in adenosine 3',5'-cyclic phosphate; however, no hydrolysis studies have appeared. The synthesis of non-ionic diesters of nucleoside cyclic phosphonates is discussed in the following section.

231

230 R = CH$_3$ or CH$_3$CH$_2$

5. Nucleoside Cyclic Phosphonates

The synthesis of the isosteric phosphonate analogues of nucleoside 5'-phosphates (**232**) and nucleoside 3'-phosphates (**233**) has provided a route to the synthesis of nucleoside 3',5'-cyclic phosphonates where the phosphate oxygen has been replaced by a methylene group (Jones and Moffatt, 1969; Jones et al., 1970). Facile cyclization of the nucleoside 5'-phosphonates **232a** to the nucleoside 3',6'-cyclic phosphonates (**234a**) with dicyclohexylcarbodiimide occurred in high yields even in aqueous pyridine. The nucleoside 3',6'-cyclic phosphonates (**234a**) were also obtained by hydrolysis of the intermediate aryl nucleoside 3',6'-cyclic phosphonates (**234b**) which were formed by base-catalyzed transesterification of the diaryl nucleoside 5'-phosphonates (**232b**). Prior blocking of the 2'-position of the nucleoside 3'-phosphonate (**233**) was necessary for cyclization of **233** to the nucleoside 3',5'-cyclic phosphonate (**235**).

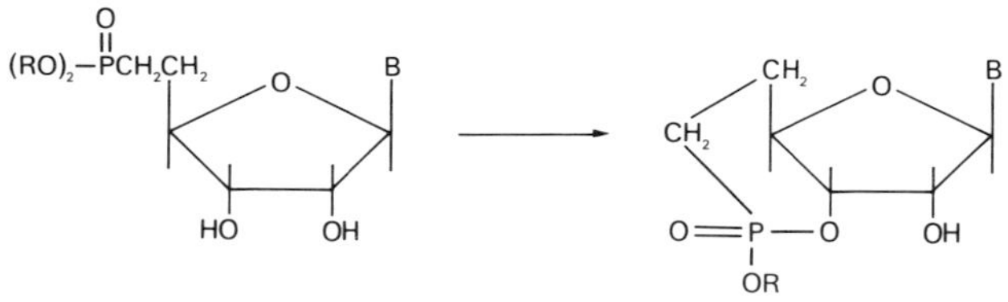

232 a, R = H
b, R = Aryl

234 a, R = H
b, R = Aryl

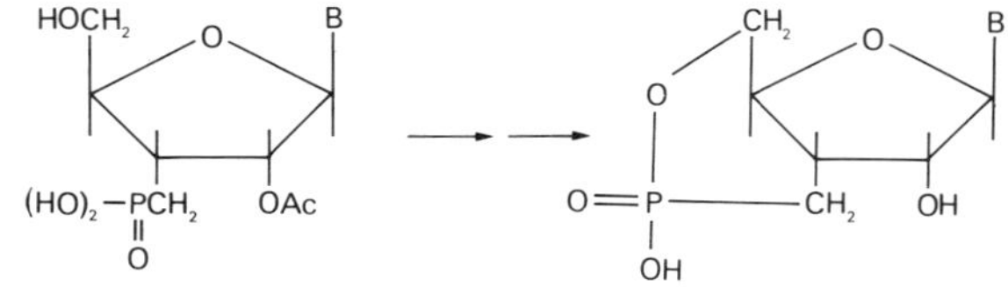

233

235

B = Adenine or other heterocycles
Ac = Acetyl

During the selective iodination of thymidine with methyltriphenoxyphosphonium iodide, a low yield of thymidine 3',5'-cyclic methylphosphonate (**236**) was isolated (Verheyden and Moffatt, 1970). The mechanism of this reaction was suggested to occur by intramolecular attack of the 3'-hydroxyl upon the phosphorus atom of the 5'-diphenoxymethylphosphonium intermediate **237** giving a cyclic phosphonium salt **238** or related phosphorane which decomposed to thymidine 3',5'-cyclic methylphosphonate (**236**).

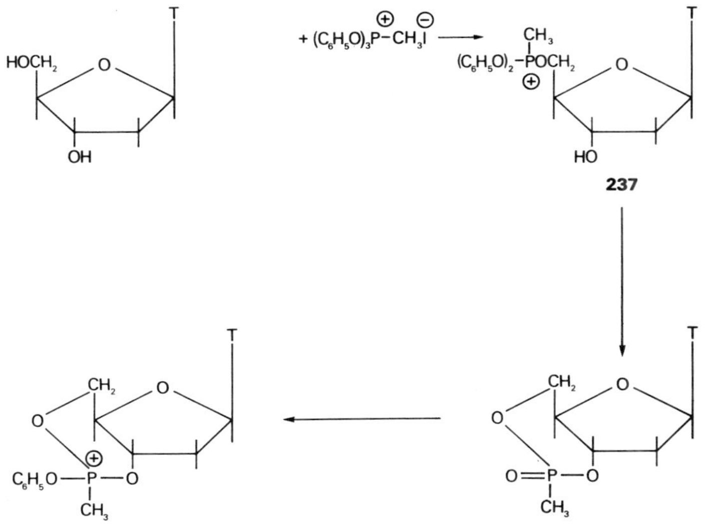

T = Thymine

6. Nucleoside Cyclic Pyrophosphates

A further modification of the cyclic phosphate moiety is illustrated by the 2',5'- and 3',5'-cyclic pyrophosphates of adenosine (**239** and **240**) formed during barium hydroxide degradation of adenosine 5'-triphosphate (Lipkin et al., 1959a,b). Adenosine 3',5'-cyclic pyrophosphate (**240**) has also been isolated from red algae *Porphyra perforata* (Su and Hassid, 1962).

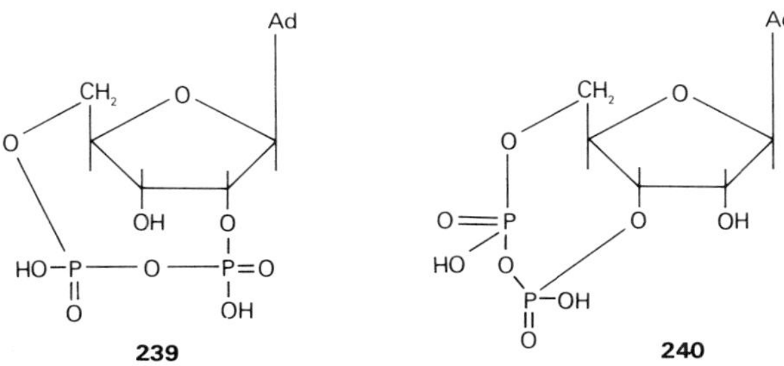

Ad = Adenine

When cytidine (**241**) was heated at 60°C with partially hydrolyzed phosphorus oxychloride, the $O^2,2'$-cyclocytidine 3',5'-cyclic pyrophosphate (**242**) was formed which, on further hydrolysis, gave the *ara*-cytidine 3',5'-cyclic pyrophosphate **243** (Kanai and Ichino, 1971).

D. Addenda

The reaction of chloroacetaldehyde with adenosine 3',5'-cyclic phosphate gave the highly fluorescent compound, 3-β-D-ribofuranosylimidazo[2,3-i]purine 3',5'-cyclic phosphate (1,N^6-etheneoadenosine 3',5'-cyclic phosphate) (**244**) Secrist et al., 1972).

Lake et al. (1972) have reported the synthesis of 9-(4-thio-β-D-ribofuranosyl)adenine 3′,5′-cyclic phosphate (**245**a) by dicyclohexylcarbodiimide cyclization of the respective 5′-nucleotide. The $N^6,2'$-O-dibutyryl derivative **245**a and the N^6-(n-butyl) derivative **245**b were also prepared.

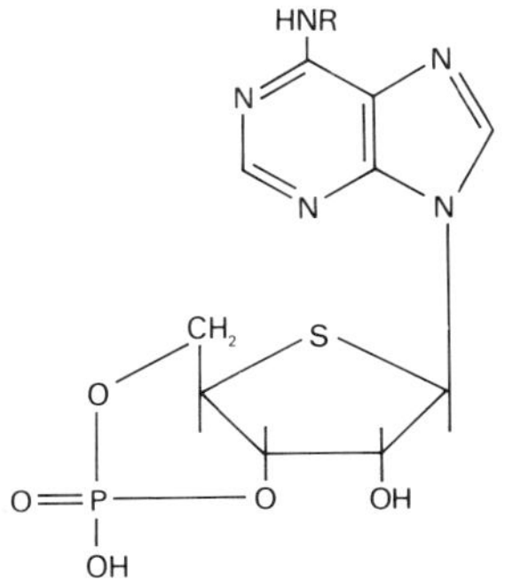

245 <u>R</u>
a, H
b, CH$_3$CH$_2$CH$_2$CH$_2$

The remainder of this chapter will be devoted to the biological activities of various cyclic nucleotides as compared to those of the two known naturally occurring 3′,5′-cyclic nucleotides, cyclic AMP and cyclic GMP. Unfortunately, biological activity has not been reported for all the aforementioned cyclic nucleotides. The following discussion is an attempt to formulate structure-activity relationships for nucleoside 3′,5′-cyclic phosphates from the published data known to the authors and from data generously communicated to us.

IV. INTRODUCTION TO THE BIOLOGICAL ACTIVITY OF ANALOGUES OF NUCLEOSIDE 3′,5′-CYCLIC PHOSPHATES

The role of cyclic AMP (cAMP) as a second messenger mediating the action of a large number of hormones and other agents has been well established and reviewed in a number of books and articles (Greengard and Costa, 1970; Robison et al., 1971; Liddle and Hardman, 1971; Hardman et al., 1971), as well as in some recently published and soon to be published symposia (Robison and Sutherland, 1971) and congresses (Greengard et al., 1972). The list of the number and types of different systems in which either cAMP or cyclic GMP (cGMP) can be shown to have a biological effect continues to grow. Table 4 contains a compilation of some biological processes upon which cAMP has recently been shown to exert an effect. Cyclic AMP (as the $N^6,2'$-O-dibutyryl derivative, dbcAMP) has been shown to inhibit the release of histamine from leukocytes obtained from allergic donors (Lichtenstein and Margolis, 1968). Those results were immediately extended to the effects of dbcAMP on immunoglobulin E (IgE)-mediated histamine release from human lung fragments (Assem and Schild, 1969) and release of slow-reacting substance of anaphylaxis (SRSA) from similar preparations (Ishizaka et al., 1971; Orange et al., 1971a,b; Kaliner et al., 1971). Weissmann et al. (1971) reported that cAMP was able to inhibit the release of lysosomal enzymes from macrophages.

TABLE 4. Some biological activities in which cAMP has recently been implicated

Tissue affected	Biological effect	Reference
Rauscher murine leukemia virus	Phosphorylation of specific structural proteins (viral replication)	Strand and August, 1971
3T3-Mouse fibroblasts (SV-40 transformed and normal)	Interconversion of cAMP-PDE from high to low K_m	D'Armiento et al., 1972
Chicken liver slice	Inhibition of lipogenesis	Allred and Roehrig, 1972
Human leukocytes	Inhibition of allergic histamine release	Bourne et al., 1972
Horse lymphocytes	Stimulation of uridine incorporation into RNA	Averner et al., 1972
Phagocytes	Inhibition of release of lysosomal enzymes	Weissmann et al., 1971
Rat liver homogenate	Inhibition of cholesterol biosynthesis	Bloxham, 1971
Rat prostate, seminal vesicles	Stimulation of activity of hexokinase phosphofructokinase, GADPH, pyruvate kinase (mimicking action of androgens, e.g., testosterone)	Singhal et al., 1971
Mice marrow cells in tissue culture	Inhibition of growth of cells	Morley et al., 1971
Rats	dbcAMP increased duration of hexobarbital-induced sleeping time	Werner et al., 1972
Rats	dbcAMP inhibited mitosis of epidermal cells of new-born rats	Voorhees et al., 1972b
Bovine spermatozoa	dbcAMP and cAMP increased sperm respiration and mobility	Garbers et al., 1971

The number of different tissues in which cAMP exerts an effect, as well as the diverse nature of these tissues' physiological functions, creates the potential for evaluation of synthetic analogues of cAMP in as many different *in vitro* and *in vivo* systems as there are unique biological activities of cAMP itself. The purpose of the remainder of this review will be to present and discuss the information available to us from the literature, and through personal communications of unpublished data up to May 1972, on the biological effects of synthetic derivatives of 3′,5′-cyclic nucleotides. Because of the ubiquitous use of *one* synthetic derivative of cAMP, dbcAMP, it will be impossible to discuss every single biological activity of this substance or of cAMP itself. However, in instances where the activity of an analogue has been compared with either cAMP or dbcAMP, the comparative data will be presented or the reader will be referred to the original reference for details related to that compound.

The application of exogenous cAMP to an isolated tissue or organ at physiological concentrations (10^{-8} to 10^{-6}M) does not produce the desired response in most instances. In fact, utilization of concentrations even two orders of magnitude greater (10^{-6} to 10^{-4}M) is not sufficient to produce the expected response in some instances. Theoretically, several reasons become apparent for this lack of activity: the obvious possibility of poor transport of cAMP into whole cells, and the destruction of cAMP by either extracellular (Chang, 1968) or intracellular cyclic nucleotide phosphodiesterases (PDE's) (Thompson and Appleman, 1971).

The use of the naturally occurring cAMP or cGMP in *in vitro* or *in vivo* studies both as molecular probes and potential therapeutic agents suffers from the potential drawback of a lack of specificity, since these naturally occurring second messengers must of necessity be able to activate all cyclic nucleotide-dependent processes. Derivatives of these agents could exhibit a greater degree of specificity.

For the biologist, in general, the synthesis of derivatives of cyclic nucleotides related to the naturally occurring nucleotides has held great promise in overcoming some of the drawbacks of cAMP and cGMP discussed above. For the pharmacologist, his hope for a compound more potent than cAMP and more tissue specific in its pharmacological action might be answered. For the molecular biologist, the availability of derivatives able to induce specifically one enzyme but not another, or to regulate growth of tumor cells but not affect normal cells, or bind to and activate cAMP-dependent enzymes from one tissue but not another, might provide the molecular probes to understand fully the molecular mechanism of cAMP's action. For the biochemist, the possibility that analogues could be synthesized that might penetrate some cells to a greater degree, and at the same time be more metabolically stable than the parent nucleotide, offers a hope for the development of therapeutically useful substances whose mechanism of action might not only be understood, but whose biological activity might be predictable from in-depth structure-activity relationship (SAR) studies in many of the systems about to be discussed.

A review by Drummond and Severson (1971) covers the biological actions of cAMP analogues synthesized and studied up to early 1970.

V. BIOLOGICAL ACTIVITY (IN VITRO)

A. Effects on Isolated Enzyme Systems

1. Cyclic Nucleotide-Dependent Protein Kinases

Cyclic AMP has been shown to stimulate the adenosine 5'-triphosphate (ATP)-dependent phosphorylation of a wide variety of proteins, e.g., casein, protamine, phosphorylase-b-kinase (Walsh et al., 1968), neurotubular protein (Goodman et al., 1970), fat-cell lipase (Huttunen et al., 1970), ribosomal protein (Loeb and Blat, 1970), and sigma factor of bacterial RNA polymerase (Martelo et al., 1970). This enzyme, a cAMP-dependent protein kinase, has been isolated from cell-free extracts prepared from nearly every single tissue and organism studied (Miyamoto et al., 1969; Kuo and Greengard, 1969), and has been purified from a wide variety of sources, including bovine brain (Miyamoto et al., 1969). Indeed, it has been postulated that most of the various biological properties attributed to cAMP may be mediated through the stimulation of these protein kinases (Kuo and Greengard, 1969).

Examination of variations in biological activity utilizing analogues of cAMP has had to await the synthesis of sufficient supplies of the pure cyclic nucleotide derivatives possessing regularly varied structural modifications. Some reports have appeared in which cyclic nucleotide derivatives of the naturally occurring nucleosides have been studied as alternate activators of protein kinases. Miyamoto et al. (1969) have shown that inosine 3',5'-cyclic monophosphate (cIMP) and cGMP had 32 and 7% the activity of cAMP, respectively, at concentrations of 5×10^{-7} M, using a partially purified preparation of bovine brain protein kinase. The same authors showed that uridine 3',5'-cyclic monophosphate (cUMP) and cytidine 3',5'-cyclic monophosphate (cCMP) were approximately 7% as active as cAMP, and deoxythymidine 3',5'-cyclic monophosphate (cdTMP) was virtually inactive.

a. Activity of 6,8-Disubstituted Purine-9-(β-D-Ribofuranosyl) 3',5'-Cyclic Monophosphates (PuRcMP)

Muneyama et al. (1971) reported the synthesis of a series of 8-substituted analogues of cAMP and their relative ability to activate a cAMP-dependent protein kinase purified from bovine brain. The results presented in Table 5 indicate that all of the analogues studied were able to stimulate incorpora-

TABLE 5. Relative protein kinase activity (α) of 8-substituted cAMP analogues at 10^{-8} to 10^{-4} M [a]

R			α		
	10^{-8}	10^{-7}	10^{-6}	10^{-5}	10^{-4}
H	1.0	1.0	1.0	1.0	1.0
NHCH$_2$CH$_2$OH	0	0	0	0.18	0.16
SCH$_2$CH$_2$OH	0	0.25	0.61	0.65	0.70
SCH$_2$CH$_3$	0.88	0.93	1.14	1.07	1.09
SCH$_2$C$_6$H$_5$	0.36	1.15	0.95	0.98	0.73
NHCH$_2$C$_6$H$_5$	0	0.091	0.56	0.85	0.30
NHCH$_2$C$_4$H$_4$O	0.50	0.38	0.88	0.98	1.28
N(CH$_3$)$_2$	0.43	0.56	0.91	1.50	1.60
N$_3$	1.57	0.63	0.80	1.0	0.86
Br	0.73	0.65	0.93	0.98	1.46
SCH$_3$	2.4	1.10	0.94	1.15	0.88
OCH$_3$	0.7	0.42	0.80	1.13	0.76
NCH$_3$	0.8	0.65	0.65	1.13	1.17
SH	2.0	1.4	1.13	1.20	1.55
OH	1.0	1.22	1.04	1.13	1.57
NH$_2$	1.7	0.62	0.81	1.0	0.85

[a] Muneyama et al., 1971.

tion of ^{32}P-phosphate into histone to some extent over a widely varying range of concentrations. The ability of these derivatives to stimulate a purified cAMP-dependent protein kinase allows the assessment of the biological agonist-like activity of a cyclic nucleotide analogue without the problem of cell permeability which is encountered in using a tissue slice or whole organ preparation. In addition, it allows the evaluation and comparison of large numbers of compounds with widely varying structures using a kinetic parameter, the activation constant (K_a value). These values were determined by a double reciprocal plot of $1/v$ vs. $1/[A]$ where $[A]$ is the concentration of activator (e.g., cAMP, analogue, etc.) (Miyamoto et al., 1969; Muneyama et al., 1971). The activation of the protein kinase by cAMP has been shown to occur according to the scheme proposed in Eq. (1) (Reimann et al., 1971). The activation constant that was calculated for cAMP or for an

$$C \cdot R \xrightleftharpoons{cAMP} C + R \quad (1)$$

(inactive complex)   (active form of the enzyme)   (cAMP binding unit)

analogue was for the process of converting the inactive protein kinase to a catalytically active subunit (C) and the regulatory protein (R). The catalytic activity measured, of course, was the incorporation of ^{32}P-phosphate into a protein acceptor (histone) from the γ-phosphate of ATP. Figure 3 shows the results obtained for cAMP as well as a typical plot for one of the analogues, 8-methoxyadenosine 3',5'-cyclic monophosphate (8-methoxy-cAMP). In our laboratory, over a period of several years, K_a values for cAMP ranging from 0.9 to 5.0×10^{-8} M have been routinely obtained using a preparation of bovine brain protein kinase purified through the stage of DEAE cellulose chromatography (fraction I) (Miyamoto et al., 1969). A mean value for cAMP of $3.2 \pm 0.8 \times 10^{-8}$ M has been calculated. As a means of comparing the biological activity of a number of 8-substituted analogues, the K_a values as well as the "relative K_a values," which we denote as K_a' ($K_a' = K_a$ cAMP/K_a analogue) for 17 8-substituted analogues of cAMP are presented in Table 6. Correlation coefficients of > 0.98 were routinely obtained for a linear regression analysis of the straight lines obtained in the double reciprocal plots. Nine of the 17 compounds tested were more effective than cAMP itself ($K_a' > 1.0$) (Muneyama et al., 1971). That the enzyme prefers the presence of a sulfur atom at C-8 rather than a nitrogen atom can best be illustrated by the fact that 8-thio-cAMP was two times more active than 8-amino-cAMP (Table 6, **1** vs. **9**), and that 8-benzylthio-cAMP was approximately 80 times more active than 8-benzylamino-cAMP (Table 6, **8** vs. **17**). A comparison of the relative activity of 8-(2-hydroxy-

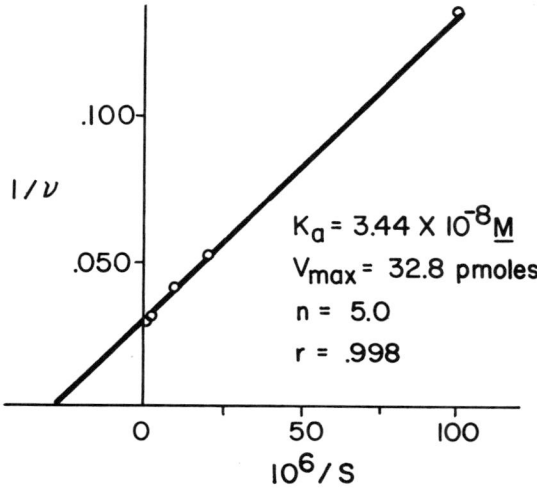

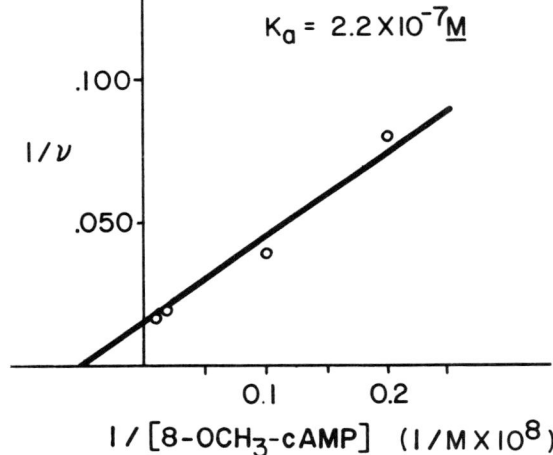

FIG. 3. *Top:* Lineweaver Burk plot for cAMP bovine brain protein kinase; *bottom:* for 8-OCH$_3$-cAMP.

ethylthio)cAMP (**12**) and 8-(2-hydroxyethylamino)cAMP (**16**) reveals a similar relationship (Simon et al., 1973).

The results with 8-methylthio- (**5**), 8-ethylthio- (**7**), and 8-benzylthio-cAMP (**8**), as well as with 8-chloro- (**2**), 8-bromo- (**3**), and 8-iodo-cAMP (**6**) also indicate that there is a fairly good degree of either bulk tolerance or

TABLE 6. Activation constants for
8-substituted-cAMP derivatives[a]

	R	$K_a^b(\times10^{-8})$	$K_a'^c$
1	SH	1.3	3.8
2	Cl	2.0	3.1
3	Br	1.6	2.9
4	OH	1.7	2.8
5	SCH_3	3.9	2.4
6	I	3.5	2.1
7	SCH_2CH_3	3.3	2.0
8	SCH_2–(phenyl)	2.8	1.7
9	NH_2	3.2	1.5
10	N_3	4.7	0.68
11	$N(CH_3)_2$	15	0.59
12	SCH_2CH_2OH	8.2	0.54
13	$HNCH_3$	12	0.38
14	OCH_3	22	0.29
15	$HNCH_2$–(furyl)	21	0.23
16	$HNCH_2CH_2OH$	27	0.11
17	$HNCH_2$–(phenyl)	58	0.023

[a] Simon et al., 1973
[b] K_a = Activation constant from Lineweaver-Burk plot
[c] $K_a' = \dfrac{K_a\text{cAMP}}{K_a\text{cmpd}}$ = Relative activation constant

preference for hydrophobic groupings at the C-8 position. The apparent low level of activity exhibited by both 8-(2-hydroxyethylamino)cAMP (**16**) and 8-(2-hydroxyethylthio)cAMP (**12**) indicates that a more polar group such as the CH_2CH_2OH is less tolerated than a nonpolar CH_2CH_3 substituent (Simon et al., 1973).

The K_a' values for a series of six 8-substituted derivatives of cAMP have been compared for cAMP-dependent kinases isolated and purified from three different tissues. The results presented in Table 7 show that

TABLE 7. *Comparison of activation constants (K_a's) of several 8-substituted cAMP derivatives using brain, heart and liver protein kinase*

	$K_a \times 10^8$ (K_a')		
R	Bovine brain[a]	Rat liver[a]	Bovine heart[b]
SH	1.28 (3.78)	1.21 (2.1)	2.0 (1.5)
OH	1.77 (2.78)	1.68 (1.51)	2.0 (1.5)
NH_2	3.15 (1.54)	4.87 (0.52)	2.0 (1.5)
SCH_2CH_2OH	8.90 (0.54)	—	3.0 (1.0)
$HNCH_2CH_2OH$	27.6 (0.17)	16.4 (0.15)	10.0 (0.3)
$HNCH_2C_6H_5$	58.1 (0.08)	16.5 (0.15)	10.0 (0.3)

[a] Bauer et al., 1971
[b] Kuo et al., 1972

when the compounds were ranked in order of decreasing potency, a similar order of activity was exhibited against all three enzymes for each of the compounds studied. A near perfect correlation was obtained.

Similar studies have been carried out utilizing a series of 8-substituted cIMP derivatives. The results presented in Table 8 indicate that cIMP

TABLE 8. *Ability of several 8-substituted cIMP derivatives to activate a cAMP-dependent (bovine brain) protein kinase*[a]

R	Bovine brain K_a'
H	0.59
SCH_2CH_3	0.32
$SCH_2C_6H_5$	0.16
Br	0.14
N_3	0.07
NH_2	0.02
H (N^1-oxide)	0.02

[a] Shuman et al., 1973a

itself was less effective than cAMP and also that substitution of a grouping at the 8 position further decreased the compound's effectiveness. It is interesting to point out that of the 8-substituted cIMP derivatives studied, the 8-alkylthio derivatives were the most active, just as in the case of the 8-substituted cAMP analogues (Shuman et al., 1973a).

Kuo et al. (1972) have studied the ability of several of the 8-substituted analogues of cAMP to inhibit binding of cAMP to the bovine brain protein kinase and to cause dissociation of the holoenzyme into a regulatory and catalytic subunit. They have been able to demonstrate that the relative ability of the analogues to dissociate the bovine brain enzyme into a catalytic subunit (3.6 S particle) was in good agreement with their relative potency in activating the bovine brain or bovine heart enzyme. They have also studied the relative ability of the 8-substituted analogues to inhibit the binding of ^{3}H-cAMP to the bovine brain protein kinase (Kuo et al., 1972). Their results indicate a lack of correlation between the analogues' effectiveness to compete with cAMP for binding and to activate the protein kinase. Similar findings have been reported by Meyer et al. (1972b) for a series of 6-substituted derivatives of cAMP. Kuo et al. (1972) suggest the possible existence of multiple binding sites on the regulatory subunit on the basis of their data, and a similar implication could be made from Meyer's experiments as well (1972b).

b. 2,6,8-Trisubstituted PuRcMP Derivatives

The synthesis and biological activity of very few 2,6,8-trisubstituted analogues have been described. The data presented in Table 9 show the relative ineffectiveness of cGMP and five of its 8-substituted derivatives in activating the bovine brain protein kinase. Although the activity is extremely low, it does appear that both 8-methylthio- and 8-benzylthio-cGMP are more potent than cGMP itself against the cAMP-dependent enzyme

TABLE 9. Ability of several 8-substituted cGMP derivatives to activate a cAMP-dependent (bovine brain) protein kinase

R	Protein kinase[a] K_a' bovine brain
H	0.005
SCH$_2$–⟨phenyl⟩	0.03
SCH$_3$	0.02
Br	0.003
HN—CH$_3$	0.001
HN—CH$_2$–⟨phenyl⟩	0.0001

[a] Shuman et al., 1973a

(Shuman et al., 1973a). Of the compounds synthesized, only 8-bromo- and 8-benzylamino-cGMP were evaluated against the *cGMP-dependent* protein kinase purified from lobster muscle. 8-Bromo-cGMP was found to be approximately 100 times more potent than 8-benzylamino-cGMP against the *cGMP-dependent* enzyme (Kuo et al., 1972; Shuman et al., 1973a). It is interesting to note that 8-bromo-cAMP was approximately 100 times more effective than 8-benzylamino-cAMP vs. the *cAMP-dependent* enzyme (Table 6).

c. 6-Substituted PuRcMP Derivatives

In examining the relative activity of several naturally occurring 2- and 6-substituted PuRcMP derivatives with the cAMP-dependent and cGMP-

TABLE 10. *The relative ability of some 6-substituted purine ribonucleoside 3',5'-cycle phosphates to activate a purified cAMP-dependent (bovine brain) protein kinase*[a]

IV

R	α^b				
	10^{-8}	10^{-7}	10^{-6}	10^{-5}	10^{-4}
NH_2(cAMP)	1.0 (26.1)[c]	1.0 (112)	1.0 (169)	1.0 (175)	1.0 (115)
NHC_2H_5	0.66	0.95	1.05	0.92	2.2
$N(C_2H_5)_2$	0.98	1.0	1.08	1.3	2.4
SH	0.48	0.52	0.90	0.97	1.48
OCH_3	0.68	0.77	0.99	1.07	1.63
OH	0.35	0.31	0.75	1.06	1.60
SCH_3	1.2	1.0	1.1	1.45	2.0

[a] Meyer et al., 1972b

[b] α = Ratio of $\dfrac{\text{pmoles incorporated in presence of test compound}}{\text{pmoles incorporated in absence of test compound}}$

[c] pmoles ^{32}P incorporated into histone

dependent protein kinases, it became obvious that substitution of an oxo function at C-6 (cIMP) lowered the ability of the compound to activate the cAMP-dependent kinase. When cIMP was compared with cAMP with regard to its ability to activate the cGMP-dependent kinase, it was found to be 10 to 20 times more effective than cAMP, although it was still less potent than cGMP (Tables 10 and 11). The effect of several 6-substituted purine derivatives (Table 11, Y = SH, SCH_3, SC_2H_5, $SCH_2C_6H_5$) on the cGMP-dependent protein kinase was examined and compared with their effect on the cAMP-dependent protein kinase. The results with these analogues dramatically demonstrate that the cGMP- and cAMP-dependent kinases have vastly different specificities with regard to structural alterations

TABLE 11. *Activation of cGMP-dependent protein kinase from lobster muscle*[a]

R	Y	Z	K_a(M × 10^8)	K_a'[b]	r[c]
cGMP	OH	NH_2	3.16	1.0	0.99
	NH_2	H	43.5	0.07	0.99
	OH	H	12.10	0.26	0.94
	SH	H	1.70	1.85	0.99
	OCH_3	H	7.69	0.41	0.99
	NHC_2H_5	H	2.21	1.41	0.95
	$N(C_2H_5)_2$	H	14.30	0.22	0.98
	SCH_3	H	7.5	0.42	0.99
	$SCH_2C_6H_5$	H	11.7	0.27	0.94
	SC_2H_5	H	7.95	0.40	0.96

[a] Meyer et al., 1972b

[b] $K_a' = \dfrac{K_a \text{ for cGMP}}{K_a \text{ for analogue}}$

[c] r = Correlation coefficient for linear regression analysis of Lineweaver-Burk double reciprocal plots

at the 6 position of the purine nucleus and their relative ability to activate cAMP-dependent and cGMP-dependent enzymes. In fact, one derivative of cGMP lacking the 2-amino group [Table 11, Y = SH, 9-(β-D-ribofuranosyl)purin-6-thione-3',5'-cyclic monophosphate], was more effective than cGMP itself in the cGMP-dependent system, whereas it was less active than cAMP vs. the cAMP-dependent enzyme (Table 11). It should be pointed out that 6-methylthio-PuRcMP behaved more nearly like a cAMP analogue than a cGMP analogue, since it was more active than cAMP using the cAMP-dependent enzyme, and less active than cGMP using the lobster kinase. The ability of 6-thiopurine derivatives to act more like guanosine analogues has been reported previously with regard to the mechanism of action of 6-mercaptopurine (6-MP) and 6-methylthiopurine (6-MMP) as

TABLE 12. K_a' (relative affinity constants) values for 6-substituted purine 3',5'-cyclic nucleotides against cAMP-dependent bovine brain protein kinase[a]

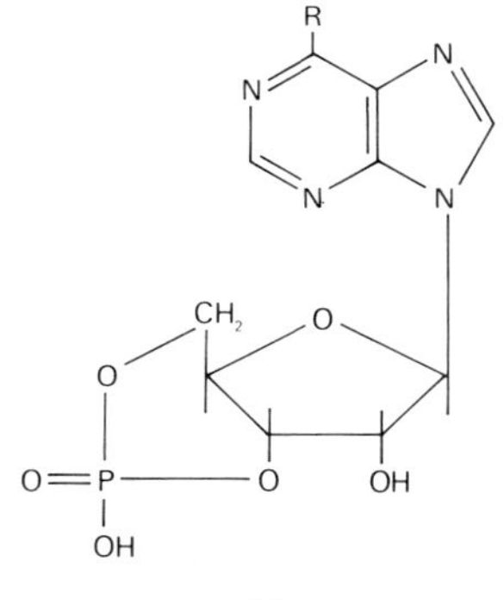

IV

R	K_a'[b]
NHC_2H_5	0.50
SH	0.56
OH	0.59
$N(C_2H_5)_2$	0.66
OCH_3	0.66
$SCH_2C_6H_5$	1.25
SCH_3	1.45
SC_2H_5	2.0

[a] Meyer et al., 1972b

[b] $K_a' = \dfrac{K_a \text{ for cAMP}}{K_a \text{ for analogue}}$

antitumor agents (Ho, 1971; Montgomery, 1970), and the relative ability of the 9-β-D-ribofuranosyl derivatives to be phosphorylated by adenosine kinase.

The presence of a methylthio, ethylthio, or benzylthio moiety at the 6 position actually increases the activity of the compound for a cAMP-dependent protein kinase as compared to the parent nucleotide (cAMP) (Table 12). The presence of the same substituents causes a *decrease* in activity for the cGMP-dependent kinase from lobster muscle. The substitution of a sulfur atom for oxygen actually enhances the ability to activate the cGMP-dependent enzyme (Meyer et al., 1972b), while decreasing the ability of the compound to activate the cAMP-dependent enzyme. These results could be explained on the basis of the difference in the tautomeric forms of the 6-alkylthio derivatives compared to the 6-thione-9-(β-D-ribofuranosyl)purine 3',5'-cyclic monophosphate, or variation in electron density at C-6 between the 6-alkylthio and 6-thione derivative (Meyer, 1972b).

d. Miscellaneous Derivatives

Although a systematic synthesis of analogues of cyclic nucleotides varying in structure in the carbohydrate portion of the molecule has not appeared, many isolated studies have been reported in which several unique structural analogues of cAMP have been synthesized and evaluated as alternate activators of a purified protein kinase. The results presented in Table 13 are a compilation of K_a' and α values (protein kinase activity in presence of analog at [0.1 μM]/protein kinase activity in presence of cAMP at [0.1 μM]) for 17 different cyclic nucleotide derivatives. Several generalizations can be made regarding SAR by examination of Table 13.

(1) Substitution at C-2' or change of configuration of the 2'-OH results in nearly complete loss of activity for protein kinase (**1** vs. **2, 7, 8**).
(2) Formation of anhydronucleoside (8,2'-anhydro-8-thio- or 8,2-anhydro-8-oxo-arabino 3',5'-cyclic monophosphate) abolishes activity (**17 and 18**).
(3) Change in configuration at C-3' also abolishes activity (**1** vs. **6**).
(4) The substitution of a methylene group at C-3' or C-5' as in the homophosphonate analogues nearly abolishes activity (**1** vs. **9, 10**).
(5) The substitution of a sulfur in place of oxygen in the exocyclic phosphate ring abolishes activity (**1** vs. **13**).
(6) The substitution of a sulfur in place of oxygen at C-5' of the ribose does *not* have a great effect on protein kinase activity (**1** vs. **12**).

(7) The replacement of the ionizable OH on the phosphorus of the 3′,5′-cyclic ring with a di- or amethylamino group abolishes activity for protein kinase (Table 13, **16,** and Meyer et al., 1973).

(8) Replacement of N^7 of the purine ring with a C-H or C-CN moiety has little effect on protein kinase activity (**1** vs. **14, 15**).

TABLE 13. *Effect of some cyclic nucleotide analogues on bovine brain protein kinase (cAMP dependent)*

	Analogue	K_a'	$\alpha(0.1\ \mu M\ [S])$
1	cAMP	1.0	1.0
2	Ara-cAMP	0.001[a]	0.046[a]
3	8-Thio-ara-cAMP	—	0.035[a]
4	8-Amino-ara-cAMP	0.01[a]	0.065[a]
5	8-Oxo-ara-cAMP	0.01[a]	0.053[a]
6	6-Amino-9-β-D-xylofuranosylpurine 3′,5′-cyclic phosphate	—	<0.01[b]
7	2′-O-Methyl-cAMP	0.002[a]	—
8	2′-Deoxy-cAMP	0.01[a]	
9	3′-Cyclic ester of 5′-Deoxy-5′-(dehydroxyphosphinyl-methyl)adenosine	0.002[c]	0.09[c]
10	5′-Cyclic ester of 3′-Deoxy-3′-(dehydroxyphosphinyl-methyl)adenosine	—	0.06[d]
11	5′-Deoxy-5′-thioinosine 3′,5′-cyclic phosphorothioate	0.005[e]	
12	5′-Deoxy-5′-thioadenosine 3′,5′-cyclic phosphorothioate	0.48[e]	0.80[e]
13	Adenosine 3′,5′-cyclic phosphorothioate	0.003[b]	
14	Toyocamycin 3′,5′-cyclic phosphate	0.51[f]	0.86[f]
15	Tubercidin 3′,5′-cyclic phosphate	0.69[b]	0.82[b]
16	N^6-Dimethylamino-9-β-D-ribofuranosylpurine 3′,5′-cyclic N,N-dimethyl phosphoramidate	<0.005[g]	Inactive
17	8,2′-Anhydro-8-thio-9-β-D-arabinofuranosyladenine 3′,5′-cyclic phosphate	<0.005[a]	Inactive
18	8,2′-Anhydro-8-oxo-9-β-D-arabinofuranosyladenine 3′,5′-cyclic phosphate	<0.005[a]	Inactive

[a] Khwaja et al., 1973
[b] Determined using activation of phosphorylase-b-kinase (Drummond and Powell, 1970)
[c] Concentration of analogue and cAMP, both at 0.5 μM (Kuo and Greengard, 1970b)
[d] Concentration of analogue and cAMP, both at 5 μM (Kuo and Greengard, 1970b)
[e] Shuman et al., 1973b
[f] Shuman et al., 1972c
[g] Meyer et al., 1973

2. cAMP Receptor Protein (E. coli)

Recently, the interaction of several analogues of cAMP with the cAMP receptor protein (CRP) from *E. coli* has been described (Anderson et al., 1972). In *E. coli,* the level of cAMP regulates a number of inducible enzymes (Pastan and Perlman, 1970). In order to mediate the synthesis of these inducible enzymes under catabolite control, cAMP must associate with a specific protein referred to as the CRP or the catabolite gene activator. The results of extensive study with the *lac* and *gal* operons suggest that the complex of cAMP and CRP stimulates messenger RNA synthesis by enhancing binding of CRP to the promoter region of DNA.

In an attempt to establish what structural elements in the cAMP molecule were responsible for cAMP binding to CRP, Anderson and his colleagues examined the ability of several cyclic nucleotide analogues to inhibit ^{3}H-cAMP binding to CRP. It was interesting to note that tubercidin 3′,5′-cyclic monophosphate (TucMP) was as effective as cold cAMP in inhibiting binding of ^{3}H-cAMP to CRP (Anderson et al., 1972). This analogue, in which a carbon atom was substituted for nitrogen at position 7 of the purine ring, was also the only one of the analogues tested that was as effective as cAMP in stimulating *gal* operon transcription (Anderson et al., 1972). TucMP was the most effective of the compounds tested in inhibiting binding of ^{3}H-cAMP to CRP. It is interesting to note that 8-thio-cAMP and 8-bromo-cAMP were next in order of potency after TucMP. All three compounds were quite active as stimulators of the cAMP-dependent protein kinase from bovine brain (Table 6). Anderson et al. (1972) suggested that cAMP was quite specific in its ability to combine with CRP in a manner that resulted in enhanced transcription, since only one of these derivatives, TucMP, was effective in promoting transcription, whereas the two other analogues, 8-thio- and 8-bromo-cAMP, were as active as TucMP as inhibitors of cAMP binding to CRP and were more active as protein kinase activators in a mammalian system. 6-Amino-9-(β-D-arabinofuranosyl)-purine 3′,5′-cyclic monophosphate (ara-cAMP) was only 10% as effective as 8-bromo-cAMP in inhibiting binding of ^{3}H-cAMP to CRP. The 5′-octylamino-5′-deoxy-cAMP was also 10% as effective as 8-bromo-cAMP, whereas 7-methyl-cIMP was only 1% as active as the bromo derivative. As further indication that different structural features in the cAMP molecule are responsible for binding to CRP and stimulating or inhibiting DNA transcription, Anderson et al. (1972) showed that one of the least potent inhibitors of cAMP binding to CRP was the most potent inhibitor of cAMP-stimulated transcription *in vitro,* i.e., the 5′-octylamino-5′-deoxy-cAMP.

In several recent studies, Schweizer and Robins (1972) showed that 8-bromo- and 8-thio-cAMP existed almost exclusively in the *syn* conformation, whereas cAMP could exist in either the *syn* or *anti* forms. They have also shown that tubercidin can exist in either conformation. It is possible that, as Anderson suggests, the inability of the cyclic nucleotide analogues to exist in the *anti* conformation (as in 8-bromo-cAMP or 8-thio-cAMP) may prevent the conformational change in CRP that allows for transcription to occur upon interaction with *E. coli* DNA.

3. Cyclic Nucleotide Phosphodiesterases

The main importance of 3',5'-cyclic nucleotide enzymes responsible for cyclic nucleotide degradation have related in the past to controlling endogenous levels of cAMP within the biological system being examined. The implications of this control with regard to production of useful pharmacologic agents is well appreciated when one views the actions of such cAMP PDE inhibitors as theophylline (Grollman and Grollman, 1970), RO 20-1724 (Sheppard and Wiggan, 1971), SQ 20009 and Librium® (Beer et al., 1972), quazodine (Amer and Browder, 1971), papaverine (Kukovetz and Pöch, 1970), and a large number of experimental drug substances purported to exert their effects *via* inhibition of destruction of endogenous cAMP. In addition, these inhibitors of cAMP PDE may exert tissue specificity by virtue of relatively specific inhibition on an enzyme(s) from one tissue and not another.

The details of the biochemistry, enzymology, and drug interactions with cyclic nucleotide PDE's constitute the subject of a review in this volume by Appleman, Thompson, and Russell. The interaction of cyclic nucleotide analogues and PDE's is covered in this section and represents an extremely important consideration within the framework of the biological activity of these cyclic nucleotide derivatives.

As discussed in the previous section, these analogues have demonstrated the ability to act as cAMP-like agonists. Extended duration of activity might exist *in vivo* if the analogues were resistant to cleavage by the enzymes responsible for the degradation of cAMP and cGMP. In addition to acting as cAMP agonists, the analogues might exert biological activity by inhibiting PDE, thereby raising endogenous levels of cAMP.

a. Analogues as Substrates of PDE's: 6-Substituted- and 6,8-Disubstituted- 9-(β-D-Ribofuranosyl)-Purine 3',5'-Cyclic Monophosphates

The determination of whether or not analogues of cAMP act as substrates for PDE has been studied mainly by using a coupled enzymatic reaction. This involves the measurement of the inorganic phosphate that is released from the adenosine 5'-monophosphate (5'-AMP) or the 5'-AMP derivative [5'-AMP-X, Eq. (3)], which is produced upon cleavage of the 3',5'-cyclic phosphate or 3',5'-cyclic phosphate analogue, by PDE in the primary reaction [Eq. (2)].

$$cAMP-X \xrightarrow{PDE} 5'-AMP-X \qquad (2)$$

$$5'-AMP-X \xrightarrow[\text{alkaline phosphatase}]{} Adenosine-X + P_i \qquad (3)$$

Michal et al. (1970a) reported on the substrate activity of a series of 6- and 8-substituted purine derivatives using a commercial enzyme from beef heart and an S-100 fraction prepared from rat adipose tissue homogenate. Incubations were allowed to occur for 30 to 90 min in the presence and absence of enzyme and the amount of inorganic phosphate present in the incubated blank was subtracted from the incubated sample. Comparisons were made at 0.65 mM concentrations of test compound for the heart enzyme, and 0.3 mM for the adipose tissue. These values (0.3 to 0.65 mM) are very close to the K_m values for cAMP using the dog heart enzyme and adipose tissue enzyme, respectively. As can be seen Table 14), only the 8-bromo-cAMP was split at any appreciable rate by the enzyme isolated from dog heart. In two instances the adipose tissue enzyme apparently had a lower degree of specificity.

Muneyama et al. (1971) examined a wider range of 8-substituted compounds and, in addition, checked the results of the coupled assay described above by utilizing a chromatographic method that separated the 5'-AMP analogue from the 3',5'-cyclic phosphate derivative using paper chromatography in isopropanol:NH_4OH. Although the results are similar to those obtained by Michal et al. (1970a) in that none of the 8-alkylamino derivatives were cleaved by PDE in either study, the free 8-amino-cAMP was found by Muneyama et al. (1971) to be cleaved at a rate that was 66% that of cAMP for both the kidney and pig brain enzyme. Comparable data on this compound are not available from Michal's study. While specific α (relative rate of hydrolysis, see Table 15) values for such derivatives as 8-thio- and 8-bromo-cAMP differ between the two studies (Table 14; Muneyama et al., 1971), they are surprisingly similar, considering that enzymes of varying degrees of purity and from different tissue sources were used. The studies reported by Muneyama et al. (1971) were carried out at a concentration (1 mM) of cAMP that was three to four times greater than that used by Michal et al. (1970a). This value (1 mM) was approximately five to 10 times the K_m value of cAMP for kidney and brain PDE, and should have readily allowed for measurement of activity even if a K_m 10-fold greater existed for the analogues as compared to cAMP.

Substitution of the amino group at the 6 position appears to have a less dramatic effect on the ability of the enzyme to cleave the resulting analogue 6-Chloro-PuRcMP has been reported to be cleaved at a rate that was 35% that of cAMP and 6-methoxy-PuRcMP was cleaved at a rate that was 10% that of cAMP by the dog heart enzyme. A series of 6-alkylamino purine derivatives were relatively resistant to cleavage (Michal et al., 1970a) (Table 16). In a later study (Meyer et al., 1972b), it was demonstrated that 6-thio-, 6-alkylthio-, 6-alkoxy-, and 6-alkylamino-PuRcMP were much less

TABLE 14. *Rate of cleavage of 8-substituted cAMP and cIMP derivatives with dog heart and rat adipose tissue PDE[a]*

VI

R	Y	% Hydrolysis relative to cAMP	
		Heart PDE ($[S]^b = 0.65$ mM)	Adipose PDE ($[S]^b = 0.3$ mM)
Br	NH_2	8–10	—
SH	NH_2	0.5–0.8	—
$HNCH_2C_6H_5$	NH_2	2–3	< 16
$HNCH_2C_6H_5(2\text{-}CH_3)$	NH_2	0.4–0.7	< 7
$HNCH_2C_6H_5(2\text{-}Cl)$	NH_2	0.3–0.8	—
$HNCH_2CH_2C_6H_5$	NH_2	0.1–0.3	
$HN(CH_2)_4CH_3$	NH_2	< 0.5	
(morpholine, tetramethyl)	NH_2	1.4	
Br	OH	2.6	< 10
$HNCH_2C_6H_5$	OH	0.2–0.5	0
(morpholine, tetramethyl)	OH	< 2.5	0
(piperidine, tetramethyl)	OH	< 3.5	—

[a] Michal et al., 1970b
[b] [S] = Concentration of analogue in assay

TABLE 15. *Effect of cAMP PDE's on some 6-substituted purine ribonucleoside 3',5'-monophosphates*[a]

IV

R	α[b]	
	Pig brain frontal cortex	Rabbit kidney
NH_2	1.0	1.0
OH	0.46	—
SH	1.04	1.06
OCH_3	0.57	0.53
SCH_3	0.52	0.41
SC_2H_5	—	0.57
$SCH_2C_6H_5$	—	0.54
HNC_2H_5	0.32	0.20
$N(C_2H_5)_2$	0.10	0.09

[a] Meyer et al., 1972b

[b] $\alpha = \dfrac{\text{rate of hydrolysis of test compounds}}{\text{rate of hydrolysis of cAMP}}$

resistant to cleavage by kidney and brain PDE than the corresponding 8-thio, 8-alkylthio, 8-alkoxy, and 8-alkylamino derivatives of cAMP (Muneyama et al., 1971) (Table 15).

(*i*) *Alterations in the carbohydrate or phosphate moiety of cAMP.* The synthesis and enzymatic cleavage of 3',5'-cyclic phosphate derivatives of the naturally occurring deoxyribonucleosides was first reported by Drummond et al. (1964), who demonstrated that 2'-deoxyadenosine 3',5'-cyclic monophosphate (cdAMP) was cleaved by a beef brain PDE at 50% the rate of cAMP and by dog heart at the same rate as cAMP. Michal et al. (1970a) reported that cdAMP was cleaved at 80% the rate of cAMP by rat adipose

TABLE 16. *Effect of dog heart and rat adipose tissue cAMP PDE on the hydrolysis of some 6-substituted purine ribonucleoside 3',5'-monophosphates*[a]

[Structure IV: 6-R-substituted purine ribonucleoside 3',5'-monophosphate]

R	α (% hydrolysis relative to cAMP)	
	Heart (PDE [S][b] = 0.65 mM)	Adipose Tissue (PDE [S][b] = 0.3 mM)
Cl	35	—
OCH$_3$	10–12	—
HNCOC$_3$H$_7$	1–3	—
HNCOC$_6$H$_5$	0.2–0.5	< 5
HNCH$_2$C$_6$H$_5$	6–8	—
HNCH$_2$C$_6$H$_5$(2-CH$_3$)	< 2	< 10
HNCH$_2$C$_6$H$_5$(2-Cl)	4–6	—
HNCH$_2$CH$_2$C$_6$H$_5$	0.6	—
HNCH(C$_6$H$_5$)CH$_3$	4	—
HN(CH$_2$)$_4$CH$_3$	4	< 14
HNCH=CHCH$_3$	4–5	—
morpholino (N-linked)	0.5–1.3	—
piperidino (N-linked)	4–4.6	—
N(CH$_3$)$_2$	6	—

[a] Michal et al., 1970*b*
[b] [S] = Concentration of analogue in assay

tissue PDE. Recent studies in our own laboratories using several derivatives of cAMP substituted at the 2' position have indicated that this position is critical to the hydrolysis of the analogue by PDE.

(ii) *2'-Derivatives of cAMP as substrates for cAMP-PDE.* Table 17 gives the results of a study comparing the susceptibility of the 2'-derivatives of cAMP to cAMP PDE's purified from rabbit lung, rabbit kidney, beef heart, and beef brain. These hydrolysis rates were determined by measuring the phosphate removed from the 5'-nucleotide product with bacterial alkaline phosphatase. Essentially the same results were obtained when the hydrolysis rates were determined by chromatographic separation and quantitation of the resultant 5'-nucleotide.

TABLE 17. 2'-Derivatives of cAMP as substrates for high K_m cAMP PDE's[a]

Compound	Relative rate of hydrolysis[b]			
	Beef heart	Beef brain	Rabbit lung	Rabbit kidney
cAMP	100	100	100	100
2'-O-Acetyl-cAMP	58	55	43	50
2'-O-Methyl-cAMP	49	39	20	30
2'-O-(2,4-Dinitrophenyl)-cAMP	17	6	12	11
2'-O-Butyryl-cAMP	81	71	22	37
N^6-Butyryl-cAMP	8	7	16	3
dbcAMP	7	3	1	3
Ara-cAMP	18	15	25	17
cdAMP	92	59	38	67

[a] Miller et al., 1973
[b] The rates of hydrolysis by each enzyme are expressed relative to that of cAMP. The actual rates of cAMP hydrolysis (nmoles 5'-nucleotide formed/min) were 12, 22, 12, and 30 by the heart, brain, lung, and kidney enzymes, respectively. The rates were determined by colorimetric determination of the phosphate produced by the action of bacterial alkaline phosphatase on the 5'-nucleotide product of PDE hydrolysis.

Most strikingly, N^6-butyryl-cAMP and dbcAMP were found to be very poor substrates for each of the four enzymes. In comparison, 2'-O-butyryl-cAMP, 2'-O-acetyl-cAMP, and 2'-O-methyl-cAMP, which do not contain an N^6-substitution, were quite good substrates for cAMP PDE, as was cdAMP. 2'-O-(2,4-Dinitrophenyl)-cAMP, a 2'-O ether containing a relatively large steric and electron-withdrawing substituent at the 2'-position, was a poor substrate. Ara-cAMP was also hydrolyzed at a relatively slow rate compared to cAMP. There were significant variations in tissue specificity; for example, while 2'-O-acetyl-cAMP and ara-cAMP were each hydrolyzed at essentially the same rate by all four enzymes (50 and 25%, respectively), 2'-O-butyryl-cAMP was split four times faster by the heart enzyme than by the lung enzyme (81% vs. 22%). Of the four cAMP PDE preparations investigated, the beef heart PDE appeared to be the least sensitive to modifications of cAMP in the 2'-position. With the exception of ara-cAMP, the heart enzyme preparation was able to hydrolyze each of the 2'-derivatives of cAMP at a faster rate than any of the three other enzyme preparations. By comparison, the lung enzyme preparation was the most sensitive to changes in the cAMP molecule at the 2'-position, and hydrolyzed most of the derivatives at a slower rate than did the other three enzyme preparations.

For those derivatives that were substrates, the nature of the products of PDE action was also investigated. The product in all cases migrated like a nucleoside monophosphate upon electrophoresis in phosphate buffer at pH 7.0. The product of the action of PDE on 2'-O-methyl-cAMP was shown to be exclusively 2'-O-methyladenosine 5'-phosphate by chromatographic comparison with an authentic sample of the material. With 2'-O-acetyl-cAMP as substrate, the product was 2'-O-acetyladenosine 5'-phosphate. This material was chromatographically distinct from 2'-O-acetyl-cAMP, cAMP and 5'-AMP, and migrated as a diionic species upon electrophoresis at pH 7.0. When treated with 5 N ammonium hydroxide for 1 hr at 50°C (to remove the 2'-O-acetyl group), the product was chromatographically identical to 5'-AMP. By the same criteria, 2'-O-butyryl-cAMP was shown to be hydrolyzed by PDE to 2'-O-butyryl-5'-AMP. Under the conditions used in the PDE incubation, there was no detectable deacylation of the 2'-O-acyl-cAMP derivatives to cAMP by any of the four enzyme preparations. The ether linkage of the 2'-O-methyl and 2'-O-(2,4-dinitrophenyl) substituents were also stable under the incubation conditions used.

Several other analogues have been reported with substitutions at the 2'-carbon of the ribose moiety. Besides ourselves (Miller et al., 1973), Jastorff and Freist (1972) reported on the synthesis of the 2'-O-(2,4-dinitrophenyl) ether of cAMP. They indicated that this derivative was split by beef heart PDE to a considerable extent, although there was no comparison with the rate at which cAMP was cleaved. The authors also reported that equimolar concentrations of this compound and cAMP caused a 50% inhibition in the rate of cAMP hydrolysis; however, the concentration of cAMP in the reaction mixture was not stated. Another 2'-O ether, the 2'-O-tetrahydropyranyl-cGMP (2'-O-THPcGMP), was synthesized by Straus and Fresco (1965) and reported to be cleaved by a rabbit brain PDE. Cyclic GMP was split at 16% the rate of cAMP in this study, and the 2'-O-THPcGMP was split at 1.7% the rate of cAMP. The N^2-benzoyl-cGMP was split about equally as well as cGMP, whereas the N^2-benzoyl-2'-O-THPcGMP was cleaved at the same rate as the 2'-O-THP derivative.

Perhaps one of the most critical structural features within the purine 3',5'-cyclic phosphate is, as would be expected, the 3',5'-cyclic phosphate ring itself. Drummond and Powell (1970) have demonstrated that two phosphonate analogues synthesized by Jones and co-workers (1970), the 3',5'-cyclic ester of 9-(5'-deoxy-5'-dehydroxyphosphinylmethyl-β-D-ribofuranosyl)adenine (Table 13, 9) and the 5'-cyclic ester of 9-(3'-deoxy-3'-dehydroxyphosphinyl-methyl-β-D-ribofuranosyl)adenine (Table 13, 10), were both relatively inactive toward PDE. As one might expect, based on the relative chemical stability of the C-P linkage, 9 was able to be cleaved to a slight extent (4%), whereas 10 was completely inactive. Finally, several phosphorothioate analogues of cAMP have been reported; one of these (Table 13, 13), was found to be completely stable to hydrolysis by brain PDE (Drummond and Powell, 1970) and beef heart PDE (Eckstein and Bär, 1969). However, the second sulfur analogue (12), in which the sulfur atom was not in the exocyclic position, was cleaved at a rate that was 30% that of cAMP (Shuman et al., 1973b).

In an interesting unpublished study, Holy and Fruhaufova (1972) observed that beef heart cAMP PDE does not split L-adenosine 3',5'-cyclic phosphate, whereas the rat kidney PDE will apparently cleave this derivative at a rate comparable to the natural D-isomer.

Several papers on the existence of cyclic nucleotide PDE's (from rat kidney, adipocytes, ox heart, and beef heart) specific for pyrimidine cyclic phosphates have appeared (Hardman and Sutherland, 1965; Klotz and Stock, 1971). A specificity appears to exist with regard to cUMP, since cCMP and thymidine 3',5'-cyclic monophosphate (cTMP) were poor substrates. Long et al. (1972) have recently reported on the synthesis of 1-(β-D-arabinofuranosyl)cytosine 3',5'-cyclic monophosphate (ara-cCMP), 1-(β-D-arabinofuranosyl)uracil 3',5'-cyclic monophosphate (ara-cUMP), and several of their 4-substituted derivatives. Under conditions in which cUMP was cleaved at a rate 17% that of cAMP by a rabbit kidney PDE, ara-cUMP was split at only 1% the rate, whereas ara-cCMP was cleaved 4% as fast as cAMP.

In summary, the studies described to date indicate the rather narrow specificity with regard to substitution of any functional group (SH, OH, SR, NR, NR_2OR), except amino, at position 8 of the purine nucleus. Although the presence of a substituent at N^6 drastically reduces the ability of the cAMP derivative to serve as a substrate (N^6-benzoyl, N^6-alkylamino), the substitution of either oxo or thio or alkoxy or alkylthio groupings at C-6 has relatively little effect on substrate activity.

To some extent, *substitutions* at the C-2' of the ribose moiety are less critical with regard to the ability of the resulting compounds to serve as

PDE substrates than changes in *configuration* about C-2' or C-3'. Presence of a sulfur atom in the exocyclic phosphate ring eliminates substrate activity, but the 5'-deoxy-5'-thio analogue of cAMP was a good substrate.

b. Analogues as Inhibitors of PDE's

Intracellular concentrations of cAMP can be increased by either the enhanced activity of the enzyme adenylate cyclase or the reduced activity of the enzyme cAMP PDE. Levels of yet another naturally occurring cyclic purine nucleotide (cGMP) may indirectly alter the intracellular concentration of cAMP by inhibiting its degradation (Rosen, 1970a,b). Rosen has further reported that cIMP was a potent inhibitor of frog erythrocyte PDE, whereas the 3',5'-cyclic nucleotides of pyrimidine nucleoside were inactive.

Harris, in some preliminary studies (Harris et al., 1971), reported that cIMP, 8-bromo-cAMP, and 8-methylthio-cAMP were potent inhibitors of cat heart PDE (I_{50} values $= 10^{-7}$ to 10^{-5}M), whereas dbcAMP and purine 2',3'-phosphates were poor inhibitors ($I_{50} = 10^{-4}$M). Muneyama et al. (1971) showed that 8-thio-cAMP was less effective than theophylline (I_{50} theophylline $= 2.7 \times 10^{-4}$M, I_{50} 8-thio-cAMP $= 4.5 \times 10^{-4}$M). The addition of an alkyl group to the 8-thio moiety increased the inhibitory activity about five to 10 times (I_{50} 8-methylthio $= 9.4 \times 10^{-5}$M, I_{50} 8-benzylthio $= 5 \times 10^{-5}$M). These studies were carried out at high cAMP concentrations (1×10^{-4}M), and most likely represent activity of these analogues against the low affinity cAMP PDE.

Harris et al. (1972) have carried out inhibition studies utilizing partially purified preparations of PDE from cat heart and rat brain. The concentration of cAMP in the assay mixtures used to determine I_{50} values was 1.6×10^{-7}M. Under these conditions, the high affinity (low K_m) cAMP PDE activity was probably being measured. Enzymes were prepared by homogenization of the appropriate tissue in 5 to 10 volumes of 0.05 M imidazole buffer (pH 7.5) containing 5 mM dithiothreitol. The homogenates were centrifuged at 39,000 $\times$ g, and ammonium sulfate was added to 50% saturation. The precipitates were collected and dialyzed against 20 volumes of buffer.

Harris (1972) was able to demonstrate that, with the exception of 8-thio-cAMP, derivatives of cAMP with sulfur in the 8 position were effective inhibitors of both cat heart and rat brain PDE. 8-Methylthio-cAMP had an I_{50} value of 3.9×10^{-5}M vs. cat heart PDE and 8-benzylthio-cAMP had a value of 2.4×10^{-5}M vs. cat heart PDE and 4.0×10^{-5}M vs. rat brain PDE. One other 8-substituted derivative, 8-amino-cAMP, was also a potent inhibitor of cat heart PDE, giving an I_{50} value of 2.3×10^{-5}M. It does not appear, however, that the ability of a compound to be utilized as a substrate is predictive of the inhibitory activity of the compound, since 8-benzylthio-cAMP was inactive as a substrate for the kidney enzyme, but was as good an inhibitor as the 8-amino-cAMP, which was a substrate. The results presented in Table 18 summarize the I_{50} values for the 8-substituted derivatives studied.

TABLE 18. *Inhibition of cAMP PDE activity by 8-substituted derivatives of cAMP*[a]

cAMP Derivatives[b]	I_{50}[c] (μM)	
	Cat heart PDE	Rat brain PDE
8-Thio-cAMP	330	540
8-Methylthio-cAMP	39	125
8-Ethylthio-cAMP	41	52
8-(2-Hydroxyethyl)thio-cAMP	44	200
8-Benzylthio-cAMP	24	40
8-Amino-cAMP	23	400
8-Azido-cAMP	90	150
8-Methylamino-cAMP	160	700
8-Dimethylamino-cAMP	3,300	2,300
8-(2-Hydroxyethyl)amino-cAMP	270	4,400
8-Hydroxy-cAMP	Inactive[d]	Inactive[d]
8-Methoxy-cAMP	130	600
8-Bromo-cAMP	16	67
dbcAMP	100	650

[a] Harris et al., 1972
[b] Assays were performed at a cAMP concentration of 1.6×10^{-7}M. The 8-substituted derivatives of cAMP were dissolved in 0.01 N NaOH.
[c] Concentration causing a 50% inhibition of enzymatic activity
[d] Inactive = $I_{50} > 10^{-2}$M

One of the most effective inhibitors of both the cat heart and rat brain enzymes was 8-bromo-cAMP, with I_{50} values of 16 μM for cat heart PDE and 67 μM for rat brain PDE. Harris et al. (1972) have shown that when a Dixon plot was constructed using analogues at two different substrate concentrations, competitive type kinetics were obtained for the PDE from cat heart with a K_i value of 20 μM.

Several 2'-substituted derivatives of cAMP have been studied for their relative ability to inhibit partially purified cAMP PDE's from rabbit lung and rabbit kidney, beef heart and beef brain. Table 19 gives a summary of this data. The four PDE's used all had low K_m values for cAMP of about 10^{-7}M, and the cAMP concentration used was 1.7×10^{-7}M. Whereas the I_{50} values for theophylline for all four enzymes were similar, the I_{50} values for the 2'-derivatives of cAMP showed distinct differences in tissue specificity. The beef heart and rabbit lung were found to be the most sensitive to inhibition by all the derivatives. As in the substrate studies, the conformation and presence of the 2'-hydroxyl is important for interaction of the cyclic nucleotide with the low K_m PDE enzymes as inhibitors.

TABLE 19. Inhibition of low K_m cAMP PDE's by 2'-derivatives of cAMP[a]

Compound	I_{50} (μM)[b]			
	Beef heart	Beef brain	Rabbit lung	Rabbit kidney
2'-O-Acetyl-cAMP	10	150	5.0	110
2'-O-Methyl-cAMP	8.7	230	4.8	41
2'-O-(2,4-Dinitrophenyl)-cAMP	26	57	25	89
2'-O-Butyryl-cAMP	5.6	23	5.6	36
N^6-Butyryl-cAMP	37	230	46	180
dbcAMP	37	770	230	320
Ara-cAMP	24	1200	190	800
cdAMP	21	210	11	180
Theophylline	130	230	250	160

[a] Miller et al., 1973
[b] I_{50} = the concentration of compound that produces 50% of maximal inhibition. The concentration of cAMP used was 1.7×10^{-7}M.

B. Crude Tissue Homogenates or Subcellular Fractions

1. *Glycogenolysis, Activation of Phosphorylase*

The utilization of either a 12,000 × g supernatant (S-12) prepared from rat liver homogenate or an S-30 fraction prepared from rabbit skeletal muscle has been described by Michal et al. (1970b) and DuPlooy et al. (1971) to study the ability of several base and sugar substituted analogues of cAMP to activate liver and muscle phosphorylase b kinase kinase (i.e., protein kinase). The principles of this assay system are described below:

(1) phosphorylase b kinase kinase (inactive) $\xrightarrow{\text{cAMP}/\text{Mg}}$ phosphorylase b kinase kinase (active)

(2) phosphorylase b kinase (inactive) $\xrightarrow{\text{active phosphorylase b kinase kinase}/\text{ATP}}$ phosphorylase b kinase (active)

(3) phosphorylase b $\xrightarrow[\text{ATP}]{\text{active phosphorylase b kinase}}$ phosphorylase a
(muscle or liver)

(4) glycogen $\xrightarrow[\text{PO}_4]{\text{phosphorylase a}}$ glucose-1-PO$_4$

(a) glucose-1-PO$_4$ $\xrightarrow{\text{phosphoglucomutase}}$ glucose-6-PO$_4$

(b) glucose-6-PO$_4$ $\xrightarrow[\text{dehydrogenase}]{\text{glucose-6-phosphate}}$ 6-phosphogluconic acid

NAD ⌢ NAD·H

In theory, the utilization of a crude S-12 fraction in the case of liver, or S-30 fraction in the case of muscle, provides the necessary components for each of these reactions, and the addition of a cyclophosphate derivative in place of cAMP at varying concentrations allows the calculation of a K_m value for the activation of phosphorylase b kinase kinase in reaction (1). In our opinion, this system suffers from the disadvantage that crude homogenates are utilized in a series of coupled biochemical reactions. This provides a less rigorous means of comparing the cAMP-like activity of these analogues than the use of a purified protein kinase as described in the studies of Muneyama and co-workers (1971) and Kuo and Greengard (1970).

The data presented in Table 20 list the K_m values (concentration of cyclophosphate giving half-maximal activation) for a series of 8-substituted cAMP and cIMP analogues as determined by DuPlooy et al. (1971). In order to compare the activity of each compound relative to cAMP, we have calculated an additional value (K_m') which is the ratio of the K_m of cAMP/K_m of analogue.

The results in Table 20 do not appear to agree in some instances with the results recorded elsewhere by the Boehringer Mannheim group (1971); in the latter study, e.g., 8-bromo-cAMP was found to have a K_m of 50 × 10^{-8}M compared to a K_m for cAMP of 7 × 10^{-8}M. This would indicate that 8-bromo-cAMP was less effective than cAMP in stimulating the activation of phosphorylase b kinase kinase. The value reported for the K_m' for 8-bromo-cAMP, based on DuPlooy's data (1971), of 1.5 (Table 20) is more nearly comparable to the K_a' of 2.9 obtained for this compound against a purified cAMP-dependent protein kinase (Muneyama et al., 1971 or Table 6 of this review). In general, the results reported by DuPlooy are consistent with those reported in Table 6, since 8-alkylamino derivatives of cAMP

TABLE 20. *Activation constants for several 8-substituted cAMP and cIMP derivatives for liver glycogen phosphorylase*

8-Substituted derivatives of cAMP	Liver system[a]	
	$K_m \times 10^8$	K_m'
cAMP	1.5	1.0
8-Bromo-cAMP	1.0	1.5
8-Allylamino-cAMP	1.0	1.5
8-Piperidino-cAMP	4.0	0.37
8-(2'-Chlorobenzylamino)-cAMP	6.0	0.25
8-Morpholino-cAMP	6.0	0.25
8-Benzylamino-cAMP	7.0	0.21
8-(2-Methylbenzylamino)-cAMP	8.0	0.17
8-(4-Methylbenzylamino)-cAMP	10.0	0.16
cIMP	20.0	0.075
8-Bromo-cIMP	15.0	0.10
8-Anilino-cIMP	20.0	0.075
8-Piperidino-cIMP	30.0	0.050
8-Cyclohexylamino-cIMP	50.0	0.030
8-(2-Chlorobenzylamino)-cIMP	60.0	0.025
8-Benzylamino-cIMP	80.0	0.017
8-(4-Methylbenzylamino)-cIMP	80.0	0.017

[a] Michal et al., 1970a

were less potent than cAMP itself using the purified cAMP-dependent protein kinase.

In two studies (Michal et al., 1970b; DuPlooy et al., 1971), several 6-substituted derivatives of PuRcMP were also examined. The data obtained using a liver S-12 fraction and a muscle S-30 fraction are presented in Table 21. In general, the 6-alkylamino-PuRcMP's were found to be more active than the corresponding 8-alkylamino-cAMP derivatives as activators of phosphorylase.

The only compound that could be directly compared as an activator of both the liver phosphorylase and purified bovine brain protein kinase was the 6-methoxy-cAMP derivative (Table 11). In both studies this analogue was 50% as effective as cAMP (Tables 11 and 21).

2. *Lipolysis, Activation of Lipases*

In a recent article, the Boehringer Mannheim group (1971) studied the ability of cyclophosphates to stimulate lipolysis using a cell-free system. Measurement of lipolysis was performed according to the method of Rizack (1964), using homogenized fat cells. In the Boehringer Mannheim article, the ratio of the concentration of cyclic nucleotide required to give half-maximal activation of phosphorylase was calculated. The authors pointed

TABLE 21. *Activation constants for several 6-substituted purine ribonucleoside 3',5'-monophosphates for liver and muscle glycogen phosphorylase*

IV

R	Liver[a]		Muscle[a]	
	$K_m \times 10^8$	K_m'	$K_m \times 10^8$	K_m'
NH_2	1.5	1.0	10.0	
dbcAMP	100.0	0.015		
Butyrylamino	2.5	0.60	20.0	0.5
Benzoylamino	6.0	0.25	10.0	1.0
2-Chlorobenzylamino	0.4	3.75		
1-Phenethylamino	0.8	1.87		
Benzylamino	1.0	1.5	3.0	3.33
4-Methylbenzylamino	1.0	1.5		
Piperidino	1.5	1.0		
Allylamino	2.0	0.75		
1-Pentylamino	2.0	0.75		
Methoxy	3.0	0.5		
Morpholino	3.0	0.5		
Chloro	4.0	0.375		
2-Methylbenzylamino	1.5	1.0	6.0	1.66

[a] Michal et al., 1970a

out that cAMP, for example, gave a ratio of about 500 ($4 \times 10^{-5}/7 \times 10^{-8}$M). We feel that although this is an interesting demonstration of tissue specificity, a more valid comparison would be obtained if the ratio of concentration of cyclic nucleotide giving half-maximal activation were compared in each case. In a detailed paper soon to be published, and presented last year at the First International Congress of cAMP held in Milan, Italy, July, 1971, Michal et al. (1972) determined the K_a values (concentration giving half-maximal activation of lipolysis) for a number of 6- and 6-8-substituted-PuRcMP derivatives of 5-amino-1-(β-D-ribofuranosyl)imidazole-4-carboxamide (AICAR). Using these K_a values, we can now compare the K_a

for lipolysis and the K_m for glycogenolysis. The results presented in Table 22 are for activation of a cell-free lipolysis system utilizing a series of 6- and 6,8-substituted-PuRcMP derivatives. K_a is as previously described. We have calculated the ratio of K_a for lipolysis/K_m for activation of liver phosphorylase kinase. This ratio utilizes the concentrations giving half-maximal activation for both lipolysis and glycogenolysis.

TABLE 22. Activation of lipolysis in a cell-free system by several 6- and 8-substituted derivatives of cAMP and cIMP[a]

Compound	K_a (M)	K_a Lipolysis/K_m phosphorylase
cAMP	1.5×10^{-5}	—
dcAMP	2×10^{-5}	—
cIMP	1.5×10^{-6}	7.5
cGMP	1.5×10^{-5}	—
cCMP	1.3×10^{-5}	—
cUMP	2.5×10^{-5}	—
Adenosine 3'-5'-c-phosphorothioate	6×10^{-6}	—
N^6-Benzoyl-cAMP	2.5×10^{-7}	4.1
2'-O-Butyryl-cIMP	5×10^{-7}	—
N^6, 2'-O-Dibutyryl-cAMP	1.2×10^{-6}	1.2
N^2, 2'-O-Dibutyryl-cGMP	1.1×10^{-6}	—
6-Thio-PuRcMP	2×10^{-6}	—
6-(1-Phenylethylamino)-PuRcMP	10^{-7}	12.3
6-Benzylamino-PuRcMP	4×10^{-8}	4
6-(4-methylbenzylamino)-PuRcMP	2×10^{-7}	20
6-Morpholino-PuRcMP	10^{-7}	3.3
8-Bromo-cAMP	2×10^{-6}	200
8-Benzylamino-cAMP	3×10^{-8}	0.43
8-(4-Methylbenzylamino)-cAMP	10^{-7}	1.0
8-Thio-cAMP	4×10^{-6}	—
8-Benzyloxy-cAMP	2×10^{-7}	—
8-Bromo-cIMP	5×10^{-8}	0.3
8-Benzylamino-cIMP	2×10^{-8}	0.025
8-(4-Methylbenzylamino)-cIMP	4×10^{-8}	0.050

[a] Taken in part from a paper presented at the 1st International Congress of cAMP in Milan, Italy, July, 1971, by G. Michal (with the courtesy of the authors).

A comment should be made regarding the relatively high degree of activity observed utilizing the 3',5'-cyclic phosphates of the naturally occurring nucleosides, as well as the sulfur analogue adenosine 3',5'-phosphorothioate, in which sulfur replaces oxygen in the exocyclic position of the phosphate ring. Cyclic UMP and cdAMP were found to be slightly less effective than cAMP. Because of these results, and the known poor activity of cUMP in stimulating protein kinase, one wonders whether the activation observed with this crude cell-free system is indicative of the specific biochemical activity of cAMP, namely, activation of a purified protein kinase, or whether it represents a process or processes other than activation of a protein kinase.

Regardless of the molecular mechanism that is operative, the results utilizing the cyclic phosphates described in Table 22 are interesting in that every derivative studied (except dcAMP and cUMP) was more active than cAMP itself (K_a analogue $<$ K_a cAMP). At the concentrations tested, however, none of these analogues (with the exception of dbcAMP) was capable of causing the same degree of maximal stimulation as cAMP. An extremely interesting observation was made by Michal (1972) in this series of experiments utilizing some derivatives of AICAR 3',5'-cyclic phosphate. The data indicated that AICAR 3',5'-cyclic phosphate was incapable of stimulating lipolysis in the cell-free system, but that at 5×10^{-6}M concentration, it inhibited lipolysis to an extent of 43%. AICAR 5'-phosphate was at least 10 times more potent than the 3',5'-cyclic phosphate as an inhibitor of lipolysis both *in vitro* and *in vivo*.

Finally, the authors raised a note of caution with regard to their cell-free lipase preparation, indicating that recent attempts to obtain a preparation capable of being activated to the same degree with dbcAMP have not been successful and that maximal activation ratios of 2.0 have recently been obtained, compared with a value of 4.5 observed in their earlier experiments.

C. Whole-Cell Suspensions or Tissue Slices

The utilization of a whole-cell or tissue-slice system, a biological system more closely related to the whole-animal system, offers some advantages over the crude homogenate or purified enzyme system. These systems give some insight into the transport of the cyclic nucleotide derivatives across the cell's plasma membrane. In our opinion, utilization of the crude homogenates discussed in the previous section offers little advantage, if any, over the purified cAMP-dependent protein kinase. Furthermore, the use of crude-cell homogenates, in contrast to intact-cell preparations, is less desirable since no insight into cyclic nucleotide membrane transport can be obtained with the former. For processes *known* to involve activation of protein kinase as a primary mechanism of action by cAMP, the purified enzyme system should be utilized. For those processes which cAMP does appear to regulate, but in which activation of a protein kinase may not be the primary site of action, a tissue-slice or whole-cell suspension undoubtedly proves more useful than the purified protein kinase *until* the molecular mechanism has been established.

1. *Glycogenolysis*

Michal and co-workers (1970*b*) and DuPlooy et al. (1971) have shown that a series of cAMP derivatives bearing mainly 8-alkylamino functions were capable of stimulating conversion of glycogen phosphorylase b to glycogen phosphorylase a in a liver homogenate. In a recent paper (Bauer et al., 1971), an attempt has been made to demonstrate that a series of 8-substituted cAMP analogues were capable of activating the glycogenolytic system in rat liver slices and, further, that the relative order of potency of these derivatives was the same for stimulation of glycogenolysis as it was for the relative ability of these analogues to activate a purified liver protein kinase. The data presented in Table 23 indicate that when the analogues

TABLE 23. *Activation of rat liver protein kinase and glycogenolysis in rat liver slices by 8-substituted derivatives of cAMP*[a]

R	Liver protein kinase ($K_a \times 10^8$)	Liver glycogenolysis ($A_{50} \times 10^6$)
H (cAMP)	2.54	100.0
SCH$_2$–C$_6$H$_5$	1.06	2.66
SH	1.21	1.0
OH	1.68	1.28
NH$_2$	4.87	4.60
NHCH$_2$CH$_2$OH	16.4	15.80
NHCH$_2$–C$_6$H$_5$	16.5	—
dbcAMP	—	5.0

[a] Bauer et al., 1971

were ranked in order of potency based on their A_{50} values (concentration of analogue giving half-maximal stimulation of glycogenolysis), there was, with the exception of cAMP itself, a near perfect correlation with their order of potency (as determined from K_a values) as protein kinase activators. Although the quantitative values of K_a and A_{50} were not the same, it is interesting to note that they consistently differed by a factor of about 100, and that 8-amino-cAMP was one-fourth as active as the 8-thio-cAMP in stimulating both liver protein kinase and liver glycogenolysis in the liver slice preparation. Similarly, 8-(2-hydroxyethylamino)cAMP was about one-fifteenth as active as the 8-thio derivative in both activation of protein kinase and stimulation of glycogenolysis (Bauer et al., 1971).

2. Lipolysis, Adrenal Steroidogenesis

The lipolytic effects of several 6- and 8-substituted analogues of cAMP on intact collagenase-treated lipocytes prepared by the method of Rodbell (1964) was reported by Michal et al. (1972). Cyclic AMP was found to cause half-maximal activation at a concentration of approximately 5×10^{-2}M, while dbcAMP had a K_a value of 1.6×10^{-3}M. The N^6-benzoyl-PuRcMP derivative was approximately 150 times more active than dbcAMP. The N^6-(4-methylbenzylamino)PuRcMP was 500 times more potent than dbcAMP.

Two 8-substituted purine derivatives, 8-piperidino-cAMP and 8-benzylamino-cIMP, were evaluated for their activity. The former compound was 80 times more potent and the latter 30 times more active than dbcAMP. Consistent with the results obtained in the cell-free system, both AICAR 3',5'-cyclic phosphate and AICAR 5'-phosphate were quite potent inhibitors of lipolysis using depression of epinephrine-stimulated lipolysis as a measure in whole-cell studies, and dbcAMP-stimulated lipolysis in the cell-free system.

In a relatively complete study, Free et al. (1971) examined 13 different 8-substituted derivatives of cAMP for their ability to activate both steroidogenesis in adrenal cells and lipolysis in the isolated rat epididymal lipocyte. Cyclic AMP, dbcAMP, and most of the 8-substituted derivatives were capable of activating the cells to the same extent that epinephrine or ACTH did using the lipocyte, or to the extent that ACTH activated steroidogenesis in the adrenal cell. Stimulatory potency of the derivatives was compared on the basis of the concentration required to give half-maximal activation of steroidogenesis or lipolysis (A_{50} values), and these are listed in Table 24. Most of the 13 8-substituted derivatives were many times (up to 50-fold) more potent than cAMP as stimulators of steroidogenesis and

TABLE 24. *Correlation of potency of 13 8-substituted cAMP analogues as activators of lipolysis and adrenal steroidogenesis with ability to activate a purified cAMP-dependent protein kinase*

	Y_1 Lipolysis[a]		X Protein kinase[b]		Y_2 Steroidogenesis[a]	
Analogue	Rank	Potency	Rank	Potency	Rank	Potency
SCH$_3$	1	(180)	4	(2.4)	1	(65)
Br	5	(440)	2	(2.92)	2.5	(85)
N$_3$	8.5	(1,440)	8	(0.68)	2.5	(85)
OH	4	(260)	3	(2.80)	4	(90)
SCH$_2$CH$_3$	2	(230)	5	(2.0)	5	(110)
N(CH$_3$)$_2$	11	(3,000)	9	(0.59)	6	(130)
SCH$_2$CH$_2$OH	6	(850)	10	(0.54)	8	(150)
NH$_2$	7	(1,200)	7	(1.50)	8	(150)
OCH$_3$	10	(1,600)	12	(0.29)	8	(150)
SCH$_2$C$_6$H$_5$	8.5	(1,400)	6	(1.70)	10	(190)
SH	3	(250)	1	(3.80)	11	(380)
NHCH$_3$	12	(3,300)	11	(0.38)	12	(460)
NHCH$_2$CH$_2$OH	13	(14,000)	13	(0.11)	13	(1,000)

Rank correlations:
Y_1 vs. X $Y_1 = 12.77 - 0.8242$ x; $R = -0.8253$; $N = 13$; $p < 0.001$
Y_2 vs. X $Y_2 = 10.35 - 0.4780$ x; $R = -0.4813$; $N = 13$; $p < 0.1$
Y_1 vs. Y_2 $Y = 2.060 + 0.5771$ x; $R = 0.5803$; $N = 13$; $p < 0.02$

[a] Free et al., 1971. The analogues were ranked on a basis of 1 to 13, from least to greatest degree of potency. The values listed under Potency are A_{50} values, (the mM concentration of analogue giving 50% stimulation of lipolysis or steroidogenesis).

[b] Muneyama et al., 1971. The protein kinase was purified from bovine brain. The analogues were ranked on a basis of 1 to 13, from greatest to least degree of potency. Potency is defined as the K_a'. The greater the K_a', the greater the potency.

lipolysis. Several of the derivatives were also more potent than dbcAMP. One group of compounds (8-oxo-, 8-methylthio-, 8-ethylthio-, and 8-bromo-cAMP) displayed relatively high potency in both lipocytes and adrenal cells.

Although these 8-substituted derivatives are by and large potent activators of the cAMP-dependent protein kinase, their activity as inhibitors of PDE does raise some question as to the mechanism by which they are exerting their action at the cellular level. Free et al. (1971) pointed to three main lines of evidence to suggest that their effects are directly on activation of protein kinase.

(1) All of the 13 8-substituted derivatives examined by Free, except one, demonstrated protein kinase activation over the range of 1 to 100 μM, similar to that seen with cAMP, and had K_a values that were similar to cAMP. The single exception, 8-(2-hydroxyethyl)amino-cAMP, showed less than 10% of the kinase activation shown by cAMP and was also the only derivative that showed less activity than cAMP in the adrenal cell and lipocyte assay.

(2) Three compounds that displayed little or no inhibitory activity toward PDE, 8-oxo-cAMP, 8-thio-cAMP, and 8-dimethylamino-cAMP (Table 18), were still potent activators of lipolysis and steroidogenesis.

(3) Free et al. (1971) observed that potent inhibitors of PDE, such as SQ 20009, that potentiate cAMP activity in adrenal and fat cells, have little or no activity on their own. The cyclic nucleotide derivatives produced maximal activation in the absence of exogenous cAMP; it is therefore unlikely that the cyclic nucleotide analogues are acting mainly as PDE inhibitors.

The fact that these derivatives are extremely resistant to PDE may in part account for their enhanced potency in these cell types. This, however, cannot be the sole reason for their potency, since one of these 8-substituted derivatives, 8-amino-cAMP, was nearly as good a substrate for PDE as cAMP, but was 20 times more active than cAMP in adrenal cells and five times more potent than cAMP in lipocytes. Since this compound was only about 1.5 times more potent than cAMP as an activator of protein kinase, part of the enhanced potency might result from enhanced penetrability into the cell. It is possible in only two cases, those of cAMP and dbcAMP, to compare the concentrations giving half-maximal activation for lipolysis in the studies described by Michal and co-workers (1972) with those of Free (1971). In both cases, Free obtained values lower than those obtained by Michal (8.5×10^{-3}M vs. 50×10^{-3}M for cAMP and 5×10^{-4}M vs. 20×10^{-4}M for dbcAMP). Both laboratories reported utilization of Rodbell's method (1964) for preparation of their whole-cell suspensions.

Finally, perhaps the most convincing demonstration that activation of lipolysis is correlated with activation of a protein kinase can be seen by comparison of their rank in order of potency as protein kinase activators with their rank in order of potency as lipolytic agents in fat cells. A correlation coefficient (r) of 0.825 was obtained, with a p value of $< .001$ (Table 24), thus indicating a highly significant degree of correlation. When similar calculations were done comparing the stimulation of adrenal cell corticosteroid production with protein kinase activation, an r value of 0.48 was obtained, with a p value of < 0.1.

3. Regulation of Tumor Cell Growth, Antiviral Activity, Enzyme Induction in Tissue Culture

Changes in the growth rate of tumor cells, correlation of growth rate of cells with levels of cAMP in the cells, changes in morphology of Chinese hamster cells by dbcAMP (Hsie and Puck, 1971), and restoration of morphological characteristics of normal fibroblasts to sarcoma cells treated with cAMP and dbcAMP (Johnson et al., 1971) have all recently been reported. Prasad (1972a) has recently reported on the effects of dbcAMP and several agents (prostaglandins and PDE inhibitors) capable of raising levels of endogenous cAMP on the growth of a neural tumor (mouse neuroblastoma) cell in tissue culture.

In a recent paper, Prasad (1972a) reported on the effect of several substances including dbcAMP, 8-benzylthio-cAMP, prostaglandin E_1 (PGE_1), and X-ray on the morphological differentiation in two neuroblastoma clones. One clone, the NBDB$^-$, was found to be insensitive to dbcAMP, RO 20-1724 [4-(3'-butoxy-4'-methoxybenzyl)imidazolidin-2-one], and 8-benzylthio-cAMP; however, the NBA2 clone was quite sensitive to PGE_1. The 8-benzylthio-cAMP was capable of producing 66% differentiated cells. The same concentration (1 mM) of dbcAMP produced only 42% differentiated cells.

In a second series of experiments (Prasad, 1972b), 8-benzylthio-cAMP was shown to produce an effect as early as 24 hr after treatment of the cells in culture, with a maximal effect seen 3 days after treatment. When the drug was removed from the medium and fresh growth medium added 1 day after treatment, the drug-induced differentiation as measured 5 days later was reversible. If the drug was removed 3 days after treatment, the number of differentiated cells as measured 2 days later was $60.0 \pm 7.0\%$ compared to $67.0 \pm 5.0\%$, and the effects were still reversible. If the drug was left in the culture for 5 days continuously, the effect was not reversible. These results are consistent with those obtained with previous studies using dbcAMP, RO 20-1724 and PGE_1 (Prasad, 1972a) in the same cell lines.

The author felt that 8-benzylthio-cAMP had some advantages over dbcAMP in that it was readily soluble in aqueous solution (pH 8.3) and was relatively less toxic, more stable, and more potent in causing morphological differentiation than dbcAMP. Several other analogues [8-thio-, 8-amino-, 8-(2-hydroxyethylthio)-, and 8-methylamino-cAMP] were relatively incapable of causing an increase in the number of differentiating cells; however, 8-thio-cAMP caused a three-fold increase in the percent of differentiated cells in treated cultures vs. untreated cultures (31 vs. 10%). Each of these compounds, with the exceptions of 8-benzylthio-cAMP and 8-(2-hydroxyethylthio)-cAMP, were quite lethal to the cells.

In some recent studies, van Wijk et al. (1972) have shown that 8-methylthio-cAMP was extremely effective in causing large process formation in a glial-cell tumor line, a response very similar to that seen with dbcAMP. A concentration of 3×10^{-5}M was capable of 60% inhibition of 7-day growth compared to a 20% inhibition of 7-day growth by dbcAMP at the same concentration. 8-Amino-cAMP produced similar effects on the growth of glial cell tumors and neuroblastoma tumors. Although this compound was not capable of causing process formation, a concentration as low as 10^{-6}M was 40% inhibitory to cell growth. The effects of several of the analogues on growth of hepatoma cells in culture (H35) and upon induction of tyrosine aminotransferase (TAT) and phosphoenol pyruvate (PEP) carboxykinase activity were also examined. The data shown in Table 25 indicate $8\text{-SCH}_3\text{-} > 8\text{-Br-} > 8\text{-SC}_2\text{H}_5\text{-} > 8\text{-SH-} > 8\text{-SCH}_2\text{C}_6\text{H}_5\text{-} > 8\text{-SCH}_2\text{CH}_2\text{OH-} > 8\text{-NHCH}_2\text{C}_6\text{H}_5\text{-cAMP}$ as inducers of TAT. A rank correlation analysis of this data gave a correlation coefficient of $r = 0.76$ with a p value of < 0.05 when TAT-inducing activity was compared with protein kinase activation. The anti-tumor effects of these derivatives were in all probability due to their action as cyclic nucleotide analogues, rather than as a metabolic product, since nearly all of the compounds were resistant to several different PDE's (Muneyama et al., 1971).

The utilization of cyclic nucleotide analogues of anti-tumor nucleoside derivatives was studied by Thomas and Montgomery (1968). They reported on the synthesis of the 3′,5′-cyclic phosphate of 9-(β-D-ribofuranosyl)-purin-6-thione (6-MPR) and its cytotoxicity against HEp-2/0 and HEp-2/MP [a cell line resistant to 6-MP because of its lack of the hypoxanthine-guanine phosphoribosyl pyrophosphate (HX-G PRPP) transferase necessary to convert 6-MP to the active 5′-nucleotide derivative]. The 3′,5′-cyclic phosphate of 6-MPR was nearly as active as 6-MP against the nonresistant cell line, but was inactive against the resistant cell line. Thomas suggested that the lack of activity of 6-MPR 3′,5′-cyclic phosphate in their cell line was due to extracellular cleavage to the 5′-phosphate.

TABLE 25. *Correlation of potency of eight 8-substituted cAMP analogues as activators of liver hepatoma tyrosine transaminase[a] and stimulation of protein kinase[b]*

R	Y_1[a]	X[b]
SCH_3	1 (206)	3 (2.40)
SH	4 (153)	1 (3.80)
NH_2	8 (19)	6 (1.50)
Br	2 (157)	2 (2.92)
SC_2H_5	3 (156)	4 (2.0)
$SCH_2C_6H_5$	5 (120)	5 (1.70)
SCH_2CH_2OH	6 (100)	7 (0.54)
$NHCH_2C_6H_5$	7 (74)	8 (0.11)

$Y_1 = 1.071 + 0.762\ X$
$r = 0.762; N = 8$
$p < 0.05$

[a] van Wijk et al., 1972. The compounds were ranked on a scale of 1 to 8, from highest to least potency. Potency is defined as the % stimulation of TAT induction at a single concentration of test compound.
[b] Muneyama et al., 1971. Protein kinase purified from bovine brain. Order of potency and rank are as described in Table 24.

Meyer et al. (1972b) have recently demonstrated that cAMP PDE isolated from animal tissue was capable of cleaving the 6-MPR 3′,5′-cyclic phosphate at a rate even greater than that of cAMP.

LePage and Hersh (1972) have recently evaluated the cyclic phosphate of this same derivative, as well as the cyclic phosphates of methylmercaptopurine riboside (6-methylthio-PuRcMP) and ara-cAMP, against: (a) L1210R, a lymphoid leukemia line resistant to 6-MP because of its lack of the HX-G PRPP transferase; (b) an Ehrlich carcinoma line, Ehrlich TGRII, lacking the HX-G PRPP; and (c) an Ehrlich carcinoma line, Ehrlich/MMPR, lacking adenosine kinase and therefore resistant to 6-methylthio-PuRcMP, *in vivo*.

Ara-cAMP apparently was able to penetrate L1210R cells, since it produced significant inhibition of the L1210R tumor at the highest concentration tested. LePage and Hersh (1972) indicated that the tests against Ehrlich TGRII were "compromised by inhibition of the immune response to this tumor (not a compatible host)." The 3′,5′-cyclic phosphate of 6-MPR was inhibitory to the Ehrlich/MMPR line, whereas the methylmercaptopurine riboside-treated "controls" were immunosuppressed and outgrew controls.

Biochemical conversions of 9-(β-D-arabinofuranosyl)adenine (ara-A) and ara-cAMP were studied using intact L1210R cells. Over a period of 30 min, ara-A was almost completely converted to 9-(β-D-arabinofuranosyl)-hypoxanthine (ara-H) by the intact L1210R cells. Ara-cAMP was unchanged under the conditions of the incubation. When similar studies were performed utilizing homogenates of L1210 cells, ara-A was converted to ara-H (65% by 15 min, and 95% by 30 min). With ara-cAMP as a substrate, the homogenate gave 85% conversion to ara-A 5′-monophosphate in 30 min. All three cyclic nucleotides were able to penetrate intact growing cells sufficiently to produce significant anti-tumor effects. In the case of nucleoside analogues that must first be converted to the active nucleotide form by nucleoside kinases, there may be special significance to the use of these types of derivatives because they would bypass the need for cells to be in a growth cycle with high nucleoside kinase levels, and such analogues might be resistant to deactivation (e.g., deamination) after having penetrated the cell intact.

Using similar reasoning, Kreis and Wechter (1972) reported on the utilization of ara-cCMP as an inhibitor of P815 and P815/ara-C cell growth in culture. Ara-cCMP and N^6-octanoyl-ara-cCMP were three and 10 times less effective than 1-(β-D-arabinofuranosyl)cytidine (ara-C) against both sensitive and resistant cell lines. In some *in vivo* studies, they were able to identify the di- and triphosphates of 1-(β-D-arabinofuranosyl)uracil (ara-U) in animals injected with the N^6-octanoyl derivative along with ara-C, ara-U, and 1-(β-D-arabinofuranosyl)uracil 5′-phosphate (ara-UMP).

Long and co-workers (1972) have reported on the details of a synthetic procedure for ara-cCMP and its *in vitro* and *in vivo* antiviral activity against several DNA viruses. In each instance, ara-cCMP was at least as effective as the corresponding nucleoside derivative.

Sidwell and co-workers (1972a), in a study reported at the VIIth International Congress of Chemotherapy in Prague, 1971, showed that cAMP, 8-methylthio-cAMP, 8-bromo-cAMP, and cAMP-N^1-oxide were moderately active against both DNA and RNA viruses in KB cells. Pretreatment of the cells or treatment early in the infection appeared to increase the activity of 8-methylthio-cAMP, suggesting inhibition of some key early intracellular mechanism. No *in vivo* anti-herpes simplex virus (HSV) activity could be seen in hamster or rabbit keratitis tests, but definite activity was demonstrated when the compounds were topically applied to HSV-induced lesions in mouse tails. The authors suspect that the type of virus-infected cell may be a key factor in this antiviral activity, since limited reports by others and preliminary experiments in their own laboratory using other cell lines demonstrated a wide variation in the degree of antiviral activity of cAMP and its derivatives.

It should be remembered that at least one of these derivatives that was found to possess antiviral activity (8-methylthio-cAMP) was also shown by van Wijk et al. (1972) to be an extremely potent inhibitor of tumor cell growth. Sidwell (1972b) has also demonstrated that several 6-thio- and 6-alkylthio-PuRcMp derivatives exhibited antiviral activity. Significant inhibition of HSV-induced cytopathogenic effect (CPE) was seen using 6-thio, 6-methylthio, 6-benzylthio, and 6-hydroxylamino derivatives of PuRcMP. The 6-thio derivative also strongly inhibited an RNA virus (parainfluenza). Each of these derivatives was also 100 to 1,000 times more toxic to tumor cell lines than to normal cells.

Finally, the 3′,5′-cyclic phosphate of a recently reported broad spectrum antiviral agent, Virazole ICN 1229 (1-β-D-ribofuranosyl-1,2,4-triazole-3-carboxamide), was also shown to inhibit replication of both DNA and RNA viruses in tissue culture, although the spectrum of activity and relative potency was not as great as that of Virazole. Virazole 3′,5′ cyclic phosphate has been shown to be easily cleaved to its 5′-phosphate by kidney and brain PDE (Shuman et al., 1972a,b).

D. Whole Tissue, Organ Perfusion, Tissue Fragment

1. *Anterior Pituitary Hormone Release*

It was previously demonstrated (Bowers et al., 1968) that cAMP could stimulate the release of certain pituitary hormones. In 1962 and 1968, Posternak and his colleagues reported on the synthesis of analogues of cAMP with acyl moieties at the N^6 and $2'$-O positions (Posternak et al., 1962; Cehovic et al., 1968). Within the past several years, these same authors have synthesized and studied the biological activity of several N^6-alkylamino and C-8 analogs of cAMP. In recent articles, Posternak and Cehovic reviewed their own studies relating to the synthesis (Posternak, 1971) and effects of these derivatives (Posternak and Cehovic, 1971), as well as 6-amino-3-(β-D-ribofuranosyl)purine $3',5'$-cyclic phosphate (iso-cAMP) on the release of growth hormone (Posternak et al., 1971), thyroid-stimulating hormone (TSH) (Cehovic et al., 1968), and prolactin (Cehovic et al., 1970) by rat pituitary glands (Posternak and Cehovic, 1971). Of those compounds studied, the $N^6,2'O$-dibutyryl derivative of 8-thio-cAMP was the most active in stimulating growth hormone release, whereas 8-bromo- and 8-thio-cAMP were the next most active. The following list represents the order of decreasing potency of the 8- and 6-substituted analogues of cAMP on growth hormone release *in vitro* (Posternak and Cehovic, 1971).

I. $N^6,2'$-O-Dibutyryl-8-thio-cAMP
II. N^6-n-Butyl-cAMP
   8-Bromo-cAMP
   8-Thio-cAMP
   dbcAMP
III. 8-Hydroxy-cAMP
   8-Thiomethyl-cAMP
   N^6-Dimethyl-cAMP
IV. N^6-Monomethyl-cAMP
   8-Amino-cAMP
   2-Amino-cAMP
V. N^6-t-Butyl-cAMP
   cAMP

Cehovic has reported that none of the derivatives cited above enhance the release of prolactin with the exception of $N^6,2'$-O-dibutyryl-8-thio-cAMP (Posternak and Cehovic, 1971). Iso-cAMP, which was extremely potent in enhancing TSH release, was relatively inactive on growth hormone release. The compounds evaluated exhibited the following order of potency as stimulators of TSH release: iso-cAMP > 6-N-butyl-cAMP > 6-N-dimethyl-cAMP > 6-N-monomethyl-cAMP > N^6-t-butyl-cAMP. Effects of 8-substituted derivatives of cAMP on TSH release have not been reported; however, from the fact that iso-cAMP stimulates TSH release while having little effect on growth hormone release, it may be possible to develop agents with discrete specificity of action.

2. Mediator Release from Sensitized Lung Fragments

The elegant studies on the regulation of mediator release (histamine, SRSA) from both human (Orange et al., 1971a,b) and rat lung fragments (Ishizaka et al., 1971) sensitized to IgE have offered a unique opportunity to study the regulation of these potent mediators of the allergic phenomenon by cyclic nucleotides.

Effects of dbcAMP and theophylline on release of SRSA and histamine from human lung fragments sensitized to IgE was reported (Orange et al., 1971a,b). It was demonstrated that theophylline caused an increase in cAMP levels and a concomitant decrease in SRSA and histamine release. Further, dbcAMP was effective in inhibiting release of these mediators in IgE-stimulated cells. Initial studies indicated that cAMP levels dropped to below baseline values in sensitized fragments challenged with IgE. Addition of isoproteronol or epinephrine at time of challenge caused a pronounced stimulation of cAMP levels in the fragments and also prevented mediator release.

Kaliner et al. (1972b) postulated that cholinergic mechanisms were involved in enhancement of mediator release in these same sensitized lung fragments and that atropine, an anticholinergic drug, was capable of blocking acetylcholine-enhanced mediator release in these segments. They further postulated that cGMP might regulate cholinergic effects on mediator release. 8-Bromo-cGMP has been previously demonstrated to be twice as potent as cGMP as an activator of cGMP-dependent lobster muscle protein kinase (Table 9). The results obtained by Kaliner and his colleagues demonstrate that low concentrations (10^{-5} and 10^{-6}M) of 8-bromo-cGMP enhanced release of SRSA by 140 and 25% in one experiment and at 10^{-6}M it enhanced SRSA release by 200% in a second experiment. 8-Bromo-cGMP stimulated histamine release by 20 to 40% at 10^{-6} and 10^{-5}M. These studies represent a classic demonstration on the use of analogues, in this case, cyclic nucleotide analogues, to help elucidate the biological significance of a naturally occurring nucleotide (cGMP) in the important physiological process of cholinergic transmission in the human lung. Studies such as these illustrate the rational approach behind the search for novel therapeutic agents.

3. Arterial or Tracheal Chain Smooth Muscle Relaxation

Cyclic AMP, along with at least two of the enzymes (cAMP PDE and adenyl cyclase) regulating its tissue levels, may be implicated in the regulation of smooth muscle function. Compounds that inhibit PDE (e.g., theophylline), or stimulate adenyl cyclase (e.g., epinephrine) allow an accumulation of cAMP and also cause relaxation of smooth muscle. Here, both the biochemical and functional events are dose dependent and quantitatively correlated (Triner et al., 1971), although this may be an instance where the molecular mechanism of cAMP may not involve effects on protein kinase.

Perhaps one of the more satisfying criteria that cAMP is involved in smooth muscle relaxation would be a direct demonstration that cAMP or several of its derivatives could cause relaxation of smooth muscle. It was found that in addition to increasing levels of cAMP within the cell, application of dbcAMP was accompanied by relaxation of arterial smooth muscle (Triner et al., 1971). Rubin et al. (1971) presented findings that analogues of cAMP, capable of mimicking cAMP's function in other cell systems, could cause relaxation of several different smooth muscle preparations. They have studied the action of a series of 12 8-substituted analogues on the contraction of guinea pig trachea and rat portal vein smooth muscle. Their studies indicated that 8-benzylthio-cAMP was nearly 10 times more active than dbcAMP in terms of the IC_{50} value (concentration producing 50% relaxation of tracheal chain or decrease in amplitude of portal vein). Several other derivatives (8-methylthio-, 8-thio-, 8-ethylthio-, 8-methoxy-, and 8-bromo-cAMP) were among those more effective than dbcAMP against the tracheal chain preparation. Although these results would *a priori* suggest that a cAMP analogue was capable of producing a cAMP-like biological effect, and in the case in point, caused relaxation of smooth muscle *directly,* an alternate explanation is possible, since 8-benzylthio-cAMP was one of the more potent inhibitors of PDE isolated from cat heart, rat brain, rabbit lung, rabbit kidney, and bovine heart. If inhibition of PDE were the main mechanism by which the analogues exerted their effect, however, one would have expected that 8-bromo-cAMP and 8-methylthio-cAMP (nearly as potent PDE inhibitors as the 8-benzylthio derivative vs. cat heart and rat brain PDE) might have been equally as potent as 8-benzylthio-cAMP against the smooth muscle preparation. They were not. The relaxant activities described by Rubin of these several analogues *in vitro* indicated that 8-thio derivatives were > 8-oxo > 8-amino and 8-methylamino derivatives. A similar order of activity was found by Muneyama et al. (1971) in assessing them as activators of purified bovine brain protein kinase. Until these analogues are evaluated as activators of a smooth muscle protein kinase or smooth muscle PDE and assayed for transport into smooth muscle cells, it will be difficult to attempt a correlation of either of these activities with their effects on smooth muscle function.

4. Glycogenolysis in Perfused Liver

As might be expected, the literature on the use of analogues of cAMP in a perfused organ system or in whole animal systems *in vivo* is at this stage of development quite sketchy and confined almost exclusively to use in laboratory animals or perfused organs therefrom.

The glycogenolytic activity of cAMP, as well as dbcAMP and N^6-monobutyryl-cAMP was reported in perfused rat livers by Levine and Washington (1970). Henion et al. (1967) reported that N^6-monobutyryl-cAMP, dbcAMP, and N^6-monooctanoyl-cAMP were all more effective in raising blood glucose in anesthetized dogs than the 2'-O-butyryl and 2'-O-monooctanoyl derivatives. Cyclic GMP, cTMP, cUMP, cCMP, cIMP, 2'-O-butyryl-cIMP, and TucMP were infused in equimolar dosages (7.8 mmoles/hr) using either whole rat blood or Ringer's solution (BLD) for 4 hr, or Krebs-Ringer bicarbonate (KRB) for 50 min. The increases in blood glucose were rapid (all significantly increased at 30 min) for TucMP, cIMP, and 2'-O-butyryl-cIMP. Maximal changes occurred within the first 2 hr (Levine and Washington, 1970). Maximal hyperglycemia was greater than that observed for cAMP but less than that observed for dbcAMP by the same authors (Levine and Washington, 1970). Using a KRB perfusion media, the rank in order of potency for the compounds was dbcAMP > cIMP > TucMP. There were, however, no attempts made to determine dose-response relationships, nor to measure a minimum glycogenolytic dose. Relative potencies were based only on the maximum glycogenolytic response (glucose release) at a given (7.38 mM) dose.

VI. BIOLOGICAL EFFECTS *IN VIVO*

Hanze (1968) reported that TucMP was absorbed by human red blood cells to an extent of 40% that of tubercidin, whereas tubercidin was absorbed to an extent of 100% of the administered dose in the same cells. Tubercidin 5'-phosphate was also absorbed to an extent of 100%. He has also studied the LD_{50} of TucMP and tubercidin in mice and reported a value of 18 to 40 mg/kg for both compounds.

An N^6-substituted derivative of cAMP has recently been reported to have some interesting central effects. 6-Piperidino-9-(β-D-ribofuranosyl)-purine 3',5'-cyclic phosphate (HD 233) has been shown by Vargiu and Spano (1971) to prolong the sleeping time induced by sodium hexobarbitol from 25 to 123 min when administered i.p. to mice. The levels of hexobarbitol were reported to be doubled in the brain after HD 233 administration, whereas dbcAMP had no effect. No other details were given. Henion reported as early as 1967 (Henion et al., 1967) that cAMP and dbcAMP given to mice at 50 mg/kg produced slight drowsiness within 10 min.

Several of the 8-substituted derivatives were examined by Braun (1972) for their effect in augmenting synthesis of antibody to sheep red blood cells in CFW mice assayed 48 hr after immunization. The number of plaque-forming cells per 10^8 spleen cells was determined. Cyclic AMP or dbcAMP were shown to double or triple the number of antibody-forming cells when used in concentrations of 200 to 500 µg per mouse given by i.p. route (Ishizaka et al., 1971). Of all the analogues studied (8-thio-, 8-oxo-, 8-amino-, and 8-benzylamino-cAMP), only 8-amino-cAMP at 400 µg i.p. per mouse and 8-benzylamino-cAMP at 100 µg i.p. per mouse enhanced the number of antibody-forming cells by 60 and 25%, respectively (Braun, 1972).

Michal et al. (1972) studied the effects of i.p. administration of several 6- and 6,8-substituted cAMP analogues on blood glucose levels, blood steroid levels, and further examined the effects of these and AICAR cyclic nucleotide derivatives on lipolysis *in vivo*. AICAR 5'-monophosphate caused a decrease in free fatty acid production from 0.45 to 0.24 meq/hr when 100 mg/kg was given by i.v. administration. This drop was statistically significant ($p < 0.01$). Administration of AICAR 3',5'-cyclic phosphate caused a drop from 0.4 to 0.3 meq/hr, which was not statistically significant.

The administration of dbcAMP and N^6,2'-*O*-dibutyryl-cGMP caused a 0.79 and 1.1 mg/ml increase in blood glucose at 20 and 80 mg/kg dose i.p. of each compound. Of the derivatives tested, 6-benzylamino-PuRcMP caused a similar degree of stimulation at 80 mg/kg as dbcAMP, whereas the 6-(4'-methylbenzylamino)PuRcMP produced a 1.56 and 1.92 mg/ml increase at 20 and 80 mg/kg, respectively. This effect was nearly twice that produced by dbcAMP. The same compound was the most active of the derivatives studied for increasing levels of blood steroids at 20 mg/kg; however, at 80 mg/kg, no further increase was observed. The effect of the cyclic nucleotide analogues on alteration of blood steroid levels appeared less reproducible than effects on blood glucose (Michal et al., 1972).

To our knowledge, none of the synthetic analogues discussed in this chapter have been administered to human volunteers with the exception of cAMP, cGMP, and dbcAMP. Levine and co-workers reported that rapid i.v. administration of cAMP in doses of 4 to 10 mg/kg at a rate of 0.3 to 0.5 mg/kg/min induced cardiac acceleration, anti-diuresis, and increases in plasma glucose, insulin, and corticol and growth hormone concentrations. These responses were often accompanied by headache and abdominal pain (Levine, 1968a,b, 1969, 1970; Levine et al., 1968).

In a separate study, it was demonstrated that dbcAMP produced a greater physiological response than cAMP at similar doses; plasma glucose increased from 48 to 150 mg%, heart rate from 30 to 52 beats/min, plasma insulin levels from 31 to 38 μU/ml, and plasma growth hormone levels from 9 to 12 ng/ml. Incidence of side effects such as testicular pain, fatigue, and headache also decreased or disappeared dramatically in dbcAMP- vs. cAMP-treated patients (Levine, 1970).

Finally, kinetic parameters and renal clearances of plasma cAMP and cGMP were evaluated in normal subjects using ^{3}H-labeled cyclic nucleotide. The tracer was administered by a single rapid i.v. injection and by constant i.v. infusion. Approximately 85% of the elimination of the cyclic nucleotides from the circulation was due to extrarenal clearance (Broadus et al., 1970). It is interesting that cyclic nucleotides are the only nucleotides which have been detected in urine. Virtually all the cGMP in human urine appears to be filtered from plasma; no nephrogenous cGMP was identified. Urinary cAMP was derived from both plasma and kidney, and under appropriate hormonal stimulation the contribution from either source may be altered (Broadus et al., 1970).

The synthesis of derivatives of the biologically important second messenger, cAMP, appears to have been productive from the standpoint of learning more about the molecular mechanisms whereby cAMP exerts its effects. What then are the therapeutic areas and what are the pathophysiological conditions in which cAMP or its more biologically active analogs could participate?

Clearly, any condition in which lower levels of cAMP appear to produce a pathological condition could lend itself to treatment with cAMP or its analogues. In a recent series of articles (Voorhees et al., 1971a,b, 1972a; Voorhees and Duell, 1971), it has been suggested that the incomplete differentiation, accelerated proliferation, and glycogen accumulation in psoriatic lesions could be due to deficient amounts of cAMP. Voorhees and co-workers (1972b) have found a statistically significant decrease in cAMP in portions of skin from involved psoriatic lesions compared to that of samples of skin from uninvolved areas of the patient as well as from control subjects. The suggestion that topical application of cAMP analogues might be used therapeutically either to normalize differentiation or to prevent increased proliferation from occurring appears to be well founded and subject to experimental proof.

The demonstration by Orange and Kaliner (Orange et al., 1971a,b; Kaliner et al., 1971) that enhanced levels of cAMP in lung fragments prepared from human lungs sensitized to IgE prevent release of SRSA and histamine gives a strong impetus for the possible evaluation of cAMP analogues in the treatment of bronchial asthma or other atopic allergies.

The literature abounds with suggestions, both indirect and direct, that cAMP plays a role in the growth and differentiation of tumor cell lines. Indeed, Prasad (1972a,b) has shown that several analogues of cAMP were capable of preventing growth of tumor cells and causing morphological alteration into normal fibroblasts. The area of antiviral chemotherapy also holds some promise for these new chemotherapeutic agents (Sidwell et al., 1972a,b).

Finally, pharmacological actions related to the many effects of cAMP on such diverse organs as the pituitary and adrenal glands, heart, liver, lung, brain, GI tract, and smooth muscle suggest great therapeutic potential for specific synthetic cyclic nucleotide derivatives. One such novel molecular biological approach has been described by Weinryb et al. (1972) who have recently reported that several different analogues of cAMP were capable of inhibiting rat lung adenylate cyclase. The discovery of analogues of cAMP that were unable to substitute for cAMP in its many biological reactions and yet could regulate adenylate cyclase might prove invaluable as a novel type of β-adrenergic blocking agent, acting at a site beyond hormone-membrane interaction.

VII. ACKNOWLEDGMENTS

We wish to thank Drs. Jon P. Miller and Rich B. Meyer, Jr. for their many helpful and stimulating comments during the preparation and proofreading of this manuscript. We thank Dr. Merrill Camien for his statistical analysis of some of the data.

VIII. LIST OF ABBREVIATIONS USED

1. AICAR = 5-amino-1-(β-D-ribofuranosyl)imidazole-4-carboxamide
2. 5'-AMP = adenosine 5'-monophosphate
3. ara-A = 9-(β-D-arabinofuranosyl)adenine
4. ara-C = 1-(β-D-arabinofuranosyl)cytidine
5. ara-cAMP = 6-amino-9-(β-D-arabinofuranosyl)purine 3',5'-cyclic monophosphate
6. ara-cCMP = 1-(β-D-arabinofuranosyl)cytosine 3',5'-cyclic monophosphate
7. ara-cUMP = 1-(β-D-arabinofuranosyl)uracil 3',5'-cyclic monophosphate
8. ara-H = 9-(β-D-arabinofuranosyl)hypoxanthine
9. ara-U = 1-(β-D-arabinofuranosyl)uracil
10. ara-UMP = 1-(β-D-arabinofuranosyl)uracil 5'-phosphate
11. ATP = adenosine 5'-triphosphate
12. cAMP = adenosine 3',5'-cyclic monophosphate
13. cCMP = cytidine 3',5'-cyclic monophosphate
14. cdAMP = 2'-deoxyadenosine 3',5'-cyclic monophosphate

15. cdTMP = deoxythymidine 3′,5′-cyclic monophosphate
16. cGMP = guanosine 3′,5′-cyclic monophosphate
17. cIMP = inosine 3′,5′-cyclic monophosphate
18. CPE = cytopathogenic effect
19. CRP = cAMP receptor protein
20. cTMP = thymidine 3′,5′-cyclic monophosphate
21. cUMP = uridine 3′,5′-cyclic monophosphate
22. dbcAMP = N^6,2′-O-dibutyryladenosine 3′,5′-cyclic monophosphate
23. HSV = herpes simplex virus
24. HX-G PRPP = hypoxanthine-guanine phosphoribosyl pyrophosphate
25. IgE = immunoglobulin E
26. iso-cAMP = 6-amino-3-(β-D-ribofuranosyl)purine 3′,5′-cyclic phosphate
27. KRB = Krebs-Ringer Bicarbonate
28. 6-MP = 6-mercaptopurine
29. 8-methoxy-cAMP = 8-methoxyadenosine 3′,5′-cyclic monophosphate
30. 6-methylthio-PuRcMP = 6-methylthiopurine-9-(β-D-ribofuranosyl) 3′,5′-cyclic monophosphate
31. 6-MMP = 6 methylthiopurine
32. 6-MPR = 9-(β-D-ribofuranosyl)purine-6-thione
33. 2′-O-THPcGMP = 2′-O-tetrahydropyranyl-cGMP
34. PDE = phosphodiesterase
35. PEP = phosphoenol pyruvate
36. PGE = prostaglandin E
37. PO_4 = inorganic phosphate
38. PuRcMP = purine-9-(β-D-ribofuranosyl) 3′,5′-cyclic monophosphate
39. SAR = structure-activity relationship
40. SRSA = slow reacting substance of anaphylaxis
41. TAT = tyrosine aminotransferase
42. TSH = thyroid-stimulating hormone
43. TucMP = tubercidin 3′,5′-cyclic monophosphate

IX. REFERENCES

Akerfeldt, S. (1962): Further studies on S-substituted phosphorothioic acids II. Synthesis and certain properties of some potential antiradiation drugs. *Acta Chemica Scandinavica*, 16:1897–1907.

Allred, J. B., and Roehrig, K. L. (1972): Inhibition of hepatic lipogenesis by cyclic 3′,5′-nucleotide monophosphates. *Biochemical and Biophysical Research Communications*, 46:1135–1139.

Amer, M. S., and Browder, H. P. (1971): Effect of quazodine on phosphodiesterase. *Proceedings of the Society for Experimental Biology and Medicine*, 136:750–752.

Anderson, W. B., Perlman, R. L., and Pastan, I. (1972): Effect of adenosine 3′,5′-monophosphate analogs on the activity of the cyclic adenosine 3′,5′-monophosphate receptor in *Escherichia coli*. *Journal of Biological Chemistry*, 247:2717–2722.

Assem, E. S. K., and Schild, H. O. (1969): Inhibition by sympathomimetic amines of histamine release induced by antigen in passively sensitized human lung. *Nature*, 224:1028–1029.

Averner, M. J., Brock, M. L., and Jost, J. P. (1972): Stimulation of ribonucleic acid synthesis

in horse lymphocytes by exogenous cyclic adenosine 3′,5′-monophosphate. *Journal of Biochemistry,* 247:413–417.
Baddiley, J., Buchanan, J. G., and Szabo, L. (1954): Sugar phosphates. Part I. Derivatives of glucose 4:6-(hydrogen phosphate). *Journal of the Chemical Society,* 3826–3832.
Baddiley, J., and Thain, E. M. (1953): Coenzyme A. Part VII. Pantetheine 2′ and 2′,4′-phosphates and a new method for the synthesis of cyclic phosphates. *Journal of the Chemical Society,* 903–906.
Bailly, O. (1921a): Sur l'action de l'epichlorhydrine sur le phosphate monoacide de sodium en solution aqueuse et sur la stabilité d'un diether monoglyceromonophosphorique. *Bulletin de la Société Chimique de France,* 29:274–280.
Bailly, O. (1921b): Action du brome sur les allylphosphates en solution aqueuse: Passage de la acide monoallylphosphorique au diether $\alpha\gamma$ monoglyceromonophosphorique. *Bulletin de la Société Chimique de France,* 29:280–283.
Bailly, O. (1922): Sur l'action de l'epichlorhydrine sur le phosphate neutre de sodium en solution aqueuse et sur la stabilité d'un diether diglyceromonophosphorique. *Bulletin de la Société Chimique de France,* 31:848–862.
Bald, R. W., and Holý, A. (1971): Nucleic acid components and their analogues CXLII. Preparation of 3-(β-D-ribofuranosyl)uracil 2′,3′-cyclic phosphate and related compounds and their behaviour towards pancreatic ribonuclease and ribonuclease T2. *Collection of Czechoslovak Chemical Communications,* 36:3657–3669.
Barker, G. R., and Foll, G. E. (1957): Biosynthesis of polynucleotides. Part II. The synthesis and properties of phosphoryl derivatives of adenine glucoside. *Journal of the Chemical Society,* 3794–3798.
Bartell, L. S., Su, L. S., and Yow, H. (1970): Lengths of phosphorus-oxygen and sulfur-oxygen bonds: An extended Hückel molecular orbital examination of Cruickshank's d-p picture. *Inorganic Chemistry,* 9:1903–1912.
Bauer, R. J., Swiatek, K. R., Robins, R. K., and Simon, L. N. (1971): Adenosine 3′,5′-cyclic monophosphate derivatives. II. Biological activity of some 8-substituted analogs. *Biophysical and Biochemical Research Communications,* 45:526–531.
Beer, B., Chasin, M., Clody, D., Vogel, J. R., and Horovitz, Z. P. (1972): Cyclic adenosine monophosphate phosphodiesterase in brain: Effect on anxiety. *Science,* 176:428–430.
Berger, N. A., and Eichhorn, G. L. (1971): Interaction of metal ions with polynucleotides and related compounds. XIV. Nuclear magnetic resonance studies of the binding of copper (II) to adenine nucleotides. *Biochemistry,* 10:1847–1857.
Bergmeyer, H. U., Michal, G., Nelboeck-Hochstetter, M., Stork, H., and Weimann, G., Boehringer Mannheim GMBH (1971). Imidazole-ribosyl-cyclophosphates. West German Patent No. 2,026,040.
Blackburn, G. M., Brown, M. J., Harris, M. R., and Shire, D. (1969): Synthetic studies of nucleic acids on polymer supports. Part II. Mechanisms of phosphorylation with carbodiimides and arenesulphonyl chlorides. *Journal of the Chemical Society* (C), 676–683.
Blecher, M., Ro'Ane, J. T., and Flynn, P. D. (1970): Metabolism of dibutyryl cyclic adenosine 3′,5′-monophosphate during its regulation of lipolysis and glucose oxidation in isolated rat epididymal adipocytes. *Journal of Biological Chemistry,* 245:1867–1870.
Bloxham, D. P. (1971): Studies on the control of cholesterol biosynthesis: The adenosine 3′,5′-cyclic monophosphate-dependent accumulation of a steroid carboxylic acid. *Biochemical Journal,* 123:275–278.
Boehringer Mannheim GmbH (1970a): Cyclofosfaten en werkwijze voor hun bereiding. Nederland Patent No. 6,913,671.
Boehringer Mannheim GmbH (1970b): Werkwijze voor het bereiden van de op de 8-planta gemodificocorde purineribofuranoside 3′,5′-cyclofosfaten. Nederland Patent No. 7,003,222.
Boehringer Mannheim GmbH (1971): Purineribofuranoside 3′,5′-cyclophosphates. British Patent No. 1,257,546.
Borden, R. K., and Smith, M. (1966a): Nucleotide synthesis. III. Preparation of nucleoside 3′,5′-cyclic phosphates in strong base. *Journal of Organic Chemistry,* 31:3247–3253.
Borden, R. K., and Smith, M. (1966b): Nucleotide synthesis. II. Nucleotide *p*-nitrophenyl and 2,4-dinitrophenyl esters. *Journal of Organic Chemistry,* 31:3241–3246.

Bourne, H. R., Lichtenstein, L. M., and Melmon, K. L. (1972): Pharmacologic control of allergic histamine release *in vitro:* Evidence for an inhibitory role of 3′,5′-adenosine monophosphate in human leukocytes. *Journal of Immunology,* 108:695–705.

Bowers, C. Y., Robison, G. A., Lee, K. L., Verster, F. D. B., and Schally, A. V. (1968): Program of the Annual Meeting of the Thyroid Society.

Boyd, D. B. (1969): Mechanism of hydrolysis of cyclic phosphate esters. *Journal of the American Chemical Society,* 91:1200–1205.

Braun, W. (1972): *personal communication.*

Broadus, A. E., Kaminsky, N. I., Hardman, J. G., Sutherland, E. W., and Liddle, G. W. (1970): Kinetic parameters and renal clearances of plasma adenosine 3′,5′-monophosphate and guanosine 3′,5′-monophosphate in man. *Journal of Clinical Investigation,* 49:2222–2236.

Bromilow, R. H., and Kirby, A. J. (1972): Intramolecular general acid catalysis of phosphate monoester hydrolysis. The hydrolysis of salicyl phosphate. *Journal of the Chemical Society Perkin II,* 149–155.

Brown, D. M., McGrath, D. I., and Todd, A. R. (1952): Nucleotides. Part XII. The preparation of cyclic 2′:3′-phosphates of adenosine, cytidine and uridine. *Journal of the Chemical Society,* 2708–2714.

Brown, D. M., and Higson, H. M. (1957): Phospholipids. Part I. The hydrolysis of some esters of cyclohexanediol phosphates. *Journal of the Chemical Society,* 2034–2041.

Bruice, T. C., and Benkovic, S. (1966): Phosphate esters. Intramolecular nucleophilic catalysis. In: *Bioorganic Mechanisms, Vol. 2,* pp. 35–48. W. A. Benjamin, Inc., New York.

Brunswick, D. J., and Cooperman, B. S. (1971): Photo-affinity labels for adenosine 3′,5′-cyclic monophosphate. *Proceedings of the National Academy of Sciences,* 68:1801–1804.

Bunton, C. A. (1970): Hydrolysis of orthophosphate esters. *Accounts of Chemical Research,* 3:257–265.

Cawley, T. N., and Letters, R. (1971): Phosphate migration in some phosphate monoesters and diesters of methyl-α-D-mannopyranoside. *Carbohydrate Research,* 19:373–382.

Cehovic, G., Marcus, I. Vengadabady, S., and Posternak, T. (1968): Sur la préparation de l'acide iso-adénosine 3′,5′-phosphorique (iso-AMP cyclicque) et sur certaines de ses propriétés biologiques. *Comptes Rendus des Séances de la Société de Physique et d'Histoire Naturelle de Gèneve,* 3:135–139.

Cehovic, G., Marcus, I., Gabbai, A., and Posternak, T. (1970): Étude de l'action de certaines nouveaux analogues de l'AMP cyclique sur la liberation de l'hormone de croissance et de la prolactine *in vitro. Comptes Rendus des Séances de l'Académie des Sciences (D),* 271:1399–1401.

Chang, Y. Y. (1968): Cyclic 3′,5′-adenosine monophosphate phosphodiesterase produced by the slime mold *Dictyostelium discoideum. Science,* 161:57–59.

Chládek, S., and Nagyvary, J. (1972): Nucleophilic reaction of some nucleoside phosphorothioates. *Journal of the American Chemical Society,* 94:2079–2085.

Collin, R. L. (1966): The electronic structure of phosphate esters. *Journal of the American Chemical Society,* 88:3281–3287.

Collin, R. L. (1969): Hückel and extended Hückel calculations on bonding in phosphate diesters. *Annals of the N.Y. Academy of Sciences,* 158:50–64.

Cook, W. H., Lipkin, D., and Markham, R. (1957): The formation of a cyclic dianhydrodiadenylic acid (I) by the alkaline degradation of adenosine 5′-triphosphoric acid (II). *Journal of the American Chemical Society,* 79:3607–3608.

Coulter, C. L. (1968): Cyclic uridine 3′,5′-phosphate: Molecular structure. *Science,* 159:888–889.

Coulter, C. L., and Greaves, M. L. (1970): Cyclic cytidine 2′,3′-phosphate: Molecular structure. *Science,* 169:1097–1098.

Cox, Jr., J. R., and Ramsay, O. B. (1964): Mechanisms of nucleophilic substitution in phosphate esters. *Chemical Reviews,* 64:317–352.

D'Armiento, M., Johnson, G. S., and Pastan, I. (1972): Regulation of adenosine 3′,5′-cyclic monophosphate phosphodiesterase activity in fibroblasts by intracellular concentrations of cyclic adenosine monophosphate. *Proceedings of the National Academy of Sciences,* 69:459–462.

Dekker, C. A., and Khorana, H. G. (1954): Carbodiimides. VI. The reaction of dicyclohexylcarbodiimide with yeast adenylic acid. A new method for the preparation of monoesters of ribonucleoside 2'- and 3'-phosphates. *Journal of the American Chemical Society,* 76: 3522–3527.
Drummond, G. I., Gilgan, M. W., Reiner, E. J., and Smith, M. (1964): Deoxyribonucleoside 3',5'-cyclic phosphates. Synthesis and acid-catalyzed and enzymic hydrolysis. *Journal of the American Chemical Society,* 86:1626–1630.
Drummond, G. I., and Powell, C. A. (1970): Analogues of adenosine 3',5'-cyclic phosphate as activators of phosphorylase-b-kinase and as substrates for cyclic 3',5'-nucleotide phosphodiesterase. *Molecular Pharmacology,* 6:24–30.
Drummond, G. I., and Severson, D. L. (1971): Biological actions of cyclic AMP analogs. In: *Annual Reports in Medicinal Chemistry,* edited by C. K. Cain. Academic Press, New York.
DuPlooy, M., Michal, G., Weiman, G., Nelböck, M., and Paoletti, R. (1971): Cyclophosphates. I. Effect of various cyclophosphates on phosphorylase-b-kinase activation. *Biochimica et Biophysica Acta,* 230:30–39.
Eckstein, F., and Bär, H. P. (1969): Enzymatic hydrolysis of adenosine 3',5'-cyclic phosphorothioate. *Biochimica et Biophysica Acta,* 191:316–321.
Eckstein, F. (1970): Nucleoside phosphorothioates. *Journal of the American Chemical Society,* 92:4718–4723.
Falbriard, J. G., Posternak, T., and Sutherland, E. W. (1967): Preparation of derivatives of adenosine 3',5'-phosphate. *Biochimica et Biophysica Acta,* 148:99–105.
Fischer, E. (1914): Über phosphorsäureester des methylglucosides und theophyllin-glucosides. *Chemische Berichte,* 47:3193–3205.
Forrest, H. S., and Todd, A. R. (1950): Nucleotides. Part V. Riboflavin 5'-phosphate. *Journal of the Chemical Society,* 3295–3299.
Free, C. A., Chasin, M., Paik, V. S., and Hess, S. M. (1971): Steroidogenic and lipolytic activities of 8-substituted derivatives of cyclic 3',5'-adenosine monophosphate. *Biochemistry,* 10: 3785–3789.
Friedrich, V. W. (1963): Zur kenntnis der phosphorsäureester der benzimidazolriboside. *Zeitschrift für Naturforschung,* 18b:455–462.
Fujimoto, Y., and Naruse, M. (1968): Nucleoside 3',5'-cyclic phosphates. Japanese Patent No. 6,816,988.
Furth, J. J., and Cohen, S. S. (1967): Inhibition of mammalian DNA polymerase by the 5'-triphosphate of 9-β-D-arabinofuranosyladenine. *Cancer Research,* 27:1528–1533.
Garbers, D. L., Lust, W. D., First, N. L., and Lardy, H. A. (1971): Effects of phosphodiesterase inhibitors and cyclic nucleotides on sperm respiration and motility. *Biochemistry,* 10:1825–1831.
Gerster, J. F., and Robins, R. K. (1966): Purine nucleosides XIII. The synthesis of 2-fluoro- and 2-chloroinosine and certain derived purine nucleosides. *Journal of Organic Chemistry,* 31:3258–3262.
Gibson, D. W., Beer, M., and Barrnett, R. J. (1971): Gold (III) complexes of adenine nucleotides. *Biochemistry,* 10:3669–3679.
Goodman, D. B. P., Rasmussen, H., DiBella, F., and Guthro, E. C., Jr. (1970): Cyclic adenosine 3',5'-monophosphate-stimulated phosphorylation of isolated neurotubule subunits. *Proceedings of the National Academy of Sciences,* 67:652–659.
Greengard, P. (1971): On the reactivity and mechanism of action of cyclic nucleotides. *Annals of the N.Y. Academy of Sciences,* 185:18.
Greengard, P. and Costa, E., editors (1970): *Advances in Biochemical Psychopharmacology, Vol. 3, Role of Cyclic AMP in Cell Function.* Raven Press, New York.
Greengard, P., Paoletti, R., and Robison, G. A. (1972): *Advances in Cyclic Nucleotide Research, Vol. 1. Physiology and Pharmacology of Cyclic AMP.* Raven Press, New York.
Grollman, A., and Grollman, E. F. (1970): *Pharmacology and Therapeutics, 7th Edition.* Lea and Febiger, Philadelphia.
Haga, K., Kainosho, M., and Yoshikawa, M. (1971): Studies of phosphorylation. V. The synthesis of inosine 5'-thiophosphates. *Bulletin of the Chemical Society of Japan,* 44:460–463.

Hampton, A., Brow, L. W., and Bayer, M. (1969): Analogs of inosine 5'-phosphate with phosphorus-nitrogen and phosphorus-sulfur bonds. Binding and kinetic studies with inosine 5'-phosphate dehyrogenase. *Biochemistry,* 8:2303–2311.

Hanze, A. R. (1968): Nucleic acids. V. Nucleotide derivatives of tubercidin (7-deazaadenosine). *Biochemistry* 7:923–939.

Hardman, J. G., Robison, G. A., and Sutherland, E. W. (1971): Cyclic nucleotides. In: *Annual Reviews of Physiology,* edited by V. E. Hall. Annual Reviews, Inc., Palo Alto.

Hardman, J. G., and Sutherland, E. W. (1965): A cyclic 3',5'-nucleotide phosphodiesterase from heart with specificity for uridine 3',5'-phosphate. *Journal of Biological Chemistry,* 240:3704–3705.

Harris, D. N., Chasin, M., Phillips, M. B., Goldenberg, H., Samaniego, S., and Hess, S. M. (1972): Effect of cyclic nucleotides on activity of cyclic 3',5'-adenosine monophosphate phosphodiesterases. *Biochemical Pharmacology* (in press).

Harris, D. N., Phillips, M. B., and Goldenberg, H. J. (1971): Interaction of cyclic nucleotides with cyclic nucleotide phosphodiesterases of the cat heart. *Federation Proceedings,* 30:219.

Henion, W. F., Sutherland, E. W., and Posternak, T. (1967): Effects of derivatives of adenosine 3',5'-phosphate on liver slices and intact animals. *Biochimica et Biophysica Acta,* 148:106–113.

Hevesi, L., Wolfson-Davidson, E., Nagy, J. B., and Bruylants, A. (1972): Contribution to the mechanism of the acid-catalyzed hydrolysis of purine nucleosides. *Journal of the American Chemical Society,* 94:4715–4720.

Hirata, M., and Hayaishi, O. (1966): Enzymic formation of deoxyadenosine 3',5'-phosphate. *Biochemical and Biophysical Research Communications,* 24:360–364.

Ho, D. H. W. (1971): Metabolism of 6-methylthiopurine ribonucleoside 5'-phosphate. *Biochemical Pharmacology,* 20:3538–3539.

Holý, A. (1968): Oligonucleotidic compounds XXIV. Synthesis of 2',3'-cyclic phosphates of inosine, xanthosine, 9-(β-D-ribofuranosyl)-6-mercaptopurine, and 9-(β-D-ribofuranosyl)-2-amino-6-mercaptopurine. *Collection of Czechoslovak Chemical Communications,* 33:2245–2256.

Holý, A. (1969): Nucleic acid components and their analogues CXXVIII. Phosphorylation of some thymine 1-pentopyranosyl and 1-hexopyranosyl derivatives by the reaction with triethyl phosphite and hexachloroacetone. *Collection of Czechoslovak Chemical Communications,* 34:3510–3522.

Holý, A. (1970): Nucleic acid components and their analogues CXXX. Preparation of nucleotide derivatives of 1'-homouridine and their behaviour towards some nucleolytic enzymes. *Collection of Czechoslovak Chemical Communications,* 35:81–88.

Holý, A., and Bald, R. (1971): Nucleic acid components and their analogues CXXXVIII. Synthesis of 2',3'-cyclic phosphates derived from some pyrimidine ribonucleosides and their behaviour towards pancreatic ribonuclease and ribonuclease T2. *Collection of Czechoslovak Chemical Communications,* 36:2809–2823.

Holý, A., Chládek, S., and Žemlička, J. (1969): Oligonucleotidic compounds XXIX. Reactions of ribonucleoside 2'(3')-phosphates with dimethylformamide acetals. *Collection of Czechoslovak Chemical Communications,* 34:253–271.

Holý, A., and Smrt, J. (1966): Oligonucleotidic compounds XI. Synthesis of ribonucleoside 2',3'-cyclophosphates from nucleosides via nucleoside 2',3'-phosphites. *Collection of Czechoslovak Chemical Communications,* 31:1528–1534.

Holý, A., Smrt, J., and Šorm, F. (1965): Nucleic acid components and their analogues LXXI. Oxidation of nucleoside 5'-phosphites on treatment with halo acid derivatives and hexchloroacetone. *Collection of Czechoslovak Chemical Communications,* 30:3309–3319.

Holý, A., and Šorm, F. (1966a): Nucleic acid components and their analogues LXXXI. A selective synthesis of ribonucleoside 2'(3')-phosphites. *Collection of Czechoslovak Chemical Communications,* 31:1562–1568.

Holý, A., and Šorm, F. (1966b): Nucleic acid components and their analogues LXXX. Preparation of nucleoside phosphites by reaction of nucleosides with triphenyl phosphite. *Collection of Czechoslovak Chemical Communications,* 31:1544–1561.

Holý, A., and Sörm, F. (1969): Oligonucleotidic compounds XXXII. Phosphorylation of 1-lyxofuranosyl, 1-xylofuranosyl and 1-arabinofuranosyl derivatives of uracil and thymine with triethyl phosphite and hexachloroacetone. *Collection of Czechoslovak Chemical Communications*, 34:1929–1953.

Honjo, M., Furukawa, Y., Yoshioka, Y., Imada, A., Fujii, S., Ootsu, K., Kimura, T., Komeda, T., and Matsumoto, T. (1968): Synthesis of 6-mercaptopurine 2'-deoxyribonucleoside and its related compounds, and their biological activities. *Annals of the Reports of the Takeda Research Laboratory*, 27:1–19.

Hsie, A. W., and Puck, T. T. (1971): Morphological transformation of Chinese hamster cells by dibutyryl adenosine cyclic 3',5'-monophosphate and testosterone. *Proceedings of the National Academy of Sciences*, 68:358–361.

Hubert-Habart, M., and Goodman, L. (1969): The direct formation of a 3',5'-cyclic mononucleotide from an adenine nucleoside. *Journal of the Chemical Society*, 740–741.

Hutchinson, D. W. (1970): Phosphates and phosphonates of biochemical interest. In: *Organophosphorus Chemistry*, Vol. 1, edited by S. Trippett, pp. 141–175. John Wright and Sons Ltd., Bristol.

Hutchinson, D. W. (1971): Phosphates and phosphonates of biochemical interest. In: *Organophosphorus Chemistry*, Vol. 2, edited by S. Trippett, pp. 119–155. John Wright and Sons Ltd., Bristol.

Huttunen, J. K., Steinberg, D., and Mayer, S. E. (1970): ATP-Dependent and cyclic AMP-dependent activation of rat adipose tissue lipase by protein kinase from rabbit skeletal muscle. *Proceedings of the National Academy of Sciences*, 67:290–295.

Ikehara, M., and Ohtsuka, E. (1963a): Studies of nucleosides and nucleotides. XXI. A new synthesis of thymidine 5'-triphosphate and the use of P^1,P^2-di-(2-cyanoethyl)pyrophosphate in the nucleoside triphosphate synthesis. *Chemical and Pharmaceutical Bulletin*, 11:1358–1363.

Ikehara, M., and Ohtsuka, E. (1963b): Analogs. XVI. Synthesis of 9-D-erythrityladenine and its phosphates. *Chemical and Pharmaceutical Bulletin*, 11:1095–1101.

Ikehara, M., and Ohtsuka, E. (1963c): Studies on coenzyme analogs. XIX. Further investigations of phosphorylation using morpholinophosphorodichloridate and P^1-diphenyl P^2-morpholino pyrophosphorochloridate. *Chemical and Pharmaceutical Bulletin*, 11:1353–1358.

Ikehara, M., and Tzawa, I. (1966): Studies of nucleosides and nucleotides. XXIX. Direct synthesis of nucleoside 2',3'-cyclic phosphates. *Journal of Organic Chemistry*, 31:819–821.

Ikehara, M., and Uesugi, S. (1969): Studies of nucleosides and nucleotides. XXXVIII. Synthesis of 8-bromoadenosine nucleotides. *Chemical and Pharmaceutical Bulletin*, 17:348–354.

Imai, K., Fujii, S., Takanohashi, K., Furukawa, Y., Masuda, T., and Honjo, M. (1969): Studies on phosphorylation. IV. Selective phosphorylation of the primary hydroxyl group in nucleosides. *Journal of Organic Chemistry*, 34:1547–1550.

Ishizaka, T., Ishizaka, K., Orange, R. D., and Austen, K. F. (1971): Pharmacologic inhibition of the antigen-induced release of histamine and slow reacting substance of anaphylaxis (SRSA) from monkey lung tissues mediated by human IgE. *Journal of Immunology*, 106:1267–1273.

Jacob, T. M., and Khorana, H. G. (1964): Studies on polynucleotides. XXX. A comparative study of reagents for the synthesis of the C_3'-C_5' internucleotidic linkage. *Journal of the American Chemical Society*, 86:1630–1635.

Jardetzky, C. D. (1962): Proton magnetic resonance of nucleotides IV. Ribose conformation. *Journal of the American Chemical Society*, 84:62–66.

Jastorff, B., and Freist, W. (1972): New analogs of adenosine-3',5'-cyclophosphate. *Angewandte Chemie*, 11:713.

Jastorff, B., and Krebs, T. (1972): Analoge des adenosine-(3',5')-cyclophosphats mit stickstoff und schwefelatomen im phosphatring. *Chemische Berichte*, 105:3192–3202.

Johnson, G. S., Friedman, R. M., and Pastan, I. (1971): Restoration of several morphological characteristics of normal fibroblasts in sarcoma cells treated with adenosine 3',5'-cyclic

monophosphate and its derivatives. *Proceedings of the National Academy of Sciences,* 68:425–429.
Jones, G. H., Albrecht, H. P., Damadaran, N. P., and Moffatt, J. G. (1970): Synthesis of isosteric phosphonate analogs of some biologically important phosphodiesters. *Journal of the American Chemical Society,* 92:5510–5511.
Jones, G. H., and Moffatt, J. G. (1969): 3'-Cyclic esters of 5'-deoxy-5'-(dihydroxyphosphinylmenthyl)-nucleosides. United States Patent No. 3,446,793.
Kaiser, E. T., and Lee, T. W. S. (1971): Structure and enzymatic reactivity of an aromatic five-membered cyclic phosphate diester. Biological implications. *Journal of the American Chemical Society,* 93:2351–2353.
Kaliner, M. A., La Raia, P. J., Orange, R. P., and Austen, K. F. (1972a): Cyclic 3',5'-adenosine monophosphate and modulation of the immunologic release of histamine and slow reacting substance of anaphylaxis in human lung. In: *Advances in Cyclic Nucleotide Research, Vol. 1: The Physiology and Pharmacology of Cyclic AMP,* edited by P. Greengard, R. Paoletti, and G. A. Robison, p. 577 (abstract). Raven Press, New York.
Kaliner, M., Orange, R. P., and Austen, K. F. (1972b): Immunological release of histamine and slow reacting substance of anaphylaxis from human lung. IV. Enhancement by cholinergic and α-adrenergic stimulation. *Journal of Experimental Medicine,* 136:556–567.
Kanai, T., and Ichino, M. (1971): Some phosphate esters of cyclocytidine and aracytidine. *Tetrahedron Letters,* 1965–1968.
Khan, S. A., and Kirby, A. J. (1970): The reactivity of phosphate esters. Multiple structure-reactivity correlation for the reactions of triesters with nucleophiles. *Journal of the Chemical Society (B),* 1172–1182.
Khan, S. A., Kirby, A. J., Wakselman, M., Horning, D. P., and Lawlor, J. M. (1970): Intramolecular catalysis of phosphate diester hydrolysis. Nucleophilic catalysis by the neighbouring carboxy-group of the hydrolysis of aryl 2-carboxyphenyl phosphates. *Journal of the Chemical Society (B),* 1182–1187.
Khorana, H. G. (1961): Cyclic phosphate formation and its role in the chemistry of phosphate esters of biological interest. In: *Some Recent Developments in the Chemistry of Phosphate Esters of Biological Interest.* John Wiley and Sons, Inc., New York.
Khwaja, T. A., Harris, R., Bauer, R. J., Simon, L. N., and Robins, R. K. (1973): Synthesis and biological activity of derivatives of 9-β-D-arabinofuranosylpurine 3',5'-cyclic phosphate (*unpublished results*).
Khwaja, T. A., Harris, R., and Robins, R. K. (1972b): Synthesis of 9-β-D-arabinofuranosylpurine 3',5'-cyclic phosphates from adenosine 3',5'-cyclic phosphate. *Tetrahedron Letters,* 4681–4684.
Khwaja, T. A., and Robins, R. K. (1972): *personal communication.*
Kikkoman Shoyu Company (1971a): 3',5'-Cyclic adenylic acid production. United States Patent No. 3,630,842.
Kikkoman Shoyu Company (1971b): 3',5'-Cyclic deoxyadenylic acid. Japanese Patent No. 7,142,958.
Kirby, A. J., and Younas, M. (1970a): The reactivity of phosphate esters. Reactions of diesters with nucleophiles. *Journal of the Chemical Society (B),* 1165–1171.
Kirby, A. J., and Younas, M. (1970b): Base catalysis of nucleophilic aromatic substitution. The reactions of primary and secondary amines with nitrophenyl phosphates. *Journal of the Chemical Society (B),* 1187–1189.
Klee, W. A., and Mudd, S. H. (1967): The conformation of ribonucleosides in solution. The effect of structure on the orientation of the base. *Biochemistry,* 6:988–998.
Klotz, U., and Stock, K. (1971): Evidence for a cyclic nucleotide phosphodiesterase with high specificity for cyclic uridine 3',5'-monophosphate in rat adipose tissue. *Naunyn-Schmiederbergs Archiv für Pharmakologie,* 269:117–120.
Kochetkov, N. K., Budovskii, E. I., Sverdlov, E. D., Simukova, N. A., Turchinskii, M. F., and Shibaev, V. N. (1972): Cleavage of phosphoester bonds and some other reactions of phosphate groups of nucleic acids and their components. In: *Organic Chemistry of Nucleic Acids, Part B,* edited by N. K. Kochetkov and E. I. Budovskii. Plenum Press, London.
Kreis, W., and Wechter, W. J. (1972): Biological activity of 3',5'-cyclic phosphates of 1-β-D-

arabinofuranosylcytosine (ara-C). *Proceedings of the American Association for Cancer Research,* 13:62.

Kukovetz, W. R., and Pöch, G. (1970): Inhibition of cyclic 3',5'-nucleotide phosphodiesterase as a possible mode of action of papaverine and similarly acting drugs. *Naunyn-Schmiedebergs Archiv für Pharmakologie,* 267:189–194.

Kumamoto, J., Cox, J. R., Jr., and Westheimer, F. H. (1956): Barium ethylene phosphate. *Journal of the American Chemical Society,* 78:4858–4860.

Kuo, J. F., and Greengard, P. (1969): Cyclic nucleotide-dependent protein kinases. IV. Widespread occurrence of adenosine 3',5'-monophosphate-dependent protein kinase in various tissues and phyla of the animal kingdom. *Proceedings of the National Academy of Sciences,* 64:1349–1355.

Kuo, J. F., and Greengard, P. (1970): Stimulation of adenosine 3',5'-monophosphate-dependent and guanosine 3',5'-monophosphate-dependent protein kinases by some analogs of adenosine 3',5'-monophosphate. *Biochemical and Biophysical Research Communications,* 40:1032–1038.

Kuo, J. F., Miyamoto, E., Reyes, P., and Greengard, P. (1972): Reactivity of various cyclic nucleotide analogs on adenosine 3',5'-monophosphate-dependent and guanosine 3',5'-monophosphate-dependent protein kinases (*in preparation*).

Lake, W. C., Anisuzzaman, A. K. M., and Whistler, R. L. (1972): 4'-Thioadenosine 3',5'-cyclic phosphate and derivatives: Chemical synthesis and hydrolysis by phosphodiesterase. Abstracts 164th American Chemical Society Meeting, New York City, August, 1972.

Lapper, R. D., Mantsch, H. H., and Smith, I. C. P. (1972): A carbon-13 and hydrogen-1 nuclear magnetic resonance study of the conformations of 3',5'- and 2',3'-cyclic nucleotides. A demonstration of the angular dependence of three-bond spin-spin couplings between carbon and phosphorus. *Journal of the American Chemical Society,* 94:6243–6244.

Lawson, A. M., Stillwell, R. N., Tacker, M. M., Tsuboyama, K., and McCloskey, J. A. (1971): Mass spectrometry of nucleic acid components. Trimethylsilyl derivatives of nucleotides. *Journal of the American Chemical Society,* 93:1014–1023.

Lecoq, J. (1956): Cyclisation des chloroalkyl phosphates. *Comptes Rendus des Séances de l'Académie des Sciences,* 242:1902–1903.

Lee, W. W., Fisher, L. V., and Goodman, L. (1971): 9-(β-D-Arabinofuranosyl)adenine 3',5'-cyclic phosphate. *Journal of Heterocyclic Chemistry,* 8:179–180.

LePage, G. A., and Hersh, E. (1972): Cyclic nucleotides as carcinostatic agents. *Biophysical and Biochemical Research Communications,* 46:1918–1922.

Levine, R. A. (1968a): Antidiuretic responses to exogenous adenosine 3',5'-monophosphate in man. *Clinical Science,* 34:253–260.

Levine, R. A. (1968b): Effects of exogenous adenosine 3',5'-monophosphate in man. II. Glucose, non-esterified fatty acid and cortisol responses. *Metabolism, Clinical Experimental,* 17:34–45.

Levine, R. A. (1969): Stimulation of plasma insulin and growth hormone in man by cyclic 3',5'-AMP. Presented at Protein and Polypeptide Hormones 3, Amsterdam International Congress. In: *Excerpta Medica Foundation Series No. 161,* pp. 879–881.

Levine, R. A. (1970): Effects of exogenous adenosine 3',5'-monophosphate in man. *Clinical Pharmacology and Therapeutics,* 11:238–243.

Levine, R. A., Dixon, L. M., and Franklin, R. B. (1968): Effects of exogenous adenosine 3',5'-monophosphate in man. I. Cardiovascular responses. *Clinical Pharmacology and Therapeutics,* 9:168–179.

Levine, R. A., and Washington, A. (1970): Glycogenolytic activity of cyclic 3',5'-monophosphates in perfused rat liver. *Endocrinology,* 87:377–382.

Lichtenstein, L. M., and Margolis, S. (1968): Histamine release *in vitro:* inhibition by catecholamines and methylxanthines. *Science,* 161:902–903.

Liddle, G. W., and Hardman, J. G. (1971): Cyclic adenosine monophosphate as a mediator of hormone action. *New England Journal of Medicine,* 285:560–566.

Lipkin, D., Cook, W. H., and Markham, R. (1959a): Adenosine 3',5'-phosphoric acid: A proof of structure. *Journal of the American Chemical Society,* 81:6198–6203.

Lipkin, D., Markham, R., and Cook, W. H. (1959b): The degradation of adenosine 5'-triphosphoric acid (ATP) by means of aqueous barium hydroxide. *Journal of the American Chemical Society,* 81:6075–6080.

Loeb, J. E., and Blat, C. (1970): Phosphorylation of some rat liver ribosomal proteins and its activation by cyclic AMP. *Federation of European Biochemistry Societies Letters,* 10:105–108.

Long, R. A., Szekeres, G. L., Khwaja, T. A., Sidwell, R. W., Simon, L. N., and Robins, R. K. (1972): Synthesis and antitumor and antiviral activity of 1-β-D-arabinofuranosylpyrimidine 3′,5′-cyclic phosphates. *Journal of Medicinal Chemistry,* 15:1215–1218.

Markham, R. (1957): The preparation and assay of cyclic nucleotides. *Methods in Enzymology,* 3:805.

Markham, R., and Smith, J. D. (1952): The structure of ribonucleic acids. 1. Cyclic nucleotides produced by ribonuclease and by alkaline hydrolysis. *Biochemical Journal,* 52:552–557.

Martelo, O. J., Woo, S. L. C., Reimann, E. M., and Davie, E. W. (1970): Effect of protein kinase on ribonucleic acid polymerase. *Biochemistry,* 9:4807–4813.

Mauret, P., and Fayet, J. (1969): Dipole moments of OP(XYZ) and P(XYZ) molecules. Polarity of the different bonds. *Bulletin de la Société Chimique de France,* 7:2363–2365.

Mauret, P., Fayet, J., Voigt, D., Labarre, M., and Labarre, J. (1968): Dipole moments of P(XYZ) and OP(XYZ) type molecules; a study of the nature of the phosphorus-oxygen bond. *Journal de Chimie Physique et de Physicochimie Biologique,* 65:549–561.

Meyer, R. B., Shuman, D. A., and Robins, R. K. (1972a): The synthesis of 2,6-disubstituted-9-β-D-ribofuranosylpurine 3′,5′-cyclic phosphates (*unpublished results*).

Meyer, R. B., Shuman, D. A., Robins, R. K., Bauer, R. J., Dimmitt, M. K., and Simon, L. N. (1972b): Synthesis and biological activity of several 6-substituted 9-β-D-ribofuranosylpurine 3′,5′-cyclic phosphates. *Biochemistry* 11:2704–2709.

Meyer, R. B., Shuman, D. A., and Robins, R. K. (1972c): *in preparation.*

Meyer, R. B., Shuman, D. A., and Robins, R. K. (1973): Synthesis of purine nucleoside 3′,5′-cyclic phosphoramidates. *Tetrahedron Letters,* 269–272.

Michal, G., DuPlooy, M., Woschee, M., Nelböck, M., and Weiman, G. (1970a): Cyclophosphate. II. Messung der aktivierung des phosphorylasesystems durch cyclophosphate. *Zeitschrift für Analytical Chemie,* 252:183–188.

Michal, G., Nelböck, M., and Weiman, G. (1970b): Cyclophosphate. III. Spaltung vershiedener cyclophosphate durch phosphodiesterase aus herz und fettgewebe. *Zeitschrift für Analytical Chemie,* 252:189–193.

Michal, G., Nelböck, M., Paoletti, R., Weiman, G., and Wunderwald, P. (1972): Lipolytic and antilipolytic effects of cyclophosphates. In: *Advances in Cyclic Nucleotide Research, Vol. 1: Physiology and Pharmacology of Cyclic AMP,* edited by P. Greengard, R. Paoletti, and G. A. Robison, p. 582 (abstract). Raven Press, New York.

Michelson, A. M. (1958): Synthesis of nucleoside cyclic phosphates. *Chemistry and Industry,* 70–71.

Michelson, A. M. (1959): Polynucleotides. Part II. *Journal of the Chemical Society,* 3655–3669.

Mikolajczyk, M. (1966): Sterochemie der reaktion von thiophosphorsäuren mit dicyclohexylcarbodiimid (DCCI). *Chemische Berichte,* 99:2083–2090.

Miller, J. P., Shuman, D. A., Scholten, M. B., Dimmitt, M. K., Stewart, C. M., Khwaja, T. A., Robins, R. K., and Simon, L. N. (1973): Synthesis and biological activity of some 2′ derivatives of adenosine 3′,5′-cyclic phosphate. *Biochemistry,* 12:1010.

Miyamoto, E., Kuo, J. F., and Greengard, P. (1969): Cyclic nucleotide-dependent protein kinases. III. Purification and properties of adenosine 3′,5′-monophosphate-dependent protein kinase from bovine brain. *Journal of Biological Chemistry,* 244:6395–6402.

Moffatt, J. G., and Khorana, H. G. (1957): Phosphorylated sugars. IV. The synthesis of D-xylose 3-phosphate *via* 1,2-O-isopropylidene-D-xylofuranose 3′,5′-cyclic phosphate. *Journal of the American Chemical Society,* 79:1194–1200.

Montgomery, J. A. (1970): The biochemical basis for the drug action of purines. In: *Progress in Medicinal Chemistry,* Vol. 7, edited by G. P. Ellis and G. B. West. Appleton Century-Crofts, New York.

Morley, A., Quesenberry, P., Garrity, M., and Stohlman, F., Jr. (1971): Inhibition of marrow

growth by cyclic AMP. *Proceedings for the Society of Experimental Biology and Medicine,* 138:57–59.
Mosher, H. S., Reinhart, J., and Prosser, H. C. (1953): The phosphorylation of chloromycetin. *Journal of the American Chemical Society,* 75:4899–4901.
Muneyama, K., Bauer, R. J., Shuman, D. A., Robins, R. K., and Simon, L. N. (1971): Chemical synthesis and biological activity of 8-substituted adenosine 3′,5′-cyclic monophosphate derivatives. *Biochemistry,* 10:2390–2395.
Murayama, A., Jastorff, B., Cramer, F., and Hettler, H. (1971): 5′-Amido analogs of adenosine 3′,5′-cyclic monophosphate. *Journal of Organic Chemistry,* 36:3029–3033.
Nagyvary, J. (1969): Arabinonucleotides II. The synthesis of $O^2,2'$-anhydrocytidine 3′-phosphate, a precursor of 1-β-D-arabinosylcytosine. *Journal of the American Chemical Society* 91:5409–5410.
Naito, T. and Sano, M. (1965): Adenosine cyclic 3′,5′-phosphate. Japanese Patent No. 9063.
Naylor, R. and Gilham, P. T. (1966): Studies on some interactions and reactions of oligonucleotides in aqueous solution. *Biochemistry* 5:2722–2728.
Nohara, A., Imai, K., and Honjo, M. (1966): Synthesis of the glucose analogs of inosine 5′-phosphate. *Chemical and Pharmaceutical Bulletin* (Tokyo) 14:491–495.
Orange, R. P., Austen, W. G., and Austen, K. F. (1971a): Immunological release of histamine and slow reacting substance of anaphylaxis from human lung. I. *Journal of Experimental Medicine,* 134:136s–148s.
Orange, R. P., Kaliner, M. A. LaRaia, P. J. and Austen, K. F. (1971b): Immunological release of histamine and slow reacting substance of anaphylaxis from human lung. II. Influence of cellular levels of cyclic AMP. *Federation Proceedings,* 30:1725–1729.
Ortiz, P. J. (1972): The inhibition of *E. coli* adenyl cyclase by ara ATP. *Biochemical and Biophysical Research Communications,* 46:1728–1733.
Panzica, R. P., Rousseau, R. J., Robins, R. K., and Townsend, L. B. (1972): A study on the relative stability and a quantitative approach to the reaction mechanism of the acid-catalyzed hydrolysis of certain 7- and 9-β-D-ribofuranosylpurines. *Journal of the American Chemical Society,* 94:4708–4714.
Pastan, I., and Perlman, R. (1970): Cyclic adenosine monophosphate in bacteria. *Science,* 169:339–344.
Posternak, T. (1971): Chemistry of cyclic nucleoside phosphates and synthesis of analogs. In: *Cyclic AMP,* edited by G. A. Robison, R. W. Butcher, and E. W. Sutherland. Academic Press, New York.
Posternak, T., and Cehovic, G. (1971): Derivatives and analogues of cyclic nucleotides. *Annals of the N.Y. Academy of Sciences,* 185:42–49.
Posternak, T., and Falbriard, J. G. (1971): Preparation of cyclic GMP. In: *Cyclic AMP,* edited by G. A. Robison, R. W. Butcher and E. W. Sutherland. Academic Press, New York.
Posternak, T., Marcus, I., and Cehovic, G. (1971): Preparation de nouveaux derivés de l'AMPc (substitutes en position C-8 et C-2) et étude de leur action sur la liberation de l'hormone de croissance. *Comptes Rendus des Séances de l'Académie des Sciences* (D), 272:622–625.
Posternak, T., Marcus, I., Gabbai, A., and Cehovic, G. (1969): Endocrinologie. Preparation et etude de quelques propriétés biologiques d'analogues de l'acide adénosine 3′,5′-phosphorique. *Comptes Rendus des Séances de l'Académie des Sciences* (D), 269:2409–2412.
Posternak, T., Sutherland, E. W., and Henion, W. F. (1962): Derivatives of cyclic 3′,5′-adenosine monophosphate. *Biochimica et Biophysica Acta,* 65:558–560.
Prasad, K. N. (1972a): Neuroblastoma clones: Prostaglandin *vs.* dibutyryl cyclic AMP, benzylthio cyclic AMP, phosphodiesterase inhibitors and X-ray. *Proceedings of the Society for Experimental Biology and Medicine,* 140:126–129.
Prasad, K. N. (1972b): 8-Substituted analogs of 3′,5′-cyclic AMP induce morphological differentiation of mouse neuroblastoma cells in culture. *Cancer Research* (*submitted for publication*).
Preobrazhenskaya, N. N., Shabarova, Z. A., and Profof'ev, M. A. (1967): The synthesis and hydrolytic stability of amino acid derivatives of uridine cyclophosphates. *Doklady Akademii Nauk SSSR,* 174:100–103 (Chemical Abstracts 68:13339w).

Rall, T. W., and Sutherland, E. W. (1958): Formation of cyclic adenine ribonucleotide by tissue particles. *Journal of Biological Chemistry,* 232:1065–1076.

Rall, T. W., Sutherland, E. W., and Berthet, J. (1957): The relationship of epinephrine and glucagon to liver phosphorylase. *Journal of Biological Chemistry,* 224:463–475.

Reimann, E. M., Brostrom, C. O., Corbin, J. D., King, C. A., and Krebs, E. G. (1971): Separation of regulatory and catalytic subunits of the cyclic 3',5'-adenosine monophosphate-dependent protein kinase(s) of rabbit skeletal muscle. *Biochemical and Biophysical Research Communications,* 42:187–194.

Rizack, M. A. (1964): Activation of an epinephrine-sensitive lipolytic activity from adipose tissue by adenosine 3',5'-phosphate. *Journal of Biological Chemistry,* 239:392–395.

Robison, G. A., Butcher, R. W., and Sutherland, E. W. (1971): *Cyclic AMP.* Academic Press, New York.

Robison, G. A., and Sutherland, E. W. (1971): Cyclic AMP and the function of eukaryotic cells: An introduction. *Annals of the New York Academy of Sciences,* 185:5–9.

Rodbell, M. (1964): Metabolism of isolated fat cells. I. Effects of hormones on glucose metabolism and lipolysis. *Journal of Biological Chemistry,* 239:375–380.

Rosen, O. M. (1970a): Preparation and properties of a cyclic 3',5'-nucleotide phosphodiesterase isolated from frog erythrocytes. *Archives of Biochemistry and Biophysics,* 137:435–441.

Rosen, O. M. (1970b): Interaction of cyclic GMP and cyclic AMP with a cyclic nucleotide phosphodiesterase of the frog erythrocyte. *Archives of Biochemistry and Biophysics,* 139:447–449.

Rubin, B., O'Keefe, E. H., Waugh, M. H., Kotler, D. G., De Maio, D. A., and Horovitz, Z. P. (1971): Activities *in vitro* of 8-substituted derivatives of adenosine 3',5'-cyclic monophosphate on guinea pig trachea and rat portal vein. *Proceedings for the Society of Experimental Biology and Medicine,* 137:1244–1248.

Russell, A. F., and Moffatt, J. G. (1969): Synthesis of some nucleotides derived from 3'-deoxythymidine. *Biochemistry,* 8:4889–4896.

Schweizer, M. P., Broom, A. D., Ts'o, P. O. P., and Hollis, D. P. (1968): Studies of inter- and intramolecular interaction in mononucleotides by proton magnetic resonance. *Journal of the American Chemical Society,* 90:1042–1055.

Schweizer, M. P., and Robins, R. K. (1973): NMR studies on the conformation of nucleosides and 3',5'-cyclic nucleotides. *Conformation of Biological Molecules and Polymers, Proceedings of the Fifth Jerusalem Symposium on Quantum Chemistry and Biochemistry,* edited by B. Pullman and E. Bergmann. Academic Press, New York (*in press*).

Secrist, J. A., III, Barrio, J. R., Leonard, N. J., Villar-Palasi, C., and Gilman, A. G. (1972): Fluorescent modification of adenosine 3',5'-monophosphate: spectroscopic properties and activity in enzyme systems. *Science,* 177:279–280.

Serafini, A., Labarre, J. F., Veillard, A., and Vinot, G. (1971): Bonding in phosphorus trifluoride and phosphorus oxyfluoride. An *ab initio* SCF-LCAO-MO study. *Journal of the Chemical Society* (D), 996–998.

Shapiro, R., and Danzig, M. (1972): Acidic hydrolysis of deoxycytidine and deoxyuridine derivatives. The general mechanims of deoxyribonucleoside hydrolysis. *Biochemistry,* 11:23–29.

Shen, T. Y. (1970): Nucleosides and nucleotides as potential therapeutic agents. *Angewandte Chemie International Edition,* 9:678–688.

Sheppard, H., and Wiggan, G. (1971): Analogues of 4-(3,4-dimethoxybenzyl)-2-imidazolidinone as potent inhibitors of rat erythrocyte adenosine cyclic 3',5'-phosphate phosphodiesterase. *Molecular Pharmacology,* 7:111–115.

Shuman, D. A., Miller, J. P., Boswell, K. H., Muneyama, K., Robins, R. K., and Simon, L. N. (1973a): Synthesis and biological activity of several 8-substituted derivatives of cyclic IMP and cyclic GMP (*in preparation*).

Shuman, D. A., Miller, J. P., Robins, R. K., and Simon, L. N. (1973b): Synthesis and biological activity of 5'-thio-5'-deoxynucleoside 3',5'-cyclic phosphorothioates. *Biochemistry* (*in press*).

Shuman, D. A., Simon, L. N., and Robins, R. K. (1972): *Unpublished results.*
Sidwell, R. W., Huffman, J., Meyer, R. B., Shuman, D. A., Simon, L. N., and Robins, R. K. (1972b): Antiviral activity and inhibition of tumor cell growth by a series of 6-substituted-9-(β-D-ribofuranosyl)purine 3′,5′-monophosphate derivatives. Abstracts Vth International Congress on Pharmacology, San Francisco, July, 1972.
Sidwell, R. W., Huffman, J., Shuman, D. A., Muneyama, K., and Robins, R. K. (1972a): Cyclic AMP derivatives as antiviral agents. Proceedings of the VIIth International Congress of Chemotherapy, Prague, August, 1971. Avicenum Press, Prague.
Simon, L. N., Swiatek, K. R., Bauer, R. J., Free, C. F., Miller, J. P., Shuman, D. A., Boswell, K. H., and Robins, R. K. (1973): The correlation of activation of lipolysis, adrenal steroidogenesis, glycogenolysis with activation of a cyclic AMP-dependent protein kinase, by 6-8-disubstituted analogs of adenosine 3′,5′-cyclic phosphate (*in preparation*).
Singhal, R. L., Pasulekar, M. R., and Vijayvargiya, R. (1971): Metabolic control mechanisms in mammalian systems. Involvement of adenosine 3′,5′-cyclic monophosphate in androgen action. *Biochemical Journal,* 125:329–342.
Smith, I. C. P., Mantsch, H. H., Lapper, R. D., Deslauriers, R., and Schleich, T. (1973): A study of the conformations of nucleic acids by carbon-13 and hydrogen nuclear magnetic resonance spectoscopy. Vth Jerusalem Symposium on Quantum Chemistry and Biochemistry, edited by B. Pullman and E. Bergmann. Academic Press, New York (*in press*).
Smith, M., Moffatt, J. G., and Khorana, H. G. (1958): Carbodiimides VIII. Observations on the reactions of carbodiimides with acids and some new applications in the synthesis of phosphoric acid esters. *Journal of the American Chemical Society,* 80:6204–6212.
Smith, M., Drummond, G. I., and Khorana, H. G. (1961): Cyclic phosphates. IV. Ribonucleoside 3′,5′-cyclic phosphates. A general method of synthesis and some properties. *Journal of the American Chemical Society,* 83:698–706.
Smith, M., and Jardetzky, C. D. (1968): The NMR spectra and conformation of the nucleotide 3′,5′-cyclic phosphates. *Journal of Molecular Spectroscopy,* 28:70–80.
Smith, M., and Khorana, H. G. (1959): Specific synthesis of the C_5'-C_3' interribonucleotide linkage: The synthesis of uridylyl-5′-3′-uridine. *Journal of the American Chemical Society,* 81:2911–2912.
Sober, H. A. (1970): *Handbook of Biochemistry, Second Edition.* Chemical Rubber Company, Cleveland.
Steiner, A. L., Ferrendelli, J. A., and Kipnis, D. M. (1972a): Radioimmunoassay for cyclic nucleotides. III. Effect of ischemia, changes during development and regional distribution of adenosine 3′,5′-monophosphate and guanosine 3′,5′-monophosphate in mouse brain. *Journal of Biological Chemistry,* 247:1121–1124.
Steiner, A. L., Pagliara, A. S., Chase, L. R., and Kipnis, D. M. (1972b): Radioimmunoassay for cyclic nucleotides. II. Adenosine 3′,5′-monophosphate and guanosine 3′,5′-monophosphate in mammalian tissues and body fluids. *Journal of Biological Chemistry,* 247:1114–1120.
Steiner, A. L., Parker, C. W., and Kipnis, D. M. (1972c): Radioimmunoassay for cyclic nucleotides. I. Preparation of antibodies and iodinated cyclic nucleotides. *Journal of Biological Chemistry,* 247:1106–1113.
Strand, M., and August, J. T. (1971): Protein kinase and phosphate acceptor proteins in Rauscher murine leukemia virus. *Nature New Biology,* 233:137–140.
Straus, D. B., and Fresco, J. R. (1965): Synthesis of *N*-benzoyl-2′-*O*-tetrahydropyranylguanosine 5′-phosphate, an intermediate in the chemical synthesis of polyriboguanylic acid. *Journal of the American Chemical Society,* 87:1364–1374.
Su, J., and Hassid, W. Z. (1962): Carbohydrates and nucleotides in the red alga *Porphyra perforata.* II. Separation and identification of nucleotides. *Biochemistry,* 1:474–480.
Sundaralingam, M., and Abola, J. (1972): Molecular conformation of adenosine 3′,5′-monophosphonate monohydrate. *Nature New Biology,* 235:244–245.
Sutherland, E. W., and Rall, T. W. (1957): The properties of an adenine ribonucleotide produced with cellular particles, ATP, Mg^{++} and epinephrine or glucagon. *Journal of the American Chemical Society,* 79:3608.

Sutherland, E. W., and Rall, T. W. (1958): Fractionation and characterization of a cyclic adenine ribonucleotide formed by tissue particles. *Journal of Biological Chemistry*, 232: 1077-1091.
Swislocki, N. I. (1970): Decomposition of dibutyryl cyclic AMP in aqueous buffers. *Analytical Biochemistry*, 38:260-269.
Tazawa, I., Tazawa, S., and Ts'o, P. O. P. (1972): *Biochemistry*, 11:4931.
Tener, G. M., Khorana, H. G., Markham, R., and Pol, E. H. (1958): Studies on polynucleotides. II. The synthesis and characterization of linear and cyclic thymidine oligonucleotides. *Journal of the American Chemical Society*, 80:6223-6230.
Tesser, G. I., Fisch, H., and Schwyzer, R. (1972): Limitations of affinity chromatography: Sepharose-bound cyclic 3',5'-adenosine monophosphate. *Federation of European Biochemistry Societies Letters*, 23:56-58.
Thomas, H. J., and Montgomery, J. A. (1968): Derivatives and analogs of 6-mercaptopurine ribonucleotide. *Journal of Medicinal Chemistry*, 11:44-48.
Thompson, W. J. and Appleman, M. M. (1971): Cyclic nucleotide phosphodiesterase and cyclic AMP. *Annals of the N.Y. Academy of Sciences*, 185:36-41.
Tokyo Tanabe Seiyaku Company (1972): Riboflavin dibutyryl. Japanese Patent No. 7,208,316.
Triner, L., Nahas, G. G., Vulliemoz, Y., Overweg, N. I. A., Verosky, M., Habif, D. V., and Ngai, S. H. (1971): Cyclic AMP and smooth muscle function. *Annals of the N.Y. Academy of Sciences*, 185:458-476.
Ts'o, P. O. P. (1972): *personal communication*.
Ueda, T., and Fox, J. J. (1967): The mononucleotides. In: *Advances in Carbohydrate Chemistry*, edited by M. L. Wolfron and R. S. Tipson. Academic Press, New York.
Ueda, T., and Kawai, I. (1970): A convenient synthesis of ribonucleoside 2',3'-cyclic phosphates from ribonucleosides and ribonucleotides. *Chemical and Pharmaceutical Bulletin (Tokyo)*, 18:2303-2308.
Ukita, T., and Hayatsu, H. (1961): Organic phosphates. XVIII. Synthesis of lyxouridine 2',3'-cyclic phosphate and related compounds. *Chemical and Pharmaceutical Bulletin (Tokyo)*, 9:1000-1005.
Usher, D. A., Richardson, D. I., Jr., and Oakenfull, D. G. (1970): Models of ribonuclease action II. Specific acid, specific base, and neutral pathways for hydrolysis of a nucleotide diester analog. *Journal of the American Chemical Society*, 92:4699-4712.
van Wijk, R., Wicks, W. D., and Clay, K. (1972): Effects of derivatives of cyclic 3',5'-adenosine monophosphate on the growth, morphology, and gene expression of hepatoma cells in culture. *Cancer Research*, 32:1905-1911.
Vargiu, L., and Spano, P. F. (1971): Some central effects of a new derivative of cyclic 3',5'-adenosine monophosphate. *Naunyn Schmiedebergs Archiv für Pharmakologie*, 269:410.
Verheyden, J. P. H., and Moffatt, J. G. (1970): Halo sugar nucleosides. II. Iodination of secondary hydroxyl groups of nucleosides with methyltriphenoxyphosphonium iodide. *Journal of Organic Chemistry*, 35:2868-2877.
Voorhees, J. J., and Duell, E. A. (1971): Psoriasis as a possible defect of the adenyl cyclase-cyclic AMP cascade. *Archives of Dermatology*, 104:352-358.
Voorhees, J. J., Duell, E. A., Bass, L. J., and Kelsey, W. H. (1971a): Inhibition of epidermal cell division by isoproterenol, dibutyryl cyclic AMP and theophylline. *Clinical Research*, 19:682.
Voorhees, J. J., Duell, E. A., Bass, L. J., Powell, J. A., and Harrell, E. R. (1972a): Decreased cyclic AMP in the epidermis of lesions of psoriasis. *Archives of Dermatology*, 105:695-701.
Voorhees, J. J., Duell, E. A., and Kelsey, W. H. (1972b): Dibutyryl cyclic AMP inhibition of epidermal cell division. *Archives of Dermatology*, 105:384-386.
Voorhees, J. J., Duell, E. A., Kelsey, W., Engelhard, K., and Sibrack, L. (1971b): The epinephrine-chalone, adenyl cyclase, phosphodiesterase cascade in the *in vitro* control of epidermal growth. *Clinical Research*, 19:365.
Wagner, E. L. (1963): Calculated bond characters in phosphoryl compounds. *Journal of the American Chemical Society*, 85:161-164.

Walsh, D. A., Perkins, J. P., and Krebs, E. G. (1968): An adenosine 3',5'-monophosphate-dependent protein kinase from rabbit skeletal muscle. *Journal of Biological Chemistry,* 243:3763-3765.
Watenpaugh, K., Dow, J., Jensen, L. H., and Furberg, S. (1968): Crystal and molecular structure of adenosine 3',5'-cyclic phosphate. *Science,* 159:206-207.
Wechter, W. J. (1967): Nucleic acids. I. The synthesis of nucleotides and dinucleoside phosphates containing *ara*-cytidine. *Journal of Medicinal Chemistry,* 10:762-773.
Wechter, W. J. (1969): Nucleic acids. VIII. Synthesis and chemistry of *ara*-cytidine 2',5'-cyclic phosphate. Phosphate anisotropy. *Journal of Organic Chemistry,* 34:244-247.
Weimann, G., and Khorana, G. H. (1962): Studies on polynucleotides. XVII. On the mechanism of internucleotide bond synthesis by the carbodiimide method. *Journal of the American Chemical Society,* 84:4329-4341.
Weinryb, I., Michel, I. M., and Hess, S. M. (1972): Characterization of adenylate cyclase from guinea pig lung and measurement of its inhibition by substrate analogs and cyclic nucletides. Abstracts of the Vth International Congress on Pharmacology, San Francisco, July, 1972.
Weissmann, G., Dukor, P., and Zurier, R. B. (1971): Effects of cAMP on release of lysosomal enzymes from phagocytes. *Nature,* 231:131-135.
Werner, M., Buterbaugh, G. G., and Blake, D. A. (1972): Inhibition of hepatic drug metabolism by cyclic 3',5'-adenosine monophosphate. *Research Communications in Chemical Pathology and Pharmacology,* 3:249-263.
Westheimer, F. H. (1968): Pseudo-rotation in the hydrolysis of phosphate esters. *Accounts of Chemical Research,* 1:70-78.
Wilchek, M., Salomon, Y., Lowe, M., and Selinger, Z. (1971): Conversion of protein kinase to a cyclic AMP independent form by affinity chromatography on N^6-caproyl 3',5'-cyclic adenosine monophosphate-sepharose. *Biochemical and Biophysical Research Communications,* 45:1177-1183.
Yoshikawa, M., Kato, T., and Takenishi, T. (1969): Studies of phosphorylation III. Selective phosphorylation of unprotected nucleosides. *Bulletin of the Chemical Society of Japan,* 42:3505-3508.
Yurkevich, A. M., Kolodkina, I. I., Varshavskaya, L. S., Borodulina-Shvetz, V. I., Rudakova, I. P., and Preobrazhenski, N. A. (1969): The reaction of phenylboronic acid with nucleosides and mononucleotides. *Tetrahedron Letters,* 25:477-484.
Zmudzka, B., Szer, W., and Shugar, D. (1962): Preparation and chemical and enzymic properties of phosphate esters of 1-(β-D-glucopyranosyl)-uracil and thymine. *Acta Biochimica Polonica,* 9:321-341.
Zoltewicz, J. A., and Clark, D. F. (1972): Kinetics and mechanism of the hydrolysis of guanosine and 7-methylguanosine nucleosides in perchloric acid. *Journal of Organic Chemistry,* 37:1193-1197.
Zoltewicz, J. A., Clark, D. F., Sharpless, T. W., and Grahe, G. (1970): Kinetics and mechanisms of the acid-catalyzed hydrolysis of some purine nucleosides. *Journal of the American Chemical Society,* 92:1741-1750.

Clinical Studies and Applications of Cyclic Nucleotides

Ferid Murad

OUTLINE

I. Introduction	356
II. Examination of Cyclic Nucleotide Levels in Extracellular Fluids and other Clinical Specimens	357
A. Urinary Cyclic Nucleotides	357
B. Plasma Cyclic Nucleotides	358
C. Other Extracellular Fluids	359
D. Tissues and Formed Elements	359
III. Basal Cyclic Nucleotide Levels in Extracellular Fluids	360
A. Urine	360
B. Plasma	363
C. Miscellaneous Fluids	364
IV. Effects of Hormones and Other Agents on Cyclic Nucleotide Levels in Plasma and Urine	365
A. Parathyroid Hormone and Calcitonin	365
B. Glucagon	367
C. Antidiuretic Hormone	368
D. Catecholamines	368
E. Growth Hormone and Insulin	369
F. Other Hormones	369
V. Cyclic Nucleotides in Disease States	369
A. Calcium Disorders	369
B. Diabetes Insipidus	374
C. Affective Disorders	374
D. Psoriasis	374
E. Asthma and Atopic Dermatitis	374

VI. Summary and Speculation 376
VII. Acknowledgments 377
VIII. References 377
    Addendum 381

I. INTRODUCTION

Although a considerable amount of information is available concerning the regulation of cyclic nucleotide metabolism in animal systems, relatively little has been reported dealing with its application in areas of clinical medicine. Although this lag in the clinical application of basic investigation is a common phenomenon, it is likely that a major factor in this delay has been the seemingly difficult and tedious techniques and methods in cyclic nucleotide quantification which were available to the clinical laboratory and investigator. However, with the development of relatively sensitive, simple methods for the estimation of cyclases, phosphodiesterases, and cyclic nucleotides, we will undoubtedly see more interest in clinical areas than we have during the past several years.

At our present state of knowledge, it appears very likely that cyclic nucleotides offer the endocrinologist, clinical pharmacologist, cardiologist, and other subspecialists a powerful tool in several areas of investigation: (1) evaluating the pathophysiology of many endocrine and metabolic disorders, (2) "rationally" designing and testing new pharmacologic and therapeutic agents, (3) diagnosing endocrine and metabolic disorders, and (4) evaluating surgical and/or medical therapy in several diseases.

Some of the studies in patients reviewed below represent extensions of previously described studies in *in vivo* and *in vitro* animal systems. Obviously, it is important that many levels of investigation are undertaken (cell-free systems, intact cell systems, and *in vivo* studies in animals) to test hypotheses in relatively simple systems as well as in more organized and complex systems. Presumably, investigation then would develop toward human studies for the purpose of understanding disease states and their therapy. Frequently, new ideas and concepts arise from clinical studies which may be best pursued in more controllable animal systems. However, in some instances as discussed below (pseudohypoparathyroidism, cystic fibrosis, idiopathic osteoporosis, nephrolithiasis, atopic dermatitis, psoriasis, affective disorders, etc.), suitable animal models for investigation are unavailable.

In this chapter many of the clinical studies with cyclic nucleotides in normal subjects and patients are summarized. Much of this chapter describes factors which influence cyclic nucleotide levels in extracellular fluids, particularly plasma and urine, because of their availability to the clinical investigator. Readers are also referred to another recent review of this topic (Broadus et al., 1971). Some emphasis (and at times speculation) has been made in those areas that may be useful to the clinician. I hope this chapter will be of interest also to the many basic investigators who have helped provide the core of information from which rational and useful clinical studies can be launched.

II. EXAMINATION OF CYCLIC NUCLEOTIDE LEVELS IN EXTRACELLULAR FLUIDS AND OTHER CLINICAL SPECIMENS

A. Urinary Cyclic Nucleotides

Urine specimens have proved to be a useful source of specimens for clinical studies with cyclic nucleotides. This is attributable to the relatively high concentrations of cyclic AMP and cyclic GMP in urine and the ease with which individual or serial specimens can be obtained. Under basal conditions the cyclic AMP concentration in human urine is of the order of several micromolar (Butcher and Sutherland, 1962; Takahashi et al., 1966; Chase et al., 1969b). The concentration of cyclic GMP in human urine is about one-fifth that of cyclic AMP (Price et al., 1967; Broadus et al., 1970a; Steiner et al., 1970; Murad and Pak, 1972a,b; Williams et al., 1972). The original isolation and characterization of cyclic GMP as a natural compound occurred with the studies of Ashman et al. (1963), working with rat urine. These observations undoubtedly led to the search for enzymes which synthesize and degrade the nucleotide. This area is reviewed by Goldberg, O'Dea, and Haddox in this volume.

Only minor preparation of urine specimens prior to assay appears necessary. To decrease bacterial growth, various laboratories have collected urine specimens with HCl (Hardman et al., 1966), acetic acid and thymol (Broadus et al., 1971), sodium metabisulfite (Paul et al., 1971), and chloroform (Abdulla and Hamadah, 1970). Other laboratories, including our own, collect urine samples in refrigerated containers (4°C) and, at the termination of the collection period (up to 24 hr), aliquots are stored at −20°C for assay. We have seen no significant alteration in cyclic AMP or cyclic GMP levels when human urine is stored at −20°C for 1 to 2 years. However, storage of unacidified urine samples at room temperature for more than 2 to 3 days or at −4°C for more than several weeks has resulted in losses in cyclic nucleotides, which we have attributed to bacterial growth and hydrolysis.

It has also been possible to assay the cyclic AMP and cyclic GMP content in urine with several methods that do not require prior purification. With random samples, similar values for cyclic AMP content (Murad et al.,

1971a) have been obtained in unpurified and purified (anion-exchange chromatography) samples when the protein binding assay of Gilman (1970) or a modification (Murad et al., 1969c) of the liver phosphorylase activation assay described by Butcher et al. (1965) was used. Unpurified and purified urines also gave similar values when the two assay methods were compared (Murad et al., 1971a). Similarly, several laboratories with cyclic AMP radioimmunoassays have not purified urine samples prior to assay (Steiner et al., 1970, 1972; Munson et al., 1971; Williams et al., 1972). Also, with the cyclic GMP protein binding assay (Murad et al., 1971a; Murad and Gilman, 1971) or the cyclic GMP radioimmunoassay (Steiner et al., 1970, 1972), it has not been necessary to purify urines prior to cyclic GMP quantification. Goldberg et al. (1969) also found it unnecessary to purify urine prior to assay of cyclic AMP or cyclic GMP with their recycling method. The ability to assay crude urine samples has obviously permitted processing more samples in clinical screening studies. The ability to assay cyclic AMP and cyclic GMP content in unpurified urine samples simultaneously with a double isotope-binding assay (Murad and Gilman, 1971) or radioimmunoassay (Wehmann et al., 1972) has also decreased technical time considerably.

B. Plasma Cyclic Nucleotides

Although another readily available source for specimens is plasma, studies with plasma have been considerably fewer. This is probably due to the relatively low concentration of cyclic nucleotides in plasma (about two orders of magnitude lower than urine) (Murad et al., 1969c; Broadus et al., 1970a,b; Steiner et al., 1970) and the need to purify samples prior to assay with most of the current methods. One laboratory employing a radioimmunoassay for cyclic AMP found that purification of plasma but not urine prior to assay was necessary due to the presence of interfering materials (Munson et al., 1971). However, with the antibody preparation used by Steiner et al. (1972), no interference was observed with deproteinized plasma samples.

Whereas urinary cyclic nucleotides can be determined with a few microliters of urine, it is usually necessary to process several milliliters of plasma due to low levels and losses with purification. We obtain this amount of plasma from approximately 7 ml of anticoagulated blood in order to allow some excess plasma over the buffy coat after centrifugation. Although the cyclic AMP concentration is relatively low in plasma and higher in leukocytes (Stossel et al., 1970) and platelets (Cole et al., 1971), systematic studies with comparison of plasma and serum levels of cyclic nucleotides and effects of, for example, hemolysis, platelet disruption, and temperature, during collection have not been reported. Studies by Broadus et al. (1970a) have shown that cyclic AMP and cyclic GMP are apparently not bound to plasma proteins and that plasma cyclic nucleotides are not derived to any significant degree from formed elements.

Phosphodiesterase activity in the blood has been described in several species (Davoren and Sutherland, 1963; Broadus et al., 1970a; Patterson et al., 1971) and is apparently derived from formed elements. When blood is collected in 6 to 7 mM EDTA (the concentration achieved in EDTA-containing vacutainer tubes), chilled in ice, and the plasma separated after centrifugation, hydrolysis of cyclic AMP and cyclic GMP is not apparent (R. Weitzman and F. Murad, *unpublished observations*). We have inadvertently left blood samples collected in EDTA at room temperature for 30 to 60 min before centrifugation without finding any alteration in plasma cyclic AMP or cyclic GMP content. Similar observations have been noted by Patterson et al. (1971). These investigators have also found that in some plasma samples stored at $-20°C$ without EDTA the cyclic nucleotide levels gradually decline. This may be attributable to a phosphodiesterase which apparently possesses some activity at low temperatures (O'Dea et al., 1971).

C. Other Extracellular Fluids

Cyclic AMP has also been found in spinal fluid (Murad et al., 1969c; Robison et al., 1970; Broadus et al., 1970a), milk (Kobata et al., 1961), and amniotic fluid (Broadus et al., 1971). In addition to urine and plasma, low levels of cyclic GMP have been found in spinal fluid (Broadus et al., 1970a). One report described very high concentrations of cyclic AMP in saliva (Stefanovich and Wells, 1971) which could not be confirmed (T. W. Rall, *personal communication*).

D. Tissues and Formed Elements

Tissue biopsy specimens for cyclic nucleotide levels and activities of cyclases and phosphodiesterases and the profile of hormone responsiveness of the system should also prove useful. This has been the case with leukocytes from patients with hay fever (Lichtenstein and Margoulis, 1968) and asthma (Smith and Parker, 1970) and with skin biopsies from patients with atopic dermatitis (Mier and Urselmann, 1970) and psoriasis (Voorhees et al., 1972). The examination of cyclic nucleotide metabolism in such readily obtainable specimens as platelets and leukocytes may provide us with some additional understanding of hematological disorders. Examination of cyclic nucleotide content in biopsy specimens after hormonal or pharmacological challenge may prove valuable in examination of organ function and pathophysiology of disease. Obviously, the tissue source for biopsy and the necessity to fix samples rapidly to prevent possible alteration in cyclic nucleotide content will limit this approach considerably. However, the hormonal response profile of intact tissue or adenyl cyclase from biopsy specimens seems attainable and warrants attention. The ability to culture various cell lines from patients would also permit such studies and, in addition, aid investigation of possible genetic deficiencies in the adenyl cyclase-phosphodiesterase systems.

III. BASAL CYCLIC NUCLEOTIDE LEVELS IN EXTRACELLULAR FLUIDS

A. Urine

The studies of Broadus et al. (1970a) in normal volunteers under basal conditions indicate that about one-half of the cyclic AMP in urine is derived from plasma load to the kidney and glomerular filtration. The remainder is contributed primarily by the proximal nephron under the influence of parathyroid hormone (Chase and Aurbach, 1967; Kaminsky et al., 1970b). The values for plasma and renal contribution to urinary cyclic AMP were determined by comparing inulin, endogenous cyclic AMP, and ^{3}H-cyclic AMP clearances (Broadus et al. 1970a). In contrast, urinary cyclic GMP appears to be derived entirely from glomerular filtration under basal conditions (Broadus et al., 1970a). Administration of agents such as glucagon and parathyroid hormone can alter the ratio of renal to plasma contribution to urinary cyclic AMP as described below.

Several laboratories have noted that the daily basal excretion rates of cyclic AMP in an individual are generally very stable. In many patients who were examined for several days or weeks, we found daily excretion rates of cyclic AMP to vary as little as 10 to 20% (Murad and Pak, 1972a,b). Although daily excretion rates for cyclic GMP have been found to be stable in some patients, in general they have varied as much as 50 to 100% for unexplained reasons (Murad and Pak, 1972a,b).

In 12 hospitalized adult normal volunteers on a house diet and normal activity, we observed 24-hr excretion rates of 2.50 ± 0.13 μmoles cyclic AMP/g creatinine (mean ± SE) and 0.55 ± 0.08 μmoles cyclic GMP/g creatinine; the ranges were 1.7 to 3.2 and 0.1 to 1.0, respectively (Murad and Pak, 1972a,b). In these normal subjects, the daily excretion rates when not normalized for creatinine were 3.20 ± 0.18 μmoles cyclic AMP (range 2.2 to 4.2) and 0.90 ± 0.12 μmoles cyclic GMP (range 0.16 to 1.38). The values for cyclic AMP excretion rates are somewhat lower with a narrower range than those reported from other laboratories (Chase and Aurbach, 1969b; Broadus et al., 1970a; Broadus et al., 1970b; Taylor et al., 1970; Estep et al., 1970). This may be attributable to our studies with 24-hr urine samples to eliminate effects of diurnal variations (see below) and to the fact that the patients were hospitalized. In a small series of five nonhospitalized normal subjects, we found 24-hr excretion rates of 3.76 ± 0.38 μmoles cyclic AMP/g creatinine and 0.50 ± 0.18 μmoles cyclic GMP/g creatinine. The reason for the higher cyclic AMP excretion rate in nonhospitalized normal subjects is not apparent.

Kaminsky et al. (1970b) have reported a positive correlation between the cyclic AMP level in urine and the creatinine content. Taylor et al. (1970) reported a positive correlation with cyclic AMP excretion and body surface area. Patients with reduced renal function and diminished endogenous creatinine clearances have decreased cyclic AMP excretion rates (Taylor et al., 1970; Estep et al., 1970). This may be due to decreased glomerular filtration of cyclic AMP and/or decreased renal mass and responsiveness to PTH. Whether cyclic GMP excretion rates are diminished because of renal disease and reduced glomerular filtration is unknown. However, this would be expected since urinary cyclic GMP appears to be derived primarily from glomerular filtration (Broadus et al., 1970a).

We and others have found that urinary cyclic AMP and cyclic GMP concentrations are inversely related to urine volume so that the total quantity of nucleotide excreted is independent of urine volume. This relationship has been observed with wide variations in fluid intake and urine volume and osmolarity. As discussed below, this would suggest that the effects of ADH on cyclic nucleotide excretion would be absent or relatively small.

Most patients demonstrate a diurnal pattern in cyclic AMP excretion (Fig. 1) (Murad and Pak, 1972a,b). When looked for in a small series of patients, two other laboratories failed to see a diurnal pattern in cyclic AMP excretion (Chase et al., 1969b; Broadus et al., 1971). However, another laboratory has also observed diurnal patterns in cyclic AMP excretion (T. W. Rall, *personal communication*). Many of our patients had a peak excretion rate in the afternoon. However, some patients demonstrated peak excretion rates in the morning or evening. When patients were examined with 4 hourly or 6 hourly urines for several days to 2 weeks, the peaks and the lows of an individual patient occurred at similar times each day (i.e., the daily excretion pattern in a particular patient was generally consistent). The degree of excursion from peak to low during the course of a day was usually 30 to 50% but in some instances was as much as two- to

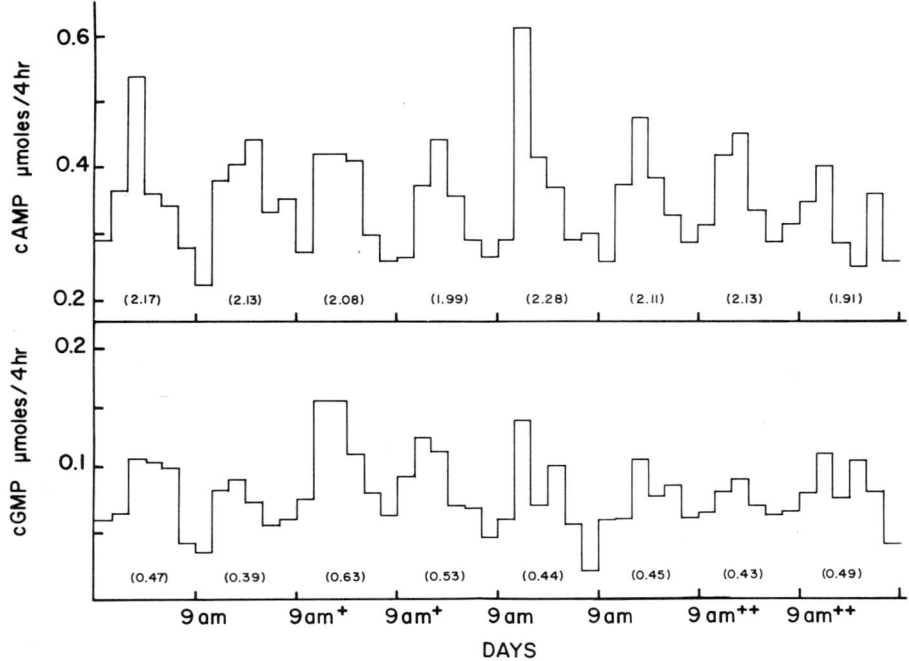

FIG. 1. Diurnal variation in cyclic nucleotide excretion and effects of caffeine and exercise (from Murad and Pak, 1972b). Urinary cyclic nucleotide levels were examined in a hospitalized normal volunteer on a metabolic diet which excluded methylxanthines. On the days indicated, the patient was given 6 cups of coffee/day (approximately 600 to 900 mg caffeine) (+) or exercised with stair climbing three times a day (++). Daily excretion rates (μmoles/24 hr) are included in parentheses.

threefold. Since we observed small diurnal fluctuations in endogenous creatinine clearances (higher during the day), excretion rates of cyclic AMP were normalized for the amount of creatinine in the urine as well as examined with absolute excretion rates. With either method prominent, diurnal patterns were observed (Murad and Pak, 1972b). The presence of methylxanthines in the diet (six cups of coffee/day or the equivalent of 600 to 900 mg caffeine/day) did not alter the circadian pattern or the absolute quantities of cyclic AMP in the urine (Fig. 1) (Murad and Pak, 1972b). Strenuous exercise did not alter cyclic AMP excretion rates or the diurnal variation (Fig. 1). Others have reported similar observations with methylxanthines (Williams et al., 1972) and exercise (Paul et al., 1971) on cyclic AMP excretion rates.

Although we have also observed a diurnal variation in urinary cyclic GMP with peak excretion rates in the afternoon (Fig. 1), many of our patients demonstrated a relatively stable excretion of cyclic GMP. Exercise had no effect on cyclic GMP excretion; however, the administration of caffeine produced small (25 to 40%) increases in cyclic GMP excretion (Fig. 1) (Murad and Pak, 1972b). The circadian rhythm of urinary cyclic nucleotides has also been observed in hypoparathyroid subjects, suggesting that factors other than fluctuating levels of serum parathyroid hormone are responsible for this phenomenon (Murad and Pak, 1972b). Whether the circadian pattern in urine is reflecting plasma variations and altered filtration and/or renal synthesis and contribution is unknown.

In several patients, we examined the circadian pattern in cyclic AMP and cyclic GMP excretion and phase-shifted the subjects so that they were awake during the night and slept during the day (G. Harbert and F. Murad, *unpublished observations*). These preliminary studies demonstrated a phase shift in urinary excretion of cyclic AMP and cyclic GMP.

In unpublished studies, we found that the excretion rates (μmoles/day) of cyclic AMP and cyclic GMP in children, ages 6 through 18, increased with age (W. Moss, A. Johanson, and F. Murad, *unpublished observations*). When values were normalized for urinary creatinine (μmoles nucleotide/g creatinine), values were considerably higher in young children and decreased with age. Hardman et al. (1969) found that excretion rates (relative to body weight) were higher in younger than older rats. However, Paul et al. (1971) stated that excretion rates (μmoles/day) of cyclic AMP in children were no different than in adults.

B. Plasma

The tissue sources of basal levels of plasma cyclic nucleotides are unknown. Liver and kidney appear to be the major sources of plasma cyclic AMP under the influence of glucagon and parathyroid hormone, respectively. Administration of glucagon in hepatectomized dogs failed to increase plasma or urinary cyclic AMP (Broadus et al., 1970b). Similarly in nephrectomized patients maintained on dialysis, parathyroid hormone failed to increase plasma cyclic AMP (Kaminsky et al., 1970b). Recently, Blonde et al. (1972) reported that, in catheterized anesthetized dogs, superior mesenteric vein plasma cyclic GMP was significantly higher than that in hepatic vein or aorta, indicating that the small intestine and/or colon adds some cyclic GMP to the circulation and that most of this is removed by the liver. Plasma cyclic AMP levels in these three vessels were similar as were those in the superior vena cava, coronary sinus, pulmonary artery, and femoral vein. These studies suggest that under basal conditions many tissues can contribute small quantities of cyclic nucleotides to plasma.

Incubation and perfusion media from *in vitro* or *in situ* studies with erythrocytes (Davoren and Sutherland, 1963), platelets (Cole et al., 1971), fat cells (Butcher et al., 1968; Kuo and DeRenzo, 1969; Murad et al., 1969*b*), leukocytes (Stossel et al., 1970), cerebellar slices (Kakiuchi and Rall, 1968), liver (Broadus et al., 1970*b*), *E. coli* (Makman and Sutherland, 1965), and slime mold (Konijn et al., 1968) contain variable quantities of cyclic AMP. In general, hormonal stimulation of cyclic AMP accumulation in intact cell systems has resulted in accumulation of nucleotide in the medium. The rate of this process has varied in different systems and, except for pigeon erythrocytes (Davoren and Sutherland, 1963), the intracellular concentration has greatly exceeded that of the medium. The appearance of cyclic AMP in the perfusion medium with glucagon in rat liver *in situ* has been found to occur under conditions in which changes in intracellular concentrations were undetectable (Broadus et al., 1970*b*). Presumably release, leakage, or transport of cyclic AMP from cells to interstitial fluid could be one method of regulating intracellular cyclic AMP levels.

Although extracellular cyclic AMP may be a mechanism for intercellular communication as has been demonstrated with the slime mold (Konijn et al., 1968), its importance in plasma remains to be established. The low concentration in plasma and the relatively high concentrations required for effects in intact cell systems would suggest it to be unimportant unless mechanisms exist which permit its concentration.

Except for the recent studies of Blonde et al. (1972), no reports are available describing the release of cyclic GMP from tissues.

C. Miscellaneous Fluids

Several laboratories have reported cyclic AMP in human cerebrospinal fluid with levels in the range of 10 to 40 nM (Murad et al., 1969*c*; Robison et al., 1970; Broadus et al., 1970*a*). In addition, one laboratory has reported low levels of cyclic GMP in spinal fluid (0 to 7 nM) (Broadus et al., 1970*a*). Except for one study reporting similar spinal fluid concentrations of cyclic AMP in several normal, epileptic, and depressed patients (Robison et al., 1970), this area has received little attention.

Very high concentrations (30 μM) of cyclic AMP have been reported in salivary secretions (Stefanovich and Wells, 1971). However, another laboratory observed very low levels (2 to 13 pmoles/mg protein or approximately 10 to 60 nM) (T. W. Rall, *personal communication*). The reason for the large discrepancy is unknown. To date there have been no further reports with reference to either continued basic or clinical investigation with these observations. Similarly cyclic AMP has been described in human milk (Kobata et al., 1961), amniotic fluid (Broadus et al., 1971), and seminal fluid (Gray, 1970) and rat bile (Levine et al., 1969). Nothing is known about factors which regulate its concentration in these fluids. Amniotic fluid is derived in part from fetal urine (Chez et al., 1964); however, it is not known what other sources, if any, contribute to the presence of cyclic AMP.

IV. EFFECTS OF HORMONES AND OTHER AGENTS ON CYCLIC NUCLEOTIDE LEVELS IN PLASMA AND URINE

A. Parathyroid Hormone and Calcitonin

The studies of Chase and Aurbach (1967) and Chase et al. (1969b) with parathyroid hormone administration demonstrated pronounced rapid increases in urinary cyclic AMP levels in rats and humans. These effects preceded the well-known phosphaturic effect of parathyroid hormone. These studies subsequently led to the demonstration of parathyroid hormone stimulation of renal cortical and bone adenyl cyclase (Chase and Aurbach, 1968; Chase et al., 1969a). The effects of parathyroid hormone on urinary cyclic AMP and renal and bone adenyl cyclase have been confirmed in several laboratories. In addition, the synthetic biologically active 1–34 amino acid fragment of parathyroid hormone can mimic the effects of the native polypeptide on both cyclase activation and induced cyclic AMP excretion (Potts et al., 1971). The effects of intravenous parathyroid hormone on urinary cyclic AMP occur rapidly, and excretion rates may increase as much as 200-fold (Table 1). Whereas large doses of parathyroid hormone (200 U) also increase urinary excretion of cyclic GMP (Table 1), low doses increase only cyclic AMP excretion (Kaminsky et al., 1970b). Parathyroid hormone can also increase plasma levels of both cyclic nucleotides (Table 1). The effects of parathyroid hormone on plasma cyclic AMP are presumably of renal origin since they were not observed after its administration in nephrectomized patients (Kaminsky et al., 1970b). In the latter study, plasma cyclic GMP was not examined. Studies summarized below suggest that parathyroid hormone is the most potent agent influencing the kidney's contribution to urinary cyclic AMP. The ability of this hormone to produce large increases in urinary cyclic AMP excretion has been used by some laboratories to evaluate patients with parathyroid and other calcium disorders (see below).

Intravenous EDTA infusions have been administered to patients to decrease serum calcium and enhance parathyroid hormone secretion. In

TABLE 1. Effects of parathyroid hormone on plasma and urine cyclic nucleotides

Plasma			Urine				
				cAMP		cGMP	
Time (min)	cAMP	cGMP	Time (min)	nmoles/ min	µmoles/ g creat.	nmoles/ min	µmoles/ g creat.
	(nM)						
0	9.9	2.7	−100−0	1.8	1.6	0.47	0.41
30	46.6	10.9	0−30	393.6	303.0	16.30	12.56
60	27.7	4.5	30−60	14.1	12.4	3.80	3.33
90	18.0	3.5	60−90	4.8	4.1	0.43	0.36
120	9.9	−	90−120	2.4	2.3	0.15	0.23

A patient with surgical hypoparathyroidism resulting from parathyroid removal at the time of thyroidectomy was given 200 U of PTH intravenously (zero time). Purified plasma samples and urines were analyzed for cyclic nucleotide content with protein-binding methods (Gilman, 1970; Murad et al., 1971a).
From Murad and Weitzman, 1972.

similar studies EDTA has produced increases in cyclic AMP excretion rates (Kaminsky et al., 1970b; Estep et al., 1970). Some laboratories have also observed that calcium infusions decrease cyclic AMP excretion in man (Kaminsky et al., 1970b; Estep et al., 1970) and rat (Chase and Aurbach, 1967). Although the latter has been used as evidence to indicate that normally much of the kidney's contribution to urinary cyclic AMP is under the influence of parathyroid hormone, we have found that the effects of calcium infusion on cyclic AMP and cyclic GMP excretion are quite variable (Murad and Pak, 1972b).

Although calcitonin can antagonize the effects of parathyroid hormone on bone resorption, it is capable of stimulating renal cortical and bone adenyl cyclase and enhancing cyclic AMP accumulation in these tissues (Murad et al., 1969a, 1970; Marcus et al., 1971). The effects of calcitonin on cyclic AMP accumulation do not appear to result from decreased phosphodiesterase activity (Murad et al., 1970). Evidence has been previously presented to suggest that parathyroid hormone and calcitonin increase cyclic AMP accumulation in different cell types in bone and perhaps kidney to account for their apparent physiological antagonism (Murad et al., 1970; Murad and Weitzman, 1972). Calcitonin can increase urinary excretion of cyclic AMP in humans (Murad and Pak, 1972b) and dogs (F. Murad, H. B. Brewer, and C. Y. Pak, *unpublished observations*). However, in parathyroidectomized rats (Chase and Aurbach, 1967), dogs, and humans (Murad and Pak, 1972b), no effects are observed. Presumably the effects of calcitonin are secondary to parathyroid hormone secretion as a result of its hypocalcemic effect. However, recent studies have also demonstrated direct effects of calcitonin on parathyroid hormone release *in vitro* which are probably mediated by cyclic AMP (Sherwood and Abe, 1972).

B. Glucagon

Glucagon has also been shown to increase plasma (Broadus et al., 1970b) and urinary (Broadus et al., 1970b; Taylor et al., 1970; Williams et al., 1972) cyclic AMP levels (Table 2). The studies of Broadus and his colleagues indicate that the increase in urinary cyclic AMP with glucagon is due to elevated plasma levels and increased glomerular filtration with little or no direct renal contribution. One laboratory reported that glucagon can increase urine levels of cyclic GMP (Williams et al., 1972); no effects on cyclic GMP were observed by another group (Broadus et al., 1970b). We have observed small effects (approximately a doubling) of glucagon on urinary cyclic GMP levels without changes in plasma levels (Table 2). These studies suggest that some urinary cyclic GMP may be of renal origin under the influence of glucagon.

Glucagon is capable of stimulating adenyl cyclase or cyclic AMP accumulation in a number of tissues including liver (Rall and Sutherland, 1958), heart (Murad, 1968; Murad and Vaughan, 1969; Levey et al., 1969, 1970), pancreatic islets (Turtle and Kipnis, 1967), fat (Butcher et al., 1968), skeletal muscle (Murad et al., 1972), kidney (Marcus and Aurbach, 1969; Murad et al., 1970), and parathyroid glands (Gitelman et al., 1972). However, the isolated studies with hepatectomized dogs suggest that the major or sole source of increased plasma cyclic AMP with glucagon is the liver (Broadus et al., 1970b).

TABLE 2. *Effects of glucagon on plasma and urine cyclic nucleotides*

	Plasma			Urine			
					cAMP		cGMP	
Time (min)	cAMP	cGMP	Time (min)	nmoles/ min	μmoles/ g creat.	nmoles/ min	μmoles/ g creat.
	(nM)						
0	26.8	8.8	−90−0	6.1	6.7	1.0	1.10
20	290.7	7.6	0−30	25.7	24.3	1.7	1.65
80	40.2	5.3	30−90	27.4	28.3	2.1	2.13
140	23.5	6.6	90−150	15.5	15.2	1.2	1.15

A hyperparathyroid patient was given 1 mg of glucagon intravenously (zero time). Plasma and urinary cyclic nucleotides were evaluated as in Table 1.

C. Antidiuretic Hormone

The relatively constant cyclic AMP excretion rates observed by us and others with variable levels of fluid intake, urine volume, and plasma and urine osmolarity suggest that antidiuretic hormone normally has little or no effect on cyclic AMP excretion. Relatively large doses of vasopressin have produced small effects in rats and humans (less than two- to threefold) in several laboratories (Takahashi et al., 1966; Chase and Aurbach, 1967; Taylor et al., 1970; Bell et al., 1971; Fichman and Brooker, 1970, 1971). However, other laboratories including our own have failed to confirm these observations (Kaminsky et al., 1970b; Williams et al., 1972). The inability to resolve this simple issue is inexplicable and frustrating, particularly since several laboratories have spent considerable time in examining cyclic AMP excretion rates and effects of antidiuretic hormone in pituitary diabetes insipidus, nephrogenic diabetes insipidus, and inappropriate ADH syndromes (see below). The data thus far suggest that physiologic levels of antidiuretic hormone have little or no effect on cyclic AMP excretion. However, large doses of the hormone are capable of producing relatively small and variable effects on cyclic AMP excretion.

One laboratory has reported that the small effects of antidiuretic hormone on cyclic AMP excretion can be augmented after the administration of chlorpropamide (Fichman and Brooker, 1970). This effect of chlorpropamide has been attributed to its ability to inhibit renal phosphodiesterase (Brooker and Fichman, 1971).

Several hormones (parathyroid hormone, calcitonin, antidiuretic hormone, glucagon, and catecholamines) are capable of acting on the kidney to influence electrolyte and/or water metabolism. Presumably all are affecting different cell types or at least cells distinguishable from those affected by parathyroid hormone, since only parathyroid hormone produces a marked direct increase in cyclic AMP secretion by the nephron.

D. Catecholamines

It has been reported that catecholamines increase levels of plasma (Kaminsky et al., 1970a) and urinary (Kaminsky et al., 1970a; Williams et al., 1972) cyclic AMP and cyclic GMP. The increases in cyclic AMP can be prevented with beta-adrenergic blockers whereas the increases in cyclic GMP are prevented with alpha blockers (Kaminsky et al., 1970a). Whether these effects are direct or indirect is unknown. It is of interest that stimulatory effects of acetylcholine have been observed on cyclic GMP accumulation in rat heart (George et al., 1970) and rat ductus deferens preparations (Schultz et al., 1972). Increased compensatory cholinergic activity may contribute to the observed changes in cyclic GMP levels with catecholamines. Also, catecholamines can release parathyroid hormone; however, this appears to be a beta-adrenergic effect (Sherwood and Abe, 1972). Studies with catecholamines on cyclic nucleotide levels after parathyroid ablation or in the presence of atropine have not been reported.

E. Growth Hormone and Insulin

Acute administration of growth hormone does not alter cyclic AMP excretion (Taylor et al., 1970; A. Johanson and F. Murad, *unpublished observations*). Although direct studies with insulin have not been reported, the levels of cyclic AMP and cyclic GMP in urine during an oral glucose tolerance test suggest it to have no effect (Broadus et al., 1970b). Urine levels of cyclic AMP and cyclic GMP were unchanged except for a slight rise in cyclic AMP 3 to 4 hr after glucose ingestion when blood glucose levels decreased below basal levels. The late effect described may be secondary to glucagon release.

F. Other Hormones

Hardman et al. (1966) reported that in rats hypophysectomy decreased urinary cyclic GMP without altering cyclic AMP. The excretion of cyclic GMP could be restored by the administration of a mixture of luteinizing hormone, follicle-stimulating hormone, thyroid-stimulating hormone, prolactin, growth hormone, ACTH, and hydrocortisone; however, hydrocortisone and thyroxine together were also partially effective. Thyroidectomized rats excreted less cyclic GMP; however, large doses of thyroxine restored the level. Williams et al. (1972) found no effects of ACTH, hydrocortisone, or triodothyronine on cyclic AMP excretion; effects of these agents on cyclic GMP excretion were not reported.

Although these studies suggest that gonadotropins and steroids have little or no effect on cyclic AMP excretion, Taylor et al. (1970) have reported increases in cyclic AMP excretion during pregnancy and in the midst of the menstrual cycle (approximately 40 to 50% increases in each instance). The reasons for these small increases are not known.

V. CYCLIC NUCLEOTIDES IN DISEASE STATES

A. Calcium Disorders

The first report of altered cyclic nucleotide metabolism in human disease was that of Chase et al. (1969b). These investigators found that patients with pseudohypoparathyroidism failed to increase (less than two- to fourfold increases) their urinary excretion of cyclic AMP after administration of parathyroid hormone unlike normals, hyperparathyroid patients, idiopathic or surgical hypoparathyroid patients, or pseudohypoparathyroid patients. Previously the diagnosis of pseudohypoparathyroidism had been made in patients by the lack of a phosphaturic and/or hypercalcemic response to parathyroid hormone administration; the effects in normals, however, are frequently small and quite variable. Because of the relatively large increase in cyclic AMP excretion observed normally and the small effect seen with pseudohypoparathyroid patients (Table 3), the examination of urinary cyclic AMP before and after parathyroid hormone is a much better diagnostic test for the disorder, as suggested by Chase et al. (1969b). Patients with pseudohypoparathyroidism also demonstrated little or no increase in plasma cyclic AMP and a smaller increase in urinary and plasma cyclic GMP with parathyroid hormone than normal patients (Table 3) (Murad and Weitzman, 1972).

TABLE 3. *Effect of parathyroid hormone on plasma and urinary cyclic nucleotides in a pseudohypoparathyroid patient*

	Plasma			Urine			
	cyclic AMP	cyclic GMP		cyclic AMP		cyclic GMP	
Time (min)	(nM)		Time (min)	nmoles/ min	μmoles/ g creat.	nmoles/ min	μmoles/ g creat.
0	8.7	6.5	−60–0	0.8	0.7	0.7	0.66
30	7.8	7.0	0–30	7.9	5.8	0.7	0.51
60	13.9	13.1	30–60	1.7	1.3	0.9	0.69
90	12.4	7.4	60–90	2.1	1.9	1.0	0.90
120	8.6	—	90–120	1.1	1.5	0.4	0.46

A patient with pseudohypoparathyroidism was given 200 U of parathyroid hormone intravenously (zero time). Cyclic nucleotides were determined as in Table 1.
From Murad and Weitzman, 1972.

Several laboratories have also reported that hyperparathyroid patients excrete increased amounts and hypoparathyroid patients decreased amounts of cyclic AMP (Chase et al., 1969b; Kaminsky et al., 1970b; Taylor et al., 1970; Estep et al., 1970; Murad et al., 1971b; Murad and Pak, 1972a,b). Because of diurnal cyclic AMP excretion, basal cyclic AMP excretion rates were examined in one or more 24-hr urine samples in patients with normal endogenous creatinine clearances (greater than 90 ml/min) (Murad et al., 1971b; Murad and Pak, 1972a,b; Murad and Weitzman, 1972).

Hypercalcemic and normocalcemic hyperparathyroid patients excreted significantly larger amounts of cyclic AMP than did normal subjects (Fig. 2 and Table 4). Hypoparathyroid patients (idiopathic, post surgical, and pseudo-) excreted significantly less cyclic AMP than normals. Only one

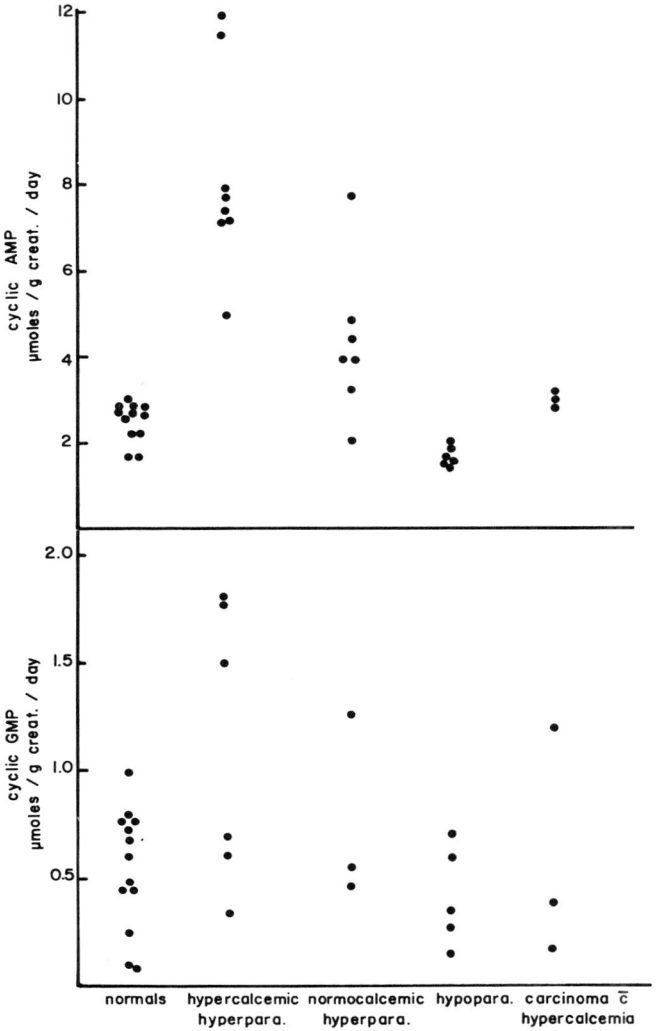

FIG. 2. Daily excretion rates of cyclic nucleotides (from Murad and Pak, 1972b). From one to five 24-hr urine samples were collected from each patient during hospitalization. The points represent the mean value for each patient.

patient with hyperparathyroidism and three with hypoparathyroidism excreted quantities of cyclic AMP that fell within the normal range. On the other hand, patients with nonparathyroid carcinoma and hypercalcemia, idiopathic osteoporosis, and nephrolithiasis excreted normal quantities of cyclic AMP (Table 4) (Murad and Pak, 1972a,b; Murad and Weitzman, 1972). One patient that we examined with sarcoidosis and hypercalcemia had a low cyclic AMP excretion rate of 1.26 μmoles/g creatinine. Dohan et al. (1971) reported that patients with nonparathyroid carcinoma and hypercalcemia have low cyclic AMP excretion rates. Studies to date indicate that cyclic AMP excretion rates can be successfully used to evaluate parathyroid function in either normocalcemic or hypercalcemic patients.

TABLE 4. *Cyclic nucleotide excretion rates with various calcium disorders*

Diagnosis	Urinary cyclic AMP	Urinary cyclic GMP
	(μmoles/g creat. $\pm$ S.E.)	
Normal	2.50 $\pm$ 0.13 (12)	0.55 $\pm$ 0.08 (13)
Hypercalcemic hyperparathyroidism	8.19 $\pm$ 0.82 (8)[a]	1.12 $\pm$ 0.26 (6)[a]
Normocalcemic hyperparathyroidism	4.29 $\pm$ 0.66 (7)[a]	0.75 $\pm$ 0.25 (3)
After surgery for hyperparathyroidism	3.13 $\pm$ 0.32 (15)	0.62 $\pm$ 0.09 (13)
Hypoparathyroidism	1.48 $\pm$ 0.10 (10)[a]	0.42 $\pm$ 0.10 (5)
Idiopathic osteoporosis	3.29 $\pm$ 0.36 (7)	0.73 $\pm$ 0.12 (7)
Nephrolithiasis	2.80 $\pm$ 0.17 (20)	0.56 $\pm$ 0.10 (17)
Malignancy and hypercalcemia	3.08 $\pm$ 0.07 (3)	0.58 $\pm$ 0.11 (3)

One to five 24-hr urine specimens from each hospitalized normal subject and patient with the indicated disorders were analyzed for cyclic nucleotide content. The number of patients examined are included in parentheses. Values are taken from reports of Murad and Pak (1972a,b) and Murad and Weitzman (1972).
[a] Values are significantly different than normals ($p < 0.025$).

The specificity of elevated cyclic AMP excretion rates for hyperparathyroidism remains to be established with additional studies. As discussed below, preliminary studies with several patients with pheochromocytoma (N. Atuk and F. Murad, *unpublished observations*), cystic fibrosis (Simopoulos et al., 1971), and inappropriate ADH syndrome (R. Weitzman and F. Murad, *unpublished observations;* Murad and Pak, 1972b) have demonstrated increased cyclic AMP excretion rates. False negative tests may also be observed. For example, one of our patients with a parathyroid adenoma and hypercalcemia with decreased renal function (endogenous creatinine clearance of 25 ml/min) had a normal cyclic AMP excretion rate (Murad and Pak, 1972b). Patients with decreased renal function have been reported to have low cyclic AMP excretion rates (Taylor et al., 1970; Estep et al., 1970) and decreased cyclic AMP excretion with parathyroid hormone (Estep et al., 1970).

The hypocalcemia and decreased response to parathyroid hormone in hypomagnesemia (alcoholism, gastrointestinal malabsorptive states, etc.) (Estep et al., 1969) suggest that patients with hyperparathyroidism and hypomagnesemia would have normal or decreased (falsely negative) cyclic AMP excretion rates. However, this remains to be established.

Presumably with additional knowledge of factors that influence cyclic nucleotide excretion and the activity of parathyroid hormone, false positive and negative tests with cyclic AMP excretion in parathyroid disorders can be avoided. Although other factors and diseases can alter cyclic AMP excretion, cyclic AMP excretion rates can be used currently in conjunction with other laboratory tests and the clinical setting to evaluate parathyroid function.

Of the patients that we examined with various calcium disorders only patients with hypercalcemic hyperparathyroidism were found to have significantly elevated cyclic GMP excretion rates (Fig. 2, Table 4). However, some of these patients had values that fell within the normal range. It is of interest that high doses of parathyroid hormone can increase plasma and urinary cyclic GMP levels (Table 1) (Kaminsky et al., 1970b; Williams et al., 1972). Low doses of parathyroid hormone increase only cyclic AMP levels in these fluids (Kaminsky et al., 1970b).

Upon removal of hyperplastic or adenomatous parathyroid glands (Table 4, Fig. 3), the excretion rates of cyclic AMP and cyclic GMP declined and returned to normal with few exceptions (Murad et al., 1971b; Murad and Pak, 1972a,b). Kaminsky et al. (1970b) and Taylor et al. (1970) also reported decreased cyclic AMP excretion rates after parathyroid surgery for hyperparathyroidism. Kaminsky and his colleagues also reported that a patient with a "PTH-secreting" pulmonary tumor had a high cyclic AMP excretion rate which decreased after radiotherapy.

We have also found in preliminary studies that patients with cystic fibrosis have elevated cyclic AMP excretion rates which may be attributable to underlying hyperparathyroidism (Simopoulos et al., 1971). Additional studies with cystic fibrosis are currently in progress in our laboratory.

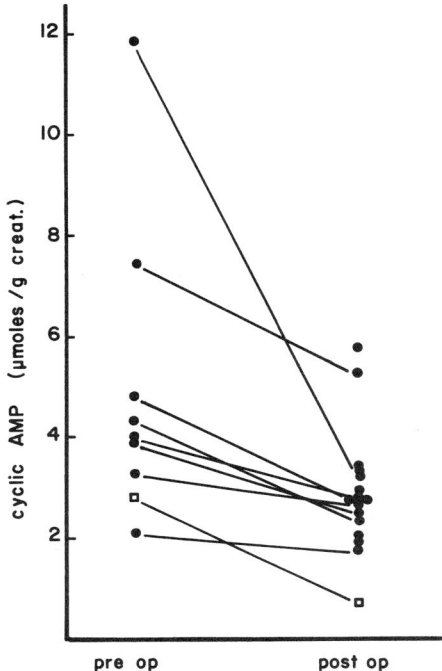

FIG. 3. Cyclic AMP excretion in hyperparathyroidism pre- and postoperatively (from Murad and Pak, 1972b). Cyclic AMP excretion rates were examined in nine hyperparathyroid subjects before and after parathyroid surgery. An additional seven patients were studied only postoperatively and are included. The patient represented by the open square had decreased renal function to account for her normal excretion rate.

B. Diabetes Insipidus

Several laboratories have reported that patients with diabetes insipidus respond to vasopressin with small increases in cyclic AMP excretion (Bell et al., 1971; Fichman and Brooker, 1971). These investigators also reported that patients with nephrogenic diabetes insipidus fail to increase their cyclic AMP excretion with vasopressin infusion. Fichman and Brooker (1971) found that the latter patients also had a smaller increase in urinary cyclic AMP with parathyroid or glucagon administration. These patients failed to demonstrate increased urine osmolarity with cyclic AMP infusion unlike normal patients or patients with pituitary diabetes insipidus (Bell et al., 1971). These studies suggest that patients with nephrogenic diabetes insipidus have multiple defects: (1) inability to generate and/or excrete cyclic

AMP with vasopressin, (2) decreased response to other hormones such as glucagon and parathyroid hormone, and (3) diminished renal response to cyclic AMP.

Whereas we found three patients with an inappropriate ADH syndrome due to "ADH-secreting" tumors to have elevated urinary excretion rates of cyclic AMP (R. Weitzman and F. Murad, *unpublished observations;* Murad and Pak, 1972*b*), we and others have not seen significant changes in urinary cyclic AMP with vasopressin administration.

C. Affective Disorders

Elevated and decreased cyclic AMP excretion rates have been reported with mania and depressive disease, respectively (Abdulla and Hamadah, 1970; Paul et al., 1971). With drug therapy (i.e., tricyclic antidepressants, L-DOPA, or lithium carbonate), the improvement in affect in these patients was associated with cyclic AMP excretion rates that returned to normal. The utilization of cyclic AMP excretion rates for diagnosis and/or examination of the pathophysiology in affective disorders must await additional studies. To date, tissue sources of the altered urinary cyclic AMP are unknown; the lack of spinal fluid changes in cyclic AMP with depression (Robison et al., 1970) suggests that it is not from the central nervous system directly. Levels of catecholamines and their metabolites in urine are altered in affective illnesses (Weil-Malherbe and Ström-Olson, 1958) and may account in part for these observations.

D. Psoriasis

Recently, Voorhees and his associates (1972) reported that epidermis from psoriatic skin contains less adenyl cyclase activity than normal epidermis. In addition, increased cyclic AMP levels in epidermis are associated with decreased mitotic activity and proliferation. These observations are analogous to the report of increased cyclic AMP levels during the stationary phase of growth of contact-inhibited fibroblasts (Otten et al., 1971). Such studies may provide models for designing and examining new therapeutic agents in psoriasis.

E. Asthma and Atopic Dermatitis

It has been postulated that asthma represents a defective or diminished beta-adrenergic response in bronchial smooth muscle and perhaps in other tissues as well (Szentivanyi, 1968). Recent studies have lent some additional support to this theory by demonstrating that asthmatic patients have less of an increase in urinary cyclic AMP after epinephrine but not glucagon administration than normal patients (Bernstein et al., 1972). Also, it has been recently reported that leukocytes from asthmatic patients (Smith and Parker, 1970) or patients with atopic dermatitis (Parker and Eisen, 1972) accumulated less cyclic AMP in the presence of catecholamine than leukocytes from normal patients. Adenyl cyclase activities in skin biopsies from involved and uninvolved areas of skin in patients with atopic dermatitis were similar (Mier and Urselmann, 1970). Effects of hormones were not examined in this study. These studies will undoubtedly lead to examination of cyclase and phosphodiesterase activities in biopsy or postmortem bronchial smooth muscle preparations from normal and asthmatic patients.

VI. SUMMARY AND SPECULATION

Considerable headway has been made during the past few years with the incorporation of cyclic nucleotide metabolism in clinical studies. Clearly cyclic nucleotide levels in urine and plasma have proved to be useful additional diagnostic tools in the evaluation of several disorders. However, it is also quite apparent that we have only begun to scratch the surface in terms of clinical application.

It is not so farfetched that one could envisage genetic deficiencies in cyclase(s) or one of the phosphodiesterases analogous to the various deficiencies in glycogen storage diseases. Indeed, protein kinase deficiency has recently been described (Hug et al., 1970). Studies by Chase et al. (1969b) suggest that pseudohypoparathyroidism represents one such genetic defect in the adenyl cyclase–cyclic AMP system. At present there are not sufficient data to establish whether this disease represents a defect in parathyroid hormone receptors in kidney (and perhaps also bone) or a deficiency in adenyl cyclase or both. With the availability of patients and specimens our current methodology is sufficient to resolve such questions. Answers to such questions will undoubtedly lead to new models for drug testing and new forms of therapy.

The studies with adenyl cyclase from failing hearts also indicate that disease processes (e.g., congestive heart failure) can result in secondary changes in the receptor–adenyl cyclase system. Levey et al. (1970) found that adenyl cyclase from failing myocardium had a diminished response to glucagon; the effects of fluoride ion and norepinephrine were unaltered. These observations helped resolve some of the controversy regarding the therapeutic efficacy of glucagon as an inotropic agent in normal and failing hearts.

The studies in the past several years of the effects of hypothalamic-releasing factors on the release of various pituitary hormones suggest that the administration of cyclic nucleotide analogues and phosphodiesterase inhibitors may prove useful in evaluating pituitary reserve and distinguishing hypothalamic disorders.

Examination of arterial-venous cyclic nucleotide differences may prove to be one of the more direct assessments of an organ's response to an appropriate hormone.

At present there are numerous examples of simple as well as complex questions in clinical medicine in which our current methodology with cyclic nucleotides can be applied to obtain additional information regarding metabolic regulation, pathophysiology, diagnosis, and therapy.

VII. ACKNOWLEDGMENTS

Some of these studies were supported by grants from the U.S. Public Health Service (AM 15316, Clinical Research Center grant RR-304), the National Cystic Fibrosis Research Foundation, and the Virginia Heart Association. The author is a recipient of U.S. Public Health Service Research Career Development Award, AM-70,456.

VIII. REFERENCES

Abdulla, Y. H., and Hamadah, K. H. (1970): 3',5' Cyclic adenosine monophosphate in depression and mania. *Lancet,* Feb. 21:378–381.

Ashman, D. F., Lipton, R. L., Melicow, M. M., and Price, T. D. (1963): Isolation of adenosine 3',5' monophosphate and guanosine 3',5' monophosphate from rat urine. *Biochemical and Biophysical Research Communications,* 11:330–334.

Bell, N. H., Clark, C. M., Avery, S., Sinha, T., Trygstad, C., and Allen, D. O. (1971): Demonstration of defects in the formation of and response to cyclic 3',5' adenosine monophosphate in vasopressin-resistant diabetes insipidus. *Clinical Research,* 19:685.

Bernstein, R. A., Linarelli, L., Facktor, M. A., Friday, G. A., Drash, A., and Fireman, P. (1972): Decreased urinary cyclic 3',5'-adenosine monophosphate after epinephrine in asthmatic patients. *Journal of Allergy and Clinical Immunology,* 49:86.

Blonde, L., Wehmann, R. E., and Steiner, A. L. (1972): Regulation of cyclic nucleotide levels in canine plasma. *Clinical Research,* 20:541.

Broadus, A. E., Hardman, J. G., Kaminsky, N. I., Ball, J. H., Sutherland, E. W., and Liddle, G. W. (1971): Extracellular cyclic nucleotides. *Annals of the N.Y. Academy of Sciences,* 185:50–66.

Broadus, A. E., Kaminsky, N. I., Hardman, J. G., Sutherland, E. W., and Liddle, G. W. (1970a): Kinetic parameters and renal clearances of plasma adenosine 3',5'-monophosphate and guanosine 3',5'-monophosphate in man. *Journal of Clinical Investigation,* 49:2222–2236.

Broadus, A. E., Kaminsky, N. I., Northcutt, R. C., Hardman, J. G., Sutherland, E. W., and Liddle, G. W. (1970b): Effects of glucagon on adenosine 3',5'-monophosphate and guanosine 3',5'-monophosphate in human plasma and urine. *Journal of Clinical Investigation,* 49:2237–2245.

Brooker, G., and Fichman, M. (1971): Chlorpropamide and tolbutamide inhibition of adenosine

3',5'-cyclic monophosphate phosphodiesterase. *Biochemical and Biophysical Research Communications*, 42:824-828.

Butcher, R. W., Baird, C. E., and Sutherland, E. W. (1968): Effects of lipolytic and antilipolytic substances on adenosine 3',5'-monophosphate levels in isolated fat cells. *Journal of Biological Chemistry*, 243:1705-1712.

Butcher, R. W., Ho, R. J., Meng, H. C., and Sutherland, E. W. (1965): The measurement of adenosine 3',5'-monophosphate in tissues and the role of the cyclic nucleotide in the lipolytic response of fat to epinephrine. *Journal of Biological Chemistry*, 240:4515-4523.

Butcher, R. W., and Sutherland, E. W. (1962): Purification and properties of cyclic 3',5'-nucleotide phosphodiesterase and use of this enzyme to characterize adenosine 3',5'-phosphate in human urine. *Journal of Biological Chemistry*, 237:1244-1250.

Chase, L. R., and Aurbach, G. D. (1967): Parathyroid function and the excretion of 3',5'-adenylic acid. *Proceedings of the National Academy of Sciences*, 58:518-525.

Chase, L. R., and Aurbach, G. D. (1968): Renal adenyl cyclase: Anatomically separate sites for parathyroid hormone and vasopressin. *Science*, 159:545-547.

Chase, L. R., Fedak, S. A., and Aurbach, G. D. (1969a): Activation of skeletal adenyl cyclase by parathyroid hormone in vitro. *Endocrinology*, 84:761-768.

Chase, L. R., Melson, G. L., and Aurbach, G. D. (1969b): Pseudohypoparathyroidism: Defective excretion of 3',5'-AMP in response to parathyroid hormone. *Journal of Clinical Investigation*, 48:1832-1844.

Chez, R. A., Smith, F. G., and Hutchinson, D. L. (1964): Renal function in the intrauterine primate fetus. *American Journal of Obstetrics and Gynecology*, 90:128-131.

Cole, B., Robison, G. A., and Hartman, R. C. (1971): Studies on the role of cyclic AMP in platelet function. *Annals of the N.Y. Academy of Sciences*, 185:477-487.

Davoren, P. R., and Sutherland, E. W. (1963): The effects of L-epinephrine and other agents on the synthesis and release of adenosine 3',5'-phosphate by whole pigeon erythrocytes. *Journal of Biological Chemistry*, 238:3009-3015.

Dohan, P., Yamashita, K., Larsen, R., Davis, B., Deftos, L., and Field, J. (1971): Urinary cyclic AMP and parathyroid hormone in the differential diagnosis of hypercalcemia. *Clinical Research*, 19:474.

Estep, H., Fratkin, M., Moser, A., and Robinson, F. (1970): Cyclic AMP excretion in hypercalcemic states. *Clinical Research*, 18:358.

Estep, H., Shaw, W. A., Watlington, C., Hobe, R., Holland, W., and Tucker, S. G. (1969): Hypocalcemia due to hypomagnesemia and reversible parathyroid hormone unresponsiveness. *Journal of Clinical Endocrinology and Metabolism*, 29:842-848.

Fichman, M., and Brooker, G. (1970): Cyclic AMP and the antidiuretic effect of chlorpropamide in diabetes insipidus. *Clinical Research*, 18:121.

Fichman, M., and Brooker, G. (1971): Deficient renal cyclic AMP production in nephrogenic diabetes insipidus. Presented at the 53rd annual meeting of the Endocrine Society, San Francisco, California, June.

George, W. J., Polson, J. P., O'Toole, A. G., and Goldberg, N. D. (1970): Elevation of guanosine 3',5'-cyclic phosphate in rat heart after perfusion with acetylcholine. *Proceedings of the National Academy of Sciences*, 66:398-403.

Gilman, A. G. (1970): A protein binding assay for adenosine 3',5'-phosphate. *Proceedings of the National Academy of Sciences*, 67:305-312.

Gitelman, H. J., Alderman, F. R., and Dufresne, L. R. (1972): In vivo regulation of parathyroid gland cyclic 3',5'-AMP content. *Clinical Research*, 20:35.

Goldberg, N. D., Dietz, S. B., and O'Toole, A. G. (1969): Cyclic guanosine 3',5'-monophosphate in mammalian tissues and urine. *Journal of Biological Chemistry*, 244:4458-4466.

Gray, J. P. (1970): Ph.D. dissertation. Vanderbilt University, Nashville, Tennessee.

Hardman, J. G., Davis, J. W., and Sutherland, E. W. (1966): Measurement of guanosine 3',5'-monophosphate and other cyclic nucleotides. *Journal of Biological Chemistry*, 241:4812-4815.

Hardman, J. G., Davis, J. W., and Sutherland, E. W. (1969): Effects of some hormonal and other factors on the excretion of guanosine 3′,5′-monophosphate and adenosine 3′,5′-monophosphate in rat urine. *Journal of Biological Chemistry*, 244:6354–6362.

Hug, G., Schubert, W. K., and Chuck, G. (1970): Loss of cyclic 3′,5′-AMP dependent kinase and reduction of phosphorylase kinase in skeletal muscle of a girl with deactivated phosphorylase and glycogenosis of liver and muscle. *Biochemical and Biophysical Research Communications*, 40:982–988.

Kakiuchi, S., and Rall, T. W. (1968): The influence of chemical agents on the accumulation of adenosine 3′,5′-phosphate in slices of rabbit cerebellum. *Molecular Pharmacology*, 4:367–378.

Kaminsky, N. I., Ball, J. H., Broadus, A. E., Hardman, J. G., Sutherland, E. W., and Liddle, G. W. (1970a): Hormonal effects on extracellular cyclic nucleotides in man. *Transactions of the Association of American Physicians*, 83:235–244.

Kaminsky, N. I., Broadus, A. E., Hardman, J. G., Jones, D. J., Ball, J. H., Sutherland, E. W., and Liddle, G. W. (1970b): Effects of parathyroid hormone on plasma and urinary adenosine 3′,5′-monophosphate in man. *Journal of Clinical Investigation*, 49:2387–2394.

Kobata, A., Kida, M., and Ziro, S. (1961): Occurrence of 3′,5′-cyclic AMP in milk. *Journal of Biochemistry*, 50:275–276.

Konijn, T. M., Barkley, D. S., Chang, Y. Y., and Bonner, J. T. (1968): Cyclic AMP: A naturally occurring acrasin in the cellular slime molds. *American Naturalist*, 102:225–233.

Kuo, J. F., and DeRenzo, E. C. (1969): A comparison of the effects of lipolytic and antilipolytic agents on adenosine 3′,5′-monophosphate levels in adipose cells as determined by prior labelling with adenine-8-^{14}C. *Journal of Biological Chemistry*, 244:2252–2260.

Levey, G. S., and Epstein, S. E. (1969): Activation of adenyl cyclase by glucagon in cat and human heart homogenates. *Circulation Research*, 24:151–156.

Levey, G. S., Prindle, K. H., and Epstein, S. E. (1970): Effects of glucagon on adenyl cyclase activity in the left and right ventricles of heart and in experimentally produced isolated right ventricular failure. *Journal of Molecular Cellular Cardiology*, 1:403–410.

Levine, R. A., Lewis, S. E., Shulman, J., and Washington, A. (1969): Metabolism of cyclic adenosine 3′,5′-monophosphate-8-^{14}C by isolated, perfused rat liver. *Journal of Biological Chemistry*, 244:4017–4022.

Lichtenstein, L. M., and Margoulis, S. (1968): Histamine release *in vitro:* Inhibition by catecholamines and methylxanthines. *Science*, 161:902–903.

Makman, R. S., and Sutherland, E. W. (1965): Adenosine 3′,5′ phosphate in Escherichia coli. *Journal of Biological Chemistry*, 240:1309–1314.

Marcus, R., and Aurbach, G. D. (1969): Bioassay of parathyroid hormone in vitro with a stable preparation of adenyl cyclase from rat kidney. *Endocrinology*, 85:801–810.

Marcus, R. J., Heersche, N. M., and Aurbach, G. D. (1971): Effects of calcitonin on formation of 3′,5′-AMP in bone and kidney. Presented at the 53rd annual meeting of the Endocrine Society, San Francisco, California, June.

Mier, P. D., and Urselmann, E. (1970): The adenyl cyclase of skin: Adenyl cyclase levels in atopic dermatitis. *British Journal of Dermatology*, 83:364–366.

Munson, W. A., Island, D. P., and Liddle, G. W. (1971): Validation of a clinically useful radioimmunoassay of adenosine 3′,5′-phosphate. *Clinical Research*, 19:31.

Murad, F. (1968): Effect of glucagon on heart. *New England Journal of Medicine*, 279:434–435.

Murad, F., Brewer, H. B., and Vaughan, M. (1969a): Effect of thyrocalcitonin on cyclic AMP formation by rat kidney. *Clinical Research*, 17:591.

Murad, F., Brewer, H. B., and Vaughan, M. (1970): Effect of thyrocalcitonin on adenosine 3′,5′-cyclic phosphate formation by rat kidney and bone. *Proceedings of the National Academy of Sciences*, 65:446–453.

Murad, F., and Gilman, A. G. (1971): Adenosine 3′,5′monophosphate and guanosine 3′,5′-monophosphate: A simultaneous protein binding assay. *Biochimica et Biophysica Acta*, 252:397–400.

Murad, F., Manganiello, V., and Vaughan, M. (1969b): Effects of guanosine 3′,5′-monophosphate on glycerol production and accumulation of adenosine 3',5'-monophosphate by fat cells. *Journal of Biological Chemistry,* 245:3352-3360.

Murad, F., Manganiello, V., and Vaughan, M. (1971a): A simple sensitive protein-binding assay for guanosine 3',5'-monophosphate. *Proceedings of the National Academy of Sciences,* 68:736-739.

Murad, F., and Pak, C. Y. (1972a): Clinical applications of cyclic AMP levels. Presented at the Fifth International Congress on Pharmacology, San Francisco, California, July (*in press*).

Murad, F., and Pak, C. Y. (1972b): Urinary excretion of adenosine 3',5'-monophosphate and guanosine 3',5'-monophosphate. *New England Journal of Medicine,* 286:1382-1387.

Murad, F., Pak, C., and Thomas, E. (1971b): Urinary excretion of cyclic AMP as a diagnostic test for altered parathyroid function. Presented at the International Conference on the Physiology and Pharmacology of cyclic AMP, Milan, Italy, July.

Murad, F., Rall, T. W., and Vaughan, M. (1969c): Conditions for the formation, partial purification and assay of an inhibitor of adenosine 3',5'-monophosphate. *Biochimica et Biophysica Acta,* 192:430-445.

Murad, F., Shen, L. C., and Larner, J. (1972): Effect of glucagon on rat diaphragms. *Federation Proceedings,* 31:889.

Murad, F., and Vaughan, M. (1969): Effect of glucagon on rat heart adenyl cyclase. *Biochemical Pharmacology,* 18:1053-1059.

Murad, F., and Weitzman, R. (1972): Effects of parathyroid hormone and calcitonin on cyclic AMP metabolism. Presented at the Fourth International Congress of Endocrinology, Washington, D.C., June (*in press*).

O'Dea, R. F., Haddox, M. K., and Goldberg, N. D. (1971): Interaction with phosphodiesterase of free and kinase-complexed cyclic adenosine 3',5'-monophosphate. *Journal of Biological Chemistry,* 246:6183-6190.

Otten, J., Johnson, G. S., and Pastan, I. (1971): Cyclic AMP levels in fibroblasts: Relationship to growth rate and contact inhibition of growth. *Biochemical and Biophysical Research Communications,* 44:1192-1198.

Parker, C. W., and Eisen, A. Z. (1972): Altered cyclic AMP metabolism in atopic eczema. *Clinical Research,* 20:418.

Patterson, W. D., Hardman, J. G., and Sutherland, E. W. (1971): Metabolism of cyclic nucleotides in rat blood. *Federation Proceedings,* 30:220.

Paul, M. J., Cramer, H., and Goodwin, F. K. (1971): Urinary cyclic AMP excretion in depression and mania. *Archives of General Psychiatry,* 24:327-333.

Potts, J. T., Tregar, G. W., Keutmann, H. T., Niall, H. D., Sauer, R., Deftos, L. J., Dawson, B. F., Hogan, M. L., and Aurbach, G. D. (1971): Synthesis of a biologically active N-terminal tetratriconta peptide of parathyroid hormone. *Proceedings of the National Academy of Sciences,* 68:63-68.

Price, T. D., Ashman, D. F., and Melicow, M. M. (1967): Organophosphates of urine, including adenosine 3',5'-monophosphate and guanosine 3',5' monophosphate. *Biochimica et Biophysica Acta,* 138:452-465.

Rall, T. W., and Sutherland, E. W. (1958): Formation of a cyclic adenine ribonucleotide by tissue particles. *Journal of Biological Chemistry,* 232:1065-1076.

Robison, G. A., Copper, A. J., Whyebrow, P. C., and Prange, A. J. (1970): Cyclic AMP in affective disorders. *Lancet,* 2:1028-1029.

Schultz, G., Hardman, J. G., Davis, J. W., Schultz, K., and Sutherland, E. W. (1972): Determination of cyclic GMP by a new enzymatic method. *Federation Proceedings,* 31:440 Abs.

Sherwood, L. M., and Abe, M. (1972): Adrenergic receptors and the release of parathyroid hormone. *Journal of Clinical Investigation,* 51:88a.

Simopoulos, A., Pak, C., Kattwinkel, J., Murad, F., diSant Agnese, P., and Bartter, F. (1971): Calcium metabolism and parathyroid function in cystic fibrosis. Presented at the American Pediatric Society Meeting, Atlantic City, New Jersey, May.

Smith, J. W., and Parker, C. W. (1970): The responsiveness of leukocyte cyclic adenosine

monophosphate to adrenergic agents in patients with asthma. *Journal of Laboratory Clinical Medicine*, 76:993–994.
Stefanovich, V., and Wells, H. (1971): Cyclic AMP in human saliva. *Federation Proceedings*, 30:565.
Steiner, A. L., Parker, C. W., and Kipnis, D. M. (1970): The measurement of cyclic nucleotides by radioimmunoassay. In: *Advances in Biochemical Psychopharmacology, Vol. 3: Role of Cyclic AMP in Cell Function*, edited by P. Greengard and E. Costa, pp. 89–111. Raven Press, New York.
Steiner, A. L., Pagliara, A. S., Chase, L. R., and Kipnis, D. M. (1972): Adenosine 3',5'-monophosphate and guanosine 3',5'-monophosphate in mammalian tissues and body fluids. *Journal of Biological Chemistry*, 247:1114–1120.
Stossel, T., Murad, F., and Vaughan, M. (1970): Regulation of glycogen metabolism in polymorphonuclear leukocytes. *Journal of Biological Chemistry*, 245:6228–6234.
Szentivanyi, A. (1968): The beta adrenergic theory of the atopic abnormality in bronchial asthma. *Journal of Allergy*, 42:203.
Takahashi, K., Kamimura, M., Shinko, T., and Tsuji, S. (1966): Effects of vasopressin and water-load on urinary adenosine 3',5'-cyclic monophosphate. *Lancet*, 2:967.
Taylor, A. L., Davis, B. B., Pawlson, L. G., Josimovich, J. B., and Mintz, D. H. (1970): Factors influencing the urinary excretion of 3',5'-adenosine monophosphate in humans. *Journal of Clinical Endocrinology and Metabolism*, 30:316–323.
Turtle, J. R., and Kipnis, D. M. (1967): An adrenergic receptor mechanism for the control of cyclic 3',5'-adenosine monophosphate synthesis in tissues. *Biochemical and Biophysical Research Communications*, 28:797–802.
Voorhees, J. J., and Duell, E. A., Kelsey, W. H., and Hayes, E. (1972): Effects of alpha and beta adrenergic stimulation on cyclic AMP formation and mitosis in epidermis. *Clinical Research*, 20:419.
Wehmann, R. E., Blonde, L., and Steiner, A. L. (1972): Simultaneous radioimmunoassay for the measurement of adenosine 3',5'-monophosphate and guanosine 3',5'-monophosphate. *Endocrinology*, 90:330–335.
Weil-Malherbe, H., and Ström-Olsen, R. (1958): Humoral changes in manic-depressive psychosis with particular reference to the excretion of catecholamines in urine. *Journal of Mental Science*, 104:696–704.
Williams, R. H., Barish, J., and Ensinck, J. W. (1972): Hormone effects upon cyclic nucleotide excretion in man. *Proceedings of the Society for Experimental Biology and Medicine*, 139:447–454.

ADDENDUM

Since this chapter was written, additional reports have appeared in which cyclic AMP and cyclic GMP levels were examined in patients. Many of these studies have dealt with urinary excretion of cyclic nucleotides and are for the most part preliminary reports in abstract form. Complete reports should be forthcoming in the near future.

As referred to above in this chapter, Moss et al. (1973) reported that absolute urinary excretion rates of cyclic AMP and cyclic GMP (μmoles/

day) in children increased with age. When values were related to body weight or urinary creatinine, the ratios were found to decline with age to approach adult values. However, the ratio of cyclic AMP to cyclic GMP excretion was not significantly altered with age. Children with cystic fibrosis had lower excretion rates of cyclic AMP and higher excretion rates of cyclic GMP; their ratio of cyclic AMP to cyclic GMP excretion was about one-half that of normals (Moss et al., 1973). Kopp et al. (1972) reported that urinary cyclic AMP (μmoles/g urinary creatinine) in adults was higher in normal females than normal males which may be a reflection of the effects of ovulation on cyclic AMP excretion (see above).

Since the preparation of this manuscript we have been studying 30 patients with primary hyperparathyroidism and normal renal function. In all patients the diagnosis was confirmed histologically and/or with elevated serum parathyroid hormone levels. All but three patients had cyclic AMP excretion rates greater than our hospitalized normals which is 2.82 ± 0.14 μmoles/g creatinine (mean $\pm$ SE, n = 25). In addition to the previously mentioned studies, several other laboratories have reported similar observations with hyperparathyroidism (Tsang et al., 1972; Neelon et al., 1973; Pak et al., 1973). Clearly, cyclic AMP excretion rates will be a valid and useful adjunct to the diagnosis of hyperparathyroidism in normocalcemic or hypercalcemic patients. Eight additional patients with "ectopic-PTH producing tumors" have been studied and found to have elevated cyclic AMP excretion rates (Pak et al., 1973; Murad, *unpublished observations*).

Since cyclic AMP is an intermediate in many hormone- and drug-induced responses and since other factors influence cyclic AMP excretion (see above), elevated urinary cyclic AMP will not be confined to patients with hyperparathyroidism. For example, Rosen (1972) reported that patients with hyperthyroidism also had elevated cyclic AMP excretion rates. Several laboratories have confirmed these findings (Kopp et al., 1973; Estep et al., 1973). In addition, we have also examined one patient with a carcinoid tumor and found excretion rates of 5.39 and 3.02 μmoles/g creatinine for cyclic AMP and cyclic GMP, respectively, both of which are distinctly elevated.

Although oral administration of methylxanthines failed to alter urinary cyclic AMP excretion (see above), intravenous infusions of aminophylline increased urinary cyclic AMP 25 to 50% (Weitzman and Murad, 1973*a*). Administration of aminophylline or chlorpropamide to several normal patients and pseudohypoparathyroid patients failed to alter significantly their cyclic nucleotide responses to parathyroid extract administration (Weitzman and Murad, 1973*b*). However, one patient with pseudohypoparathyroidism had normal or supernormal increases in plasma and urinary cyclic nucleotides after parathyroid extract and phosphodiesterase inhibitors, suggesting that some patients with this disorder may have accelerated renal hydrolysis of cyclic nucleotides rather than decreased synthesis (Weitzman and Murad, 1973b; Weitzman et al., 1973).

REFERENCES

Estep, H., Fratkin, M., Campbell, G., and Smith, E. P. (1973): Hypercalcemia and cyclic AMP excretion in hyperthyroidism. *Clinical Research,* 21:490.
Kopp, L. E., Lin, T., and Tucci, J. R. (1972): Factors affecting cyclic AMP excretion in normal human subjects. *Clinical Research,* 20:866.
Kopp, L. E., Lin, T., and Tucci, J. R. (1973): Urinary cyclic AMP in patients with hyperthyroidism. *Clinical Research,* 21:496.
Moss, W. W., Johanson, A. J., Murad, F., and Selden, R. (1973): Urinary excretion of cyclic AMP and cyclic GMP in normal children and in those with cystic fibrosis. *Clinical Research,* 21:45.
Neelon, F. A., Birch, B. M., Drezner, M., and Lebovitz, H. E. (1973): Urinary cyclic adenosine monophosphate as an aid in the diagnosis of hyperparathyroidism. *Lancet,* 1:631–634.
Pak, C. Y., Ohata, M., Chakmakjian, Z., Lawrence, E. C., Silverstein, R., Waters, O., and Stewart, A. (1973): Elevation of urinary adenosine 3',5'-monophosphate in primary hyperparathyroidism. *Journal of Clinical Endocrinology (in press).*
Rosen, O. M. (1972): Urinary cyclic AMP in Grave's disease. *New England Journal of Medicine,* 287:670–671.
Tsang, C. P., Lehotay, D. C., and Murphy, B. E. (1972): Competitive binding assay for adenosine 3',5'-monophosphate employing a bovine adrenal protein: Application to urine, plasma and tissues. *Journal of Clinical Endocrinology,* 35:809–817.
Weitzman, R., and Murad, F. (1973a): Effects of phosphodiesterase inhibition on urinary cyclic AMP and phosphate after parathyroid extract in pseudohypoparathyroidism. *Clinical Research,* 21:48.
Weitzman, R., and Murad, F. (1973b): Effects of aminophylline, chlorpropamide and parathyroid extract on plasma and urinary cyclic AMP in pseudohypoparathyroidism. *Clinical Research,* 21:89.
Weitzman, R., Murad, F., and Owen, J. (1973): Cyclic AMP and cyclic GMP levels in pseudohypoparathyroidism. Proc. of the 55th Annual Meeting of the Endocrine Soc., Chicago, Ill., June.

AUTHOR INDEX

Abdulla, Y.H., 84, 357, 375
Abe, M., 367, 369
Abola, J., 239
Abrass, I.B., 124, 125, 135
Adiga, P.R., 107
Adler, A.J., 133
Agren, G., 102
Akerfeldt, S., 281
Alberici, M., 9, 22
Albert, S., 101
Alexander, R. L., 110, 121, 122, 131
Allen, D.O., 8, 84, 86
Allfrey, V.G., 102
Allred, J.B., 292
Ambesi-Impiombato, F.S., 108, 133
Amer, M.S., 88, 308
Anderson, I., 102
Anderson, S.L., 102
Anderson, W.B., 106, 107, 188, 261, 306, 307
Antoni, F., 120
Appleman, M.M., 65, 67, 69, 70, 72, 74, 76, 77, 81, 113, 114, 169, 176, 186, 293
Ariens, E.J., 46
Armiento, M. d', 71, 78, 79, 87
Arnaiz, G.R.D.L., 9, 22
Ashby, C.D., 113, 114, 115
Ashcroft, S.J.H., 70
Ashman, D.F., 157, 168, 357
Assem, E.S.K., 292
Astwood, E.B., 104, 131
August, J.T., 137, 292
Aurbach, G.D., 7, 19, 21, 34, 37-39, 118, 161, 162, 164-166, 199, 360, 361, 365, 366-368
Averner, M.J., 292
Baddiley, J., 230, 231, 277
Bader, H., 102
Baggio, B., 102
Bailey, C., 138
Bailly, O., 237
Baird, C.E., 9, 33, 40
Bald, R.W., 232
Balhorn, R., 134
Ball, J.H., 159, 191, 193, 198
Bar, H.P., 4, 9, 18, 23, 24, 33, 38, 315
Barker, G.R., 231, 277
Bartell, L.S., 226
Bass, A.D., 180
Bastide, F., 102, 124, 184
Bastomsky, C.-H., 70, 89

Bauer, R.J., 299, 324, 325
Bdolah, A., 181
Beall, R.J., 38
Beavo, J.A., 67, 70, 74, 75, 77, 80, 81, 169-171, 176, 187, 209, 211
Beck, W.T., 134
Beckwith, J., 106
Beer, B., 83, 308
Begmeyer, H.U., 262
Bell, N.H., 368, 374
Bellantone, R.A., 134
Belocopitow, E., 104
Benkovic, S., 228
Beraud, G., 135
Berger, J.E., 104, 131
Berger, N.A., 240
Bernard, M., 70
Bernlohr, R.W., 161, 162, 164, 166, 170, 206
Bernstein, R.A., 376
Berridge, M.J., 105
Berthet, J., 100
Berti, R., 179
Birnbaumer, L., 2, 4, 5, 9-11, 17, 19, 21, 23-25, 28, 30, 33, 37-39, 42, 43, 45, 46, 52, 53, 113, 114, 165
Bishop, J.S., 130, 176, 177, 210
Bitensky, N.W., 40, 43
Blackburn, G.M., 236
Blanchette-Mackie, J., 8, 25, 27, 30, 41
Blanco, M., 40
Blat, C., 135, 294
Blecher, M., 10, 70, 84, 86, 251
Blomquist, C., 159, 193, 205
Blonde, L., 159, 172, 363, 364
Bloxham, D.P., 292
Boehringer Mannheim Gmbh, 249, 253, 255, 259, 272, 319, 320
Bohme, E., 161
Bonner, J.J., 78
Borden, R.K., 231, 232, 248, 252, 262, 272, 277
Bos, C.J., 8
Boucher, R., 125, 137
Bourgoignie, J., 181
Bourne, H.R., 292
Boutwell, R.K., 81
Bowers, C.Y., 332
Boyd, D.B., 227, 241
Boyer, P.D., 102
Bradham, L.S., 21, 22
Bratvold, G.E., 100, 104, 113, 115, 128, 129

Braun, T., 48, 177
Braun, W., 337
Bravard, I.J., 78
Brecher, P.I., 10
Breckenridge, B. McL., 70, 77
Brewer, H.B., 366
Bridger, W.A., 102
Brisson, G.R., 80
Broadus, A.E., 159, 172, 173, 191, 338, 357-361, 363-365, 367, 369
Brodie, B.B., 4, 5
Brody, T.M., 211
Bromilow, R.H., 227
Bron, T., 178
Brooker, G., 67, 68, 70, 83, 168, 368, 374
Brostrom, C.O., 104, 106, 109-111, 114, 116, 121, 122, 128, 129, 131, 138
Brostom, M.A., 108, 116
Browder, H.P., 308
Brown, D.M., 230, 234
Brown, H.D., 23
Brown, J.D., 83
Brown, N.E., 104, 130
Bruice, T.C., 228
Brunswick, D.J., 250, 272
Budovski, E.I., 244
Bunton, C.A., 227
Burdon, R.A., 126
Burger, M.M., 49
Burk, R.R., 8
Burke, G., 18, 21, 24, 25, 39
Burnett, G., 100, 102
Butcher, R.W., 2, 9, 11, 13, 22, 24, 32, 33, 39, 40, 52, 67, 68, 70, 79, 80, 84, 85, 104, 168, 183, 197, 208, 357, 358, 364, 367.
Caldwell, A., 89
Calkins, D., 113, 114
Canellakis, E.S., 134
Carlson, S.F., 161
Carpenedo, F., 82
Cashel, M., 188
Casillas, E.R., 89, 103
Cawley, T.N., 229
Cehovic, G., 180, 201, 202, 260, 333
Cehovic, G.D., 180
Cerasi, E., 80
Chalkley, R., 134
Chambaut, A.M., 123
Chambers, D.A., 106
Chang, Y.Y., 71, 78, 293
Charollais, E., 180

Chase, L.R., 7, 118, 199, 357, 360, 361, 365, 366, 368-370, 376
Chasin, M., 40, 178
Chassey, B.M., 71
Chattopadhyay, S., 23
Chaudhuri, T.K., 83
Chelala, C.A., 105, 210
Chen, B., 106, 107
Chen, L.-J., 119, 120, 123, 124, 126, 127, 131
Cheung, W.Y., 5, 16, 67, 68, 70, 75, 76, 80, 84, 168, 171, 179
Chez, R.A., 365
Chi, Y.M., 2, 34, 39, 41
Chiang, M.,-H., 16
Chladek, S., 238, 281, 282
Christian, W., 23
Chuah, C.-C., 134
Cirillo, V.J., 40
Clark, D.F., 243
Clark, J.B., 84, 86
Clark, L.J., 86, 88
Clark, R.B., 35, 50
Clark, V.L., 161, 162, 164, 166, 170
Cohen, K.L., 40
Cohen, S.S., 265
Cole, B., 198, 208, 358, 364
Cole, R.D., 132
Collins, R.L., 227, 241
Colowick, S.P., 16, 111
Conaway, C., 23
Conn, H.O., 175
Conway, A., 72, 73
Cook, W.H., 5, 247
Cooper, R.A., 100, 128, 129
Cooperman, B.S., 250, 272
Corbin, J.D., 102, 104, 109, 111, 114, 116, 121, 122, 131, 184
Costa, E., 2, 5, 291
Coulter, C.L., 238, 239
Courte, C., 110, 113, 119, 120, 122
Cox, B., 194
Cox, J.R., Jr., 227, 228
Craig, J.W., 104, 196, 208
Creveling, C.R., 33, 43, 49
Crittenden, E.R.S., 130
Crofford, O.B., 134
Cuatrecasas, P., 40, 55, 85, 87
Dalton, C., 81
Daly, J., 33, 43, 48, 49, 53
Danzig, M., 244
D'Armiento, M., 292
Daughaday, D.H., 180

Davidson, J.N., 101
Davie, E.W., 106, 139
Davies, G.E., 80
Davis, J.W., 157, 191
Davis, W.W., 134
Davoren, P.R., 8, 359, 364
de Crombrugghe, B., 106, 107
De Haen, C., 19
Dekker, C.A., 234, 236
De Lange, R.J., 67, 100, 104, 128, 129
De Long, A., 159, 199
DeRenzo, E.C., 364
DeRobertis, E., 9, 22, 77
Desalles, L., 4, 10
Desbuquois, B., 55
DeSombre, E.R., 10
Dettbarn, W.-D., 28, 47, 180
DeVerdier, C.-H., 102
Dexter, R.N., 8
Diamond, J., 211
DiBella, F., 137
Dietze, G., 105
Dixon, G.H., 102, 134
Doerfler, W., 138
Dohan, P., 372
Dokas, L.A., 107, 133
Donnelly, T.E., Jr., 114
Dousa, T., 21, 47, 70, 80
Drummond, G.I., 4, 5, 10, 17, 19, 23, 24, 25-28, 30, 67, 68, 104, 106, 165, 242, 243, 262, 272, 294, 306, 315
Duell, E.A., 338
Duncan, L., 4, 5, 17, 19, 23-25, 28, 30, 165
Dunn, A., 70, 74
DuPlooy, M., 318, 319, 320, 324
Eckstein, F., 237, 283, 315
Edel, R., 23, 24, 28, 38
Eichhorn, G.L., 240
Eil, C., 135
Eisen, A.Z., 376
Elbert, R., 80
Emmelot, P., 8
Emmer, M., 106, 107, 187, 188, 211
England, P.J., 139
Ensinck, J.W., 80
Entman, M.L., 10
Enz, A., 83
Epstein, S.E., 10
Erlichman, J., 111, 113, 120, 121-125
Estep, H., 361, 366, 370, 373, 382

Ewart, R.B.L., 181
Ewing, L.L., 110, 122, 123, 138
Exton, J.H., 85, 133, 173, 175, 176
Fain, J.N., 83, 86, 88, 89
Fakunding, J., 135
Falbriard, J.G., 250-252, 260, 273, 287
Farago, A., 120
Fasman, G.D., 133
Feinstein, H., 10, 42
Ferguson, J.J., Jr., 134
Ferrendelli, J.A., 159, 193-195, 204
Fichman, M., 83, 368, 374
Field, M., 40
Fischer, E.H., 100, 113-115, 130
Fisher, E., 277
Fleischer, N., 80
Florendo, N.T., 77
Foll, G.E., 231, 277
Fontana, J.A., 110, 111, 123, 133
Forn, J., 21, 33, 39
Forrest, H.S., 230, 280
Forte, J.B., 76
Fox, J.J., 234, 243
Frank, W., 79
Franks, D.J., 71, 75
Frazer, S.C., 101
Free, C.A., 325, 326, 327
Freeman, S., 102
Freidberg, S.L., 2, 34
Freist, W., 253, 277, 314
Fresco, J.R., 252, 275, 314
Friedman, D.L., 101, 104
Friedman, N., 177
Friedman, R.M., 8
Friedrich, V.W., 260
Froesch, E.R., 120
Froscio, M., 137
Fujimoto, Y., 248, 252, 262, 272
Furchgott, R.F., 44, 47
Furth, J.J., 265
Gaballah, S., 77, 118, 125
Galsky, A.G., 160
Garbers, D.L., 159, 202, 292
Garren, L.D., 108-110, 113, 120-125, 134, 135, 182, 184
Gauthier, M., 134
George, W.J., 159, 179, 190-194, 201, 208, 368
Gerisch, G., 78
Gershon, E., 40

Gerster, J.F., 254
Gessa, G.L., 33
Gibson, D.W., 240
Gilden, R.V., 103, 138
Gilham, P.T., 234
Gill, G.N., 108, 109, 110, 113, 121-125, 135, 184
Gill, T.H., 34
Gilman, A.G., 33, 184, 185, 187, 200, 358, 366
Giorgio, J., 10
Gitelman, H.J., 367
Glimcher, M.J., 102
Glinsmann, W.H., 175, 178, 184
Glomset, J., 102
Goidl, J.A., 134
Goldberg, A.R., 49
Goldberg, N.D., 14, 80-82, 84, 115, 155, 157, 159, 160, 164, 168, 172, 176, 177, 179, 181, 185, 190-206, 208, 210, 358
Goldfine, I., 71, 83
Gonzales, C., 113, 114
Goodman, A.D., 181
Goodman, D.B.P., 137, 294
Goodman, H.M., 84, 88
Goodman, L., 264, 269
Goodsell, E.B., 70, 80
Goren, E.N., 8, 68, 70, 76, 170
Gorman, R.E., 40, 43
Gottesman, M.E., 106, 107
Granner, D., 7, 8, 118, 134
Granner, D.K., 118
Graves, D.J., 100, 130
Gray, J.P., 159, 161, 162, 165, 166, 173, 205, 365
Greaves, M.L., 238
Greene, H.L., 70
Greengard, P., 2, 16, 34, 101-103, 110-114, 116, 117, 120-122, 125-127, 131, 132, 135, 136, 137, 174, 184-186, 210, 211, 245, 246, 291, 294, 306, 319
Grimm, 79
Grollman, A., 308
Grollman, E.F., 308
Gulyassay, P.F., 67
Guthrow, C.E., Jr., 137
Haber, E., 10, 41
Hach, B., 182
Hadden, E.M., 196, 200
Hadden, J.W., 159, 179, 193, 196, 200, 205, 208
Haddox, M.K., 14, 115, 155, 159, 185, 192, 193, 197, 198, 200, 205

Haga, K., 281
Hahn, P., 24
Halkerston, I.D.K., 11, 52, 80
Halle, W., 179
Ham, E.A., 40
Hamadah, K.H., 357, 375
Hamilton, I.R., 16, 23, 24
Hammermeister, K.E., 113
Hampton, A., 281
Hanze, A.R., 260, 336
Harbert, G., 363
Hardin, J.M., 117, 131
Hardman, J.G., 68, 156, 157, 159, 161, 162, 164-166, 176, 191, 193, 195, 198, 200, 201, 247, 252, 291, 315, 357, 363, 369
Harris, D.N., 316, 317
Harris, J.S., 110, 116, 117, 120-122
Hartle, D.K., 115, 159, 193, 197, 198
Harwood, J.P., 43, 104, 106
Haschke, R.H., 100
Hassid, W.Z., 289
Hatanaka, M., 103, 138
Hayaishi, O., 16, 17, 18, 85, 111, 272
Hayatsu, H., 270, 271
Haynes, R.C., Jr., 105
Hector, O., 4, 9, 10, 11, 18, 21, 23, 33, 38, 47, 48, 52
Hedrich, J.I., 122
Heidrick, M.L., 8, 78, 178
Heilmeyer, L.M., Jr., 100
Hein, F., 182
Hemington, J.G., 70, 74
Henderson, A., 40
Henion, W.F., 336, 337
Hepp, K.D., 23, 24, 28, 38, 40, 70, 77, 85, 105
Herman, R.H., 70
Hern, E.P., 174, 184
Hersh, E., 330, 331
Hershey, J.W.B., 135
Hess, M.E., 209
Hess, S.M., 40
Hevesi, L., 243
Hevesy, G., 101
Hickenbottom, J.P., 101, 104, 113-115, 119, 130, 131
Higson, H.M., 230
Himms-Hagen, J., 13, 39
Hirata, M., 17, 18, 272
Hirch, A.H., 8
Hirsch, A.H., 111
Ho, D.H.W., 304
Ho, E.S., 104, 110, 114, 128, 129

Hochberg, A.A., 137
Hofert, J.F., 81
Hoffman, F., 186, 211
Hoffman, P.L., 47
Hollenberg, C.H., 104, 130
Holt, D.A., 21
Holy, A., 229, 232-235, 237, 269, 270, 271, 279, 280, 315
Honda, E., 82
Honjo, M., 273
Hoskins, D.D., 89, 103
House, P.D.R., 86
Hsie, A., 134
Hsie, A.W., 78, 328
Huang, M., 33
Huang, Y.C., 70, 80, 84
Hubert-Habart, M., 264, 269
Hug, G., 376
Hughes, S.D., 134
Huijing, F., 101, 115
Humes, J.L., 40
Humphrey, G.B., 126
von Hungen, K., 34
Hunkeler, F.L., 101, 104, 106, 108, 113-115, 120, 130, 131
Hutchinson, D.W., 234
Hutchison, W.C., 101
Huttunen, J.K., 104, 121, 131, 294
Hynie, S., 24, 25, 28, 38, 41, 67
Ichino, M., 290
Ide, M., 6, 11, 23
Ikehara, M., 255, 270, 277, 281
Illiano, G., 40, 85
Imai, K., 270
Imamura, H., 82
Inoue, Y., 121
Ishi, I., 80, 83, 84
Ishikawa, E., 159, 160, 162, 202
Ishizaka, T., 292, 334, 337
Iwangoff, P., 83
Jackson, E.M., 102
Jackson, J.E., 102
Jacob, T.M., 236
Jard, S., 70, 102, 124, 184
Jardetzky, C.D., 239
Jarret, L., 159, 177
Jastorff, B., 253, 277, 287, 314
Jean, D.H., 102
Jefferson, L.S., 85, 177, 190, 208
Jensen, E.V., 10
Jergil, B., 102
Johanson, A., 363, 369
Johnson, C.B., 10

Johnson, E.M., 111, 121, 125-127, 136, 137
Johnson, G., 134
Johnson, G.S., 6, 8, 16, 78, 328
Johnson, J., 40
Johnson, R.A., 21, 31, 32
Johnson, R.E., 70, 77
Johnson, R.M., 101
Jones, A.B., 9, 21, 28, 37, 39
Jones, G.H., 288, 315
Jost, J.-P., 108, 134, 182
Kabat, D., 135
Kadowitz, P.J., 191
Kaiser, E.T., 246
Kakiuchi, S., 70, 72, 76, 169, 171, 364
Kaliner, M., 292, 334, 338
Kaminsky, N.I., 159, 191, 193, 198-200, 360, 361, 363, 365, 366, 368, 370, 373
Kanai, T., 290
Kawai, I., 235
Kawasaki, A., 197
Kebabian, J.W., 34
Kelley, J.J., 102
Kelley, L.A., 25, 26
Kemp, G., 81
Kemp, R.G., 70, 80, 84, 100, 105, 128, 129
Kennedy, E.P., 100, 101, 102
Kent, A.B., 100
Khan, S.A., 227
Khandelwal, R.L., 16, 23, 24
Khorana, G.H., 236, 241
Khorana, H.G., 229, 231, 234, 236, 275
Khwaja, T.A., 265-267, 272, 277, 306
Kikkoman Shoyu Co., 247, 272
Kimberg, D.V., 40
King, C.A., 109-111, 116
King, R., 40, 43
Kingdon, H.S., 111
Kinscherf, D.A., 204
Kipnis, D.M., 83, 158, 175, 204, 367
Kirby, A.J., 227
Kish, V.M., 111
Kitabchi, A.E., 178
Klainer, L.M., 2, 34
Klee, W.A., 240
Klein, I., 30, 37, 38, 55
Klein, M.I., 117, 118
Kleinsmith, L.J., 102, 107, 111, 133
Klotz, U., 70, 72, 75, 80, 82, 315

Kobata, A., 359, 365
Kochetkov, N.K., 241, 244
Konijn, T.M., 364
Kono, T., 87
Kopp, L.E., 382
Koritz, S.B., 25, 26
Koshland, D.E., Jr., 72, 73
Krane, S.M., 102
Krans, M.J., 4, 9, 10, 11, 28, 33
Krans, M.L., 2
Krause, E.-G., 179
Krebs, E.G., 100-106, 108-117, 119-122, 124, 128-132, 134, 135, 138, 139, 184
Krebs, T., 287
Kreis, W., 269, 331
Kreutner, W., 176
Krishna, G., 4, 5, 33, 43
Krueger, B.K., 102, 112, 113, 121
Krug, F., 55
Kuehl, F.A., 40, 67, 193, 197
Keuhn, G.D., 111
Kubovetz, W.R., 82, 179, 308
Kumamoto, J., 237
Kumar, K.S.V.S., 102
Kumon, A., 108-111, 113, 119-123, 184
Kundig, W., 102
Kuo, J.F., 101-103, 112-114, 121, 125, 131, 136, 159, 160, 167, 168, 174, 184, 185, 186, 187, 190, 192, 193, 197, 210, 211, 294, 299-302, 306, 319, 364
Kupiecke, F.P., 84, 86
Kurashina, Y., 16
Labrie, F., 110, 113, 119, 120, 122, 125-127, 134, 137, 181
Lake, R.S., 134
Lake, W.C., 291
Lang, M., 10, 42
Langan, T.A., 99, 101, 102, 108, 117, 119, 126, 127, 131, 132, 133
Lapper, R.D., 239
Lardy, H.A., 137
Larner, J., 100, 101, 104, 115, 116, 119, 127, 130, 176, 213
Lasser, M., 10, 42
Lawson, A.M., 241
Lecoq, J., 237
Lee, W.W., 240
Lefkowitz, R.J., 10 38, 39, 41, 55

Lemarie, S., 110, 113, 119, 120, 122, 125-127, 137
Lemay, A., 134
LePage, G.A., 330, 331
Leray, F., 123
Letters, R., 229
Levene, P.A., 101
Levey, G.S., 10, 30, 37-39, 55, 367, 376
Levine, L., 208
Levine, R.A., 336-338, 365
Liano, S., 10
Lichtenstein, L.M., 292, 359
Liddle, G.W., 291
Lin, A.W., 10
Ling, V., 134
Lipkin, D., 5, 245, 247, 289
Lipmann, F., 11, 16, 23, 101, 102, 104, 108, 109, 111, 119-125
Lippincott, J.A., 160
Loeb, J.E., 126, 135, 294
Long, R.A., 269, 315, 331
Lorand, L., 4, 10
Lorini, M., 102
Loten, E.G., 70, 86
Louie, A.J., 134
Love, D.S., 100, 115
Lovenberg, W., 110, 111, 123, 133
Luft, R., 80
Lust, W.D., 159, 190, 193, 194, 203
Macchia, V., 108, 134
Macintyre, E.H., 8, 35
MacKinley, A.G., 119
MacManus, J.P., 71, 75, 182
Maeno, H., 111, 121, 125-127, 132, 135-137
Mahafee, D., 178
Majumber, G.C., 113, 118, 119, 122-124, 135, 185
Makman, M.H., 117, 118
Makman, R.S., 106, 206, 364
Mallette, L.E., 133
Mamrak, F., 40
Mandel, L.R., 67, 80, 81
Manganiello, V., 71, 78, 79, 87, 88, 171, 211
Mansour, T.E., 105
Marcus, R., 19, 21, 24, 37, 38, 39, 366, 367
Margolis, S., 292
Margoulis, S., 359
Marinetti, G.V., 23, 137
Markham, R., 5, 230, 243
Marshall, N.B., 84
Martel, A.E., 19, 23

Martelo, O.J., 106, 139, 294
Marushige, K., 134
Massey, K.L., 70
Masui, H., 182
Mauret, P., 227
Mayer, S.E., 104, 106, 121, 129, 131
McAuslan, B.R., 137
McCune, R.W., 34
McKenzie, J.M., 89, 107
McKibbin, J.B., 134
McNeill, J.H., 80, 83, 84
Meetz, J., 205
Meisler, J., 4, 10
Meisler, M.H., 132
Melicow, M.M., 157
Menahan, L.A., 70, 86
Menon, K.M.J., 24
Menon, T., 2, 7, 8, 14, 19, 23, 30, 36, 37
Merlevede, W., 105, 210
Meyer, F., 139
Meyer, R.B., 247, 249, 253, 261, 283, 300, 301-306, 309, 311, 330, 339
Meyer, W.L., 100, 115
Michal, G., 309-312, 318, 320-325, 327, 337
Michel, M., 5
Michelson, A.M., 234
Mier, P.D., 70, 359, 376
Miki, N., 76, 84
Mikolajczyk, M., 236
Miller, J.P., 276, 313, 314, 318, 339
Mills, D.C.B., 82
Mitznegg, P., 182
Miyai, K., 89
Miyamoto, E., 102, 110, 112, 114, 116, 117, 120-122, 125, 184, 294, 296
Moffatt, J.G., 231, 238, 271, 272, 283, 288
Mommaerts, W.F.H.M., 10
Monod, J., 106
Montague, W., 70, 80, 84
Montgomery, J.A., 249, 305, 329
Moore, M.M., 5, 17, 18, 20, 24-29, 34, 35, 37, 40, 48-50
Moore, P.F., 82
Moret, V., 102
Morita, Y., 134
Morley, A., 292
Morris, H.P., 8, 23
Mosher, H.S., 230
Moss, W., 363
Moss, W.W., 381, 383
Moyer, R.W., 102

Mudd, S.H., 240
Muller-Oerlinghausen, B., 86
Muneyama, K., 255, 295, 296, 309, 311, 316, 319, 326, 329, 330, 335
Munshower, J., 8
Munske, K., 161
Munson, W.A., 358
Murad, F., 2, 39, 41, 159, 160, 167, 170, 172, 177, 184, 187, 191, 193, 199, 200, 208, 357-364, 366, 367, 369-375, 382, 383
Murayama, A., 285
Murray, A.W., 137
Murthy, P.V.N., 107
Muschek, L.D., 83
Nagyvary, J., 238, 269, 281, 282
Naim, E., 10, 14, 24-26, 42
Nair, K.G., 67, 68, 70, 80, 168
Naito, T., 248
Nakano, J., 80, 83, 84
Nanoune, J., 123
Naruse, M., 248, 252, 262, 272
Naylor, T., 234
Neblett, M., 133
Neblett, M.S., 134
Neelon, F.A., 382
Neufeld, A.H., 40, 43
Neville, D.M., 9
Newton, N.E., 115
Ney, R.I., 8, 33, 48
Ney, R.L., 134
Nichols, W.K., 177, 190, 208, 210
Nicolson, G.S., 51, 52, 53
Nishiyama, K., 110, 111, 113, 119, 120-123
Nishizuka, Y., 109-111, 113, 119-123
Nissley, P., 106, 107, 187, 188
Nohara, A., 277
Noland, B.J., 117, 131
Numata, M., 10
O'Dea, R.F., 14, 70, 155, 168-170, 184, 202, 359
Ohga, Y., 102
Ohtsuka, E., 270, 281
Okamoto, H., 16
Oken, R.L., 67
Oliver, I.T., 134, 182
Orange, R.P., 292, 334, 338
Ord, M.G., 126, 134
Ortiz, P.J., 265
O'Toole, A.G., 191
Otten, J., 6, 8, 204, 208, 375
Oye, I., 2, 37
Pagliara, A.S., 181

Pak, C.Y., 172, 191, 199, 357, 360-363, 366, 370-375, 382
Palmer, E.C., 28, 47
Palmer, G.C., 40
Pannbacker, R.G., 71, 78, 82
Panzica, R.P., 243
Paoletti, R., 179, 180
Park, C.R., 110
Parker, C.W., 158, 359, 376
Parks, J.S., 188
Pastan, I., 6-8, 10, 16, 25, 27, 30, 38, 39, 41, 78, 106, 107, 187, 206, 306
Paton, W.D.M., 46, 198
Patterson, W.D., 359
Paul, M., 33
Paul, M.J., 357, 362, 363, 375
Peake, G. T., 159, 180, 181, 193, 201, 202
Pearce, C.A., 126
Pelletier, G., 121, 125, 126, 127, 137
Pennington, S.N., 23, 67
Perkins, J.P., 1, 5, 8, 17, 18, 20, 24, 25, 28, 29, 34, 35, 37, 40, 48-50, 71, 78, 100-104, 110, 114, 119, 124, 128, 129
Perlman, R., 7, 78, 106, 107, 187, 206, 306
Perrot-Yee, S., 67, 68
Peterson, M.J., 84
Petzold, G., 34
Petzold, G.L., 110, 116, 117, 120-122
Pierre, M., 126
Pinna, L.A., 102
Plageman, P.G.W., 67, 80
Poch, G., 67, 70, 80, 82, 179, 308
Pohl, S.L., 2, 4, 5, 9, 10, 11, 17, 19, 21, 23-25, 28, 30, 33, 37-39, 42
Poirier, G., 125, 137
Polson, J.B., 191, 192
Ponten, J., 35
Popoff, C., 77, 118, 125
Posner, J.B., 104, 106, 113
Post, R.L., 102
Posternak, T., 180, 201, 202, 249, 252, 254, 255, 333
Potkonjak, D., 194
Potts, J.T., 365
Powell, C.A., 104, 106, 315
Prasad, K.N., 8, 328, 339
Preobrazhenskaya, N.N., 283
Price, T.D., 157, 357
Pricer, W., 8, 25, 27, 30, 41

Prince, W.T., 105
Puck, T.T., 78, 328
Puglisi, L., 179
Raben, M.S., 104, 131
Rabinowtiz, M., 4, 10, 100, 102, 124, 125
Rall, S.C., 132
Rall, T.W., 2, 3, 7, 8, 14, 16, 19, 23, 30, 33, 34-38, 41, 66, 79, 80, 100, 183, 247, 359, 361, 364, 367
Ramaley, R.F., 102
Ramsey, O.B., 227, 228
Rassmussen, H., 105, 137, 159, 199
Rathnam, P., 33
Ray, T.K., 23, 76, 85
Reddi, A.H., 110, 122, 123, 138
Reddy, W.J., 4
Reimann, E.M., 101, 104, 106, 108, 109, 111-116, 119-122, 130, 131, 139, 296
Richter, D., 124
Riddle, M., 134
Riedel, V., 78
Rieke, W.O., 134
Riggs, A.D., 187, 188
Riley, G.A., 105, 210
Riley, W.D., 35, 100, 104, 128, 129
Rivkin, I., 40, 178
Rizack, M.A., 320
Rizback, M.A., 131
Roberts, S., 34
Robertson, W., 40
Robins, R.K., 239, 240, 254, 277, 307
Robison, G.A., 2, 11, 13, 23, 28, 32, 39, 40, 47, 52, 66, 87, 104, 156, 179, 180, 183, 189, 197, 291, 359, 364, 375
Rodbell, M., 2, 4, 5, 9, 10-12, 17, 19, 21, 23-25, 28, 30, 33, 37-39, 42, 43-46, 166, 167, 325, 327
Rodnight, R., 102, 125, 136
Roehring, K.L., 292
Rosell-Perez, M., 101, 104
Roseman, S., 102
Rosen, O.M., 8, 9, 16, 18, 24, 40, 47, 67, 69, 70, 72, 76, 88, 111, 113, 120, 121-125, 350, 383
Rosen, S.M., 9, 16, 18, 24, 40, 47, 170
Rosenberg, I., 86

Roth, J., 8, 10, 38, 39, 41, 83
Rous, S., 178
Rubin, B., 335
Rubin, C.S., 113, 120-125
Ruddon, R.W., 102
Rudolph, S.A., 16
Rufeger, U., 82
Russell, A.F., 238, 271, 272, 283
Russell, T.R., 70, 72-75, 77, 87
Russell, V., 40
Ryan, L., 134
Ryan, W.L., 8, 78, 178
Rychlik, I., 70, 80
Sahib, M.K., 108
Salas, M.L., 108, 109, 111, 119-123
Salganicoff, L., 77
Salzman, E.W., 208
Salzman, N.P., 134
Samaniego, S.G., 40
Sams, D.J., 70, 80, 84
Sandra, P.S., 102
Sanes, J.R., 102, 112, 113, 121
Sanford, C., 159, 193, 197
Sanger, F., 127, 130
Sano, M., 248
Santhanam, K., 131
Sato, G.H., 8, 10, 134
Sattin, A., 33, 35, 204
Sayers, G., 38
Saxema, B.B., 8, 33
Schaeffer, L.D., 88
Schafer, D.E., 159, 192
Schaffhausen, B., 133
Schild, H.O., 292
Schimazu, T., 209
Schimmer, B., 8, 10, 50
Schlender, K.K., 101, 104, 113, 115, 119, 130, 183, 184
Schmidt, M.J., 28, 47
Schonhofer, P.S., 67, 70, 80
Schorr, I., 8, 33, 49
Schramm, M., 10, 14, 24-26, 42, 44, 110, 181
Schroder, J., 67, 70, 71, 80
Schultz, G., 85, 159-163, 193, 195, 197, 198, 202, 203, 368
Schultz, J., 48, 53
Schwabe, U., 80, 82, 83
Schwartz, D., 106
Schwarz, I.L., 47
Schweizer, M.P., 239, 240, 307
Secrist, J.A., 290

Seelig, S., 38
Selinger, Z., 110
Semenuk, N.S., 3, 99
Senft, G., 80, 83, 86, 177
Serafini, A., 227
Seraydarin, K., 10
Severson, D.L., 10, 17, 19, 24-28, 30, 294
Shanta, T.R., 77
Shapiro, B.M., 111
Shapiro, R., 244
Sharma, S.K., 107
Sharp, G.W.G., 24, 25, 28, 38, 41
Shen, L.C., 101, 115, 116, 119, 130
Shen, T.Y., 285
Shepherd, G.R., 117, 131
Sheppard, H., 5, 71, 81, 308
Sheppard, J.R., 8, 204, 208
Sherod, D., 134
Sherwood, L.M., 367, 369
Shimizu, H., 33, 43, 49
Shimomura, R., 121
Shimoyama, M., 78
Shin, S.-I., 134
Shlatz, L., 137
Shuman, D.A., 250, 253, 255, 258-260, 264, 281, 300-302, 306, 315, 332
Sidwell, R.W., 332, 339
Siebert, G., 126
Silberstein, H., 137
Siliprandi, N., 102
Sillen, I.G., 19, 23
Simon, L.N., 297, 298, 299
Simonis, A.M., 46
Simopoulos, A., 372, 373
Sims, M., 21
Singer, J.J., 51-53
Singhal, R.L., 292
Skidmore, I.R., 84
Smith, A.J., 122
Smith, C.H. 130
Smith, I.C.P., 239
Smith, J.B., 82
Smith, J.D., 230
Smith, J.W., 359, 376
Smith, L.K., 101, 102
Smith, M., 24, 231, 232, 234, 236, 239, 242-245, 248, 252, 262, 272, 275, 277
Smith, S.W., 101
Smithies, O., 134
Smrt, J., 232
Sneyd, J.G.T., 70, 86
Sobel, B.E., 67
Sober, H.A., 243

Soderling, T.R., 101, 104, 110, 113-115, 119, 130, 131, 184
Soifer, D., 10, 11
Solodkowska, W., 101
Solomon, S.S., 70, 88, 181
Sooknandan, G., 118
Sorm, F., 229, 232-234, 269-271
Spano, P.F., 336
Stadtman, E.R., 111, 210
Stansfield, D.A., 70, 84
Stefanovich, V., 359, 364
Steinberg, D., 104, 121, 131
Steiner, A.L., 158-160, 164, 166, 167, 172, 181, 274, 357, 358
Steiner, A.S., 180, 190, 191, 200
Stephens, D.T., 103
Stephenson, R.P., 44
Stern, R., 104, 106, 113
Stock, K., 75, 315
Stocken, L.A., 126, 134
Stone, D.B., 105
Stone, M.J., 102
Stossel, T., 358, 364
Strand, M., 137, 292
Stratman, F.W., 137
Straus, D.B., 252, 275, 314
Streeto, J.M., 4
Strom-Olson, R., 375
Strubelt, O., 80
Stull, J.T., 106, 138, 139
Sturtevant, J.M., 16
Su, J., 289
Sulakhe, P.V., 10, 25-28
Sulser, F., 40
Sundaralingam, M., 239
Sundararajan, T.A., 102
Sung, C.P., 70, 77
Sung, M.T., 134
Sutherland, E.W., 2, 3, 7-9, 11, 13-16, 19, 21-23, 30-32, 34, 36, 37, 39, 41, 52, 66-68, 70, 79, 80, 84, 100, 104, 106, 156, 157, 161, 162, 164-166, 169, 183, 189, 191, 197, 206, 247, 291, 315, 357, 359, 364, 367
Suzuki, C., 16
Swislocki, N.I., 251
Sy, J., 124
Szentivanyi, A., 375
Takahashi, K., 357, 368
Takai, K., 16
Takats, A., 120

Takeda, M., 102, 119, 120
Talwar, G.P., 107
Tao, M., 11, 16, 23, 108-111, 117, 119, 120-123, 138, 184, 185
Taylor, A.L., 361, 367-370, 373
Taylor, K.W., 181
Tazawa, I., 276, 277
Tenenhouse, A., 105
Tener, G.M., 245
Tesser, G.I., 255
Thain, E.M., 230
Thang, M.N., 139
Therriault, D.G., 70, 80
Thomas, H.J., 249, 329
Thompson, W.J., 65, 67, 69, 70, 72, 74, 76, 77, 81, 87-89, 164, 169, 176, 201, 293
Todd, A.R., 230, 280
Tokoyo Tanabe Seiyaku Co., 280
Tomasi, V., 23
Tomkins, G.M., 7, 118
Torres, H.N., 105, 113, 114, 210
Toson, G.C., 82
Traugh, J.A., 135
Traut, R.R., 104, 135
Trayser, K.A., 100, 115
Trauton, O.D., 8
Triner, K., 2
Triner, L., 82, 197, 335
Tsai, S.-C., 131
Tsang, C.P., 382
Turkington, R.W., 113, 118, 119, 122-124, 134, 135, 185
Turtle, J.R., 83, 367
Twiddy, E., 103, 138
Tymoczko, J.L., 10
Ueda, K., 8, 10
Ueda, T., 136, 234, 235, 243
Ueki, A., 16
Uesugi, S., 255
Ukita, T., 270, 271
Ullman, A., 106
Urselmann, E., 70, 359, 376
Usher, D.A., 228, 230
Valdecasas, F.G., 21, 39
Van Rossum, J.M., 46
Van Wijk, R., 329, 330, 332
Vargiu, L., 336
Varrone, S., 108, 133
Vatter, A.E., 8

Vaughan, M., 71, 78, 79, 87, 88, 104, 131, 171, 211, 367
Verheyden, J.P.H., 288
Villar-Palasi, C., 100, 101, 104, 113, 115, 116, 119, 130, 138
Voight, K., 33
Voorhees, J., 159
Voorhees, J.J., 292, 338, 359, 375
Waddy, C.T., 119
Waggle, S.R., 131
Wagner, E.L., 227
Walaas, D., 184
Walinder, O., 102
Walsh, D.A., 100-104, 110-117, 119-124, 126-131, 184, 294
Walter, R., 47
Walton, G.M., 110, 124, 125, 135
Wang, J.H.C., 76
Warburg, O., 23
Ward, W.I., 83
Washington, A., 336
Watenpaugh, K., 239
Watson, G., 133
Waud, D.R., 46
Weber, G., 8
Wechter, W.J., 269, 331
Wehmann, R.E., 172, 358
Wei, S.H., 101, 104, 113, 115, 130
Weil-Malherbe, H., 375
Weimann, G., 236
Weinryb, I., 5, 250, 339
Weiss, B., 2, 4, 5, 7, 24, 25, 67, 70
Weissman, G., 292
Weitzman, R., 359, 366, 370, 372, 373, 375, 382, 383
Weller, M., 102, 125, 136
Wells, H., 359, 364
Welsch, F., 180
Wenger, J.I., 116
Werner, M., 292
West, T.C., 80
Westheimer, F.H., 241, 287
White, A.A., 161, 162, 164-166
White, J., 159, 198
Whitfield, J.F., 170, 178, 182
Wicks, W.D., 134
Wieland, O., 23, 24, 28, 38
Wiggan, G., 71, 81, 308
Wilber, J.F., 181
Wilchek, M., 250, 251

Williams-Ashman, H.G., 110, 122, 123, 138
Williams, R.H., 357, 358, 362, 367-369, 373
Williamson, J.R., 179
Wilson, B.D., 107
Wines, N., 83
Winters, V.G., 70, 80
Wintersgill, C.J., 8
Wolff, J., 9, 21, 28, 37, 39
Wollenberger, A., 179
Woo, S.L.C., 106, 139
Wood, H.N., 78
Wool, I.G., 135
Wosilait, W.D., 100
Wright, R.L., 107
Wyatt, G.R., 103, 113
Yamamoto, M., 70
Yamamura, K., 102, 108-111, 113, 119, 120-123
Yeung, D., 134, 182
Yoshida, H., 76, 84
Yoshikawa, M., 269
Younos, M., 227
Yurkevich, A.M., 235
Zahlten, R.N., 137
Zapf, J., 120
Zmudzka, B., 277
Zoltewicz, J.A., 243
Zubay, G., 106, 187, 211

SUBJECT INDEX

Acetylcholine
  in cardiac function, 192
  effect on cAMP, 179, 180, 191, 368
  effect on cGMP, 167, 168, 180, 191-196, 209, 368
ACTH
  adenyl cyclase stimulation, 9, 21, 27, 30, 33, 37-39, 48-50, 190
  cAMP and cGMP dualism, 212
  effect on guanyl cyclase, 166
Actinomycin
  binding to chromatin, 133
  effect on phosphodiesterase, 79, 87
Adenine, cAMP incorporation, 14, 15
Adenosine
  effect on cAMP in brain, 33-35, 47-50
  protonation of, 243
  in xylofuranosyl nucleotide synthesis, 269, 270
Adenosine 2',3'-cyclic monophosphate, 235
Adenosine 3',5'-cyclic monophosphate
  see cAMP (listed among A's in index)
Adenosine 3',5'-cyclic phosphorofluoridate, 277
Adenosine 3',5'-cyclic phosphorothioate, 237, 283
  lipolysis activation, 322
  protein kinase activation, 306
Adenosine 2',5'-cyclic pyrophosphate, 289
Adenosine 3',5'-cyclic pyrophosphate, 289
Adenosine 5'-diphosphate, salts, 240
Adenosine (2') (3') (5') monophosphates, PMR spectra, 240
Adenosine 5'-p-nitrophenylphosphate, 248
Adenosine 5'-phosphate (AMP), 277, 283
  2'-O-acetyl deriv., 314
  8-bromo deriv., 255
  2'-O-butyl deriv., 314
  hydrolysis of, 242-245, 314
  2'-O-methyl deriv., 276, 314
  salts of, 240
Adenosine 5'-phosphorofluoridate, 277
Adenosine 5'-phosphorothioate, 237
  N^6-dimethylaminomethylene deriv., 283
Adenosine triphosphate, see ATP
Adenyl cyclase, 2-56
  ACTH effect on, 10, 21, 27, 30, 33, 37-41, 48-50
  adenosine effect on, 33-35, 47-50
  assay of, 3-6
  batrachotoxin effect, 33
  bromoacetyl-oxytocin inhibitor, 47
  calcium in, 19, 21, 38, 39, 105, 106
  cell growth regulation, 7, 8

  cholera toxin effect, 40, 41
  chlorpromazine effect, 37
  cobalt in, 19
  cobramine B, effect of, 37
  copper in, 19
  detergent treated, 30-32, 38
  digitonin effect, 37
  dihydroergotamine effect, 83
  EGTA effect on, 21, 22, 31, 32, 38
  epinephrine stimulation, 9, 33, 37
  FSH effect on, 33, 49
  glucagon stimulation, 10, 33, 37, 42
  gramicidin S, effect, 37
  GTP effect on, 42-47, 157
  histamine effect, 33, 35, 50
  hormone stimulation, 2, 7, 8-15, 25, 32-51, 190
  insulin effect, 40, 85
  LH stimulation, 33, 49
  lithium in, 21, 39
  magnesium in, 2, 5, 16-21, 23, 26-32
  manganese in, 19-21, 31, 32, 38
  messenger hypothesis, 7, 41
  norepinephrine effect, 25, 33-35, 47-50
  in nuclei, 10
  ontogenetic development, 47-51
  ouabain effect, 33
  phentolamine effect, 39
  phospholipase A, effect, 37, 38
  potassium in, 33, 39
  propranolol effect, 33, 39
  prostaglandin E_1, effect, 34, 35, 37, 40, 41, 50
  reaction details, 16-23
  receptor mechanism, 39, 40
  rubidium in, 39
  secretin stimulation, 33, 37
  serotonin effect, 33
  sodium fluoride stimulation, 2, 9, 13, 23-30, 31, 36-41, 48
  sodium ion effect, 39
  solubilization, 30-32, 37
  species distribution, 6, 7
  strontium in, 21
  structure, 51-56
  thymol effect, 37
  trypsin effect, 37
  TSH stimulation, 21, 25, 33, 37, 49
  urea effect, 37
  vasopressin effect, 41
  veratridine effect, 33
  in yeast, 124
  zinc in, 19
Adenylate cyclase see adenyl cyclase
Adenylic acid deaminase
  in phosphodiesterase assay, 67
Adenylyl cyclase see adenyl cyclase

393

5'-Adenylyl imidodiphosphate,
ATP substrate, 4, 107
Adipose tissue lipolysis see lipolysis
Adipose tissue lipase
  activation, 130, 131
  cAMP effect on, 104
Adrenal cortex binding protein
  in cAMP activation, 109
Adrenal cortex protein kinase
  distribution, 125, 126
  molecular wt., 122
  multiple forms, 120
Adrenal steroidogenesis
  cAMP derivatives effect on, 325-327
  cAMP effect on, 178, 182, 325
  cGMP effect on, 178, 182
Adrenalectomy, effect on cGMP
  excretion, 164, 200
Adrenohypophyseal secretory granules,
  protein kinase substrate, 137
Agarose beads in hormone binding, 41
Aging
  loss of protein kinase activity, 116
  protein kinase mol. wt. change, 123
Alditol cyclic nucleotides, 280, 281
Aldosterone
  effect on cGMP, 200
  effect on guanyl cyclase, 164
Alloxan, effect on cAMP, 190
2-Amino-6-chloro-9-(2,3,5-tri-O-acetyl-beta-
  D-ribofuranosyl) purine, 254
alpha-Aminoisobutyric acid, in hormone-
  receptor mechanism, 47
Aminophylline, growth hormone release,
  180, 202
5-Amino-1-beta-D-ribofuranosyl-imidazole-
  4-carboxamide-3',5'-cyclic phosphate,
  262
  lipolysis activation, 321, 337
5-Amino-1-beta-D-ribofuranosyl-imidazole-
  4-carboxamide-5'-phosphate, 262
  lipolysis activation, 331
6-Amino-3-(beta-D-ribofuranosyl)-purine 3',
  5'-cyclic phosphate (isocAMP), growth
  hormone release, 333
Amniotic fluid, cAMP in, 359, 365
cAMP (Adenosine 3',5'-cyclic phosphate)
    see ALSO entries under Adenyl cyclase
  2'-O-acetyl deriv.
    PDE inhibitor, 318
    PDE substrate, 313, 314
  acetylcholine effect on, 191
  adenine incorporation, 14, 15
  adenyl cyclase in formation of, 2
  adrenal steroidogenesis, 178, 182, 325
  N^6-allyl deriv., phosphorylase
    activation, 321
  8-allylamino deriv., phosphorylase activa-
    tion, 321
  2-amino deriv., hormone release, 333
  8-amino deriv., 255, 266
    adrenal steroidogenesis, 326, 327
    antibody synthesis, 337
    cell growth-enzyme induction, 329,
      330
    hormone release, 333
    lipolysis activation, 326
    muscle relaxation, 335
    PDE inhibitor, 316, 317
    PDE substrate, 309
    protein kinase activity, 295, 296, 298,
      299, 324, 326
  aminocaproyl deriv., 251
  8-(2-aminoethylthio) deriv., 255
  in amniotic fluid, 359, 365
  antibody synthesis, 337
  antidiuretic hormone effect on, 368
  antiviral activity, 332
  assay for adenyl cyclase, 3
  in asthma, 375
  in atopic dermatitis, 375
  8-azido deriv., 255
    adrenal steroidogenesis, 326
    lipolysis activation, 326
    PDE inhibitor, 317
    protein kinase activity, 295, 298, 326
  basal adenyl cyclase activity, 14, 15
  N^6-benzoyl deriv.
    adrenal steroidogenesis, 325
    lipolysis activation, 322
    PDE substrate, 312
    phosphorylase activation, 321
  N^6-benzyl deriv.
    lipolysis activation, 322
    PDE substrate, 312
    phosphorylase activation, 321
  8-benzylamino deriv.
    antibody synthesis, 337
    cell growth-enzyme induction, 329,
      330
    lipolysis activation, 322
    PDE substrate, 310
    phosphorylase activation, 320
    protein kinase activity, 295, 296, 299,
      324
  8-benzyloxy deriv., 255
    lipolysis activation, 322
  8-benzylthio deriv.
    adrenal steroidogenesis, 326
    cell growth-enzyme induction, 328-
      330
    lipolysis activation, 326
    muscle relaxation, 335
    PDE inhibitor, 316, 317
    protein kinase activity, 295-298, 324,
      326
  bladder permeability, 181
  in brain, 33, 34
  2-bromo deriv., 253
  8-bromo deriv., 255
    adrenal steroidogenesis, 326, 327
    antiviral activity, 332
    cell growth-enzyme induction, 329,
      330
    CRP binding inhibition, 307
    hormone release, 333
    lipolysis activation, 322, 326
    muscle relaxation, 335
    PDE inhibitor, 316, 317
    PDE substrate, 309, 310

cAMP (continued)
  8-bromo deriv. (continued)
    phosphorylase activation, 319, 320
    protein kinase activity, 295, 297, 298, 302, 326
  8-bromo-2'-O-tosyl deriv., 265, 266
  N^6-n-butyl deriv.
    hormone release, 333
  N^6-t-butyl deriv.
    hormone release, 333
  N^6-butyryl deriv.
    glycogenolysis, 336
    PDE inhibitor, 318
    PDE substrate, 313
    phosphorylase activation, 321
  2'-O-butyryl deriv.
    PDE inhibitor, 318
    PDE substrate, 313, 314
  calcitonin effect on, 369
  in calcium disorders, 369
  cell growth regulation, 7, 8, 178, 204, 328-332
  2-chloro deriv., 253
  8-chloro deriv., 255
    protein kinase activity, 297, 298
  chloroacetaldehyde, reaction with, 290
  N^6-(2-chlorobenzyl) deriv.
    PDE substrate, 312
    phosphorylase activation, 321
  8-(2-chlorobenzylamino) deriv.
    PDE substrate, 310
    phosphorylase activation, 320
  cholera toxin, effect on, 40, 41, 190
  cholinergic antagonist, 179, 180, 368
  cis-trans isomerism, 229
  clinical studies, 356-377, 381, 382
  cyclohexyl carbamoyl deriv., 250
  in cystic fibrosis, 382
  in depressive disease, 375
  in diabetes insipidus, 374
  diacetyl deriv., 250, 283
  diacetyl-8-oxo deriv., 255
  diadamantoyl deriv., 250
  di(aminocaproyl) deriv., 250
  dibenzoyl deriv., 250
  dibutyryl deriv. (dbcAMP)
    antibody formation, 337
    cardiac cell contractions, 179
    cell growth regulation, 328
    effect on phosphodiesterase, 79, 313, 316-318
    glucose in blood, effect on, 337
    glycogenolysis, 336
    histamine release, 294
    histone phorphorylation, effect on, 132
    human response, 337
    cGMP mimic, 181
    lipolysis activation, 322, 325
    mediator release, 334
    muscle relaxation, 335
    phosphorylase activation, 321
    sleep, effect on, 336, 337
    synthesis, 250
  dibutyryl-8-thio deriv.
    hormone release, 333
  di(diazomalonyl) deriv., 250, 273
  dihexanoyl deriv., 250
  dilauroyl deriv., 250
  N^6-dimethyl deriv., hormone release, 333
  8-dimethylamino deriv.
    adrenal steroidogenesis, 326, 327
    lipolysis activation, 326
    PDE inhibitor, 317
    protein kinase activity, 295, 298, 326
  2'-O-(2,4-dinitrophenyl) deriv., 277
    PDE inhibitor, 318
    PDE substrate, 313, 314
  dioctanoyl deriv., 250
  distearoyl deriv., 250
  ethyl ester, 287
  8-ethylthio deriv.
    adrenal steroidogenesis, 326, 327
    cell growth-enzyme induction, 329, 330
    lipolysis activation, 326
    muscle relaxation, 335
    PDE inhibitor, 317
    protein kinase activity, 295, 297, 298, 326
  gene transcription, 106
  glucagon effect on, 104, 190, 191, 367
  glucose effect on, 178, 320
  glycogenolytic activity, 336
  cGMP effect on, 170, 171, 176, 177
  cGMP similarity with, 175-179, 207
  growth hormone release, 180, 181, 332, 333
  hormone effect on, 365-369
  hydrolysis of, 74, 75, 242-245
  2-hydroxy deriv., 265
  8-hydroxy deriv.
    adrenal steroidogenesis, 326, 327
    hormone release, 333
    lipolysis activation, 326
    PDE inhibitor, 317
    protein kinase activity, 295, 298, 299, 324, 326
  8-(2-hydroxyethylamino) deriv.
    adrenal steroidogenesis, 326, 327
    lipolysis activation, 326
    PDE inhibitor, 317
    protein kinase activity, 295, 297, 298, 299, 324, 326
  8-(2-hydroxyethylthio) deriv.
    adrenal steroidogenesis, 326
    cell growth-enzyme induction, 329, 330
    lipolysis activation, 326
    PDE inhibitor, 317
    protein kinase activity, 295, 297-299, 326
  8-hydroxy-2'-O-tosyl deriv., 265, 266
  in hyperthyroidism, 382
  insulin effect on, 369
  8-iodo deriv., 255
    protein kinase activity, 297, 298
  kinetic regulation, 72-74
  lipolysis activation, 322
  liver phosphorylation, 117
  in mania disease, 375

cAMP (continued)
  8-methoxy deriv., 266
    adrenal steroidogenesis, 326
    lipolysis activation, 326
    muscle relaxation, 335
    PDE inhibitor, 317
    phosphorylase activation, 320
    protein kinase activity, 295, 296, 298, 326
  N^6-methyl deriv.
    hormone release, 333
  2'-O-methyl deriv., 276
    PDE inhibitor, 318
    PDE substrate, 313, 314
  8-methylamino deriv.
    adrenal steroidogenesis, 326
    cell growth-enzyme induction, 329
    lipolysis activation, 326
    muscle relaxation, 335
    PDE inhibitor, 317
    protein kinase activity, 295, 298, 326
  N^6-(2-methylbenzyl) deriv.
    PDE substrate, 312
    phosphorylase activation, 321
  N^6-(4-methylbenzyl) deriv.
    adrenal steroidogenesis, 325
    lipolysis activation, 322
    phosphorylase activation, 321
  8-(2-methylbenzylamino) deriv.
    PDE substrate, 310
    phosphorylase activation, 320
  8-(4-methylbenzylamino) deriv.
    lipolysis activation, 322
    phosphorylase activation, 320
  methyl ester, 287
  8-methylthio deriv.
    adrenal steroidogenesis, 326, 327
    antiviral activity, 332
    cell growth-enzyme induction, 329, 330, 332
    hormone release, 333
    lipolysis activation, 326
    muscle relaxation, 335
    PDE inhibitor, 316, 317
    protein kinase activity, 295, 297, 298, 326
  in milk, 359, 365
  mitogenesis, effect on, 182
  molecular structure, 239
  8-morpholino deriv.
    phosphorylase activation, 320
    muscle relaxation, 334
  N^6-octanoyl deriv., glycogenolysis, 336
  ontogenetic development, 47-51
  N-oxide deriv., 260
    antiviral activity, 332
  8-oxo deriv., 255, 266
    antibody synthesis, 337
  parathyroid hormone effect on, 369
  perfused liver level, 175, 336
  N^6-pentyl deriv.
    PDE substrate, 310
    phosphorylase activation, 321
  8-pentylamino deriv., PDE substrate, 310

  N^6-(1-phenethylamino) deriv.
    lipolysis activation, 322
    phosphorylase activation, 321
  phosphofructokinase, effect on, 104
  physiological response to, 337, 338
  8-piperidino deriv.
    adrenal steroidogenesis, 325
    phosphorylase activation, 321
  in plasma, 358, 359, 363-369
  protein kinase activation, 94-141, 294-302, 306, 324
  in psoriasis, 375
  purification of, 2, 3
  receptor protein binding, 306, 307
  RNA synthesis stimulation, 133, 188
  in saliva, 364
  species distribution, 6, 7
  spectra of, 240
  in spinal fluid, 359, 364
  steroid hormone action, role in, 11
  2'-O-succinyl deriv., 274
  synthesis of, 234, 235, 247, 248
  8-thio deriv., 255, 266
    adrenal steroidogenesis, 326, 327
    antibody synthesis, 337
    cell growth-enzyme induction, 329, 330
    CPR binding inhibition, 307
    hormone release, 333
    lipolysis activation, 322, 326
    muscle relaxation, 335
    PDE inhibitor, 317
    PDE substrate, 309, 310
    protein kinase activity, 295, 296, 298, 299, 324, 326
  tissue distribution, 7, 8, 359, 360
  in urine, 357, 358, 360-363, 365-369, 381
cAMP binding protein, 123, 124
cAMP dependent protein kinase see protein kinase
cAMP phosphodiesterase see Phosphodiesterase
cAMP receptor protein see Catabolite gene-activating protein
cdAMP see 2'-Deoxyadenosine-3',5'-cyclic phosphate
dbcAMP see cAMP, dibutyryl derivative
d-Amphetamine, effect on cGMP, 193, 204
Anterior pituitary protein kinase
  distribution, 125
  molecular wt., 122
Antibody formation,
  cAMP derivatives in, 337
Antidiuretic hormone
  effect on cAMP, 368
  effect on guanylate cyclase, 166
Antiviral activity
  cAMP and derivatives, 332
  virazole agent, 332
9-beta-D-Arabinofuranosyladenine, 264
  2-O'-benzyl deriv., 265
  cell growth regulation, 331
  5-beta-cyanoethylphosphate deriv., 265

9-beta-D-Arabinofuranosyladenine-
  2',5'-cyclic phosphate, 264, 265
  8-thio deriv., 266
9-beta-D-Arabinofuranosyladenine-3,5'-
  cyclic phosphate
  6-amino deriv. (ara-cAMP)
    cell growth regulation, 331, 332
    CRP binding inhibition, 307
    PDE inhibitor, 318
    PDE substrate, 313
    protein kinase activation, 306
  6-Amino-N^6-octanoyl deriv.
    cell growth regulation, 331
  6-amino-8-oxo deriv. (8-oxo-ara-cAMP)
    protein kinase activation, 306
  6-amino-8-thio deriv. (8-thio-ara-cAMP)
    protein kinase activation, 306
  8,2'-anhydro-8-oxo deriv.
    protein kinase activation, 306
  8,2'-anhydro-8-thio deriv., 272
    protein kinase activation, 306
  6,8-diamino deriv. (8-amino-ara-cAMP)
    protein kinase activation, 306
  molecular structure, 240
9-beta-D-Arabinofuranosyladenine-5'-
  phosphate, 264, 265
9-beta-D-Arabinofurnanosyladenine-5'-
  triphosphate, 265
1-(beta-D-Arabinofuranosyl) cytidine cell
  growth regulation, 331
1-(beta-D-Arabinofuranosyl) cytidine-3',5'-
  cyclic pyrophosphate, 290
1-beta-D-Arabinofuranosylcytosine,N'-
  benzoyl deriv., 269
1-beta-D-Arabinofuranosylcytosine-2',5'-
  cyclic phosphate, 269
1-beta-D-Arabinofuranosylcytosine-3',5'-
  cyclic phosphate (ara-cCMP),
  269
  PDE substrate, 315
1-beta-D-Arabinofuranosylcytosine-3'-
  phosphate, 269
9-(beta-D-Arabinofuranosyl)-hypoxanthine
  (ara-H), 331
1-beta-D-Arabinofuranosyl-4-methyl-thio-
  2-pyrimidone-3',5'-cyclic phosphate,
  269
1-beta-D-Arabinofuranosyl-4-thiouracil-
  3',5'-cyclic phosphate, 269
1-beta-D-Arabinofuranosyluracil, 269
  cell growth regulation, 331
1-beta-D-Arabinofuranosyluracil-3',5'-cyclic
  phosphate (ara-cUMP), 269
  PDE substrate, 315
Arthrobacter atrocyoneus
  protein kinase in, 103
Aspartic acid, phosphorylation of, 102
Assay procedures
  adenyl cyclase, 3-6
  phosphodiesterase, 66-68
  plasma, 358, 359
  urine, 357, 358
Asthma, cAMP in, 375
Atopic dermatitis, cAMP in, 375

ATP (Adenosine triphosphate)
  barium hydroxide degradation, 289
  conversion to cAMP, 2, 4, 16-23
  radioactive labeled, 4
ATPase
  activation of, 105
  adenyl cyclase contaminant, 4
  sodium fluoride effect on, 23
Atropine
  dibutyryl cGMP blocking, 180
  effect on cGMP, 190, 191, 193, 194, 20
  in mediator release, 334
6-Azauridine 3',5'-cyclic phosphate, 233

Bacteria
  cAMP regulatory capacity, 187-189
  gene transcription, 106
  cGMP regulating capacity, 187-189, 206
  guanylate cyclase in, 161, 162
  protein kinase in, 103
Batrachotoxin, effect on cAMP level, 33
6-Benzamidopurine, 265
2'-O-Benzyl-3,5-di-O-p-nitrobenzoyl-alpha-
  D-arabinofuranosyl chloride, 265
4-Benzyl-2-imidazolidones,
  phosphodiesterase inhibitor, 81, 82
Biopsy specimens, 359, 360
Bis-p-nitrophenyl phosphorochloridate, 232
Bladder permeability
  cAMP and cGMP effect, 181
Bonding in cyclic phosphates, 226
Brain protein kinase
  molecular wt., 120, 121, 122
  multiple forms, 120
  preincubation with histone, 116
Bromoacetyl-oxytocin
  adenyl cyclase inhibitor, 47
Butane-1,4-diol cyclic phosphate,
  hydrolysis, 241
4-(3'-Butoxy-4'-methoxybenzyl)-imidazolidin-
  2-one, 328

Caffeine
  effect on cGMP, 202
  steroidogenesis, effect on, 178
Calcitonin
  effect on cAMP, 369
  effect on cGMP, 193, 197
Calcium
  in ATP-cAMP conversion, 19, 21, 38, 39,
    105, 106
  in ATPase inactivation, 105
  in GTP-cGMP conversion, 165, 171
  hormone modification, 105, 106
  phosphodiesterase, effect on, 76
  troponin phosphorylation, 138
Calcium disorders, cAMP in, 369
Calf serum, phosphodiesterase activation, 79
Carbachol, effect on cGMP, 193, 195
Carbamyl choline
  dibutyryl cGMP mimics action, 179
Cardiac function
  cAMP effect on, 179
  epinephrine effect on, 179
  cGMP effect on, 179, 191, 192, 208

Casein
  phosphorylation, 100-102, 111, 119
  preincubation of protein kinase, 116
Catabolite gene-activating protein
  cAMP analogs and derivatives, 306, 307
  cAMP effect on, 106, 187
  cGMP effect on, 187, 188
Catechol cyclic phosphate, 246
Cell division, histone phosphorylation
  during, 134
Cell growth regulation
  adenyl cyclase in, 8
  cAMP in, 7, 8, 16, 178, 204, 208, 328-332
  cGMP in, 178
  PDE inhibitors in, 8
  prostaglandin E_1, effect, 328
Cell nuclei, protein kinase in, 126
Cellular extrusion of cGMP, 172, 173
Cerebral cortex
  histone phosphatase activity, 132
Chloromycetin cyclic phosphate, 230
Chloroproamide, phosphodiesterase inhibitor, 83
Chlorpromazine
  effect on adenyl cyclase, 37
  effect on cGMP, 204
Cholecystokinin
  effect on phosphodiesterase, 88
Cholera toxin
  effect on adenyl cyclase, 40, 41, 190
  effect on cGMP, 190
Chromatin
  actinomycin D, binding, 133
  phosphorylation, 133
  protein kinase in, 126
alpha-Chymotrypsin
  phosphorylation of, 246
Cis-trans isomers
  in cyclic phosphates, 229
  effect on hydrolysis, 241
Clinical studies, 356-377, 381, 382
Cobalt in ATP-cAMP conversion, 19
Cobramine B, adenyl cyclase blocking, 37
Collagen, effect on cGMP, 193, 197
Concanavalin A, effect on cGMP, 193, 205
Copper, in ATP-cAMP conversion, 19
Cortisol
  effect on adenyl cyclase activity, 10
  effect on cGMP excretion, 164, 200
Creatine phosphokinase in ATP regeneration, 4
Cyclic AMP (etc.) see cAMP (etc.)
Cyclic nucleotide phosphodiesterase see phosphodiesterase
Cyclic phosphates
  chemical and physical properties, 238-247
  cis-trans conformation, 229
Cyclic phosphonates, 288, 289
Cyclic pyrophosphates, 289, 290
$O^2,2'$-Cyclocytidine 3',5'-cyclic pyrophosphate, 290
Cystic fibrosis, cAMP in, 382
Cytidine, protonation of, 243

Cytidine 2',3-cyclic phosphate (cCMP), 235
  glycogenolysis, effect on, 336
  hydrolysis of, 242
  lipolysis activation, 322
  molecular structure, 238
  PDE substrate, 315
  protein kinase activation, 294
  synthesis, 262
Cytidine 2'(3'),5'-diphosphate, 235
Cytidine 2'(3')-phosphate, hydrolysis, 242
Cytidine 5'-phosphate
  N^4-benzoyl deriv., 262
Cycloheximide
  effect on phosphodiesterase, 79, 87
Cyclohexyl isocyanate, 248
Cytoplasm, guanylate cyclase in, 162, 163
Cytosine 2',3'-cyclic phosphate, 269
Cytosol, protein kinase in, 124

7-Deazaadenosine see tubercidin
Decapitation, effect on cAMP, cGMP, 190
Denaturation, cAMP and, 110, 111
2'-Deoxyadenosine 3',5'-cyclic phosphate (cdAMP)
  bacterial synthesis, 247
  8-bromo deriv., 272
  hydrolysis of, 242
  lipolysis activation, 322
  5'-octylamino deriv., in CRP binding, 307
  PDE inhibitor, 318
  PDE substrate, 311, 313
  protein kinase activation, 306
  synthesis, 267, 272
2'-Deoxyadenosine 5'-phosphate,
  hydrolysis of, 242
5'-Deoxyadenosine 3',5'-cyclic-phosphoramidate, 5'-amino deriv., 285
5'-Deoxyadenosine 5'-phosphoramidate,
  5'-amino deriv., 285
5'-Deoxyadenosine 3',5'-thio-phosphoramidate, 5'-amino deriv., 287
2'-Deoxycytidine 3',5'-cyclic phosphate
  hydrolysis of, 242
  synthesis of, 272
2'-Deoxycytidine 5'-phosphate,
  hydrolysis, 242
3'-Deoxy-3'-(dehydroxyphosphinylmethyl)-adenosine, 5'-cyclic ester
  PDE substrate, 315
  protein kinase activation, 306
5'-Deoxy-5'-(Dehydroxyphosphinylmethyl)-adenosine, 3'-cyclic ester
  PDE substrate, 315
  protein kinase activation, 306
1-(2-Deoxy-beta-D-glucopyranosyl)-thymine, 279
2'-Deoxyguanosine 3',5'-cyclic phosphate
  hydrolysis, 242
  synthesis, 272
2'-Deoxyguanosine 5'-phosphate
  hydrolysis, 242
2'-Deoxyinosine 3',5'-cyclic phosphate, 272, 273
3'-Deoxy-3'-iodothymidine 5'-phosphate, 271

3'-Deoxy-3'-iodothymidine-5'phosphoromorpholidate, 283
1-(2-Deoxy-beta-D-threo-pento-furanosyl) thymidine 3',5'-cyclic phosphate, 271, 272
1-(2-Deoxy-beta-D-threo-pentofuranosyl-thymine) 3',5'-cyclic phosphoromorpholidate, 283
1-(2-Deoxy-beta-D-threo-pentofuranosyl) thymine 5'-diphenylphosphate, 271
9-(2-Deoxy-beta-D-ribofuranosylpurine)-3',5'-cyclic phosphate
   6-chloro deriv., 272
   6-methylthio deriv., 273
   6-thione deriv., 273
1-(2-Deoxy-beta-D-ribopyranosyl)-thymine, 233, 279
5'-Deoxy-5-thioadenosine 3',5'-cyclic-phosphorothioate, 281
   protein kinase activation, 306
5'-Deoxy-5'-thioadenosine-5'-phosphorothioate, 281
5'-Deoxy-5'-thioinosine 3',5'-cyclic phosphorothioate, 281
   protein kinase substrate, 306
5'-Deoxy-5'-thioinosine-5'-phosphorothioate, 281
2'-Deoxy thymidine 3',5'-cyclic-phosphate (cdTMP), 272
   protein kinase activity, 294
2'-Deoxyuridine 3',5'-cyclic phosphate, 272
1-(2-Deoxy-beta-D-xylopyranosyl)-thymine, 234
Depressive disease, cAMP in, 375
Deshistidyl-glucagon membrane binding, 46
Desmethylimipramine
   effect on phosphodiesterase, 83
Detergent treated
   adenyl cyclase, 30, 31, 32, 38
   guanyl cyclase, 162
Dexamethasone
   adenyl cyclase activity, effect on, 10
   guanylate cyclase activity, effect on, 166, 201
Diabetes insipidus, cAMP in, 374
1,5-Diazabicyclo-(4.3.0.)nonene-5, 271
p'p''-Di(6-azauridine-5'-yl-)-pyrophosphate, 262
Dictyostelium discoideum
   phosphodiesterase in, 78
Digitonin, effect on adenyl cyclase, 37
Dihydroergotamine
   adenyl cyclase inhibitor, 83
   phosphodiesterase inhibitor, 83
Diisopropyl fluoro phosphate
   liver protein kinase multiple form suppression, 121
4-(3,4-Dimethoxybenzyl)-2-imidazol-idinone, phosphodiesterase inhibitor, 81
N^6-Dimethylamino-9-beta-D-ribofuranosyl-purine 3',5'-cyclic-N,N-dimethyl-phosphoramidate, 306
Dimethyl formamide acetal, 234
Dimethyl phosphate, hydrolysis, 241
2,4-Dinitrofluorobenzene, 232, 277

Diphenylphosphorochloridate, 232
DNA
   in gene transcription, 106, 188
   histone complexing, 117
   hydrolysis, 230
   thymidine incorporation, 182
Dopamine, membrane binding, 41
Dualism between cAMP and cGMP, 207

E. coli
   cAMP function in, 187, 188
   gene transcription, 106
   cGMP function in, 187-189
   phosphodiesterase in, 78
   protein kinase in, 103
EDTA, membrane binding, 43, 46
EGTA, effect on adenyl cyclase, 21, 22, 31, 32, 38
Electroconvulsive shock (ECS)
   effect on cAMP, 190, 202
   effect on cGMP, 190, 193, 194, 202
Enthalpy of hydrolysis, 246
Epinephrine
   adenyl cyclase stimulant, 9
   cAMP, effect on, 190, 208
   cardiac cell contractions, effect on, 208
   guanylate cyclase, effect on, 166, 193, 197
   membrane binding, 41, 42, 44
   protein kinase, effect on, 116
9-D-Erythrityladenine 3',4'-cyclic phosphate, 281
Erythrityl adenine monophosphoromorph-lidate, 281
9-D-Erythrityladenine 5'-triphosphate, 28
Ethylene glycol cyclic phosphate, 237
   hydrolysis, 241
Excretion
   of cAMP, 360-363
   diurnal variation, 361-363
   of cGMP, 191, 196, 199-201, 360-363

Free energy of hydrolysis, 245-247
Fructose-6-phosphate
   phosphodiesterase stimulation, 78
FSH, effect on adenyl cyclase, 33, 49

Galactokinase, Galactosidase
   cGMP effect on, 187
   ppGpp effect on, 188
Gene transcription
   cAMP and analogs in, 106, 188, 307
   cGMP in, 188
Glucagon
   adenyl cyclase stimulation, 10, 33, 37
   cAMP effect on, 104, 190, 191, 367
   effect on guanylate cyclase, 166
   cGMP effect on, 191, 208, 367
   in histone phosphorylation, 41, 42
   membrane receptor binding, 41-44, 4
Glucocorticoid, in cGMP excretion, 200
9-beta-D-Glucopyranosyladenine, 277
9-beta-D-Glucopyranosyladenine-4',6'-cy phosphate, 277

9-beta-D-glucopyranosyladenine-6'-phosphate, 277
9-beta-D-Glucopyranosyladenine-6'-phosphomorpholidate, 277
7-beta-D-Glucopyranosyltheophylline, 277
Glucose
  effect on cGMP, 193
  oxidation, 178
  production, 175, 181
Glucose-6-phosphate
  phosphodiesterase stimulation, 78
Glutamic acid, phosphorylation of, 102
Glutamine synthase, ATP activation, 111
Glycerol 1,3-cyclic phosphate, 237
Glycerol production, cAMP, cGMP, effect on, 177
Glycogen phosphorylase
  cAMP effect on, 175, 183
  activation, 100
  cGMP effect on, 175, 183
Glycogen synthetase
  cAMP effect on, 104, 111, 175, 183, 184
  epinephrine effect on, 116
  cGMP effect on, 175, 183, 184, 209
  insulin effect on, 116
  magnesium effect on, 115
  phosphorylation of, 101, 129, 130
  protein kinase moderator, 113
  reduction of, 104
Glycogen synthetase kinase
  aging effect, 116
  magnesium effect on, 115
9-beta-D-Glycopyranosyladenine, 231
cGMP (Guanosine 3',5'-cyclic phosphate)
  acetylcholine effect on, 167, 168, 191-196, 209, 368
  adrenal steroidogenesis, 178, 182
  aldosterone effect on, 200
  aminophylline effect, 202
  cAMP comparison, 175-182, 207
  d-amphetamine effect, 193, 204
  atropine effect on, 180, 191-194, 204
  bacteria regulatory function, 182-189, 206
  N^2-benzoyl deriv., PDE substrate, 314
  N^2-benzoyl-2'-O-THPc deriv., PDE substrate, 314
  8-benzylamino deriv., 259
    protein kinase activation, 301
  8-benzylthio deriv., 259
    protein kinase activation, 301
  bladder permeability, 181
  8-bromo deriv., 259
    mediator release, 334
    protein kinase activation, 259
  caffeine effect on, 202
  calcitonin effect, 193, 197
  calcium effect on, 177, 193, 195-197, 199
  carbachol effect, 193, 195
  cardiac function, effect on, 179, 191, 208
  cell-free systems, 183
  cell growth inhibition, 178
  cellular extrusion, 172, 173
  chlorpromazine effect on, 204
  cholera toxin, effect of, 190
  cholinergic activity, 178, 179, 191-196, 334, 368
  collagen effect, 193, 197
  concanavalin A, effect, 193, 205
  cortisol effect, 200
  decapitation, effect of, 190
  diacetyl deriv., 283
  dibutyryl deriv.
    atropine blocking, 180
    cholinergic mimic, 179, 180
    glucose in blood, effect on, 337
    growth hormone release, 180
  8-dimethylamino deriv., 259
  electroconvulsive shock, effect, 190, 193, 194, 204
  epinephrine effect, 193, 197
  excretion, 172, 173, 191, 196, 199, 201, 338
  exogeneous, effects of, 173-183
  glucagon, effect on, 191, 367
  glucose, effect on, 177, 178, 193
  glycogenolysis, 336
  growth hormone release, 180, 181, 201
  histamine effect, 193, 197, 204
  hormone effect, 190, 200
  human response, 337
  hydrolysis of, 74, 75, 242
  hypothalamic extract, 193
  insulin effect on, 176, 177
  level alteration, 189-207
  lipolysis activation, 322
  maaloxone effect on, 193, 195
  mechanism of action, 209
  methacholine, 193, 195, 196
  8-methylamino deriv., 259
    protein kinase activation, 301
  8-methylthio deriv., 259
    protein kinase activation, 301
  mitogenesis, effect on, 182
  molecular structure, 240
  norepinephrine effect, 190, 193, 197, 204
  oxotremorine effect on, 193, 194
  oxytocin, effect on, 193, 196
  parathyroid hormone effect, 166, 193, 197, 365, 370
  perfused liver, level in, 175
  phosphoenol pyruvate carboxykinase, effect on, 182
  phytohemagglutin effect, 193, 196, 205
  pituitary gland, removal, 191, 200
  in plasma, 363
  potassium, effect of, 193, 195
  prostaglandin F_{2a} effect on, 193, 204
  protein kinase activation, 111, 113, 174, 183-187, 210, 294, 301
  reserpine effect on, 204
  in RNA synthesis, 188, 205
  serotonin effect, 197, 204
  speculation, 207-213
  in spinal fluid, 359, 364
  2'-O-succinyl deriv., 274
  synthesis, 252
  2'-O-tetrahydropyranyl deriv., 275, 314
  thyroxine effect, 200
  in urine, 338, 357, 358, 360, 365-367, 381

cGMP dependent protein kinase see protein
   kinase
Gramicidin S, adenyl cyclase blocking, 37
Growth hormone
   cAMP effect on release, 180, 181, 332
      333
   cAMP derivatives effect on, 333
   cGMP effect on release, 180, 181, 201
   phosphodiesterase effect, 88
GTP
   effect on adenyl cyclase, 42-47, 167
   cGMP synthesis, 252
      kinetics of, 164-166
Guanosine, protonation of, 243
Guanosine 3'-diphosphate-5'-diphosphate
   effect on galactokinase and beta-
      galactosidase, 188
Guanosine monophosphate see cGMP
Guanosine 2'(3') phosphate, hydrolysis, 242
Guanosine 5'-phosphate
   N^2-benzoyl deriv., 252
   p-nitrophenyl deriv., 252
   2'-O-tetrahydropyranyl deriv., 275
Guanyl cyclase, 161-168
Hexachloroacetone, 234, 269
Hexopyranosyl cyclic nucleotides, 277-280
Histamine
   cAMP, dibutyryl deriv., effect, 294
   effect on cAMP level, 33, 35, 50
   effect on cGMP level, 193, 197, 204
   effect on phosphodiesterase, 84
Histidine, phosphorylation of, 102
Histone
   cAMP stimulation, 101
   DNA complexing, 117, 133
   phosphorylation stimulation, 132
   protein kinase substrate, 116, 131, 132
Histone kinase see protein kinase
Hormone influence
   adenyl cyclase, 2, 7, 8, 13, 25, 32-51,
      190
   binding and adenyl cyclase action, 4-47
   calcium effect, 105, 106
   on cGMP level, 190, 200
   on guanylate cyclase, 164, 166-168
   on phosphodiesterase, 85-89
   protein kinase and, 104-108
HTC cells, protein kinase activity, 118
Hydrolysis
   of cyclic phosphates, 238, 241-247
   cGMP, enzymatic, 168-172
Hyperthyroidism, cAMP in, 382
Hypophysectomy, effect on cGMP
   excretion, 164, 201, 369
Hypothalamic extract, effect on cGMP, 193

Imidazole, phosphodiesterase activator, 84,
   169
Imipramine, phosphodiesterase inhibitor,
   83
cIMP (Inosine 3',5'-cyclic phosphate)
   2'-O-acetyl deriv., 249, 283
   8-amino deriv., 258
      protein kinase activation, 300
   8-anilino deriv.
      phosphorylase activation, 320

8-azido deriv., 258
   protein kinase activation, 300
8-benzylamino deriv.
   lipolysis activation, 322
   PDE substrate, 310
   phosphorylase activation, 320
8-benzylthio deriv.
   protein kinase activation, 300
8-bromo deriv., 255
   lipolysis activation, 322
   PDE substrate, 310
   phosphorylase activation, 320
   protein kinase activation, 300
2'-O-Butyryl deriv.
   glycogenolysis, effect on, 336
   lipolysis activation, 322
8-chlorobenzylamino deriv.
   phosphorylase activation, 320
8-cyclohexylamino deriv.
   phosphorylase activation, 320
8-ethyl thio deriv.
   protein kinase activation, 300
glycogenolysis, effect on, 336
lipolysis activation, 322
7-methyl deriv., 261
   CRP binding inhibition, 307
8-(4-methylbenzylamino) deriv.
   lipolysis activation, 322
   phosphorylase activation, 320
8-morpholino, PDE substrate, 310
N'-oxide deriv.
   protein kinase activation, 300
PDE substrate and inhibitor, 311, 316
8-piperidino deriv.
   PDE inhibitor, 316
   PDE substrate, 311
protein kinase activation, 185, 294, 300,
   302-304
2'-O-succinyl deriv., 274
8-thio deriv., 255
Inosine 3',5'-cyclic phosphate see cIMP
Inosine 5'-triphosphate, hydrolysis, 247
Insulin
   effect on adenyl cyclase, 40, 185, 208
   effect on cAMP, 369
   effect on cGMP, 176, 177
   effect on phosphodiesterase, 85-87
   effect on phosphorylation, 133, 134
   effect on protein kinase, 116
   in hormone-receptor mechanism, 47
Iodate, effect on adenyl cyclase, 25
9-(2,3-O-Isopropylidene-beta-D-ribofurano-
   syl) purine, 6-chloro deriv., 250
1,2-O-Isopropylidenexylose, 231
Isoproterenol
   membrane binding, 41
   in rat heart perfusion, 190

Kinetic regulation of phosphodiesterase, 72-
   74

Lacrimal gland protein kinase,
   multiple forms, 120
LH, adenyl cyclase stimulation, 33, 49
Librium, phosphodiesterase inhibitor, 308

Lipolysis
    cAMP derivatives effect on 322, 326
    cCMP effect on, 322
    cGMP effect on, 322
    cIMP and derivatives, effect on, 322
    PuRcMP derivatives effect on, 321-323
    cUMP efect on, 322
Lithium in ATP-cAMP conversion, 21, 37
Liver protein kinase
    molecular wt., 122, 123
    multiple forms, 120
Lubrol PX effect on adenyl cyclase, 30-32, 37
1-alpha-1-alpha-D-Lyxofuranosyl thymine
    2',3'-cyclic phosphate, 271
    2',5'-cyclic phosphate, 271
    3',5'-cyclic phosphate, 271
1-(beta-D-Lyxofuranosyl) uracil
    2',3'-cyclic phosphate, 229, 230
    2',5'-cyclic phosphate, 233, 271
    3',5'-cyclic phosphate, 271
    5'-phosphate, 271
    5'-phosphite, 271
    2'(3')-phosphites, 229, 230, 270

Maaloxone, effect on cGMP, 193, 195
Magnesium
    inATP-cAMP conversion, 2, 4, 5, 16-21, 26-32
    inGTP-cGMP conversion, 165, 166
    phosphodiesterase assay, 67
    phosphodiesterase effect on, 76
Mammary gland protein kinase, molecular wt., 122
Manganese
    in ATP-cAMP conversion, 19-21, 31, 32, 38
    in GTP-cAMP conversion, 165, 166
    in phosphodiesterase assay, 67
1-(alpha-1)-Mannopyranosylthymine, 279
    bis (cyclic phosphate), 280
    2',3'-cyclic phosphate, 280
    4',6'-cyclic phosphate, 280
    2',(3')-phosphate 4',6'-cyclic-phosphate, 280
    phosphite deriv., 280
Mania, cAMP in, 375
Mechanism of cAMP action, 108, 109
Membrane
    alteration in adenyl cyclase activity, 36-38
    alteration in guanylate cyclase activity, 162
    phosphodiesterase distribution, 76-78
    protein kinase substrate, 135
    structure in adenyl cyclase system, 51-56
6-Mercaptopurine (6-MP), 304
Messenger hypothesis
    in adenyl cyclase, 7, 41
    cAMP in, 291
    cGMP in, 163
    protein kinase in, 99
Methacholine, effect on cGMP, 193, 195, 196

Methyl ethylene glycol cyclic phosphate, hydrolysis, 241
6-Methylthiopurine (6-MMP), 304
Methyl xanthine
    phosphodiesterase inhibitor, 79-81, 202
Michaelis-Menten constants for
    guanyl cyclase, 164-166
    phosphodiesterase, 70, 72-74
Microtubule protein,
    protein kinase substrate, 137
Milk, cAMP in, 359, 365
Mitochondria, protein kinase in, 125
Mitogenesis, cAMP and cGMP effect, 182
Molecular structure (X-ray)
    cAMP, 239
    9-beta-D-Arabinofuranosyladenine 3',5'-cyclic phosphate, 238
    cytidine 2',3'-cyclic phosphate, 238
    cGMP, 240
    uridine 3',5'-cyclic phosphate, 239
Molecular theory
    P-O bonds, 226
    substitution mechanism, 227
Molecular weight
    of phosphodiesterase, 169, 170
    of protein kinase subunits, 120-124
Multiple forms
    of phosphodiesterase, 68-72
    of protein kinase, 119-124
Myokinase in phosphodiesterase assay, 67, 68

Nerve endings,
    phosphodiesterase in, 77
    protein kinase in, 125
Nicotinic acid,
    phosphodiesterase activation, 84
p-Nitrophenol, 232
p-Nitrophenylguanosine 5'-phosphate, 252
Norepinephrine
    adenyl cyclase, effect on, 25, 33-35, 47-51
    effect on cGMP, 190, 193, 197, 204
    lipolytic agent, 10
    membrane binding, 41, 42

Ontogenetic development of adenyl cyclase, 47-51
Orchidectomy, effect on guanyl cyclase, 164
Ornithine decarboxylase, induction of, 134
Osmotic shock, phosphodiesterase release with, 77
Ouabain, effect on cAMP, 33
Oxalacetate, inhibitor for guanyl cyclase, 165
Oxotremorine, effect on cGMP, 193, 194
Oxytocin
    effect on cGMP, 193, 197
    uterus contraction moderated by cyclic nucleotides, 182

Pancreatic lipase, activation, 131
Panteheine-2,4-cyclic phosphate, 230
Papaverine, phosphodiesterase inhibitor, 32, 308

Paraoxon, growth hormone in plasma, 180
Parathyroid hormone (PTH)
   effect on cAMP, 361, 365, 369, 370
   effect on cGMP, 193, 197, 365, 370
   effect on guanyl cyclase, 166
PDE see phosphodiesterase
PGE see prostaglandin
Phentolamine
   effect on adenyl cyclase, 39
   effect on phosphodiesterase, 83
   membrane binding, 41
Phenyl phosphodichloridate, 277
Phosphatase, alkaline, 280
Phosphocreatine in ATP regulation, 4
Phosphodiesterase (PDE)
   actinomycin effect, 79, 87
   adenyl cyclase contaminant, 4
   adipose tissue, 69, 70, 75-77
   cAMP derivatives, effect, 79, 309-314, 316-318
   assay, 66-68
   4-benzyl-2-imidazolidones, effect, 81, 82
   brain, 69-72, 75-77
   calf serum activation, 79
   calcium effect, 76
   cell growth regulation, 8, 79
   cholecystokinin effect, 88
   chloropropamide inhibition, 83
   cycloheximide effect, 79, 87
   desmethylimipramine effect, 83
   dihydroergotamine inhibitor, 83
   gastric mucosa, 77
   cGMP hydrolysis, 168-172
   growth hormone effect, 88
   heart muscle, 69, 70, 76, 77
   histamine effect, 84
   hormone effects, 84-89
   imidazole activation, 84, 169
   imipramine inhibition, 83
   cIMP inhibitor, 316
   inhibiting effect on cGMP, 202, 316
   insulin effect, 40, 85-87
   kidney, 69, 70, 76, 77
   kinetic regulation, 72-74
   librium inhibition, 308
   liver, 69, 70, 74, 75
   in lower organisms, 78
   magnesium effect, 76
   2'-O-methyl-5'-nucleotide synthesis, 276
   methyl xanthine inhibition, 5, 79-81
   Michaelis-Menten constants, 71, 73
   multiple forms, 68-72, 169
   nicotinic acid activation, 84
   papavarine inhibition, 82, 308
   pharmacology, 79-84
   phentolamine inhibition, 83
   in plants, 78
   in plasma membrane, 76
   pronase stimulation, 81
   prostaglandin E_1 effect, 87
   puramycin inhibition, 81
   PuRcMP derivatives as substrates, 308-316
   Quazodine inhibition, 308
   RO 20-1724 inhibition, 308
   skeletal muscle, 69, 70, 77
   snake venom stimulation, 76
   solubilization, 77
   specificity, 68, 72, 76
   SQ 20009 inhibition, 308
   subcellular distribution, 76-78
   sugar stimulation, 78
   theophylline inhibition, 5, 79-87, 308, 316-318
   thyroid hormone effect, 88, 89
   tissue culture, 77-79
   tolbutamide inhibitor, 83
   trypsin stimulation, 76
   3,3',5-triiodo-L-thyronine effect, 88
Phosphoenolpyruvate
   in ATP regeneration, 4
   inhibitor of guanyl cyclase, 165
Phosphoenol puruvate carboxykinase (PEP)
   cAMP effect on, 182
   cGMP effect on, 182
   induction of, 134, 329
Phosphofructokinase, cAMP effect on, 105
Phospholipase A, effect on adenyl cyclase, 37, 38
Phosphoprotein kinase in brain, 125
Phosphoramidate nucleosides, 283-285
Phosphorothioate nucleosides, 281-283
Phosphorous-oxygen bond, 226, 227
Phosphorylase kinase
   cAMP and derivatives effect, 100, 101, 104, 175, 320-322
   autophosphorylation, 128
   cGMP effect on, 175
   cIMP and derivatives, effect on, 320
   magnesium effect on, 115
   protein kinase moderator in, 113
   protein kinase substrate, 129
Phosphorylase kinase kinase see protein kinase
Phosphorylase phosphatase
   cAMP inactivation, 105
   dephosphorylation of troponin, 138
Phosphorylation
   of adipose tissue lipase, 131
   agents for, 230, 232
   cAMP mediation, 103-107
   of aspartic acid, 102
   of casein, 100, 102
   of chromatin proteins, 133
   of alpha-chymotrypsin, 246
   of diols, 230
   in dividing cells, 134
   of glutamic acid, 102
   glycogen phosphorylase in, 100
   of glycogen synthetase, 101, 129, 130
   of histidine, 102
   of histones, 101, 117, 131, 132
   insulin effect on, 133, 134
   magnesium effect, 115
   of membrane, 135, 136
   of pancreatic lipase, 131
   of phosphorylase kinase, 128, 129
   of phosvitin, 101, 102
   of polynucleotide phosphorylase, 139

Phosphorylation (continued)
  prolactin effect on, 134
  of protamine, 100, 138
  of protein kinase moderator, 113-115
  of RNA polymerase, 106-108, 139
  RNA synthesis, effect on, 133
  of ribosomal protein, 135
  of serine residue, 101, 102, 132
  of skeletal muscle protein kinase, 131
  spermatogenesis, during, 134
  substrates, for, 127, 139
  of threonine residue, 101, 102
  of triglyceride lipase, 103, 131
  of troponin, 138
Phosvitin (Vitellinic acid),
  phosphorylation of, 101, 102, 119
Physarum polycephalum,
  protein kinase in, 111
Phytohemagglutin (PHA),
  effect on cGMP, 193, 196, 205
Pituitary gland, removal,
  effect on cGMP, 191, 200
Plants
  cGMP in, 160
  protein kinase in, 103
Plasma
  cAMP in, 358, 359, 363-369
  cGMP in, 363
Plasma membrane
  adenyl cyclase in, 8-11
  hormone binding, 41
  phosphodiesterase in, 77
  protein kinase substrate, 136
  structure, 51-56
Platelet aggregation, control of, 208
Poly-L-lysine, preincubation with protein
    kinase, 116
Polynucleotide phosphorylase, protein
    kinase substrate, 139
Potassium
  effect on cAMP, 33, 38
  effect on cGMP, 193, 195
  release from liver, cGMP effect, 175,
    177
Potato, phosphodiesterase in, 78
Pronase
  phosphodiesterase stimulation, 76
Propane-1,3-diol cyclic phosphate, hydrolysis, 241
Propranolol
  effect on adenyl cyclase, 33, 39
  membrane receptor antagonist, 41, 42, 44
Prostaglandin E_1
  cell growth regulation, 328
  effect on adenyl cyclase, 33, 35, 37, 40-42, 50
  effect on phosphodiesterase, 87
Prostaglandin F_{2a}
  effect on cGMP, 193, 204
Protamine
  phosphorylation of, 101
  preincubation with protein kinase, 116, 117
Protein kinase
  in acellular slime mold, 111
  aging effect, 116
  cAMP effect on, 100, 101, 108-113, 183-187
  cAMP derivatives effect on, 294-306
  ara-cAMP and derivatives, effect on, 306
  calcium effect, 112
  casein phosphorylation, 104-108
  cCMP effect on, 294
  epinephrine effect on, 116
  gene transcription, 106
  cGMP effect on, 111, 113, 174, 183-187, 210, 294, 301
  cGMP derivatives effect on, 301, 302
  histone phosphorylation, 101
  hormone response, 104-108
  cIMP effect on, 112, 185, 294, 303
  cIMP derivatives effect on, 299, 300
  insulin effect, 116
  magnesium effect, 115
  multiple forms, 119-124
  preincubation with substrate, 116
  PuRcMP derivatives effect, 302-305
  species distribution, 102, 103
  in spermatozoa, 103
  subcellular distribution, 124-127
  substrates for, 127-139
  subunit structure, 119-124
  tissue distribution, 103
  cdTMP effect on, 294
  cUMP effect on, 185, 294
  in virus, 103
Protein kinase moderator, 113-115
Psoriasis, cAMP in, 375
Purine-9-(beta-d-ribofuranosyl)-3',5'-cyclic
    monophosphate (PuRcMP)
  6-amino deriv. see cAMP
  2-amino-6-chloro deriv., 253
  2-amino-6-hydroxy deriv. see cGMP
  6-benzylthio deriv., 250
    antiviral activity, 332
    PDE substrate, 311
    protein kinase activation, 303-305
  6-chloro deriv., 249
    PDE substrate, 309, 312
    phosphorylase activation, 321
  2,6-diamino deriv., 254
  2,6-dichloro deriv., 253
  6-diethylamino deriv., 250
    PDE substrate, 311
    protein kinase activation, 302-304
  6-ethylamino deriv., 250
    PDE substrate, 311
    protein kinase activation, 302-304
  6-ethylthio deriv., 250
    PDE substrate, 311
    protein kinase activation, 303-305
  6-hydroxy deriv. see cIMP
  6-hydroxylamino deriv.
    antiviral activity, 332
  6-methoxy deriv., 250
    PDE substrate, 309, 311, 312
    phosphorylase activation, 321
    protein kinase activation, 302-304
  6-methylthio deriv., 250
    antiviral activity, 332
    cell growth regulation, 331

PuRcMP (continued)
   6-methylthio deriv. (continued)
      PDE substrate, 311
      protein kinase activation, 302-305
   6-morpholino deriv.
      lipolysis activation, 322
      phosphorylase activation, 321
   6-piperidino deriv.
      sleep prolongation, 336
   6-thione deriv., 249, 250
      antiviral activity, 332
      cell growth regulation, 329, 330
      lipolysis activation, 322
      PDE substrate, 309, 311
      protein kinase activation, 302-305
Puromycin, phosphodiesterase inhibitor, 81
Pyruvate kinase in ATP regeneration, 4

Quazodine, phosphodiesterase inhibitor, 308

Receptor mechanism
   adenyl cyclase, 39-47
   cAMP analogs in, 306, 307
Reserpine, effect on cGMP, 204
Reticulocyte protein kinase
   molecular wt., 122
   multiple forms, 120
Riboflavin-4',5'-cyclic phosphate, 230, 280
   2',3'-di-O-butyryl deriv., 280
Riboflavin 5'-phosphate, 280
3-beta-D-ribofuranosyladenine-3',5'-cyclic
      phosphate, 260
9-beta-L-ribofuranosyladenine-3',5'-cyclic
      phosphate, 260
9-beta-D-ribofuranosylpurine-3',5'-cyclic
      phosphate see Purine-9-(beta-D-ribo-
      furanosyl)-3',5'-cyclic monophos-
      phate
1-beta-D-ribofuranosyl-1,2,4-triazole-3-
      carboxamide,
   antiviral activity, 332
1-(D-ribopyranosyl) thymine, 279
Ribosomal protein,
   protein kinase substrate, 135
Rifampicin in RNA synthesis, 107
RNA
   hydrolysis, 230
   synthesis, 133, 188, 205, 307
RNA polymerase
   cAMP in, 106-108
   protein kinase substrate, 139
RO 20-1724
   cell growth and, 328
   phosphodiesterase inhibitor, 308
Rubidium, effect on adenyl cyclase, 39
Saliva, cAMP in, 364
Sciatic nerve, effect on cGMP, 196
Secretin, adenyl cyclase stimulation, 33, 37
Serine, phosphorylation of residue, 101, 102, 132
Serine dehydratase
   cAMP effect on, 182
   cGMP effect on, 182
   induction of, 134
   serotonin effect, 193
   tissue distribution and level, 158-160
   in urine, 157

Serotonin
   effect on cAMP, 33, 197
   effect on cGMP, 193, 197, 204
Skeletal muscle protein kinase
   molecular wt., 120
   multiple forms, 119
   phosphorylation, 131
   preincubation with casein, 116
Snake venom
   in phosphodiesterase assay, 67, 68
   phosphodiesterase stimulation, 76
Sodium fluoride
   in ATP-cAMP conversion, 2, 9, 13, 15, 23-31, 36-41
   effect on ATPase activity, 23
   in GTP-cGMP conversion, 166
Sodium ion, effect on adenyl cyclase, 39
Solubilization
   of adenyl cyclase, 30-32
   of guanylate cyclase, 162, 163
   of phosphodiesterase, 77
Sonic disruption
   phosphodiesterase release with, 77
Spare receptor theory, 44-46
Species distribution
   of adenyl cyclase, cAMP, 6, 7
   of protein kinase, 102, 103, 186
Spermatogenesis
   histone phosphorylation during, 134
Spermatozoa
   guanyl cyclase in, 161
   phosphodiesterase activity in, 89
   protein kinase in, 103
Spinal fluid
   cAMP and cGMP in, 359, 364
SQ 20009, phosphodiesterase inhibitor, 308
Stereochemistry of cyclic phosphates, 229, 240, 269
Strontium in ATP-cAMP conversion, 21
Subcellular distribution
   adenyl cyclase, 8-11
   guanylate cyclase, 161-164
   phosphodiesterase, 76-78
   protein kinase, 124-127
Subunit structure of protein kinase, 119-124
Synaptic membrane, protein kinase in, 125
Synaptosome, adenyl cyclase activity, 9

Tetra-p-nitrophenyl pyrophosphate, 232
Theophylline
   adenyl cyclase, effect on, 25
   mediator release, 334
   PDE inhibitor, 5, 79-81, 308
   steroidogenesis, effect on, 178
Thiophosphoramidate nucleosides, 287
9-(4-Thio-beta-D-ribofuranosyl)-adenine
      3',5'-cyclic phosphate, 291
   n-butyl deriv., 291
   dibutyryl deriv., 291
Thymidine, DNA incorporation, 182
Thymidine 3',5'-cyclic methylphosphonate 288, 289
Thymidine 3',5' cyclic phosphate (cTMP), 272
   glycogenolysis, effect on, 178, 336
   hydrolysis of, 242
   phosphodiesterase substrate, 315

Thymidine 5'-phosphate, hydrolysis, 242
Thymol, adenyl cyclase blocking. 37
Thyroid hormone
  effect on cGMP excretion, 164, 200
  effect on guanylate cyclase, 167
  effect on phosphodiesterase, 84, 88
Thyroid stimulating hormone (TSH)
  in ATP-cAMP conversion, 21, 25, 33, 37, 49
  release of, 333
Tissue culture
  of adenyl cyclase, cAMP, 7, 8
  phosphodiesterase in, 77-79
cTMP see thymidine 3',5'-cyclic phosphate
Tolbutamide, phosphodiesterase inhibitor, 83
Toyocamycin 3',5'-cyclic phosphate, 260
  protein kinase activation, 306
Triglyeride lipase
  cAMP effect, 184
  protein kinase activation, 130, 131
3,3',5-Triiodo-L-thyronine
  effect on phosphodiesterase, 88
Triton X-100
  effect on adenyl cyclase, 30, 37, 162
  effect on guanylate cyclase, 162
  phosphodiesterase release, 77
Troponin, phosphorylation, 138
Trypsin
  effect on adenyl cyclase, 37
  phosphodiesterase stimulation, 76
Tubercidin 3',5'-cyclic phsophate (TucMP), 260
  CRP binding inhibition, 307
  glycogenolysis, effect on, 336
  protein kinase activation, 306
Tubercidin 5'-phosphate, 260
Tyrosine aminotransferase (TAT), induction of, 329
Tyrosine transaminase, induction of, 134

Urea, effect on adenyl cyclase, 37
Uridine, protonation of, 244
Uridine 3',5'-cyclic (methyl phenylalanyl) phosphoramidate, 283

Uridine 3',5'-cyclicphosphate (cUMP), 283
  glycogenolysis, effect on, 336
  hydrolysis of, 242-245
  lipolysis activation, 322
  molecular structure, 239
  phosphodiesterase substrate, 315
  2'-O-succinyl deriv., 274
  synthesis, 262
  2'-O-tetrahydropyranyl deriv., 275
Uridine 5'-phosphate
  6-aza deriv., 262
  hydrolysis of, 242-245
  2'-O-tetrahydropyranyl deriv., 275
Urine, cAMP and cGMP in, 338, 351, 357, 358, 360-363, 365-369, 381

Vasopressin, effect on adenyl cyclase, 41
Veratridine, effect on cAMP level, 33
Vinca rosea, phosphodiesterase in, 78
Virus, protein kinase in, 103
Virus protein, protein kinase substrate, 137, 138
Vitellinic acid see phosvitin

Xanthosine 3',5'-cyclic phosphate, 253
  8-bromo deriv., 259
9-beta-D-Xylofuranosyladenine, 269
9-beta-D-Xylofuranosyladenine-3',5'-cyclic phosphate, 269, 270
9-beta-D-Xylofuranosyladenine 5'-phosphoromorpholidate
  2',3'-Di-O-acetyl deriv., 270
9-beta-Xylofuranosylhypoxanthine-3',5'-cyclic phosphate, 270
9-beta-D-Xylofuranosylpurine-3',5'-cyclic phosphate
  6-amino deriv., protein kinase activity, 306
1-beta-D-Xylofuranosylthymine-3',5'-cyclic phosphate, 238
1-beta-D-Xylofuranosyluracil, 270
1-beta-D-Xylofuranosyluracil-3',5'-cyclic phosphate, 270

Yeast, adenyl cyclase in, 124

Zinc, in ATP-cAMP conversion, 19

QP
625
N89
A36
v.3

NOV 8 1976